Plant Gasotransmitters and Molecules with Hormonal Activity

Nitric Oxide in Plant Resilience to Abiotic Stress

Edited by

Ravi Gupta

Plant Stress Physiology and Proteomics Laboratory, College of General Education, Kookmin University, Seoul, South Korea

Yiming Wang

Department of Plant Pathology, College of Plant Protection, Nanjing Agricultural University, Nanjing, P.R. China

Riyazuddin Riyazuddin

Instituto de Bioingeniería, Universidad Miguel Hernández, Elche, Spain

Series Editors

Vijay Pratap Singh

Durgesh Kumar Tripathi

Vasileios Fotopoulos

ELSEVIER

ACADEMIC PRESS

An imprint of Elsevier

Elsevier
Radarweg 29, PO Box 211, 1000 AE Amsterdam, Netherlands
125 London Wall, London EC2Y 5AS, United Kingdom
50 Hampshire Street, 5th Floor, Cambridge, MA 02139, United States

ISBN: 978-0-443-40497-9

For Information on all Elsevier publications
visit our website at https://www.elsevier.com/books-and-journals

Publisher: Andre Wolff
Acquisitions Editor: Chiara Giglio
Editorial Project Manager: Ritika Dagar
Production Project Manager: Aishwarya R.
Cover Designer: Bakyalakshmi S.

Typeset by MPS Limited, Chennai, India

Contents

12. Role of nitric oxide in the mitigation of mineral stress

13. Nitric oxide in the mitigation of ultraviolet stress in plants

List of contributors

Oluwatosin Peace Adedokun Department of Plant Biology, University of Szeged, Szeged, Hungary

Gulnaz Ashraf Department of Biotechnology, COMSATS University Islamabad, Abbottabad Campus, Abbottabad, Khyber Pakhtunkhwa, Pakistan

Abu Imran Baba Department of Forest Genetics and Plant Physiology, Umeå Plant Science Centre (UPSC), Swedish University of Agricultural Sciences, Umeå, Sweden

Beimi Cui Institute of Molecular Plant Sciences, School of Biological Sciences, The University of Edinburgh, Edinburgh, United Kingdom

Zalán Czékus Department of Plant Biology, University of Szeged, Szeged, Hungary

Nilanjana Das Department of Biotechnology, Oxidative Stress Biology Laboratory, Visva-Bharati, Siksha Bhavana, Santiniketan, West Bengal, India

Renu Deswal Department of Botany, University of Delhi, New Delhi, Delhi, India

Narottam Dey Department of Biotechnology, Rice Biotechnology Laboratory, Visva-Bharati, Siksha Bhavana, Santiniketan, West Bengal, India

Retwika Ganguly Department of Biology, Suriano Lab, Montclair State University, Montclair, NJ, United States

Praveen Gupta Department of Botany, Faculty of Science, University of Lucknow, Lucknow, Uttar Pradesh, India

Ravi Gupta Plant Stress Physiology and Proteomics Laboratory, College of General Education, Kookmin University, Seoul, South Korea

Samta Gupta Department of Botany, Acharya Narendra Dev College, University of Delhi, New Delhi, Delhi, India

Malik Tahir Hayat Department of Environmental Sciences, COMSATS University Islamabad, Abbottabad Campus, Abbottabad, Khyber Pakhtunkhwa, Pakistan

Nadeem Iqbal School of Environmental Science and Engineering, Hunan University, Changsha, Hunan, P.R. China

Hafiz Muhammad Ansab Jamil Department of Biotechnology, COMSATS University Islamabad, Abbottabad Campus, Abbottabad, Khyber Pakhtunkhwa, Pakistan

Kamal Kant Institute of Plant Biology, HUN-REN Biological Research Centre, Szeged, Hungary

Aniruddhabhai Khuman Department of Bioinformatics, Marwadi University, Rajkot, Gujarat, India

Sun Tae Kim Department of Plant Bioscience, Pusan National University, Life and Industry Convergence Research Institute, Miryang, Republic of Korea

Gábor Kocsy HUN-REN Centre for Agricultural Research, Agricultural Institute, Martonvásár, Hungary

Zsuzsanna Kolbert Department of Plant Biology, University of Szeged, Szeged, Hungary

Dóra Kondak Department of Plant Biology, University of Szeged, Szeged, Hungary

Selahattin Kondak Department of Plant Biology, University of Szeged, Szeged, Hungary

Akhil Kumar Hungarian University of Agriculture and Life Sciences, Institute of Genetics and Biotechnology, Gödöllő, Hungary

Gi Hyun Lee Department of Plant Bioscience, Pusan National University, Life and Industry Convergence Research Institute, Miryang, Republic of Korea

Clément Leonard Department of Plant Bioscience, Pusan National University, Life and Industry Convergence Research Institute, Miryang, Republic of Korea

Moniba Zahid Mahmood Department of Ecology, Friedrich Schiller University, Jena, Germany

Arun Kumar Maurya Department of Botany, Multanimal Modi College, Ghaziabad, Uttar Pradesh, India

Cheol Woo Min Department of Plant Bioscience, Pusan National University, Life and Industry Convergence Research Institute, Miryang, Republic of Korea

Shivani Mishra C.M.P. Degree College, A Constituent Post Graduate College of University of Allahabad, Prayagraj, Uttar Pradesh, India; Department of Botany, Plant Physiology Laboratory, Praygraj, Uttar Pradesh, India

Muhammad Mughees Department of Biotechnology, COMSATS University Islamabad, Abbottabad Campus, Abbottabad, Khyber Pakhtunkhwa, Pakistan

Adeela Munir Department of Plant Bioscience, Pusan National University, Life and Industry Convergence Research Institute, Miryang, Republic of Korea

Muhammad Nauman School of Metallurgical Engineering, Anhui University of Technology, Ma'anshan, Anhui, P.R. China

Attila Ördög Department of Plant Biology, University of Szeged, Szeged, Hungary

Dip Pal Department of Biotechnology, Rice Biotechnology Laboratory, Visva-Bharati, Siksha Bhavana, Santiniketan, West Bengal, India

Sandip Pal Department of Biotechnology, Rice Biotechnology Laboratory, Visva-Bharati, Siksha Bhavana, Santiniketan, West Bengal, India

Qiaona Pan Institute of Molecular Plant Sciences, School of Biological Sciences, The University of Edinburgh, Edinburgh, United Kingdom

Suraj Panja Department of Biosciences and Bioengineering, Indian Institute of Technology, Guwahati, Assam, India

Péter Poór Department of Plant Biology, University of Szeged, Szeged, Hungary

Riyazuddin Riyazuddin Instituto de Bioingeniería, Universidad Miguel Hernández, Elche, Spain

Yogendra Prasad Saksena Department of Botany, Multanimal Modi College, Ghaziabad, Uttar Pradesh, India

Pratyasha Samanta Department of Biotechnology, Rice Biotechnology Laboratory, Visva-Bharati, Siksha Bhavana, Santiniketan, West Bengal, India

Gauri Saxena Department of Botany, Faculty of Science, University of Lucknow, Lucknow, Uttar Pradesh, India

Ankita Sehrawat Department of Botany, Maitreyi College, University of Delhi, New Delhi, Delhi, India

Ambreen Tahir Shah Department of Botany, University of Kashmir, Srinagar, Jammu and Kashmir, India

Amarjeet Singh Department of Botany, Mycology and Natural Pesticide Laboratory, Post Graduate College, Ghazipur, Uttar Pradesh, India

Kalpita Singh HUN-REN Centre for Agricultural Research, Agricultural Institute, Martonvásár, Hungary

Pooja Singh C.M.P. Degree College, A Constituent Post Graduate College of University of Allahabad, Prayagraj, Uttar Pradesh, India; Department of Botany, Plant Physiology Laboratory, Praygraj, Uttar Pradesh, India

Vijay P. Singh C.M.P. Degree College, A Constituent Post Graduate College of University of Allahabad, Prayagraj, Uttar Pradesh, India; Department of Botany, Plant Physiology Laboratory, Praygraj, Uttar Pradesh, India

Dwaipayan Sinha Government General Degree College, Paschim Medinipur, West Bengal, India

Yaiphabi Sougrakpam Division of Plant Physiology, Indian Agricultural Research Institute, New Delhi, Delhi, India

Ágnes Szepesi Department of Plant Biology, University of Szeged, Szeged, Hungary

Nadra Tabassum Institute of Molecular Plant Sciences, School of Biological Sciences, The University of Edinburgh, Edinburgh, United Kingdom; Department of Botany, University of Dhaka, Dhaka, Bangladesh

Zahra Tahmasebi HUN-REN Centre for Agricultural Research, Agricultural Institute, Martonvásár, Hungary

Durgesh K. Tripathi Amity University, Crop Nanobiology and Molecular Stress Physiology Lab, Amity Institute of Organic Agriculture (AIOA), Noida, Uttar Pradesh, India

Babar Usman Department of Plant Bioscience, Pusan National University, Life and Industry Convergence Research Institute, Miryang, Republic of Korea

Section I

Biosynthesis and signaling of NO

Chapter 1

Nitric oxide: production, functional modifications, and advanced detection strategies

Samta Gupta[1], **Gauri Saxena**[2], **and Praveen Gupta**[2]
[1]*Department of Botany, Acharya Narendra Dev College, University of Delhi, New Delhi, Delhi, India,* [2]*Department of Botany, Faculty of Science, University of Lucknow, Lucknow, Uttar Pradesh, India*

1.1 Introduction

Nitric oxide (NO) is a gasotransmitter that regulates many physiological as well as molecular processes in animal and plant cells (Gupta et al., 2025). The importance of NO was recognized when it was named "Molecule of the Year" in 1992, following its identification as a physiological mediator in vascular, nervous, and immune systems (Gupta et al., 2022; Koshland, 1992). This recognition quickly expanded into plant sciences, where initial studies detected NO emissions from herbicide-treated soybean leaves (Klepper, 1979). Later research identified NO in detached plant tissues, cell suspensions, and mitochondria (Sahay & Gupta, 2017; Szuba et al., 2015). Chemically, NO is a small, uncharged, and diatomic free radical, which, due to its lipophilic nature, can quickly move across the biological membranes (Yang et al., 2021). The short half-life (usually less than 10 seconds under physiological conditions), high diffusion rate (around 50 μm s^{-1}), and water solubility (~1.9 mM at 1 ATM) make it well-suited for coordinating the intra and intercellular signal transmission in the plant cells (Gupta et al., 2022). NO exists in three interchangeable forms, viz., nitrosyl cation (NO^+), neutral radical ($NO\cdot$), and nitroxyl anion (NO^-). These forms enable NO to react with biomolecules, such as metal centers, thiols, lipids, and nucleic acids (McCleverty, 2004).

The biological functions of NO depend on its reactive derivatives, collectively known as reactive nitrogen species (RNS). These include nitrogen dioxide (NO_2), dinitrogen trioxide (N_2O_3), nitroxyl (HNO), and the strong oxidant peroxynitrite ($ONOO^-$), which forms when NO reacts with superoxide

Nitric Oxide in Plant Resilience to Abiotic Stress. DOI: https://doi.org/10.1016/B978-0-443-40497-9.00004-0

($O_2^{\cdot-}$) (Kovacic & Somanathan, 2012). NO-based posttranslational modifications (PTMs) regulate cell signaling and the activity of target proteins under normal physiological as well as stress conditions (Begara-Morales et al., 2015). These NO-induced PTMs primarily include *S*-nitrosation, metal nitrosylation, and tyrosine nitration. These PTMs play important roles in cell signaling, enzyme regulation, and cellular metabolism in plants (Corpas et al., 2014). While *S*-nitrosation affects protein localization and functioning through reversible interactions of NO with the thiol (-SH) side chain of cysteine residues (Gupta et al., 2022), tyrosine nitration involves the attachment of a nitro group to tyrosine residues, resulting in permanent changes in protein function (León, 2022).

NO has been reported to participate in the regulation of seed germination, flowering, lateral root growth, stomatal closure, pollen tube guidance, senescence, and programmed cell death in plants (Sahay & Gupta, 2017). It also triggers the antioxidant defense system for the mitigation of the negative impact of abiotic stress in plants (Gupta & Seth, 2020). The beneficial roles of NO are chiefly ascribed to the regulation of enzyme activity, hormone signaling, and gene expression. For instance, NO activates reactive oxygen species (ROS) scavenging by triggering superoxide dismutase (SOD), catalase (CAT), peroxidase, and AsA-GSH cycle enzymes in plants exposed to stress conditions (Gupta & Seth, 2020). During biotic stress, NO is a crucial mediator of hypersensitive response and systemic acquired resistance to trigger localized programmed cell death as a defense response against phytopathogens (Gupta et al., 2025). The interaction of NO with tripeptide glutathione (GSH) produces *S*-nitrosoglutathione (GSNO), which functions as a stable reservoir of NO with potential signaling roles (Wei et al., 2024). Moreover, GSNO synthesis can also be promoted from the reaction between NO and unsaturated fatty acids such as nitro-linolenic acid (NO_2-Ln) (Ergashev et al., 2025). The detection and quantification of naturally produced NO remain a challenging task due to its transient nature, rapid reactivity, and low cellular concentrations. The endogenous NO levels in plants typically range from picomolar to micromolar, with production rates influenced by tissue type, developmental stage, and growth conditions (Vishwakarma et al., 2019). This inherent variability, coupled with its short half-life and rapid reactivity, warrants sensitive detection methods that can resolve its spatial distribution within subcellular compartments. The common detection methods include electrochemical sensors, quantum cascade laser (QCL)-based spectroscopy, and electron paramagnetic resonance spin trapping (EPR), each with distinct advantages and limitations (Mur et al., 2011; Vandelle & Delledonne, 2008). EPR is specific and can detect short-lived radicals when combined with spin-trapping agents such as DETC (diethyldithiocarbamate) or MGD (N-methyl-D-glucamine dithiocarbamate). QCL absorption spectroscopy allows real-time monitoring of NO in the gas phase with minimal hindrance from other substances (Vishwakarma et al., 2019). The fluorescent dye DAF-2DA (4,5-diaminofluorescein diacetate) is commonly used in plant research for its ease of use, but it has limitations in specificity because of its reaction with nitrogenous compounds (Möller et al., 2019). Hence, accurately detecting NO and its derivatives is not just a technical goal but a biological necessity. As research advances, integrating various detection methods with genetic and pharmacological tools will be vital for building a detailed understanding of NO biology in plants. This chapter aims to explore the chemical properties, synthesis, and action of NO in plants. Various detection methods for its quantification and analysis are also discussed to create a foundation for future research on the roles of NO in plant physiology.

1.2 Nitric oxide production and regulation in plants

The origin of endogenous NO in plants is a highly debated and complex topic in plant biology, owing to the diversity of potential pathways and the lack of consensus regarding the primary enzymatic route. In general, NO production in plants occurs through both oxidative and reductive pathways in multiple subcellular compartments such as cytoplasm, chloroplasts, mitochondria, endoplasmic reticulum, peroxisomes, plasma membrane, and apoplast (Gupta et al., 2025; Sahay & Gupta, 2017). L-arginine and nitrite (NO_2^-) serve as the primary precursors, with enzymatic routes involving either nitric oxide synthase (NOS)-like activity or nitrate reductase (NR)-mediated mechanisms being widely studied (Asgher et al., 2017) (Fig. 1.1). The NOS-dependent pathway was originally proposed based on the successful application of mammalian NOS inhibitors like L-NAME (N^G-nitro-L-arginine methyl ester) and L-NNA (N^G-nitro-L-arginine) in plant tissues. This pathway involves the oxidative conversion of L-arginine to NO and L-citrulline by utilizing NAD(P)H and calmodulin as cofactors (Gupta et al., 2022). Despite these functional analogies, no homologous gene encoding animal-type NOS has been reported in plants to date, prompting researchers to refer to this activity as NOS-like rather than attributing it to a defined plant NOS gene (Sahay & Gupta, 2017). In fact, polyamines and hydroxylamines have also been implicated in NOS-like NO production through the oxidative route (Chamizo-Ampudia et al., 2017). In contrast, the reductive pathway of NO production is well-characterized and involves both enzymatic and nonenzymatic reduction of NO_2^-, primarily via the NR enzyme (Astier et al., 2018; Gupta et al., 2025). The classical role of NR is to catalyze the NAD(P)H-dependent reduction of nitrate (NO_3^-) to NO_2^-, a crucial step in nitrogen assimilation (Gupta et al., 2017). Besides, it can facilitate NO production by reducing NO_2^-, particularly under high NO_2^- and low NO_3^- concentrations (Mur et al., 2013). This alternative function is substantiated by findings from Arabidopsis NR-defective single as well as double *nia1* and *nia2* mutants, which show diminished NO accumulation (Teng et al., 2025). The reduction of NO_2^- to NO can also be mediated by nitrite:NO reductase (Ni-NOR) at the

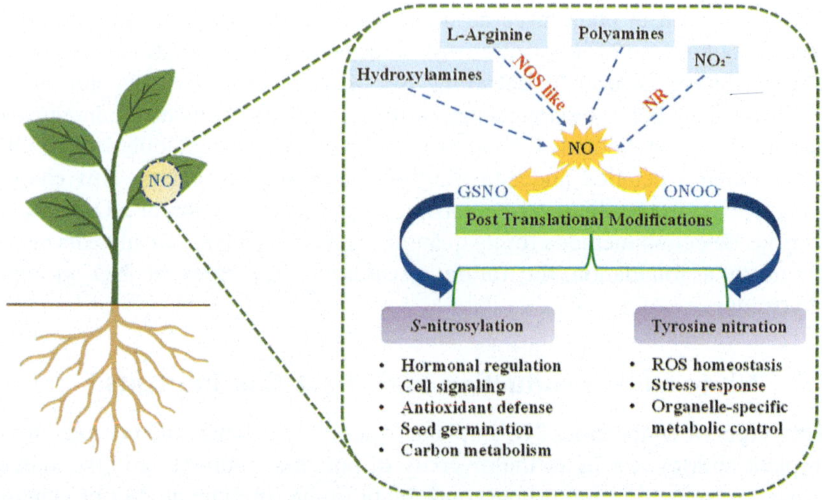

FIGURE 1.1 The biosynthesis of nitric oxide (NO) from different oxidative and reductive routes includes NOS (nitric oxide synthase) like activity and NR (nitrate reductase). NO regulates various developmental aspects in plants. NO-mediated posttranslational modifications play an important role in regulating protein function, signaling networks, and cellular balance in plants during both ideal and stressful conditions [Abbreviations: Nitrite (NO_2^-), S-nitrosoglutathione (GSNO), and peroxynitrite ($ONOO^-$)].

plasma membrane of root cells (Stöhr et al., 2001). Nonenzymatic reduction of NO_2^- to NO occurs under acidic conditions (Sahay & Gupta, 2017). In addition to NR, peroxisomal xanthine oxidoreductase, a molybdoenzyme involved in purine catabolism, has been involved in NO formation via NO_2^- reduction under hypoxic conditions (Asgher et al., 2017). Mitochondria also facilitate NO synthesis under low oxygen availability, where NO_2^- serves as an alternative electron acceptor for components of the electron transport chain, particularly complexes III and IV (Gupta et al., 2011). Chloroplasts can also mediate NO formation under light via reduction of NO_2^- involving carotenoids (Asgher et al., 2017), while ascorbate (AsA)-mediated reactions in the apoplast can reduce NO_2^- to NO (Sahay & Gupta, 2017). Similarly, in hypoxic root zones, plasma membrane-bound NR (PM-NR) and Ni-NOR have been shown to facilitate NO production (Reda et al., 2022).

The cellular levels of NO in plants are regulated by a fine-tuned homeostatic balance that involves synthesis, utilization, scavenging, and feedback regulation. This ensures optimal physiological responses while avoiding cytotoxicity. In plants, NO production is governed by the availability and metabolism of key precursors such as NO_2^- and L-arginine, which participate in reductive and oxidative pathways, respectively (Mur et al.,

2013). Under hypoxic conditions or when the photosynthetic electron transport is disrupted, the uptake of NO_3^- from the soil increases, leading to its cytosolic reduction to NO_2^-, which in turn is converted to NO. However, excess NO_2^- may be sequestered into vacuoles or effluxes from cells, thereby reducing the cytosolic NO_2^- pool and limiting NO production (Mur et al., 2013). Light exposure can also interfere with NO biosynthesis by promoting the reduction of NO_2^- to NH_4^+ in chloroplasts (Rőszer, 2012). In parallel, NR is susceptible to rapid degradation via proteolysis or inhibition by 14-3-3 proteins, thus restricting NO production under fluctuating physiological conditions (Chamizo-Ampudia et al., 2017; Reda et al., 2018). Although the arginine-dependent NO synthesis via NOS-like activity is controversial in plants, evidence suggests that photosynthesis-mediated augmentation in L-arginine levels may favor NO production under low NO_2^- conditions (He et al., 2004; Streatfield et al., 1999), while arginase activity can antagonize this route by degrading the arginine (Flores et al., 2008; Shi et al., 2013).

NO homeostasis can also be controlled by GSNO, formed via the reaction between NO and GSH. The GSNO serves as a relatively stable NO reservoir and donor during protein *S*-nitrosylation, playing a vital role in NO-mediated regulation of protein function and redox signaling (Gupta & Seth, 2023). The cytosolic enzyme *S*-nitrosoglutathione reductase (GSNOR) catalyzes the NADH-dependent degradation of GSNO into oxidized glutathione and ammonia, thereby modulating the available NO pool (Wei et al., 2024). Moreover, GSNOR activity is also subjected to *S*-nitrosylation, a feedback mechanism through which NO can modulate its own turnover by reducing GSNOR activity and preserving GSNO levels (Ergashev et al., 2025; Zhang et al., 2020). In aerobic conditions, NO reacts with oxygen to produce nitrogen dioxide (NO_2) or dinitrogen tetroxide (N_2O_4), which subsequently gets hydrolyzed to NO_2^- and NO_3^-, contributing to its turnover. NO also engages in redox reactions with $O_2{\cdot}^-$, producing $ONOO^-$, a strong nitrating agent that modifies proteins and lipids (León, 2022). Another important NO-scavenging mechanism involves hemoglobins (Hbs), a family of iron-containing globular proteins classified into nonsymbiotic (class 1), symbiotic (class 2), and truncated hemoglobins (THBs, class 3). THBs are considered as homologous to bacterial forms, while other Hbs resemble animal Hbs (Chamizo-Ampudia et al., 2017). Hbs act as NO dioxygenases, which convert NO into NO_3^- in the presence of O_2. In higher plants like *Zea mays*, NR and Hb exhibit coordinated spatial and temporal expression patterns in response to NO_3^- levels, indicating their concerted action in maintaining NO homeostasis (Trevisan et al., 2011). Thus enzymatic control, redox interactions, metabolic cross-talk, and feedback regulation maintain optimal NO homeostasis, enabling plants to effectively coordinate growth, development, and defense responses against environmental challenges.

1.3 Nitric oxide-mediated posttranslational modifications in plants

NO-mediated PTMs play an important role in regulating protein function, signaling networks, and cellular balance in plants during both ideal and stressful conditions (Fig. 1.1) (Begara-Morales et al., 2015). PTMs mainly involve NO-derived RNS like NO^+, NO^-, $ONOO^-$, peroxynitrate (O_2NOO^-), peroxynitrous acid (ONOOH), and peroxynitric acid (O_2NOOH), which modify proteins mainly through *S*-nitrosation, tyrosine nitration, and metal nitrosylation (Saddhe et al., 2019). Of these, *S*-nitrosylation and tyrosine nitration are the frequently observed NO-induced PTMs, especially in plants exposed to stress conditions (Gupta et al., 2022).

1.3.1 *S*-nitrosation

S-nitrosation is a dynamic, reversible, and redox-sensitive PTM that involves covalent addition of a NO moiety to the -SH groups of cysteine residues (Lin et al., 2025). The alteration of protein function could be attributed to the creation of steric hindrance at active sites, which affects protein function, protein–protein interactions, protein degradation, subcellular localization, and DNA-binding activity (Ergashev et al., 2025; Wei et al., 2024). Recent advancements in proteomics and imaging technologies highlighted the potential of *S*-nitrosation in regulating enzyme activity, cell signaling, ion channels, and antioxidant defense in plants (Feng et al., 2019). *S*-nitrosation is recognized as a key modulator that influences biosynthesis, perception, and signaling of phytohormones such as abscisic acid (ABA), auxin, cytokinin, salicylic acid (SA), jasmonic acid (JA), gibberellins (GA), and ethylene (Ergashev et al., 2025; Lin et al., 2025). During ABA signaling, *S*-nitrosation of ABA receptors (PYR/PYL/RCAR) and signaling kinases (SnRK2s) regulates their activity and allows plants to optimize stomatal closure and photosynthesis during water-deficit conditions (Fidler et al., 2022; García-Mata & Lamattina, 2021; Wang et al., 2022). In addition, *S*-nitrosation of negative regulators of ABA signaling, viz., protein phosphatase 2C and serine/threonine-protein kinase (SnRK), inhibits their function to alleviate salt toxicity in tomato seedlings (Wei et al., 2022). ABA growth suppressor ABI5 is also a potential target, where *S*-nitrosation at Cys153 promotes its degradation via CULLIN4-based and KEEP ON GOING E3 ligases, thereby facilitating seed germination in Arabidopsis (Albertos et al., 2015). Transcription factors, including ABA-responsive element-binding factors and dehydration-responsive element-binding proteins, undergo *S*-nitrosation that alters their DNA-binding affinity, influencing antioxidant defense, osmolyte production, and cell wall remodeling (Begara-Morales et al., 2019; Chen et al., 2020; Li et al., 2022; Zhao et al., 2024). In contrast, NO can negatively modulate ABA signaling in guard cells by *S*-nitrosating open stomata 1/sucrose nonfermenting 1-related protein kinase 2.6 (SnRK2.6), leading to ABA-insensitive stomatal closure (Wang et al., 2015). Additionally, *S*-nitrosation of MYB30 enhances its transcriptional activity to reduce ABA content during germination, thereby modulating seed dormancy in Arabidopsis (Nie et al., 2022; Zhao et al., 2024). Moreover, *S*-nitrosation can inhibit GA signaling to conserve energy for stress adaptation in plants (Chen et al., 2022). GA promotes growth by inducing proteasomal degradation of DELLA transcriptional repressors via the GA receptor-F-box protein SLY1 complex (Chen et al., 2022). However, *S*-nitrosation of the Arabidopsis DELLA protein REPRESSOR-OF-ga1--3 (RGA) at Cys-374 disrupts its interaction with SLY1, which stabilizes RGA and prevents its degradation (Chen et al., 2022; Kneeshaw et al., 2014). Auxin signaling, which regulates root initiation and development, is also influenced by *S*-nitrosation of efflux transporters, PIN-FORMED proteins, and regulatory components, including Auxin/Indole-3-Acetic Acid (Aux/IAA) repressors and Auxin response factors. Such modification alters protein localization, activity, and stability, enabling rapid reprogramming for water and nutrient uptake under abiotic stresses (Ergashev et al., 2025). For example, *S*-nitrosation of auxin receptor TRANSPORT INHIBITOR RESPONSE 1 (TIR1) at Cys10 can enhance its association with AUX/IAA proteins to facilitate their degradation and activate auxin-responsive gene expression (Terrile et al., 2012). However, modification of IAA17 at Cys70 inhibits its interaction with TIR1, thereby repressing auxin signaling (Jing et al., 2023). Additionally, *S*-nitrosation of ROP2 (a Rho-related GTPase) reduces auxin transport and root growth by altering its intracellular localization (Kenesi et al., 2023), while modification of TIR1 and SKP1 enhances protein–protein interactions in the SKP1/CULLIN1/F-Box protein (SCF)-type E3 ubiquitin ligase complex to modulate transcription of auxin-responsive genes (Iglesias et al., 2018; Terrile et al., 2022). *S*-nitrosation also impacts cytokinin signaling, where modification of AHP1 at Cys115 suppresses its phospho-relay activity (Feng et al., 2013). Moreover, auxin and cytokinin signaling can also be controlled through *S*-nitrosation of the GSNOR enzyme. Notably, NR-mediated NO production can *S*-nitrosate GSNOR at Cys10, reducing its activity and triggering an NO burst that shifts auxin and cytokinin homeostasis to regulate axillary bud growth in tomato (Yan et al., 2021).

In SA-mediated defense responses, *S*-nitrosation acts as a pivotal regulatory mechanism in modulating key signaling proteins (Wei et al., 2024). For instance, *S*-nitrosation of the immune regulator Non-expressor of PR gene 1 (NPR1) at Cys156 plays an essential role in its monomerization, activation, and nuclear translocation (Tada et al., 2008). This ensures that NPR1 remains inactive in the absence of pathogen attack, thereby preventing unnecessary immune activation. Similarly, inhibition of *Arabidopsis thaliana* salicylic acid-binding protein 3 by *S*-nitrosation at Cys280 demonstrates the important role of NO in fine-tuning the SA signaling cascade in plants (Rui et al., 2024). JA is a key phytohormone that triggers defense mechanisms against necrotrophic pathogens and osmotic stress. *S*-nitrosation of JA biosynthetic enzymes, viz., allene oxide synthase and allene oxide cyclase, alters JA production and thereby modulates defense responses in plants (Shan et al., 2025). *S*-nitrosylation also affects JA perception, as modification of its primary receptor Coronatine Insensitive1 [COI1, a core component of the Skp1-Cullin-F-box (SCF) E3 ubiquitin ligase complex] increases its affinity for the bioactive JA conjugate jasmonoyl-isoleucine (JA-Ile) (Ergashev et al., 2025). The interaction of active jasmonoyl-isoleucine (JA-Ile, the active form of JA) and the COI1-SCF complex promotes the ubiquitination and subsequent degradation of JAZ repressor proteins, releasing JA-responsive genes for activation (Pande et al., 2022). Moreover, *S*-nitrosation of *S*-adenosylmethionine synthase (ethylene biosynthesis precursor) inhibits its activity and reduces ethylene production in tomato under salt stress (Wei et al., 2022). In addition, the ethylene biosynthetic enzyme 1-aminocyclopropane-1-carboxylate oxidase homolog 4 (ACOh4) undergoes *S*-nitrosation at Cys172, resulting in conformational alterations that boost its enzymatic activity to stimulate ethylene production and improve salinity tolerance in tomato (Liu et al., 2023).

In antioxidant metabolism, ascorbate peroxidase (APX) is activated via *S*-nitrosation at Cys32 in pea plants, enhancing ROS scavenging under salinity stress (Begara-Morales et al., 2014). Monodehydroascorbate reductase (MDHAR), which is essential for AsA regeneration, is repeatedly identified as an *S*-nitrosation target in multiple species, including sunflower and tomato, where its modification boosts salt stress tolerance in plants (Begara-Morales et al., 2015; Jain et al., 2016; Tanou et al., 2012). Moreover, glutathione reductase is also reported to undergo *S*-nitrosation, potentially contributing to improved oxidative balance in transgenic tobacco (Begara-Morales et al., 2015). *S*-nitrosation of the Arabidopsis respiratory burst oxidase homolog at Cys890 resulted in the reduction of its ROS generation capacity (Yun et al., 2011), whereas *S*-nitrosation suppresses peroxiredoxin (PsPrxII F) activity in mitochondria, resulting in hydrogen peroxide (H_2O_2) accumulation in pea (Camejo et al., 2015).

Carbon metabolism of plants is also reshaped through *S*-nitrosation of respiratory enzymes (fructose-bisphosphate aldolase, glyceraldehyde-3-phosphate dehydrogenase, pyruvate kinase, citrate synthase, and malate dehydrogenase), which promote energy flux and metabolic adaptation under salinity stress (Jain et al., 2016; Wei et al., 2022). Proline synthesis enzyme Δ^1-pyrroline-5-carboxylate reductase (P5CR) can also be modified by *S*-nitrosation, which boosts proline biosynthesis and tolerance to saline-alkali and drought stress in tomato (Liu, Wei, et al., 2024). *S*-nitrosation of GABA transaminases (SlGABA-TP1/2/3) at Cys316/258/316 decreases their activity, causing GABA accumulation that stimulates malate transport via alumina-activated malate transporter 14 (*SlALMT14*) in roots and SlALMT9 in fruits, synergistically improving both stress resilience and fruit quality in tomato (Liu, Cao, et al., 2024). Therefore NO-mediated *S*-nitrosation modulates plant growth and stress responses by integrating the hormonal regulation, antioxidant defense, energy metabolism, and osmotic adjustment in plants.

1.3.2 Tyrosine nitration

Tyrosine nitration is a significant PTM that involves the covalent addition of a nitro (-NO_2) group, usually at the ortho position of the phenolic hydroxyl group of tyrosine, producing 3-nitrotyrosine (Kolbert et al., 2017). It is generally driven by $ONOO^-$ formation or by NO_2^- production via heme proteins through reactions between H_2O_2 and NO_2^-. This modification introduces a net negative charge to tyrosine residues, often impairing the function of target proteins. Therefore reducing tyrosine nitration could boost stress tolerance in plants (Sabadashka et al., 2021). In *Citrus aurantifolia*, tyrosine nitration alters the factors associated with photosynthesis, glycolysis, antioxidant metabolism, and protein folding under salt stress (Tanou et al., 2012). Confocal laser scanning microscopy (CLSM) and immunoblotting showed elevated NO_2-Tyr in *Helianthus annuus* roots under salt stress (David et al., 2015). Enzymes like peroxisomal hydroxypyruvate reductase (HPR1, photorespiratory pathway) (Corpas et al., 2013), O-acetylserine(thiol)lyase (sulfur assimilation) (Alvarez et al., 2011), and glutamine synthetase (nitrogen metabolism) (Melo et al., 2011) are reported to be inhibited by tyrosine nitration. Unlike reversible cysteine *S*-nitrosylation, tyrosine nitration is considered irreversible, making it a potent but potentially deleterious regulatory mechanism in NO signaling. Stress conditions that elevate NO often coincide with increased ROS ($O_2^{\cdot-}$, hydroxyl radicals, and H_2O_2), necessitating redox regulation via antioxidant enzymes, many of which are targeted by nitration (Begara-Morales et al., 2016; Holzmeister et al., 2015; Palma et al., 2020). In plants, SOD and CAT constitute the first line of defense against toxic ROS generated during stressful conditions. Interestingly, mitochondrial SOD, which catalyzes the conversion of $O_2^{\cdot-}$ into H_2O_2, is inhibited by tyrosine nitration (Holzmeister et al., 2015). During fruit ripening, tyrosine nitration selectively inhibits the CAT activity in the pepper (Chaki et al., 2015). Tyrosine nitration acts as a negative regulator of the AsA–GSH cycle, primarily through its impact on APX and MDHAR activities. The $ONOO^-$-mediated nitration of cytosolic APX at Tyr235 disrupts heme-group properties and thereby lowers its H_2O_2 scavenging capacity (Begara-Morales et al., 2014). Similarly, tyrosine nitration of MDHAR at Tyr345 inhibited its function for AsA regeneration (Begara-Morales et al., 2015). DHAR and GR have also been recognized as potential targets of Tyr-nitration in citrus and sunflower, respectively (Chaki et al., 2009; Tanou et al., 2012), but nitration sites remain to be characterized. Mitochondria frequently accumulate nitrated proteins due to sustained NO and ROS production, which triggers $ONOO^-$ production (Radi et al., 2002), while peroxisomes possess nitrated enzymes like serine hydroxymethyl transferase, CAT, alanine:glyoxylate aminotransferase, glycolate oxidase, and malate dehydrogenase (Lozano-Juste et al., 2011; Sandalio et al., 2019), many of which participate in ROS metabolism and redox regulation (Corpas et al., 2021). Importantly, chloroplast metabolism, including the Calvin–Benson cycle, is modulated by nitration, where enzymes such as glyceraldehyde phosphate dehydrogenase and carbonic anhydrase are inactivated (Chaki et al., 2009; Lozano-Juste et al., 2011). In addition to central metabolism enzymes, several photosystem proteins are also targeted by nitration (Galetskiy et al., 2011). Collectively, tyrosine nitration represents a multifaceted and largely irreversible modification that intricately connects NO signaling with ROS homeostasis, stress physiology, and organelle-specific metabolic control in plants.

1.4 Advanced instrumental approaches for the quantification and detection of nitric oxide in plants

Detection of endogenous levels of NO in plants faces formidable analytical challenges owing to its transient nature, low endogenous concentration, short half-life, and high reactivity with biomolecules (Vishwakarma et al., 2019). Besides, the interconversion of NO to other reactive molecules, such as NO^-, NO, and NO^+, under physiological conditions also hampers the interpretation of results in experiments. To overcome these challenges, substantial advancement has been made in analytical techniques, enabling real-time, spatially resolved, and highly sensitive measurements of NO in intact tissues, isolated cells, and cellular extracts (Li et al., 2024). In general, methods like spectrophotometry, chromatography, mass spectroscopy, fluorescent probes, EPR, chemiluminescence, oxyhemoglobin ($Hb\text{-}O_2$) assay, spin traps, and electrode-based techniques have been used for the quantification of NO in plants (Vishwakarma et al., 2019; Vitecek et al., 2008). These techniques detect either free NO or its stable oxidation products such as NO_2^- and NO_3^-. While each method offers specific advantages, their effectiveness is often influenced by the type of plant material and the experimental conditions of the study (Table 1.1).

1.4.1 Spectrophotometer-based methods

1.4.1.1 Griess reagent assay

The technique was pioneered by Johann Peter Griess (1829–1888), a German organic chemist. This method is a widely used, rapid, and inexpensive approach for quantifying NO_2^-, a stable oxidation product of NO, in biological samples. Its simplicity and inexpensive analytical feasibility have made it a popular choice across diverse scientific disciplines. Under aerobic conditions, NO is readily oxidized to NO_2^-, which can be quantified using the Griess reaction.

TABLE 1.1 Comparative overview of various NO detection methods used in plant research.

S. no.	Method	Strengths	Limitations
1.	Griess reaction	– Simple and inexpensive – Can measure NO in the gas phase	– Lower sensitivity than other methods – Not suitable for cellular localization
2.	Oxyhemoglobin assay	– Very sensitive and no specialized equipment required – Can be performed easily	– ROS may interfere by forming methemoglobin complexes, leading to false results – Sensitive to pH and temperature changes – Not suitable for cellular localization
3.	Dye-based (DAF-FM, DAF-FM-DA, DAF-2DA)	– Widely accessible and commercially available – Detects NO with high sensitivity at nanomolar levels – Enables cellular localization studies	– Sensitive to pH variations and requires the presence of oxygen – DAF-2 does not react directly with NO – Uptake of DAF-2DA varies by tissue/organelles – Generates N_2O_3 as an intermediate and can react with ascorbic/dehydroascorbic acid
4.	EPR spin trapping	– Detects NO at picomolar levels – Highly specific	– Requires semiquantitative and complex analysis – Spin traps can be costly – Specialized and expensive equipment is needed – Not suitable for localization studies
5.	Chemiluminescence	– Detects gaseous NO – Can measure NO in the ppb range – High accuracy and reliability	– Cannot localize NO in the cells and unable to measure released NO in the gas phase – Requires costly instruments and technical expertise

(Continued)

TABLE 1.1 (Continued)

S. no.	Method	Strengths	Limitations
		– Useful for in vitro enzymatic and tissue studies	
6.	Mass spectrometry	– Accurate NO quantification with ^{15}N isotopes – MIMS/RIMS allows simultaneous detection of NO, O_2, and CO_2	– Limited to gas-phase NO detection – Requires costly and advanced equipment
7.	Quantum cascade laser	– Extremely sensitive, detecting NO at the ppb range – Allows in vivo quantification	– Restricted to gas-phase NO detection – Does not provide spatial localization

NO, nitric oxide; *ROS*, reactive oxygen species; *DAF-2DA*, 4,5-diaminofluorescein diacetate.

$$2NO + O_2 \rightarrow 2NO_2$$

$$NO + NO_2 \rightarrow N_2O_3$$

$$N_2O_3 + H_2O \rightarrow 2NO_2^- + 2H^+$$

The method is based on the reaction of NO_2^- with sulfanilamide under acidic conditions, leading to the production of a diazonium compound. This intermediate subsequently couples with N-(1-naphthyl) ethylenediamine (NED) dihydrochloride to produce a stable water-soluble magenta-colored azo dye, which is quantified spectrophotometrically at 540 nm (Fig. 1.2). The resulting absorbance is directly correlated to the NO_2^- concentration, typically determined using a calibration curve ranging from 1 to 10 μM NO_2^- (Vishwakarma et al., 2019). This method has been used to determine NO content in *Brassica juncea* (Gupta & Seth, 2020), *Daucus carota* (Sun et al., 2024), *Triticum turgidum* (Aloui et al., 2024), *Oryza sativa* (Imran et al., 2024; Teng et al., 2025), and *Momordica charantia* (Kumari et al., 2025). Moreover, this assay potentially detects the endogenous NO levels in Arabidopsis NR-

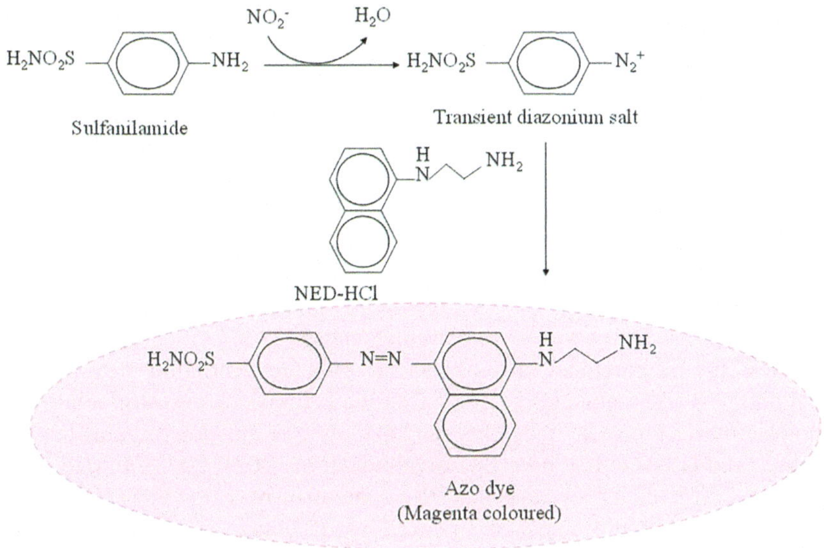

FIGURE 1.2 Detection of endogenous nitric oxide using the Griess assay. NO is oxidized to NO_2^-, which reacts with sulfanilamide to form a diazonium salt. The reaction of intermediate diazonium salt with N-(1-naphthyl) ethylenediamine produces an azo dye (magenta colored), which can be measured at 540 nm using a spectrophotometer.

deficient double mutants (*nia1nia2*) that were impaired in endogenous production (Cantrel et al., 2011; Jian et al., 2015).

Due to its two-step mechanism, variations of the Griess assay exist. In some protocols, sulfanilamide and NED are premixed before the reaction, while others involve the addition of NED after the reaction of NO_2^- with sulfanilamide. However, the most widely adopted version, particularly in commercial kits, follows a sequential method, that is, NO_2^- is first mixed with sulfanilamide, followed by the addition of NED to give the highest yield of the chromophore. Moreover, NO quantification can also be achieved using CrO_3-based oxidizer columns that convert NO to NO_2^- before Griess detection (Vitecek et al., 2008). The Saville reaction, which uses $HgCl_2$ and the Griess reagent, is also employed to determine *S*-nitrosothiols, further expanding the versatility of the Griess assay for NO measurement (Wang, Shang, et al., 2024).

1.4.1.2 Oxyhemoglobin assay

The oxyhemoglobin [Fe(II)(Hb)O_2] assay for NO determination was introduced by Feelisch and Noack (1987). The principle of this assay relies on the reaction of NO with HbO_2 to form methemoglobin [Fe(III)(Hb)] and NO_3^-, resulting in a characteristic shift in absorbance from 415–421 nm (HbO_2) to 401 nm (MetHb) (Lärfars & Gyllenhammar, 1995). NO levels are measured by monitoring the change in absorbance between 401 nm and 421 nm (Vishwakarma et al., 2019). The assay is both rapid and precise, with a sensitivity ranging from 1.3 to 2.8 nM, and has been extensively employed in studies investigating the NO signaling pathways (Murphy & Noack, 1994; Vishwakarma et al., 2019). For example, Aghdam et al. (2019) reported that melatonin-induced NO modulated the expression of key genes involved in polyamine metabolism (*ornithine decarboxylase* and *arginine decarboxylase*) and proline biosynthesis (*P5CR*, *ornithine aminotransferase*, and *proline dehydrogenase*) in tomato under chilling stress. The oxyhemoglobin assay revealed the importance of endogenous NO in restricting heavy metal (cadmium, lead, manganese, and chromium) translocation while boosting photosynthesis, antioxidant defense, and osmolyte production (glycine betaine, proline, and GSH) in bamboo (Emamverdian et al., 2021, 2023). In rice, time-course measurements revealed that NO content in root tips increased within 6 hours of aluminum exposure, peaked at 12 hours, and then stabilized, contributing to enhanced root elongation and reduced aluminum accumulation in apical tissues (Lan et al., 2021). In general, NO detection methods involve homogenization of excised leaves in an ice-cold buffer (1 M NaCl, 0.1 M sodium acetate, and 1% AsA, pH 6.0), followed by centrifugation and analysis of the supernatant for HbO_2-to-MetHb conversion. The underlying reaction mechanism can be summarized as follows

$$HbO_2 + NO \rightarrow MetHb + NO_3^-$$

A preexperiment calibration is typically performed using potassium ferricyanide or NO donors such as sodium nitroprusside (SNP), *S*-nitroso-N-acetylpenicillamine, or GSNO (Hetrick & Schoenfisch, 2009). Moreover, $ONOO^-$ is also capable of oxidizing oxyhemoglobin; hence, SOD amendment is suggested as a control to restrict the formation of $ONOO^-$ from NO and $O_2^{\cdot-}$, ensuring that the observed methemoglobin formation can be reliably attributed to NO. Despite its early popularity, the assay fell out of use due to several limitations. Firstly, the production of fresh HbO_2 is technically demanding, as it requires hemoglobin oxygenation followed by purification using chromatography. Additionally, ROS such as H_2O_2 can oxidize HbO_2 independently of NO, leading to false results. Although antioxidants such as SOD and CAT are used to mitigate ROS interference (Delledonne et al., 1998; Ma et al., 2016), the low substrate affinity of CAT (K_m = 25 mM) may still allow residual H_2O_2 to affect results. This is particularly problematic since the production of NO occurs simultaneously, or near simultaneously, with that of H_2O_2 during plant defense responses. Moreover, fluctuations in pH, temperature, and the presence of competing heme-containing proteins can alter the absorbance spectrum of Met-Hb, further complicating the measurements of endogenous NO in plants (Mur et al., 2011; Noack et al., 1992).

1.4.2 Electrode-based measurement system

Electrode-based measurement of NO offers a cost-effective and practical method for NO quantification in plants (Vishwakarma et al., 2019). A standard NO electrode is composed of a main electrode (usually Teflon/Platinum) paired with a reference electrode made of $Ag/AgCl_2$ (Besson-Bard et al., 2008; Davies & Zhang, 2008). These elements are embedded in a glass or steel tube containing an electrolyte solution made up of 30 mM NaCl or 0.3 nM HCl. The open end of the electrode is sealed with a NO-permeable membrane typically made up of chloroprene rubber, polytetrafluoroethylene, cellulose acetate, or phenylenediamine (Vishwakarma et al., 2019). The assay is based on the oxidation of NO at a potential of +0.8 to +0.9 V with respect to the standard electrode (Shibuki, 1990). Although commercially available NO sensors can be expensive, low-cost alternatives have been successfully developed. For example, Besson-Bard et al. (2008) constructed a homemade platinum/iridium microelectrode coated with nickel phthalocyanine, Nafion, and o-phenylenediamine. This configuration showed high sensitivity and selectivity in monitoring NO during cryptogein-induced responses in tobacco cell cultures. Similarly, Leshem and Pinchasov (2000) reported the suitability of using NO electrodes in whole plants by inserting the electrode directly into the fruit tissue. However, direct electrode insertion is not feasible for all tissues, limiting its utility to soft parts or cell cultures. Despite this limitation, the technique has been successfully employed in various studies, including cadmium-induced cell death in tobacco BY-2 cells (Ma et al., 2010) and NO monitoring during wounding responses in *Pelargonium* leaves (Arasimowicz et al., 2009). Agnihotri and Seth (2016) measured the endogenous NO production in tomato leaves using a sensor-based detection system. The study revealed a rapid rise in NO production in plants after NO_3^- treatment under arsenic stress. During measurement, temperature should be kept constant, considering that the solubility of NO gas is very temperature-sensitive. Microelectrodes have been developed (<1 mm) to allow direct real-time monitoring of NO in the cells. Importantly, NO electrodes are unaffected by physiologically relevant levels of NO_3^-, $O_2^{\cdot-}$, $ONOO^-$, or H_2O_2, enhancing their specificity (Besson-Bard et al., 2008). However, a key limitation is that NO emitted into the gas stream cannot be detected using this method.

1.4.3 Dye-based methods

Fluorescent probes have become indispensable tools for visualizing NO in plants due to their sensitivity, noninvasiveness, and easier application (Gupta & Seth, 2023). A variety of cell-permeable fluorescent dyes such as acridine-TEMPO-DTCS-Fe(II), FNOCT (fluorescent NO cheletropic traps), DAMBO-PH (8-(3,4-diaminophenyl)-2,6-bis(2-carboxyethyl)-4,4-difluoro-1,3,5,7-tetramethyl-4-bora-3a,4a-diaza-s-indacene), DAR-4MAM (diaminorhodamine),

DAQ (1,2-diaminoanthraquinone), DAN (2,3-diaminonaphthalene), MNIP-Cu (4-methoxy-2-(1H-naphtho[2,3-d]imidazol-2-yl)phenol-copper-II), and DAFs have been employed to visualize the distribution of NO in diverse plant tissues (Jain et al., 2016; Vishwakarma et al., 2019). These probes react with NO-derived species (dinitrogen trioxide, N_2O_3) to generate fluorescent products, such as DAN and DAF derivatives (Jain et al., 2016). Among these, DAF dyes (especially DAF-2DA) are widely employed for detecting intracellular NO in xylem (Gaupels et al., 2008), guard cells (Xu et al., 2024), pollen tubes (Jimenez-Quesada et al., 2017), root primordia (Pescador et al., 2022), ovules (Wang, Guo, et al., 2024), etc. The DAF-FM-DA probe (a derivative of DAF-2DA) offers enhanced fluorescence stability and higher sensitivity (~3 nM detection limit) across a broader pH range. These probes have enabled high-resolution imaging in plants such as *Cucumis sativus* (Janicka et al., 2022), *A. thaliana* (Pescador et al., 2022), *O. sativa* (Prakash et al., 2022), *Solanum lycopersicum* (Li et al., 2022), *B. juncea* (Gupta & Seth, 2023), and *Hordeum vulgare* (Valentovičová et al., 2025). Additionally, these dyes have allowed spatial mapping of NO production during developmental processes and stress responses, such as mechanical wounding (Huang et al., 2004), pathogen attack (Pescador et al., 2022), salinity (Gupta & Seth, 2023), drought (He et al., 2024), and heavy metals (Singh et al., 2024).

DAF-2DA diffuses into cells and is deacetylated by intracellular esterases to DAF-2, which then reacts with N_2O_3 (an NO oxidation product) to form the fluorescent triazole compound DAF-2T (λ_{exc} = 495 nm, λ_{em} = 515 nm) (Fig. 1.3). This reaction does not proceed directly with NO but rather with its oxidative intermediate (Ignarro et al., 1993), making the process dependent on the redox status of the cells (Planchet & Kaiser, 2006).

A common approach involves infiltrating leaves with 10 mM DAF-2 DA solution prepared in a buffer containing 10 mM MES-Tris, 10 mM KCl, and 0.1 mM $CaCl_2$ (pH 7.6), followed by confocal microscopy analysis one hour

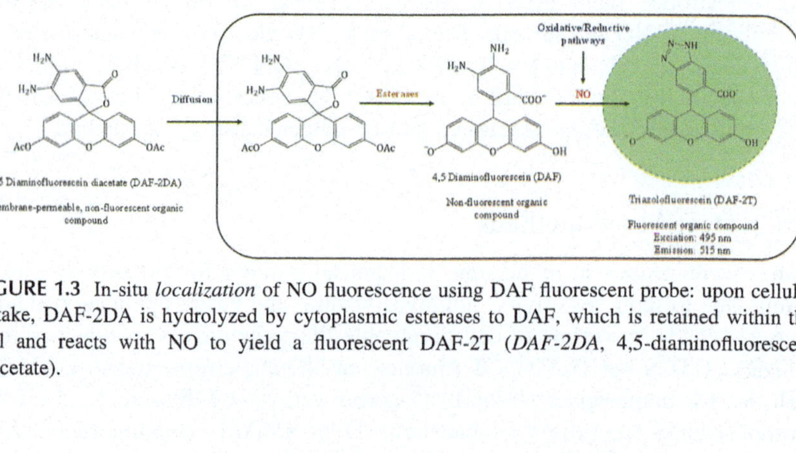

FIGURE 1.3 In-situ *localization* of NO fluorescence using DAF fluorescent probe: upon cellular uptake, DAF-2DA is hydrolyzed by cytoplasmic esterases to DAF, which is retained within the cell and reacts with NO to yield a fluorescent DAF-2T (*DAF-2DA*, 4,5-diaminofluorescein diacetate).

postinfiltration (Qu et al., 2006). Alternatively, small leaf segments (2 × 2 mm) can be incubated in 10 mM DAF-2 DA prepared in MES buffer (10 mM MES, 50 mM KCl, pH 6.15) for 10 minutes in the dark, rinsed to remove excess dye, and immediately analyzed by CLSM (Prats et al., 2005). For epidermal sections, either abaxial or adaxial leaf peels are incubated in 10 mM DAF-2 DA or 5 mM DAF-FM DA in 10 mM Tris buffer (pH 7.0–7.2) for 10–30 minutes in the dark at room temperature (Vandelle & Delledonne, 2008). After washing in dye-free buffer, samples are mounted on slides and observed using CLSM (excitation at 488 nm; emission at 500–550 nm) or epifluorescence microscopy with FITC filters (excitation at 490 nm; emission at 525 nm). NO detection in cell suspensions involves incubating cells in assay buffer containing 10–20 μM DAF-2 DA (for 15–60 minutes) in the dark under constant agitation, followed by washing to remove excess dye. Thereafter, fluorescence can be measured either directly from treated cultures using a fluorometer or from aliquots transferred into multiwell plates using a plate reader (Vandelle & Delledonne, 2008). Measurement is typically performed at 485–495 nm excitation and 510–515 nm emission. By calibrating NO standard curves, it has been observed that the detection limit of NO by DAF-2 is about 3–5 nM (Vandelle & Delledonne, 2008).

The specificity of DAF dyes has been a subject of critical scrutiny. Although they do not respond to NO_2^-, NO_3^-, $ONOO^-$, or H_2O_2 (Kojima et al., 1998), interference from dehydroascorbate and AsA, which produce DAF-like fluorescence, has been reported (Zhang et al., 2002). Ye et al. (2004) suggested freezing the probe on dry ice to minimize reactions with less mobile interfering compounds. Moreover, pH sensitivity remains a limitation, especially under stress conditions that alter intracellular pH, such as hypoxia or pathogen infection (Vitecek et al., 2008). Other potential artifacts include the influence of divalent cations, particularly calcium, which has been reported to markedly enhance the fluorescence signal of DAF-2, as well as the contribution of the excitation light itself. Moreover, repeated or prolonged exposure to excitation light does not lead to photobleaching of the dye; instead, it amplifies the fluorescence response.

Hence, shorter light periods are recommended. To confirm that fluorescence signals are indeed NO-dependent, coapplication of the NO scavenger cPTIO (2-(4-Carboxyphenyl)-4,4,5,5-tetramethylimidazoline-1-oxyl-3-oxide) is recommended. cPTIO reacts with NO to form $NO_2\cdot$ and cPTIO, effectively quenching NO-related fluorescence. As an alternative, MNIP-Cu (a copper-complexed fluorescent probe) offers enhanced specificity for NO imaging. Unlike DAF dyes, MNIP-Cu reacts directly with NO to emit blue fluorescence (excitation: 330–385 nm; emission: 420–500 nm) and performs well under both aerobic and anaerobic conditions (Keisham et al., 2019). It has been utilized for detecting auxin-induced NO generation in sunflower hypocotyls, as well as for localization studies in whole seedlings, stigma tissues, and protoplasts (Jain et al., 2016; Yadav et al., 2013). The direct reaction of

MNIP-Cu with NO makes it suitable for applications where DAF dyes may yield ambiguous results. Despite certain drawbacks, such as photobleaching and redox dependence, in DAF dyes, fluorescent NO probes remain essential for real-time visualization of NO in live cells. Integration with fluorescence and CLSM enables three-dimensional, subcellular-resolution, greatly advancing our understanding of NO signaling microdomains in plant physiology.

1.4.4 Electron paramagnetic resonance

Electron spin resonance, or electron paramagnetic resonance, is a sensitive technique employed for the detection of free radicals, including the short-lived NO radical (Galatro & Puntarulo, 2016). This method relies on the fundamental principle that unpaired electrons in a magnetic field can transition between parallel and antiparallel spin states when exposed to microwave radiation, resulting in a measurable resonance signal (Kleschyov et al., 2007). Since NO is a free radical containing an unpaired electron, it is theoretically detectable via EPR. However, its transient nature, characterized by a rapid relaxation time and extremely short half-life, renders challenges in direct detection under physiological conditions (Vishwakarma et al., 2019). To overcome this, EPR-based NO detection often employs spin trapping, wherein a chemical spin trap reacts with NO to form a more stable paramagnetic complex that can be readily detected (Steffen-Heins & Steffens, 2015). The reaction of spin trapping can be represented as

$$R_{*} + ST \rightarrow R - SA$$

where R* is the radical intermediate (e.g., NO), ST is a spin trap, and R–SA is the stable radical adduct. The effectiveness of this approach depends on multiple factors, such as the transient radical concentration, spin trap used, kinetics of adduct formation, and the stability of the resulting complex (Steffen-Heins & Steffens, 2015). When optimized, EPR detection using spin traps can identify picomolar NO levels in the samples. The major advantage of EPR lies in its specificity, that is, it only detects paramagnetic species; therefore, NO can be distinguished from nonradical nitrogen oxides such as NO_2^- or NO_3^- due to its distinct spectral fingerprint of the chemical/electronic structure around the unpaired electron (Kleschyov et al., 2007). However, the ephemeral nature of the NO radical entails using specific spin-trap chemicals, which give longevity to a NO-dependent radical signal (NO. + ST → NO-ST) (Mur et al., 2011). The spin-trapping agents have been employed in plant NO studies, which can be classified as endogenous (e.g., Hb) and exogenous [DETC and MGD, and imidazoline-derived compounds (cPTIO)]. Hbs have been widely used due to their ability to form stable NO adducts in both oxy- and deoxy-states (Vishwakarma et al., 2019). These NO-Hb complexes generate a characteristic paramagnetic signal detectable by EPR. However, iron-dithiocarbamate complexes, such as DETC and MGD, are more commonly used due to their high NO affinity and ability to form stable, EPR-detectable adducts. These adducts produce a distinct three-line EPR spectrum, making them ideal for quantifying NO in biological samples (Vandelle & Delledonne, 2008). Side-chain modifications on dithiocarbamate ligands (e.g., methyl-, ethyl-, glucamine-, and sarcosine-based groups) allow researchers to work under different environments, that is, lipophilic Fe-DETC complexes are suited for membranes, while hydrophilic Fe-MGD complexes produce ideal results in aqueous phases (Kleschyov et al., 2007). Although cPTIO can be used to trap NO, its byproducts degrade rapidly, and their use can sometimes lead to overestimation or nonspecific reactions due to simultaneous NO_2^-/NO_3^- generation (Vishwakarma et al., 2019). EPR has been effectively used to investigate NO production across various plant systems. In Arabidopsis, exposed to *Pseudomonas syringae*, NO production was measured using $Fe(MGD)_2$ spin traps combined with $^{15}NO_2^-$ labeling (Modolo et al., 2006). Similarly, EPR confirmed endogenous NO release in pollen grains (Bright et al., 2009) and embryonic axes of sorghum (Jasid et al., 2008). In *Synechococcus elongatus*, lipophilic spin traps have been used to study roles of NO in modulating the oxygen-evolving complex of photosystem II (Sarrou et al., 2003). EPR analysis of intact soybean nodules revealed the accumulation of nitrosyl-leghemoglobin ($Lb^{2+}NO$) in nodules that are defective in NO_2^- or NOR (Calvo-Begueria et al., 2018).

Despite its sensitivity, EPR is underutilized in plant biology due to the high cost of the equipment, the need for specialized expertise, and the impracticality of continuous or in vivo measurements (Mur et al., 2011). Despite its strengths, EPR spectroscopy has several limitations. One major concern is the interference from ROS, such as H_2O_2, which can oxidize Fe^{2+} or DETC and mimic NO signals. Additionally, compounds like L-NAME and certain reductants may falsely simulate NO signals or regenerate NO from NO_2^-, leading to overestimation (Vandelle & Delledonne, 2008). Moreover, continuous in planta NO monitoring is impractical, and the method's cost and technical complexity often demand expert collaboration. Nonetheless, EPR remains a benchmark method for NO detection in biological systems when precision and specificity are essential. Its continued application in plant science, particularly with innovations in spin-trap chemistry and instrumentation, could promise to deepen our understanding of the dynamic roles of NO in plant research.

1.4.5 Mass spectrometry

This technique for NO detection in plants was demonstrated by Conrath et al. (2004). Mass spectrometric approaches such as Membrane Inlet Mass Spectrometry (MIMS) and Restriction Capillary Inlet Mass Spectrometry (RIMS) offer a promising alternative to more conventional detection techniques by allowing real-time and noninvasive measurement of NO in cell cultures and whole-plant systems. MIMS is generally used to quantify NO in aqueous solution, while the RIMS approach can detect NO from the gaseous phase (Vishwakarma et al., 2019). In MIMS, a semipermeable Teflon membrane serves as an interface between the sample and mass spectrometer to allow NO to diffuse through, enabling its direct analysis based on specific mass-to-charge ratios (m/z). NO is typically identified at 30 m/z, and its abundance reflects changes in its concentration (Vandelle & Delledonne, 2008). This setup has been successfully used to detect NO production in plants such as tobacco and soybean, especially when exposed to *P. syringae* (Conrath et al., 2004). RIMS is better suited for whole plant tissues or leaves (Vandelle & Delledonne, 2008). In this method, samples are placed within a sealed cuvette, and the gas phase above the tissue is continuously mixed using a metal bellows pump. The released volatile gas diffuses through a narrow capillary tube into the mass spectrometer. A major strength of both MIMS and RIMS lies in their ability to distinguish between isotopically labeled nitrogen isotopes (Mur et al., 2011). By using ^{15}N-labeled NO_3^-, NO_2^-, arginine, or hydroxylamines, researchers can trace the biochemical origin of NO and determine the contribution of specific pathways to its production. The calibration can be achieved using standards like NO-saturated water or by reducing KNO_2 with KI (Berkels et al., 2001). In fact, calibration experiments by Conrath et al. (2004) demonstrated that one abundance unit at m/z = 30 corresponds to approximately 10–13 pmol of NO. The specificity of this technique has been validated using NO donors such as *S*-nitroso-N-acetyl-DL-penicillamine and *S*-nitroso-L-glutathione and further confirmed using NO scavengers such as cPTIO (Conrath et al., 2004). Bethke et al. (2004) also utilized a similar mass spectrometry-based approach to examine apoplastic NO production via nonenzymatic NO_2^- reduction, highlighting the broader applicability of mass spectrometric methods in plant systems. Despite its strengths, the adoption of MIMS/RIMS in plant NO research remains limited, primarily due to the high cost and technical demands of the instrumentation. Additionally, the method is generally restricted to detecting extracellular NO. Other variables, such as low oxygen levels, must be tightly controlled for optimal performance (Vandelle & Delledonne, 2008). Nevertheless, MIMS/RIMS provides a powerful platform for high-resolution, real-time detection of NO in various plant contexts, with the added advantage of isotopic discrimination, making it a valuable tool for future research in plant NO biology.

1.4.6 Chemiluminescence

Chemiluminescence is one of the robust, sensitive, and widely adopted methods among different techniques developed for the detection of NO in biological samples (Li et al., 2024). The assay is based on a well-characterized gas-phase reaction where NO interacts with ozone (O_3) to form an excited intermediate, NO_2*. As NO_2* relaxes to its ground state, it emits photons (typically above 600 nm), which are detected by a photomultiplier tube, with light intensity directly proportional to the amount of NO present in the samples (Gupta et al., 2005; Planchet et al., 2005). Moreover, gases such as ethylene and sulfur compounds can also react with O_3; however, their emission spectra fall outside this range (440–470 nm and <400 nm, respectively), making this technique highly specific for NO (Mur et al., 2011).

$$NO + O_3 \rightarrow NO_{2*} + O_2$$

$$NO_{2*} \rightarrow NO_2 + h\nu$$

A typical chemiluminescence setup includes a sample chamber, which can vary in size from whole-plant enclosures to small vials for cell or organelle analysis. A controlled gas flow is maintained using flow regulators, while the chamber is connected to a chemiluminescence detector and an O_3 generator. Calibration is routinely performed using NO-free air (0 ppben) and known concentrations of NO (1–35 ppb) obtained by diluting standard NO gas (500 ppb in nitrogen) with NO-free air (Planchet & Kaiser, 2006; Vandelle & Delledonne, 2008). This system can handle both gaseous and aqueous samples. In the latter case, NO is transferred to the gas phase by bubbling with an inert gas or gentle agitation (Gupta et al., 2005). The method offers excellent sensitivity and can detect NO at concentrations as low as 20-50 pmol (Byrnes et al., 1996; Planchet et al., 2005). The major applications include monitoring NO during anoxia, assessing NO production from hydroxylamines, and investigating NO_2^- reduction in mitochondria (Planchet et al., 2005; Wany et al., 2016). Despite its advantages, a limitation of this technique is that it detects only free, gaseous NO and not its oxidized derivatives (Planchet & Kaiser, 2006). Since NO readily converts to NO_3^- or NO_2^- in biological systems, part of the endogenous NO pool may go undetected. However, this challenge can be addressed using an indirect chemiluminescence approach. In this method, oxidized NO species in sample extracts are chemically reduced back to NO using heated acidic vanadium chloride. The regenerated NO is then measured via chemiluminescence (Rümer et al., 2009).

1.4.7 Quantum cascade laser-based system

QCL is one of the most advanced and promising technologies for the detection of cellular NO, especially in plant biology, due to its high sensitivity, precision, and capacity for real-time gas analysis (Vishwakarma et al., 2019). It effectively operates at ambient temperature and offers numerous advantages such as compact design, high spectral resolution, low power consumption, and exceptional sensitivity up to ppb levels (Montilla-Bascón et al., 2018). A QCL that emits infrared radiation at 5.3 μm (1850 cm^{-1}) is typically required for NO detection in the biological samples. There are two different lasers currently being used, namely, AllnAs/GalnAs- and GaAs/AlGaAs-based materials. In a standard setup, a gas sample is directed through a specialized multipass absorption cell, usually equipped with mirrors at both ends. This design allows the laser beam to reflect multiple times through the sample, significantly increasing the optical path length. For instance, a physical cell length of just 30 cm can achieve an effective path length of 76 m (Vishwakarma et al., 2019). The concentration of NO is determined from comparison with a gas mixture of known NO concentrations in nitrogen. To ensure stability and reproducibility, the optical system is typically operated in a temperature-controlled environment. Recent advancements in thermos-electrically cooled QCLs eliminate the need for bulky cryogenic cooling systems such as CO lasers. These lasers operate continuously in the 1847–1854 cm^{-1} region, producing output powers up to 25 mW at 245 K. Their performance can be enhanced by combining them with astigmatic multipass absorption cells and using wavelength modulation spectroscopy to reduce noise and increase sensitivity (Mur et al., 2011). The difference in strength between the light coming into the cell and the transmitted light is proportional to the NO concentration. A key outcome of this method is the ability to detect NO concentrations as low as 0.8 ppb within a second measurement interval. Such sensitivity, combined with low flow requirements (1 L h^{-1}), makes QCL ideal for applications where sample dehydration due to excessive airflow can be a concern.

One of the advantages of QCL-based NO detection is the ability to conduct long-term measurements with minimal airflow, preserving sample integrity. This makes it ideal for continuous monitoring of gaseous NO emissions under stress or during developmental transitions. However, QCL systems require specialized, often expensive equipment and considerable technical expertise to operate and maintain (Vishwakarma et al., 2019). As a result, most current QCL applications in plant science have emerged through collaborations between physicists and biologists. Nonetheless, ongoing efforts to miniaturize and simplify QCL instruments could soon make them more accessible to mainstream plant biology labs.

1.4.8 Laser photoacoustic detection

Laser photoacoustic detection (LPAD) is a highly sensitive method used for the measurement of NO emissions in plants. The technique relies on the photoacoustic effect, where infrared light absorbed by gas molecules causes pressure fluctuations that can be detected as sound waves (Vandelle & Delledonne, 2008). The magnitude of this sound is directly proportional to the concentration of the absorbing gas, such as NO. The results are usually expressed as moles of NO h^{-1} g^{-1} of fresh weight. In the setup, a photoacoustic cell is placed inside the laser cavity to enhance signal strength, typically using a CO laser that emits in the mid-infrared range (4.6-8.2 μm) (Mur et al., 2011). The sensitivity and reliability of the method can be evaluated using different concentrations (0.1 to 10 mM) of NO donor SNP, which generally gives proportional results as reported in tobacco after the first hour postinfiltration (Leshem & Pinchasov, 2000; Mur et al., 2005; Vandelle & Delledonne, 2008). The study utilized LPAD to monitor NO emissions during strawberry and avocado ripening by placing samples in a nitrogen-flushed transparent cuvette to prevent NO_2 formation. The refinement of LPAD has enabled real-time monitoring of NO emissions in tobacco and Arabidopsis during bacterial infection, with a detection sensitivity as low as 17 nmol h^{-1} (Mur et al., 2005; Mur et al., 2006). Moreover, to avoid the signals from other gases such as ethylene and NO_2, measurements are often performed across multiple laser lines (1933.4, 1921.8, 1900.0, 1856.4, and 1790.6 cm^{-1}) (Mur et al., 2011). Despite its precision and sensitivity, LPAD has several limitations. It exclusively detects gaseous NO, providing no insight into intracellular concentrations or spatial distribution within tissues. Additionally, the equipment is large, costly, and technically demanding, restricting its routine use in most of the plant biology laboratories (Mur et al., 2013; Vandelle & Delledonne, 2008).

1.5 Conclusion

NO has emerged as a versatile regulator in plant biology, orchestrating diverse processes ranging from growth and development to the regulation of gene expression and augmentation of stress tolerance. Its biological action is mediated through direct and often reversible interactions with biomolecules. Significant progress has been made in understanding the roles of NO in PTMs, particularly *S*-nitrosylation and tyrosine nitration. NO-based PTMs highlight the central role of NO in orchestrating plant responses to biotic as well as abiotic challenges, where it collaborates with ROS and key phytohormones such as ABA, SA, JA, and auxins. Yet, several questions remain unresolved, including the precise enzymatic and nonenzymatic routes of NO biosynthesis in plants, the spatio-temporal regulation of its production, and the functional consequences of its rapid interconversion into RNS. Unraveling these gaps will not only require refined detection methods but also integrative strategies that combine biochemical, genetic, and systems biology approaches. However, the short-lived and compartment-specific nature of these species makes their detection inherently challenging. Researchers have relied on a spectrum of approaches, from chemical assays that measure stable breakdown products to sophisticated instrumentation capable of capturing real-time NO dynamics. While methods such as Griess assay, chemiluminescence, fluorescence imaging with probes like DAF dyes, EPR spin trapping, and mass spectrometry have all been applied successfully, no single technique provides a complete picture. Each has inherent strengths and limitations in terms of sensitivity, specificity, spatial resolution, or compatibility with in vivo systems (Table 1.1). This makes it essential to validate findings through complementary approaches and rigorous controls, including the use of scavengers, donors, or inactive analogs. Advances in electrochemical sensing, high-throughput proteomics, and imaging now enable researchers not only to quantify NO but also to unravel its spatial and temporal homeostasis within cellular compartments. By refining detection and quantification strategies and linking them with functional outcomes, future research will continue to unlock the multifaceted roles of NO, strengthening its position as a master regulator of plant growth, development, and stress tolerance.

Acknowledgments

This work was supported by a grant from the Anusandhan National Research Foundation (ANRF) with File no.: ANRF/ECRG/2024/003069/LS.

References

Aghdam, M. S., Luo, Z., Jannatizadeh, A., Sheikh-Assadi, M., Sharafi, Y., Farmani, B., Fard, J. R., & Razavi, F. (2019). Employing exogenous melatonin applying confers chilling tolerance in tomato fruits by upregulating ZAT2/6/12 giving rise to promoting endogenous polyamines, proline, and nitric oxide accumulation by triggering arginine pathway activity. *Food Chemistry, 275*, 549–556. https://doi.org/10.1016/j.foodchem.2018.09.157, www.elsevier.com/locate/foodchem.

Agnihotri, A., & Seth, C. S. (2016). Exogenously applied nitrate improves the photosynthetic performance and nitrogen metabolism in tomato (*Solanum lycopersicum* L. cv Pusa Rohini) under arsenic (V) toxicity. *Physiology and Molecular Biology of Plants, 22*(3), 341–349. https://doi.org/10.1007/s12298-016-0370-2, http://www.springerlink.com/content/120962/.

Albertos, P., Romero-Puertas, M. C., Tatematsu, K., Mateos, I., Sánchez-Vicente, I., Nambara, E., & Lorenzo, O. (2015). S-nitrosylation triggers ABI5 degradation to promote seed germination and seedling growth. *Nature Communications, 6*. https://doi.org/10.1038/ncomms9669, http://www.nature.com/ncomms/index.html.

Aloui, N., Kharbech, O., Mahjoubi, Y., Chaoui, A., & Karmous, I. (2024). Exogenous melatonin alleviates cadmium toxicity in wheat (*Triticum turgidum* L.) by modulating endogenous nitric oxide and hydrogen sulfide metabolism. *Journal of Soil Science and Plant Nutrition, 24*(2), 2535–2552. https://doi.org/10.1007/s42729-024-01675-3, https://www.springer.com/journal/42729.

Alvarez, C., Lozano-Juste, J., Romero, L. C., García, I., Gotor, C., & León, J. (2011). Inhibition of *Arabidopsis* O-acetylserine(thiol)lyase A1 by tyrosine nitration. *Journal of Biological Chemistry, 286*(1), 578–586. https://doi.org/10.1074/jbc.M110.147678Spain, http://www.jbc.org/content/286/1/578.full.pdf+html.

Arasimowicz, M., Floryszak-Wieczorek, J., Milczarek, G., & Jelonek, T. (2009). Nitric oxide, induced by wounding, mediates redox regulation in pelargonium leaves. *Plant Biology, 11*(5), 650–663. https://doi.org/10.1111/j.1438-8677.2008.00164.x.

Asgher, M., Per, T. S., Masood, A., Fatma, M., Freschi, L., Corpas, F. J., & Khan, N. A. (2017). Nitric oxide signaling and its crosstalk with other plant growth regulators in plant responses to abiotic stress. *Environmental Science and Pollution Research, 24*(3), 2273–2285. https://doi.org/10.1007/s11356-016-7947-8, http://www.springerlink.com/content/0944-1344.

Astier, J., Gross, I., & Durner, J. (2018). Nitric oxide production in plants: An update. *Journal of Experimental Botany, 69*(14), 3401–3411.

Begara-Morales, J. C., Chaki, M., Valderrama, R., Mata-Pérez, C., Padilla, M. N., & Barroso, J. B. (2019). The function of S-nitrosothiols during abiotic stress in plants. *Journal of Experimental Botany, 70*(17), 4429–4439. https://doi.org/10.1093/jxb/erz197, http://jxb.oxfordjournals.org/.

Begara-Morales, J. C., Sánchez-Calvo, B., Chaki, M., Mata-Pérez, C., Valderrama, R., Padilla, M. N., López-Jaramillo, J., Luque, F., Corpas, F. J., & Barroso, J. B. (2015). Differential molecular response of monodehydroascorbate reductase and glutathione reductase by nitration and S-nitrosylation. *Journal of Experimental Botany, 66*(19), 5983–5996. https://doi.org/10.1093/jxb/erv306, http://jxb.oxfordjournals.org/.

Begara-Morales, J. C., Sánchez-Calvo, B., Chaki, M., Valderrama, R., Mata-Pérez, C., López-Jaramillo, J., Padilla, M. N., Carreras, A., Corpas, F. J., & Barroso, J. B. (2014). Dual regulation of cytosolic ascorbate peroxidase (APX) by tyrosine nitration and S-nitrosylation. *Journal of Experimental Botany, 65*(2), 527–538. https://doi.org/10.1093/jxb/ert396.

Begara-Morales, J. C., Sánchez-Calvo, B., Chaki, M., Valderrama, R., Mata-Pérez, C., Padilla, M. N., Corpas, F. J., & Barroso, J. B. (2016). Antioxidant systems are regulated by nitric oxide-mediated post-translational modifications (NO-PTMs). *Frontiers in Plant Science, 7*. https://doi.org/10.3389/fpls.2016.00152, https://www.frontiersin.org/journals/plant-science.

Berkels, R., Purol-Schnabel, S., & Roesen, R. (2001). A new method to measure nitrate/nitrite with a NO-sensitive electrode. *Journal of Applied Physiology, 90*(1), 317–320.

Besson-Bard, A., Griveau, S., Bedioui, F., & Wendehenne, D. (2008). Real-time electrochemical detection of extracellular nitric oxide in tobacco cells exposed to cryptogein, an elicitor of defence responses. *Journal of Experimental Botany, 59*(12), 3407–3414. https://doi.org/10.1093/jxb/ern189.

Bethke, P. C., Badger, M. R., & Jones, R. L. (2004). Apoplastic synthesis of nitric oxide by plant tissues. *The Plant Cell, 16*(2), 332–341.

Bright, J., Hiscock, S. J., James, P. E., & Hancock, J. T. (2009). Pollen generates nitric oxide and nitrite: A possible link to pollen-induced allergic responses. *Plant Physiology and Biochemistry, 47*(1), 49–55. https://doi.org/10.1016/j.plaphy.2008.09.005.

Byrnes, C. A., Bush, A., & Shinebourne, E. A. (1996). Measuring expiratory nitric oxide in humans. *Methods in Enzymology, 269*, 459–474.

Calvo-Begueria, L., Rubio, M. C., Martínez, J. I., Pérez-Rontomé, C., Delgado, M. J., Bedmar, E. J., & Becana, M. (2018). Redefining nitric oxide production in legume nodules through complementary insights from electron paramagnetic resonance spectroscopy and specific fluorescent probes. *Journal of Experimental Botany, 69*(15), 3703–3714. https://doi.org/10.1093/jxb/ery159, http://jxb.oxfordjournals.org/.

Camejo, D., Ortiz-Espín, A., Lázaro, J. J., Romero-Puertas, M. C., Lázaro-Payo, A., Sevilla, F., & Jiménez, A. (2015). Functional and structural changes in plant mitochondrial PrxII F caused by NO. *Journal of Proteomics, 119*, 112–125. https://doi.org/10.1016/j.jprot.2015.01.022, http://www.elsevier.com.

Cantrel, C., Vazquez, T., Puyaubert, J., Rezé, N., Lesch, M., Kaiser, W. M., Dutilleul, C., Guillas, I., Zachowski, A., & Baudouin, E. (2011). Nitric oxide participates in cold-responsive phosphosphingolipid formation and gene expression in *Arabidopsis thaliana*. *New Phytologist, 189*(2), 415–427. https://doi.org/10.1111/j.1469-8137.2010.03500.x, http://onlinelibrary.wiley.com/journal/10.1111/(ISSN)1469-8137.

Chaki, M., Valderrama, R., Fernández-Ocaña, A. M., Carreras, A., López-Jaramillo, J., Luque, F., Palma, J. M., Pedrajas, J. R., Begara-Morales, J. C., Sánchez-Calvo, B., Gómez-Rodríguez, M. V., Corpas, F. J., & Barroso, J. B. (2009). Protein targets of tyrosine nitration in sunflower (*Helianthus annuus* L.) hypocotyls. *Journal of Experimental Botany, 60*(15), 4221–4234. https://doi.org/10.1093/jxb/erp263, http://jxb.oxfordjournals.org/.

Chaki, M., Álvarez De Morales, P., Ruiz, C., Begara-Morales, J. C., Barroso, J. B., Corpas, F. J., & Palma, J. M. (2015). Ripening of pepper (*Capsicum annuum*) fruit is characterized by an enhancement of protein tyrosine nitration. *Annals of Botany, 116*(4), 637–647. https://doi.org/10.1093/aob/mcv016, http://aob.oxfordjournals.org/.

Chamizo-Ampudia, A., Sanz-Luque, E., Llamas, A., Galvan, A., & Fernandez, E. (2017). Nitrate reductase regulates plant nitric oxide homeostasis. *Trends in Plant Science, 22*(2), 163–174. https://doi.org/10.1016/j.tplants.2016.12.001, www.elsevier.com/inca/publications/store/3/0/9/6/0/index.htt.

Chen, L., Sun, S., Song, C. P., Zhou, J. M., Li, J., & Zuo, J. (2022). Nitric oxide negatively regulates gibberellin signaling to coordinate growth and salt tolerance in *Arabidopsis*. *Journal of Genetics and Genomics, 49*(8), 756–765. https://doi.org/10.1016/j.jgg.2022.02.023, http://www.journals.elsevier.com/journal-of-genetics-and-genomics/.

Chen, T., Tian, M., & Han, Y. (2020). Hydrogen sulfide: A multi-tasking signal molecule in the regulation of oxidative stress responses. *Journal of Experimental Botany, 71*(10), 2862–2869.

Conrath, U., Amoroso, G., Köhle, H., & Sültemeyer, D. F. (2004). Non-invasive online detection of nitric oxide from plants and some other organisms by mass spectrometry. *Plant Journal, 38*(6), 1015–1022. https://doi.org/10.1111/j.1365-313X.2004.02096.x.

Corpas, F. J., González-Gordo, S., & Palma, J. M. (2021). Nitric oxide (NO) scaffolds the peroxisomal protein–protein interaction network in higher plants. *International Journal of Molecular Sciences, 22*(5), 2444.

Corpas, F. J., Leterrier, M., Begara-Morales, J. C., Valderrama, R., Chaki, M., López-Jaramillo, J., Luque, F., Palma, J. M., Padilla, M. N., Sánchez-Calvo, B., Mata-Pérez, C., & Barroso, J. B. (2013). Inhibition of peroxisomal hydroxypyruvate reductase (HPR1) by tyrosine nitration. *Biochimica et Biophysica Acta - General Subjects, 1830*(11), 4981–4989. https://doi.org/10.1016/j.bbagen.2013.07.002.

Corpas, F. J., Begara-Morales, J. C., Sánchez-Calvo, B., Chaki, M., & Barroso, J. B. (2014). *Nitration and S-nitrosylation: Two post-translational modifications (PTMs) mediated by reactive nitrogen species (RNS) and their role in signalling processes of plant cells*. Springer Science and Business Media LLC, 267–281. 10.1007/978-3-319-10079-1_13.

David, A., Yadav, S., Baluška, F., & Bhatla, S. C. (2015). Nitric oxide accumulation and protein tyrosine nitration as a rapid and long distance signalling response to salt stress in sunflower seedlings. *Nitric Oxide: Biology and Chemistry / Official Journal of the Nitric Oxide Society, 50*, 28–37. https://doi.org/10.1016/j.niox.2015.08.003.

Davies, I. R., & Zhang, X. (2008). Nitric oxide selective electrodes. *Methods in Enzymology, 436*, 63–95. https://doi.org/10.1016/S0076-6879(08)36005-4.

Delledonne, M., Xia, Y., Dixon, R. A., & Lamb, C. (1998). Nitric oxide functions as a signal in plant disease resistance. *Nature, 394*(6693), 585–588. https://doi.org/10.1038/29087.

Emamverdian, A., Ding, Y., Barker, J., Liu, G., Li, Y., & Mokhberdoran, F. (2023). Sodium nitroprusside improves bamboo resistance under Mn and Cr toxicity with stimulation of antioxidants activity, relative water content, and metal translocation and accumulation. *International Journal of Molecular Sciences, 24*(3), 1942. https://doi.org/10.3390/ijms24031942.

Emamverdian, A., Ding, Y., Barker, J., Mokhberdoran, F., Ramakrishnan, M., Liu, G., & Li, Y. (2021). Nitric oxide ameliorates plant metal toxicity by increasing antioxidant capacity and reducing Pb and Cd translocation. *Antioxidants, 10*(12), 1981. https://doi.org/10.3390/antiox10121981.

Ergashev, U., Yang, T., Zhang, Y., Ergashev, I., Luo, L., Yu, M., & Han, Y. (2025). Protein S-nitrosylation: Roles for nitric oxide signaling in the regulation of stress-dependent phytohormones. *Physiologia Plantarum, 177*(2), e70180.

Feelisch, M., & Noack, E. A. (1987). Correlation between nitric oxide formation during degradation of organic nitrates and activation of guanylate cyclase. *European Journal of Pharmacology, 139*(1), 19–30. https://doi.org/10.1016/0014-2999(87)90493-6.

Feng, J., Chen, L., & Zuo, J. (2019). Protein S-nitrosylation in plants: Current progresses and challenges. *Journal of Integrative Plant Biology, 61*(12), 1206–1223.

Feng, J., Wang, C., Chen, Q., Chen, H., Ren, B., Li, X., & Zuo, J. (2013). S-nitrosylation of phosphotransfer proteins represses cytokinin signaling. *Nature Communications, 4*. https://doi.org/10.1038/ncomms2541.

Fidler, J., Graska, J., Gietler, M., Nykiel, M., Prabucka, B., Rybarczyk-Płońska, A., Muszyńska, E., Morkunas, I., & Labudda, M. (2022). PYR/PYL/RCAR Receptors play a vital role in the abscisic-acid-dependent responses of plants to external or internal stimuli. *Cells, 11*(8), 1352. https://doi.org/10.3390/cells11081352.

Flores, T., Todd, C. D., Tovar-Mendez, A., Dhanoa, P. K., Correa-Aragunde, N., Hoyos, M. E., Brownfield, D. M., Mullen, R. T., Lamattina, L., & Polacco, J. C. (2008). Arginase-negative mutants of *Arabidopsis* exhibit increased nitric oxide signaling in root development. *Plant Physiology, 147*(4), 1936–1946. https://doi.org/10.1104/pp.108.121459, http://www.plantphysiol.org/cgi/reprint/147/4/1936.

Galatro, A., & Puntarulo, S. (2016). Measurement of nitric oxide (NO) generation rate by chloroplasts employing electron spin resonance (ESR). In *Plant nitric oxide: Methods and protocols*. New York, NY: Springer New York, 103–112.

Galetskiy, D., Lohscheider, J. N., Kononikhin, A. S., Popov, I. A., Nikolaev, E. N., & Adamska, I. (2011). Phosphorylation and nitration levels of photosynthetic proteins are conversely regulated by light stress. *Plant Molecular Biology, 77*(4-5), 461–473. https://doi.org/10.1007/s11103-011-9824-7.

García-Mata, C., & Lamattina, L. (2021). Nitric oxide, a key molecule in the regulation of guard cell function. *Frontiers in Plant Science, 18*(10), 830.

Gaupels, F., Furch, A. C. U., Will, T., Mur, L. A. J., Kogel, K. H., & Van Bel, A. J. E. (2008). Nitric oxide generation in *Vicia faba* phloem cells reveals them to be sensitive detectors as well as possible systemic transducers of stress signals. *New Phytologist, 178*(3), 634–646. https://doi.org/10.1111/j.1469-8137.2008.02388.x.

Gupta, K. J., Fernie, A. R., Kaiser, W. M., & van Dongen, J. T. (2011). On the origins of nitric oxide. *Trends in Plant Science, 16*(3), 160–168. https://doi.org/10.1016/j.tplants.2010.11.007.

Gupta, K. J., Stoimenova, M., & Kaiser, W. M. (2005). In higher plants, only root mitochondria, but not leaf mitochondria reduce nitrite to NO, *in vitro* and *in situ*. *Journal of Experimental Botany, 56*(420), 2601–2609.

Gupta, P., Kumar, D., & Seth, C. S. (2022). Nitric oxide-mediated salinity stress tolerance in plants: Signaling and physiological perspectives. In *Advancements in developing abiotic stress-resilient plants*. CRC Press, 45–63.

Gupta, P., Saxena, G., & Gupta, R. (2025). Nitrate reductase-mediated nitric oxide synthesis in shaping stress resilience in plants. *Journal of Experimental Botany*, eraf353.

Gupta, P., & Seth, C. S. (2023). 24-Epibrassinolide regulates functional components of nitric oxide signalling and antioxidant defense pathways to alleviate salinity stress in *Brassica juncea* L. cv. Varuna. *Journal of Plant Growth Regulation, 42*(7), 4207–4222. https://doi.org/10.1007/s00344-022-10884-y, https://www.springer.com/journal/344.

Gupta, P., Srivastava, S., & Seth, C. S. (2017). 24-Epibrassinolide and sodium nitroprusside alleviate the salinity stress in *Brassica juncea* L. cv. Varuna through cross talk among proline, nitrogen metabolism and abscisic acid. *Plant and Soil, 411*(1), 483–498.

Gupta, P., & Seth, C. S. (2020). Interactive role of exogenous 24 Epibrassinolide and endogenous NO in *Brassica juncea* L. under salinity stress: Evidence for NR-dependent NO biosynthesis. *Nitric Oxide: Biology and Chemistry / Official Journal of the Nitric Oxide Society, 97*, 33–47. https://doi.org/10.1016/j.niox.2020.01.014.

He, F., Niu, M. X., Wang, T., Li, J. L., Shi, Y. J., Zhao, J. J., Li, H., Xiang, X., Yang, P., Wei, S. Y., Lin, T. T., Huang, X., Xia, X., & Wan, X. Q. (2024). The ubiquitin E3 ligase RZFP1 affects drought tolerance in poplar by mediating the degradation of the protein phosphatase PP2C-9. *Plant Physiology, 196*(4), 2936–2955. https://doi.org/10.1093/plphys/kiae497, https://academic.oup.com/plphys/article/196/4/2936/7772440.

He, Y., Tang, R. H., Hao, Y., Stevens, R. D., Cook, C. W., Ahn, S. M., Jing, L., Yang, Z., Chen, L., Guo, F., Fiorani, F., Jackson, R. B., Crawford, N. M., & Pei, Z. M. (2004). Nitric oxide represses the *Arabidopsis* floral transition. *Science (New York, N.Y.), 305*(5692), 1968–1971. https://doi.org/10.1126/science.1098837.

Hetrick, E. M., & Schoenfisch, M. H. (2009). Analytical chemistry of nitric oxide. *Annual Review of Analytical Chemistry, 2*, 409–433. https://doi.org/10.1146/annurev-anchem-060908-155146, http://arjournals.annualreviews.org/doi/pdf/10.1146/annurev-anchem-060908-155146.

Holzmeister, C., Gaupels, F., Geerlof, A., Sarioglu, H., Sattler, M., Durner, J., & Lindermayr, C. (2015). Differential inhibition of *Arabidopsis* superoxide dismutases by peroxynitrite-mediated tyrosine nitration. *Journal of Experimental Botany, 66*(3), 989–999. https://doi.org/10.1093/jxb/eru458, http://jxb.oxfordjournals.org/.

Huang, X., Stettmaier, K., Michel, C., Hutzler, P., Mueller, M. J., & Durner, J. (2004). Nitric oxide is induced by wounding and influences jasmonic acid signaling in *Arabidopsis thaliana*. *Planta, 218*(6), 938–946. https://doi.org/10.1007/s00425-003-1178-1.

Iglesias, M. J., Terrile, M. C., Correa-Aragunde, N., Colman, S. L., Izquierdo-Álvarez, A., Fiol, D. F., París, R., Sánchez-López, N., Marina, A., Calderón Villalobos, L. I. A., Estelle, M., Lamattina, L., Martínez-Ruiz, A., & Casalongué, C. A. (2018). Regulation of SCFTIR1/AFBs E3 ligase assembly by S-nitrosylation of *Arabidopsis* SKP1-like1 impacts on auxin signaling. *Redox Biology, 18*, 200–210. https://doi.org/10.1016/j.redox.2018.07.003, http://www.journals.elsevier.com/redox-biology.

Ignarro, L. J., Fukuto, J. M., Griscavage, J. M., Rogers, N. E., & Byrns, R. E. (1993). Oxidation of nitric oxide in aqueous solution to nitrite but not nitrate: Comparison with enzymatically formed nitric oxide from L-arginine. *Proceedings of the National Academy of Sciences, 90*(17), 8103–8107. https://doi.org/10.1073/pnas.90.17.8103.

Imran, M., Hussain, S., Rana, M. S., Iqbal, A., Rehman, N. U., Chen, X., & Tang, X. (2024). Exogenously applied sodium nitroprusside alleviated cadmium toxicity in different aromatic rice cultivars by improving nitric oxide accumulation and modulating oxidative metabolism. *Agronomy, 14*(8). https://doi.org/10.3390/agronomy14081857, www.mdpi.com/journal/agronomy/.

Jain, P., David, A., & Bhatla, S. C. (2016). A novel protocol for detection of nitric oxide in plants. In *Plant nitric oxide: Methods and protocols*. New York, NY: Springer New York, 69–79.

Janicka, M., Reda, M., Napieraj, N., Michalak, A., Jakubowska, D., & Kabała, K. (2022). Involvement of diamine oxidase in modification of plasma membrane proton pump activity in *Cucumis sativus* L. seedlings under cadmium stress. *International Journal of Molecular Sciences, 24*(1), 262.

Jasid, S., Simontacchi, M., & Puntarulo, S. (2008). Exposure to nitric oxide protects against oxidative damage but increases the labile iron pool in sorghum embryonic axes. *Journal of Experimental Botany, 59*(14), 3953–3962.

Jian, W., Zhang, Dw, Zhu, F., Wang, Sx, Zhu, T., Pu, Xj, Zheng, T., Feng, H., & Lin, Hh (2015). Nitrate reductase-dependent nitric oxide production is required for regulation alternative oxidase pathway involved in the resistance to *Cucumber mosaic* virus infection in *Arabidopsis*. *Plant Growth Regulation, 77*(1), 99–107. Available from https://doi.org/10.1007/s10725-015-0040-3, http://www.springeronline.com/sgw/cda/frontpage/0,11855,1-40109-70-35547532-0,00.html.

Jimenez-Quesada, M. J., Carmona, R., Lima-Cabello, E., Traverso, J.Á., Castro, A. J., Claros, M. G., & Alché, J. D. D. (2017). Generation of nitric oxide by olive (*Olea europaea* L.) pollen during in vitro germination and assessment of the S-nitroso- and nitro-proteomes by computational predictive methods. *Nitric Oxide - Biology and Chemistry, 68*, 23–37. Available from https://doi.org/10.1016/j.niox.2017.06.005, http://www.elsevier.com/inca/publications/store/6/2/2/9/2/6/index.htt.

Jing, H., Yang, X., Emenecker, R. J., Feng, J., Zhang, J., Figueiredo, M. R. A. D., Chaisupa, P., Wright, R. C., Holehouse, A. S., Strader, L. C., & Zuo, J. (2023). Nitric oxide-mediated S-nitrosylation of IAA17 protein in intrinsically disordered region represses auxin signaling. *Journal of Genetics and Genomics, 50*(7), 473–485. https://doi.org/10.1016/j.jgg.2023.05.001, http://www.journals.elsevier.com/journal-of-genetics-and-genomics/.

Keisham, M., Jain, P., Singh, N., von Toerne, C., Bhatla, S. C., & Lindermayr, C. (2019). Deciphering the nitric oxide, cyanide and iron-mediated actions of sodium nitroprusside in cotyledons of salt stressed sunflower seedlings. *Nitric Oxide: Biology and Chemistry / Official Journal of the Nitric Oxide Society, 88*, 10–26. https://doi.org/10.1016/j.niox.2019.03.008.

Kenesi, E., Kolbert, Z., Kaszler, N., Klement, É., Ménesi, D., Molnár, Á., Valkai, I., Feigl, G., Rigó, G., Cséplő, Á., Lindermayr, C., & Fehér, A. (2023). The ROP2 GTPase participates in nitric oxide (NO)-induced root shortening in *Arabidopsis*. *Plants, 12*(4), 750. https://doi.org/10.3390/plants12040750.

Klepper, L. (1979). Nitric oxide (NO) and nitrogen dioxide (NO_2) emissions from herbicide-treated soybean plants. *Atmospheric Environment (1967), 13*(4), 537–542. https://doi.org/10.1016/0004-6981(79)90148-3.

Kleschyov, A. L., Wenzel, P., & Munzel, T. (2007). Electron paramagnetic resonance (EPR) spin trapping of biological nitric oxide. *Journal of Chromatography B, 851*(1-2), 12–20.

Kneeshaw, S., Gelineau, S., Tada, Y., Loake, G. J., & Spoel, S. H. (2014). Selective protein denitrosylation activity of thioredoxin-h5 modulates plant immunity. *Molecular Cell, 56*(1), 153–162. https://doi.org/10.1016/j.molcel.2014.08.003, www.molecule.org.

Kojima, H., Nakatsubo, N., Kikuchi, K., Kawahara, S., Kirino, Y., Nagoshi, H., Hirata, Y., & Nagano, T. (1998). Detection and imaging of nitric oxide with novel fluorescent indicators: Diaminofluoresceins. *Analytical Chemistry, 70*(13), 2446–2453. https://doi.org/10.1021/ac9801723, http://pubs.acs.org/journal/ancham.

Kolbert, Z., Feigl, G., Bordé, Á., Molnár, Á., & Erdei, L. (2017). Protein tyrosine nitration in plants: Present knowledge, computational prediction and future perspectives. *Plant Physiology and Biochemistry, 113*, 56–63. https://doi.org/10.1016/j.plaphy.2017.01.028, http://www.journals.elsevier.com/plant-physiology-and-biochemistry/.

Koshland, D. E. (1992). The molecule of the year. *Science (New York, N.Y.), 258*(5090), 1861. https://doi.org/10.1126/science.1470903.

Kovacic, P., & Somanathan, R. (2012). Nitric oxide, peroxynitrite, peroxynitrous acid, nitroxyl, nitrogen dioxide, nitrous oxide: Biochemical mechanisms and bioaction. *Current Bioactive Compounds, 8*(4), 297–306.

Kumari, R., Rakhra, G., Alsahli, A. A., Bhat, J. A., & Ahmad, P. (2025). Exploring the potential of signalling molecules hydrogen sulfide and nitric oxide in augmenting salt stress resilience in bitter gourd. *BMC Plant Biology, 25*(1). https://doi.org/10.1186/s12870-025-06942-8, https://bmcplantbiol.biomedcentral.com/.

Lan, Y., Chai, Y., Xing, C., Wu, K., Wang, L., & Cai, M. (2021). Nitric oxide reduces the aluminum-binding capacity in rice root tips by regulating the cell wall composition and enhancing antioxidant enzymes. *Ecotoxicology and Environmental Safety, 208*, 111499. https://doi.org/10.1016/j.ecoenv.2020.111499.

Leshem, Y.'acov Y., & Pinchasov, Y. (2000). Non-invasive photoacoustic spectroscopic determination of relative endogenous nitric oxide and ethylene content stoichiometry during the ripening of strawberries *Fragaria anannasa* (Duch.) and avocados *Persea americana* (Mill.)1. *Journal of Experimental Botany, 51*(349), 1471–1473. https://doi.org/10.1093/jexbot/51.349.1471.

León, J. (2022). Protein tyrosine nitration in plant nitric oxide signaling. *Frontiers in Plant Science, 13*. https://doi.org/10.3389/fpls.2022.859374.

Li, J., LoBue, A., Heuser, S. K., & Cortese-Krott, M. M. (2024). Determination of nitric oxide and its metabolites in biological tissues using ozone-based chemiluminescence detection: A state-of-the-art review. *Antioxidants, 13*(2). https://doi.org/10.3390/antiox13020179, www.mdpi.com/journal/antioxidants.

Li, X., Wang, S., Chen, X., Cong, Y., Cui, J., Shi, Q., Liu, H., & Diao, M. (2022). The positive effects of exogenous sodium nitroprusside on the plant growth, photosystem II efficiency and Calvin cycle of tomato seedlings under salt stress. *Scientia Horticulturae, 299*, 111016. https://doi.org/10.1016/j.scienta.2022.111016.

Lin, W., Shang, J. X., Li, X. Y., Zhou, X. F., & Zhao, L. Q. (2025). Nitric oxide regulates multiple signal pathways in plants via protein S-nitrosylation. *Current Issues in Molecular Biology, 47*(6), Available from https://doi.org/10.3390/cimb47060407, https://www.mdpi.com/journal/cimb.

Liu, M., Wei, J. W., Liu, W., & Gong, B. (2023). S-nitrosylation of ACO homolog 4 improves ethylene synthesis and salt tolerance in tomato. *New Phytologist, 239*(1), 159–173. https://doi.org/10.1111/nph.18928, http://onlinelibrary.wiley.com/journal/10.1111/(ISSN)1469-8137.

Liu, M., Cao, B., Wei, J. W., & Gong, B. (2024). Redesigning a S-nitrosylated pyruvate-dependent GABA transaminase 1 to generate high-malate and saline–alkali-tolerant tomato. *New Phytologist, 242*(5), 2148–2162.

Liu, W., Wei, J. W., Shan, Q., Liu, M., Xu, J., & Gong, B. (2024). Genetic engineering of drought- and salt-tolerant tomato via Δ1-pyrroline-5-carboxylate reductase S-nitrosylation. *Plant Physiology, 195*(2), 1038–1052.

Lozano-Juste, J., Colom-Moreno, R., & Leon, J. (2011). *In vivo* protein tyrosine nitration in *Arabidopsis thaliana. Journal of Experimental Botany, 62*(10), 3501–3517.

Lärfars, G., & Gyllenhammar, H. (1995). Measurement of methemoglobin formation from oxyhemoglobin a real-time, continuous assay of nitric oxide release by human polymorphonuclear leukocytes. *Journal of Immunological Methods, 184*(1), 53–62. https://doi.org/10.1016/0022-1759(95)00074-K.

Ma, W., Xu, W., Xu, H., Chen, Y., He, Z., & Ma, M. (2010). Nitric oxide modulates cadmium influx during cadmium-induced programmed cell death in tobacco BY-2 cells. *Planta, 232*(2), 325–335. https://doi.org/10.1007/s00425-010-1177-y.

Ma, Z., Marsolais, F., Bykova, N. V., & Igamberdiev, A. U. (2016). Nitric oxide and reactive oxygen species mediate metabolic changes in barley seed embryo during germination. *Frontiers in Plant Science, 7*, 2016. https://doi.org/10.3389/fpls.2016.00138, http://journal.frontiersin.org/article/10.3389/fpls.2016.00138/pdf.

McCleverty, J. A. (2004). Chemistry of nitric oxide relevant to biology. *Chemical Reviews, 104*(2), 403–418. https://doi.org/10.1021/cr020623q.

Melo, P. M., Silva, L. S., Ribeiro, I., Seabra, A. R., & Carvalho, H. G. (2011). Glutamine synthetase is a molecular target of nitric oxide in root nodules of *Medicago truncatula* and is regulated by tyrosine nitration. *Plant Physiology, 157*(3), 1505–1517. https://doi.org/10.1104/pp.111.186056, http://www.plantphysiol.org/content/157/3/1505.full.pdf+html.

Modolo, L. V., Augusto, O., Almeida, I. M. G., Pinto-Maglio, C. A. F., Oliveira, H. C., Seligman, K., & Salgado, I. (2006). Decreased arginine and nitrite levels in nitrate reductase-deficient *Arabidopsis thaliana* plants impair nitric oxide synthesis and the hypersensitive response to *Pseudomonas syringae*. *Plant Science, 171*(1), 34–40. https://doi.org/10.1016/j.plantsci.2006.02.010.

Montilla-Bascón, G., Mandon, J., Harren, F. J., Mur, L. A., Cristescu, S. M., & Prats, E. (2018). Quantum cascade lasers-based detection of nitric oxide. In *Nitric oxide: Methods and protocols*. New York, NY: Springer New York, 49–57.

Mur, L. A., Carver, T. L., & Prats, E. (2006). NO way to live; the various roles of nitric oxide in plant–pathogen interactions. *Journal of Experimental Botany, 57*(3), 489–505.

Mur, L. A. J., Mandon, J., Cristescu, S. M., Harren, F. J. M., & Prats, E. (2011). Methods of nitric oxide detection in plants: A commentary. *Plant Science, 181*(5), 509–519. https://doi.org/10.1016/j.plantsci.2011.04.003.

Mur, L. A. J., Mandon, J., Persijn, S., Cristescu, S. M., Moshkov, I. E., Novikova, G. V., Hall, M. A., Harren, F. J. M., Hebelstrup, K. H., & Gupta, K. J. (2013). Nitric oxide in plants: An assessment of the current state of knowledge. *AoB Plants, 5*(0), pls052. https://doi.org/10.1093/aobpla/pls052.

Mur, L. A. J., Santosa, I. E., Laarhoven, L. J. J., Holton, N. J., Harren, F. J. M., & Smith, A. R. (2005). Laser photoacoustic detection allows in planta detection of nitric oxide in tobacco following challenge with avirulent and virulent *Pseudomonas syringae* pathovars. *Plant Physiology, 138*(3), 1247–1258. https://doi.org/10.1104/pp.104.055772, http://www.plantphysiol.org/.

Murphy, M. E., & Noack, E. (1994). Nitric oxide assay using hemoglobin method. *Methods in Enzymology, 233*(C), 240–250. https://doi.org/10.1016/S0076-6879(94)33027-1.

Möller, M. N., Rios, N., Trujillo, M., Radi, R., Denicola, A., & Alvarez, B. (2019). Detection and quantification of nitric oxide-derived oxidants in biological systems. *Journal of Biological Chemistry, 294*(40), 14776–14802. https://doi.org/10.1074/jbc.REV119.006136, http://www.jbc.org/content/294/40/14776.full.pdf.

Nie, K., Zhao, H., Wang, X., Niu, Y., Zhou, H., & Zheng, Y. (2022). The MIEL1-ABI5/MYB30 regulatory module fine tunes abscisic acid signaling during seed germination. *Journal of Integrative Plant Biology, 64*(4), 930–941. https://doi.org/10.1111/jipb.13234, http://onlinelibrary.wiley.com/journal/10.1111/(ISSN)1744-7909.

Noack, E., Kubitzek, D., & Kojda, G. (1992). Spectrophotometric determination of nitric oxide using hemoglobin. *Neuroprotocols, 1*(2), 133–139.

Palma, J. M., Mateos, R. M., López-Jaramillo, J., Rodríguez-Ruiz, M., González-Gordo, S., Lechuga-Sancho, A. M., & Corpas, F. J. (2020). Plant catalases as NO and H_2S targets. *Redox Biology, 34*. https://doi.org/10.1016/j.redox.2020.101525, http://www.journals.elsevier.com/redox-biology.

Pande, A., Mun, B. G., Rahim, W., Khan, M., Lee, D. S., Lee, G. M., Al Azzawi, T. N. I., Hussain, A., Kim, C. K., & Yun, B. W. (2022). Phytohormonal regulation through protein S-nitrosylation under stress. *Frontiers in Plant Science, 13*. https://doi.org/10.3389/fpls.2022.865542, https://www.frontiersin.org/journals/plant-science.

Pescador, L., Fernandez, I., Pozo, M. J., Romero-Puertas, M. C., Pieterse, C. M. J., & Martínez-Medina, A. (2022). Nitric oxide signalling in roots is required for MYB72-dependent systemic resistance induced by *Trichoderma* volatile compounds in *Arabidopsis*. *Journal of Experimental Botany, 73*(2), 584–595. https://doi.org/10.1093/jxb/erab294, http://jxb.oxfordjournals.org/.

Planchet, E., & Kaiser, W. M. (2006). Nitric oxide (NO) detection by DAF fluorescence and chemiluminescence: A comparison using abiotic and biotic NO sources. *Journal of Experimental Botany, 57*(12), 3043–3055. https://doi.org/10.1093/jxb/erl070, http://jxb.oxfordjournals.org/.

Planchet, E., Jagadis Gupta, K., Sonoda, M., & Kaiser, W. M. (2005). Nitric oxide emission from tobacco leaves and cell suspensions: Rate limiting factors and evidence for the involvement of mitochondrial electron transport. *The Plant Journal, 41*(5), 732–743. https://doi.org/10.1111/j.1365-313x.2005.02335.x.

Prakash, V., Rai, P., Sharma, N. C., Singh, V. P., Tripathi, D. K., Sharma, S., & Sahi, S. (2022). Application of zinc oxide nanoparticles as fertilizer boosts growth in rice plant and alleviates chromium stress by regulating genes involved in oxidative stress. *Chemosphere, 303*. https://doi.org/10.1016/j.chemosphere.2022.134554, www.elsevier.com/locate/chemosphere.

Prats, E., Mur, L. A. J., Sanderson, R., & Carver, T. L. W. (2005). Nitric oxide contributes both to papilla-based resistance and the hypersensitive response in barley attacked by *Blumeria graminis* f. sp. hordei. *Molecular Plant Pathology, 6*(1), 65–78. https://doi.org/10.1111/J.1364-3703.2004.00266.X.

Qu, Z. L., Zhong, N. Q., Wang, H. Y., Chen, A. P., Jian, G. L., & Xia, G. X. (2006). Ectopic expression of the cotton non-symbiotic hemoglobin gene GhHbd1 triggers defense responses and increases disease tolerance in *Arabidopsis*. *Plant and Cell Physiology, 47*(8), 1058–1068. https://doi.org/10.1093/pcp/pcj076.

Radi, R., Cassina, A., Hodara, R., Quijano, C., & Castro, L. (2002). Peroxynitrite reactions and formation in mitochondria. *Free Radical Biology and Medicine, 33*(11), 1451–1464. https://doi.org/10.1016/S0891-5849(02)01111-5, www.elsevier.com/locate/freeradbiomed.

Reda, M., Golicka, A., Kabała, K., & Janicka, M. (2018). Involvement of NR and PM-NR in NO biosynthesis in cucumber plants subjected to salt stress. *Plant Science, 267*, 55–64. https://doi.org/10.1016/j.plantsci.2017.11.004, http://www.journals.elsevier.com/plant-science/.

Reda, M., Janicka, M., & Kabała, K. (2022). Nitrate reductase dependent synthesis of NO in plants. In *Nitric oxide in plant biology*. Academic Press, 95–110.

Rui, L., Kang, P., Shao, J., Lu, M., Cui, B., Zhao, Y., Wang, W., Cai, H., Tang, D., Loake, G. J., Wang, M., & Shi, H. (2024). The chloroplast-localized casein kinase II α subunit, CPCK2, negatively regulates plant innate immunity through promoting S-nitrosylation of SABP3. *Plant Journal, 120*(2), 552–568. https://doi.org/10.1111/tpj.17000, http://onlinelibrary.wiley.com/journal/10.1111/(ISSN)1365-313X.

Rümer, S., Gupta, K. J., & Kaiser, W. M. (2009). Plant cells oxidize hydroxylamines to NO. *Journal of Experimental Botany, 60*(7), 2065–2072.

Rőszer, T. (2012). Nitric oxide synthesis in the chloroplast. In *The biology of subcellular nitric oxide*. Dordrecht: Springer Netherlands, 49–66.

Sabadashka, M., Nagalievska, M., & Sybirna, N. (2021). Tyrosine nitration as a key event of signal transduction that regulates functional state of the cell. *Cell Biology International, 45*(3), 481–497.

Saddhe, A. A., Malvankar, M. R., Karle, S. B., & Kumar, K. (2019). Reactive nitrogen species: Paradigms of cellular signaling and regulation of salt stress in plants. *Environmental and Experimental Botany, 161*, 86–97. https://doi.org/10.1016/j.envexpbot.2018.11.010, www.elsevier.com/inca/publications/store/2/6/7.

Sahay, S., & Gupta, M. (2017). An update on nitric oxide and its benign role in plant responses under metal stress. *Nitric Oxide: Biology and Chemistry / Official Journal of the Nitric Oxide Society, 67*, 39–52. https://doi.org/10.1016/j.niox.2017.04.011.

Sandalio, L. M., Gotor, C., Romero, L. C., & Romero-Puertas, M. C. (2019). Multilevel regulation of peroxisomal proteome by post-translational modifications. *International Journal of Molecular Sciences, 20*(19). https://doi.org/10.3390/ijms20194881, https://www.mdpi.com/1422-0067/20/19/4881/pdf.

Sarrou, J., Isgandarova, S., Kern, J., Zouni, A., Renger, G., Lubitz, W., & Messinger, J. (2003). Nitric oxide-induced formation of the S-2 state in the oxygen-evolving complex of photosystem II from *Synechococcus elongatus*. *Biochemistry, 42*(4), 1016–1023. https://doi.org/10.1021/bi026327p.

Shan, Q., Zhao, D., Cao, B., Zhu, X., Wang, C., Deng, L., Li, C., Zhang, Y., Shi, Q., & Gong, B. (2025). Jasmonic acid and nitric oxide orchestrate a hierarchical melatonin cascade for *Botrytis cinerea* resistance in tomato. *Plant Physiology, 197*(3). https://doi.org/10.1093/plphys/kiaf078, https://academic.oup.com/plphys/issue/197/3.

Shi, H., Ye, T., Chen, F., Cheng, Z., Wang, Y., Yang, P., Zhang, Y., & Chan, Z. (2013). Manipulation of arginase expression modulates abiotic stress tolerance in *Arabidopsis:* Effect on arginine metabolism and ROS accumulation. *Journal of Experimental Botany, 64*(5), 1367–1379. https://doi.org/10.1093/jxb/ers400.

Shibuki, K. (1990). An electrochemical microprobe for detecting nitric oxide release in brain tissue. *Neuroscience Research, 9*(1), 69–76. https://doi.org/10.1016/0168-0102(90)90048-j.

Singh, P., Jaiswal, S., Tripathi, D. K., & Singh, V. P. (2024). Nitric oxide acts upstream of indole-3-acetic acid in ameliorating arsenate stress in tomato seedlings. *Plant Physiology and Biochemistry, 208*. https://doi.org/10.1016/j.plaphy.2024.108461, https://www.sciencedirect.com/science/journal/09819428.

Steffen-Heins, A., & Steffens, B. (2015). EPR spectroscopy and its use in planta—a promising technique to disentangle the origin of specific ROS. *Frontiers in Environmental Science, 3*. https://doi.org/10.3389/fenvs.2015.00015.

Streatfield, S. J., Weber, A., Kinsman, E. A., Häusler, R. E., Li, J., Post-Beittenmiller, D., Kaiser, W. M., Pyke, K. A., Flügge, U. I., & Chory, J. (1999). The phosphoenolpyruvate/phosphate translocator is required for phenolic metabolism, palisade cell development, and plastid-dependent nuclear gene expression. *The Plant Cell, 11*(9), 1609–1621. https://doi.org/10.1105/tpc.11.9.1609, http://www.plantcell.org/.

Stöhr, C., Strube, F., Marx, G., Ullrich, W. R., & Rockel, P. (2001). A plasma membrane-bound enzyme of tobacco roots catalyses the formation of nitric oxide from nitrite. *Planta, 212*(5-6), 835–841. https://doi.org/10.1007/s004250000447.

Sun, M., Yang, T., Qiao, X. H., Zhao, P., Zhu, Z. P., Wang, G. L., Xu, L. L., & Xiong, A. S. (2024). Nitric oxide regulates the lignification and carotenoid biosynthesis of postharvest carrot (*Daucus carota* L.). *Postharvest Biology and Technology, 207*. https://doi.org/10.1016/j.postharvbio.2023.112593, www.elsevier.com/inca/publications/store/5/0/3/3/1/3/index.htt.

Szuba, A., Kasprowicz-Maluśki, A., & Wojtaszek, P. (2015). Nitration of plant apoplastic proteins from cell suspension cultures. *Journal of Proteomics, 120*, 158–168.

Tada, Y., Spoel, S. H., Pajerowska-Mukhtar, K., Mou, Z., Song, J., Wang, C., Zuo, J., & Dong, X. (2008). Plant immunity requires conformational charges of NPR1 via S-nitrosylation and thioredoxins. *Science (New York, N.Y.), 321*(5891), 952–956. https://doi.org/10.1126/science.1156970.

Tanou, G., Filippou, P., Belghazi, M., Job, D., Diamantidis, G., Fotopoulos, V., & Molassiotis, A. (2012). Oxidative and nitrosative-based signaling and associated post-translational modifications orchestrate the acclimation of citrus plants to salinity stress. *The Plant Journal, 72*(4), 585–599. https://doi.org/10.1111/j.1365-313x.2012.05100.x.

Teng, Z., Zheng, Q., Peng, Y., Li, Y., Meng, S., Liu, B., Peng, Y., Duan, M., Yuan, D., Zhang, J., & Ye, N. (2025). Nitrate reductase–dependent nitric oxide production mediates nitrate-conferred salt tolerance in rice seedlings. *Plant Physiology, 197*(3). https://doi.org/10.1093/plphys/kiaf080.

Terrile, M. C., París, R., Calderõn-Villalobos, L. I. A., Iglesias, M. J., Lamattina, L., Estelle, M., & Casalongué, C. A. (2012). Nitric oxide influences auxin signaling through S-nitrosylation of the *Arabidopsis* TRANSPORT INHIBITOR RESPONSE 1 auxin receptor. *Plant Journal, 70*(3), 492–500. https://doi.org/10.1111/j.1365-313X.2011.04885.x.

Terrile, M. C., Tebez, N. M., Colman, S. L., Mateos, J. L., Morato-López, E., Sánchez-López, N., Izquierdo-Álvarez, A., Marina, A., Calderón Villalobos, L. I. A., Estelle, M., Martínez-Ruiz, A., Fiol, D. F., Casalongué, C. A., & Iglesias, M. J. (2022). S-Nitrosation of E3 ubiquitin ligase complex components regulates hormonal signalings in *Arabidopsis*. *Frontiers in Plant Science, 12*. https://doi.org/10.3389/fpls.2021.794582, https://www.frontiersin.org/journals/plant-science.

Trevisan, S., Manoli, A., Begheldo, M., Nonis, A., Enna, M., Vaccaro, S., Caporale, G., Ruperti, B., & Quaggiotti, S. (2011). Transcriptome analysis reveals coordinated spatiotemporal regulation of hemoglobin and nitrate reductase in response to nitrate in maize roots. *New Phytologist, 192*(2), 338–352. https://doi.org/10.1111/j.1469-8137.2011.03822.x.

Valentovičová, K., Demecsová, L., Liptáková, Ľ., Zelinová, V., & Tamás, L. (2025). Barley root tip peroxidases convert DAF-FM and DAR-4M to an NO-independent fluorescent product using H_2O_2 derived from polyamine catabolism by polyamine oxidases. *Journal of Plant Physiology, 308*, 154477. https://doi.org/10.1016/j.jplph.2025.154477.

Vandelle, E., & Delledonne, M. (2008). Methods for nitric oxide detection during plant-pathogen interactions. *Methods in Enzymology, 437*, 575–594. https://doi.org/10.1016/S0076-6879(07)37029-8.

Vishwakarma, A., Wany, A., Pandey, S., Bulle, M., Kumari, A., Kishorekumar, R., Igamberdiev, A. U., Mur, L. A. J., & Gupta, K. J. (2019). Current approaches to measure nitric oxide in plants. *Journal of Experimental Botany, 70*(17), 4333–4343. https://doi.org/10.1093/jxb/erz242, http://jxb.oxfordjournals.org/.

Vitecek, J., Reinohl, V., & Jones, R. L. (2008). Measuring NO production by plant tissues and suspension cultured cells. *Molecular Plant, 1*, 270–284.

Wang, D., Shang, X., Ren, H., Han, P., & Liu, W. (2024). Nitric oxide is involved in the inhibition of seed germination by paclobutrazol in *Arabidopsis*. *Journal of Plant Growth Regulation, 43*(8), 2761–2770.

Wang, J., Guo, X., Chen, Y., Liu, T., Zhu, J., Xu, S., & Vierling, E. (2024). Maternal nitric oxide homeostasis impacts female gametophyte development under optimal and stress conditions. *The Plant Cell, 36*(6), 2201–2218.

Wang, L., Mu, X., Chen, X., & Han, Y. (2022). Hydrogen sulfide attenuates intracellular oxidative stress via repressing glycolate oxidase activities in *Arabidopsis thaliana*. *BMC Plant Biology, 22*(1). https://doi.org/10.1186/s12870-022-03490-3, http://www.biomedcentral.com/bmcplantbiol/.

Wang, P., Du, Y., Hou, Y. J., Zhao, Y., Hsu, C. C., Yuan, F., Zhu, X., Tao, W. A., Song, C. P., & Zhu, J. K. (2015). Nitric oxide negatively regulates abscisic acid signaling in guard cells by S-nitrosylation of OST1. *Proceedings of the National Academy of Sciences of the United States of America, 112*(2), 613–618. https://doi.org/10.1073/pnas.1423481112, http://www.pnas.org/content/112/2/613.full.pdf+html.

Wany, A., Gupta, A. K., Kumari, A., Gupta, S., Mishra, S., Jaintu, R., Pathak, P. K., & Gupta, K. J. (2016). Chemiluminescence detection of nitric oxide from roots, leaves, and root mitochondria. In *Plant nitric oxide: Methods and protocols*. New York, NY: Springer New York, 15–29.

Wei, L., Zhang, J., Wei, S., Wang, C., Deng, Y., Hu, D., Liu, H., Gong, W., Pan, Y., & Liao, W. (2022). Nitric oxide alleviates salt stress through protein S-nitrosylation and transcriptional regulation in tomato seedlings. *Planta, 256*(6). https://doi.org/10.1007/s00425-022-04015-w, https://www.springer.com/journal/425.

Wei, L., Liao, W., Zhong, Y., Tian, Y., Wei, S., & Liu, Y. (2024). NO-mediated protein S-nitrosylation under salt stress: Role and mechanism. *Plant Science, 338*, 111927. https://doi.org/10.1016/j.plantsci.2023.111927.

Xu, B., Feng, X., Piechatzek, A., Zhang, S., Konrad, K. R., Kromdijk, J., Hedrich, R., & Gilliham, M. (2024). The GABA shunt contributes to ROS homeostasis in guard cells of *Arabidopsis*. *New Phytologist, 241*(1), 73–81. https://doi.org/10.1111/nph.19390, http://onlinelibrary.wiley.com/journal/10.1111/(ISSN)1469-8137.

Yadav, S., David, A., Baluška, F., & Bhatla, S. C. (2013). Rapid auxin-induced nitric oxide accumulation and subsequent tyrosine nitration of proteins during adventitious root formation in sunflower hypocotyls. *Plant Signaling & Behavior, 8*(3), e23196.

Yan, Y., Shi, Q., & Gong, B. (2021). S-nitrosoglutathione reductase-mediated nitric oxide affects axillary buds outgrowth of *Solanum lycopersicum* L. by regulating auxin and cytokinin signaling. *Plant and Cell Physiology, 62*(3), 458–471.

Yang, Y., Huang, Z., & Li, L. L. (2021). Advanced nitric oxide donors: Chemical structure of NO drugs, NO nanomedicines and biomedical applications. *Nanoscale, 13*(2), 444–459.

Ye, X., Kim, W. S., Rubakhin, S. S., & Sweedler, J. V. (2004). Measurement of nitric oxide by 4, 5-diaminofluorescein without interferences. *Analyst, 129*(12), 1200–1205. https://doi.org/10.1039/b409394a, http://pubs.rsc.org/en/journals/journal/an.

Yun, B. W., Feechan, A., Yin, M., Saidi, N. B. B., Le Bihan, T., Yu, M., Moore, J. W., Kang, J. G., Kwon, E., Spoel, S. H., Pallas, J. A., & Loake, G. J. (2011). S-nitrosylation of NADPH oxidase regulates cell death in plant immunity. *Nature, 478*(7368), 264–268. https://doi.org/10.1038/nature10427.

Zhang, T., Ma, M., Chen, T., Zhang, L., Fan, L., Zhang, W., Wei, B., Li, S., Xuan, W., Noctor, G., & Han, Y. (2020). Glutathione-dependent denitrosation of GSNOR1 promotes oxidative signalling downstream of H_2O_2. *Plant, Cell & Environment, 43*(5), 1175–1191. https://doi.org/10.1111/pce.13727.

Zhang, X., Kim, W. S., Hatcher, N., Potgieter, K., Moroz, L. L., Gillette, R., & Sweedler, J. V. (2002). Interfering with nitric oxide measurements. 4,5-Diaminofluorescein reacts with dehydroascorbic acid and ascorbic acid. *Journal of Biological Chemistry, 277*(50), 48472–48478. https://doi.org/10.1074/jbc.M209130200, http://www.jbc.org/content/by/year.

Zhao, H., Ma, L., Shen, J., Zhou, H., & Zheng, Y. (2024). S -nitrosylation of the transcription factor MYB30 facilitates nitric oxide–promoted seed germination in *Arabidopsis*. *The Plant Cell, 36*(2), 367–382. https://doi.org/10.1093/plcell/koad276.

Chapter 2

Biosynthesis of nitric oxide under normal and challenging environmental conditions

Pooja Singh[1,2], Shivani Mishra[1,2], Amarjeet Singh[3], Durgesh K. Tripathi[4], and Vijay P. Singh[1,2]
[1]*C.M.P. Degree College, A Constituent Post Graduate College of University of Allahabad, Prayagraj, Uttar Pradesh, India,* [2]*Department of Botany, Plant Physiology Laboratory, Praygraj, Uttar Pradesh, India,* [3]*Department of Botany, Mycology and Natural Pesticide Laboratory, Post Graduate College, Ghazipur, Uttar Pradesh, India,* [4]*Amity University, Crop Nanobiology and Molecular Stress Physiology Lab, Amity Institute of Organic Agriculture (AIOA), Noida, Uttar Pradesh, India*

2.1 Introduction

Nitric oxide (NO) is an endogenous signaling molecule that plays a pivotal role in diverse physiological processes and in mediating plant responses to environmental challenges, thereby contributing significantly to advancements in plant biology and agriculture (Sun et al., 2021). It is a small, diatomic, gaseous molecule that readily diffuses across biological membranes due to its lipophilic nature (Khan, Al Azzawi, et al., 2023; Khan, Ali, et al., 2023). Remarkably, in 1992, NO was recognized as the *Molecule of the Year* by *Science* owing to its surprising biological roles in animals (Chiletti et al., 2025). In the case of plants, NO contributes to normal growth and development, including seed germination, photosynthesis, root formation, flowering, and fruit development (Gupta et al., 2011). It also interacts with other signaling molecules and phytohormones. For instance, NO promotes hydrogen gas-mediated adventitious root development in cucumber (Zhu et al., 2016). In tomato, NO is synthesized downstream of auxin (IAA) signaling and regulates cell cycle progression and cell division, which are essential for lateral root primordium formation in the pericycle region (Correa-Aragunde et al., 2004). Several studies have demonstrated that NO actively breaks seed dormancy. Pretreatment of apple seeds with the NO donor sodium nitroprusside (SNP, 5 mM) induces germination by modulating reactive oxygen species (ROS) homeostasis (Kumar & Ohri, 2023). Similarly, in *Paulownia elongata*, NO derived from SNP promotes dormancy release during early germination (Liu et al., 2019). NO also plays a crucial role in

Nitric Oxide in Plant Resilience to Abiotic Stress. DOI: https://doi.org/10.1016/B978-0-443-40497-9.00005-2

regulating flowering time to ensure reproductive success. Recent findings suggest that NO delays flowering in *Arabidopsis thaliana* by modulating the expression of *Flowering Locus C* (FLC) through epigenetic mechanisms. Specifically, NO enhances histone acetylation at the FLC by S-nitrosating histone deacetylases HDA5 and HDA6, thereby reducing their activity (Zhu et al., 2025). In tomato, NO regulates floral transition through epigenetic modifications, influencing the expression of DNA demethylation genes and flowering-related genes such as *Flowering Locus D* and *Single Flower Truss* (Hou et al., 2025).

NO also interacts with ROS to regulate the pollen tube tip growth. It also regulates intracellular calcium levels, essential for pollen tube elongation and navigation within the style (Šírová et al., 2011). Beyond developmental roles, NO is vital for plant stress tolerance and mitigates the effects of both biotic and abiotic stresses such as high temperature, salinity, heavy metals (HMs), and pathogen attack (Sarma et al., 2025). During salinity stress, NO sustains ionic balance (particularly Na^+/K^+ ratios), enhances antioxidant defenses, and improves photosynthetic efficiency (Shang et al., 2022). Evidence also indicates that early NO production protects rice roots exposed to NH_4^+ under water stress by enhancing antioxidant activity (Cao et al., 2019). In bamboo, NO confers tolerance to HMs such as lead and cadmium by regulating antioxidant defenses and photosynthetic efficiency, while preventing metal accumulation and translocation within plant tissues, thereby protecting against oxidative damage (Emamverdian et al., 2021). NO also plays a central role in plant immunity. In *A. thaliana*, the IAA-leucine resistant-like gene 6 (*AtILL6*), induced by NO, together with *AtSID2* (salicylic acid induction deficient 2), positively regulates pattern-triggered immunity, effector-triggered immunity, and systemic acquired resistance in response to both virulent and avirulent pathogens. This highlights the importance of NO-mediated induction of *AtILL6* and *AtSID2* in enhancing plant immune responses.

In this book chapter, we highlight recent advances in understanding the regulation of NO biosynthesis under both normal and stress conditions, the specific cellular signaling networks involved, and the influence of phytohormones and emerging plant growth regulators (PGRs) on NO production.

2.2 Biosynthesis of nitric oxide in plants

In animals, NO biosynthesis occurs through well-defined pathways, whereas in plants, the mechanisms underlying NO production remain incompletely understood. Several major cellular sites have been identified as contributors to NO generation in plants, including the apoplast, chloroplasts, mitochondria, peroxisomes, and biological membranes (Khan, Al Azzawi, et al., 2023; Khan, Ali, et al., 2023). NO biosynthesis in plants involves both enzymatic and nonenzymatic pathways (Fig. 2.1) (Gupta et al., 2025). The enzymatic routes can be further classified into oxidative and reductive pathways (Astier et al., 2018). The oxidative pathways produce NO through the oxidation of substrates such as L-arginine, polyamines (PAs), and hydroxylamine (HA), as well as via plant-type nitric oxide

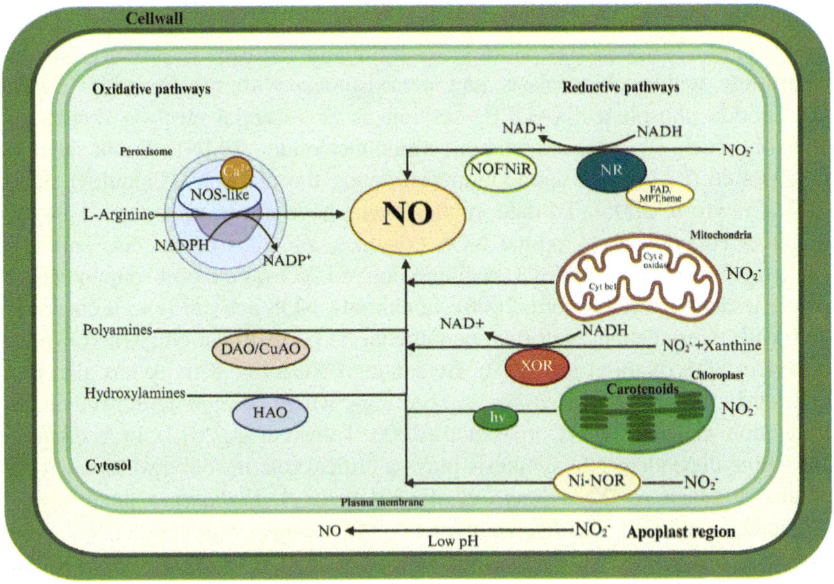

FIGURE 2.1 A schematic representation of the pathways involved in NO biosynthesis in plants. NO generation through oxidative reactions utilizing substrates such as L-arginine, polyamines, or hydroxylamine. In addition, NO is produced via reductive pathways, wherein nitrate reductase, mitochondrial components, and xanthine oxidoreductase catalyze the conversion of nitrite to NO. Furthermore, nitrite can undergo nonenzymatic conversion to NO under acidic conditions or in the presence of chemically reducing agents. Abbreviations: NO (Nitric oxide), NOS-like (nitric oxide synthase), DAO (Diamine Oxidase), CuAO (Diamine Oxidase), HAO (hydroxylamine oxidoreductase), NADH (nicotinamide adenine dinucleotide), NADPH (nicotinamide adenine dinucleotide phosphate), NOFNiR (nitric oxide-forming nitrite reductase), FAD (flavin adenine dinucleotide), MPT (molybdopterin), XOR (xanthine oxidoreductase), NiNOR (membrane-bound nitrite reductase).

synthase (NOS). In contrast, the reductive pathways primarily generate NO through the activity of nitrate reductase (NR), along with contributions from mitochondrial and plasma membrane-bound enzymes (Khan, Al Azzawi, et al., 2023; Khan, Ali, et al., 2023). Additionally, NO can be synthesized through nonenzymatic conversion of nitrite. This process is facilitated by plastid pigments (e.g., carotenoids), reducing agents such as ascorbic acid, and acidic conditions, which collectively promote the transformation of nitrite into NO (Tewari et al., 2021).

2.2.1 Oxidative pathways

2.2.1.1 From oxidation of L-arginine by nitric oxide synthase-like enzyme

NO can be synthesized from L-arginine in plants through NOS-like enzymatic activity, which resembles the mechanism observed in animals. Genomic analyses of the marine green algae *Ostreococcus tauri* and *Ostreococcus lucimarinus* revealed the presence of NOS-like genes homologous to mammalian NOS (Foresi et al., 2010). In this pathway, L-arginine is oxidized to produce NO and L-citrulline within chloroplasts and peroxisomes, with nicotinamide adenine dinucleotide phosphate (NADPH) serving as an essential electron donor, and cofactors such as calcium, flavin mononucleotide (FMN), flavin adenine dinucleotide (FAD), and calmodulin facilitating the reaction (Allagulova et al., 2023; Crawford, 2006). To date, higher plants have not been shown to contain definitive equivalents of animal NOS enzymes, and thus the precise pathways and enzymes responsible for L-arginine-derived NO production remain incompletely understood (Crawford, 2006). In animals, NOS activity is well characterized, with three distinct isoforms: endothelial NOS, neuronal NOS (nNOS), and inducible NOS (Nabi et al., 2019). Evidence of NOS-like activity has also been reported in pea and *Brassica juncea* seedlings, where L-arginine-dependent NO production was observed (Corpas et al., 2006; Talwar et al., 2012). In *A. thaliana*, L-arginine-dependent NO synthesis plays a critical role in root development and cellular architecture (Krasylenko et al., 2019). In *A. thaliana*, a gene initially identified as *AtNOS1* was later renamed *NOA1* (Nitric Oxide Associated 1) and proposed to function as a plant equivalent of NOS (Crawford, 2006). However, subsequent studies demonstrated that NOA1 lacks several characteristic sequence motifs typical of animal NOS proteins (Crawford, 2006). Instead, NOA1 exhibits similarity to a circularly permuted GTPase (Sarma et al., 2025). Research suggests that NOA1 does not directly catalyze NO synthesis but rather influences NO levels indirectly through its role in cellular metabolism. Consequently, altered NO production in NOA1-deficient plants is likely due to secondary metabolic effects rather than direct enzymatic activity (Van Ree et al., 2011). Experimental evidence further indicates that NO plays a protective role against oxidative stress, and NOS-like activity is essential for the induction of heme oxygenase under UV-B exposure. Pretreatment with SNP prevented chlorophyll degradation, reduced H_2O_2 accumulation, and minimized ion leakage in UV-B-treated soybean plants (Santa-Cruz et al., 2010). Similarly, in maize, NOS-like activity regulates the antioxidant defense system during drought stress. Exogenous application of SNP alleviates oxidative stress in maize leaves, thereby enhancing tolerance to water deficit conditions (Hao et al., 2008).

2.2.1.2 *From polyamines and hydroxylamine*

PAs, including putrescine, spermidine, and spermine, are small nitrogen-rich molecules present in all plant tissues. In addition to their established roles in regulating growth and stress tolerance, PAs also contribute to NO biosynthesis (Asija et al., 2023). Plants do not directly extract NO from PAs; rather, NO is generated through chemical transformations that occur following PA oxidation. During PA oxidation, enzymes such as diamine oxidase (DAO/CuAO) and polyamine oxidase (PAO) become highly active, catalyzing the removal of amino groups from PAs (Wimalasekera et al., 2011). The resulting products, including nitrite, can subsequently be converted into NO through both enzymatic and nonenzymatic pathways. Multiple studies provide evidence that PAs contribute to NO synthesis in plant cells, thereby supporting growth under both normal and adverse conditions. For example, in *A. thaliana*, spermine and spermidine enhance NO production in the root tip region and primary leaves (Tun et al., 2006). Similarly, DAO- and PAO-mediated PA oxidation increases NO biosynthesis in *Arabidopsis* (Wimalasekera et al., 2011).

Exogenous PA application has also been shown to elevate NO levels in soybean callus cells, where NO production depends on DAO activity and the associated generation of hydrogen peroxide. These findings suggest that PA-induced NO synthesis is not a direct effect but occurs through interconnected pathways involving PA oxidation (Yang et al., 2014). In maize, the expression of *ZmPAO6*, *ZmDAO*, and *ZmADC1* is upregulated under water stress, indicating that regulation of PA metabolism and related gene expression supports root development and enhances tolerance to waterlogging (Senthil Kumar et al., 2016). Collectively, PA-induced NO production plays a significant role in plant growth and adaptation under stress conditions. HAs, a reactive nitrogen species (RNS), represents another alternative source of NO production. Initially, it was believed that in ammonia-oxidizing bacteria (AOB), the enzyme hydroxylamine oxidoreductase (HAO) directly catalyzed the conversion of HAs to nitrite (Choi et al., 2023). However, subsequent studies revealed that NO is a mandatory intermediate in this process. Revised findings indicate that HAO performs a three-electron oxidation to produce NO, rather than a four-electron reaction (Caranto & Lancaster, 2017). Beyond AOB, certain methanotrophs also encode HAO under high-ammonia conditions. For instance, *Methylacidiphilum fumariolicum* possesses a multiheme HAO that rapidly converts HAs to NO, thereby detoxifying excess HA (Versantvoort et al., 2020). In anammox (anaerobic ammonia-oxidizing) bacteria, HAO exhibits bidirectional activity, interconverting NO and NH_2OH depending on the redox potential, thus playing a dual role in nitrogen metabolism (Irisa et al., 2014). In plants, experimental evidence suggests that the addition of HAs or salicylhydroxamate to tobacco cell suspensions increases NO production. However, the physiological relevance of this mechanism remains unclear, and further research is required to determine the conditions under which HA-derived NO contributes significantly to plant NO pools (Rümer et al., 2009).

2.2.2 Reductive pathways

2.2.2.1 By nitrate reductase

NR is widely recognized as a multifunctional enzyme localized in the cytoplasm or associated with the PM–NR, where it plays a central role in nitrogen metabolism and assimilation. It catalyzes the first and rate-limiting step of nitrate assimilation by reducing nitrate to nitrite, a process in which electrons are supplied by NADH (Astier et al., 2018; Gupta et al., 2025; Meyer et al., 2005).

Structurally, NR exists predominantly as a homodimer and requires several cofactors for its enzymatic activity, including FAD, heme, and molybdopterin (Astier et al., 2018). The enzyme contains several conserved domains: the molybdenum cofactor domain (Mo-MPT), the FAD domain, the heme-Fe-binding domain, the NAD(P)H-binding domain, and the dimer interface. During catalysis, electrons from NAD(P)H are first transferred to the FAD domain, then relayed through the heme-Fe domain to the Mo-MPT domain, where nitrate is reduced to nitrite. Under certain conditions, nitrite can be further converted to NO at the Mo-MPT domain (Gupta et al., 2025). In *A. thaliana*, two NR genes, *AtNIA1* and *AtNIA2*, encode proteins with high overall similarity (~83.5%) but differ in specific regions, particularly at the N-terminus. Double mutants (*nia1nia2*) have demonstrated that NR plays a critical role in abscisic acid (ABA)-triggered NO production in guard cells, thereby regulating stomatal movement (Wilson et al., 2008). In rice, NR contributes to NO biosynthesis, promotes lateral root development, and enhances nitrogen uptake (Sun et al., 2015). In *Medicago truncatula* nodules, which undergo symbiotic nitrogen fixation, NR isoforms *MtNR1* and *MtNR2* are involved in NO production (Berger et al., 2020). Similarly, NR-mediated NO generation has been reported in the cyanobacterium *Anabaena doliolum* (Roszer, 2014). Under specific physiological conditions, NR also catalyzes NO production in *Chlamydomonas* through an indirect mechanism involving the enzyme ARC (amidoxime reducing component), also referred to as nitric oxide-forming nitrite reductase. In this pathway, ARC utilizes electrons donated by NR to reduce nitrite to NO, highlighting NR's role as an electron donor while ARC directly forms NO (Gupta et al., 2025). Importantly, NR activity is tightly regulated by posttranslational modifications, particularly phosphorylation, to maintain appropriate levels of nitrite and NO within the cell. This regulation is essential for preserving nitrogen balance and protecting plant tissues from the potentially toxic by-products of nitrate reduction (Lillo et al., 2004).

2.2.2.2 *By plasma membrane-bound nitrite–nitrate reductase*

In plants, NO can be synthesized through multiple pathways, one of which involves a plasma membrane–bound enzyme in roots known as Ni-NOR (nitrite–NO reductase), first identified in *Nicotiana tabacum* (Roszer, 2014). Ni-NOR catalyzes the reduction of nitrite (NO_2^-) to NO, utilizing reduced cytochrome c as the electron donor. The level of NO production mediated by Ni-NOR is comparable to that of root-specific NR. However, Ni-NOR appears to represent a distinct protein, and its complete biochemical characterization as a bona fide NO-producing enzyme remains to be fully elucidated (Graska et al., 2023).

2.2.3 By the mitochondrial electron transport chain

In the plant kingdom, members of the Chlorophyceae and higher vascular plants utilize the mitochondrial electron transport chain (ETC) to produce NO by reducing nitrite under hypoxic (low-oxygen) conditions (Ermilova, 2023; Gupta et al., 2018). Mitochondrially generated NO fulfills multiple physiological functions: it decreases O_2 consumption by inhibiting cytochrome c oxidase (CcO), sustains energy generation under oxygen limitation, and contributes to processes such as seed germination, the suppression of the C_2 cycle, and the regulation of fermentative metabolism (Roszer, 2014). This pathway of NO biosynthesis has been reported in microorganisms, animal cells, and plant cells. In fungi, nitrite can serve as a terminal electron acceptor to support ATP production under hypoxia (Kumari et al., 2023). This reaction primarily occurs at complex III (cytochrome bc_6) and complex IV (CcO) of the mitochondrial ETC (Allagulova et al., 2023). In *Paracoccus denitrificans*, nitrite reduction is mediated by membrane-bound cytochromes, specifically cytochrome cd_1 and cytochrome bc, which directly catalyze nitrite reduction to NO. This membrane-associated system enables efficient electron transfer from donors to nitrite, thereby supporting energy metabolism in denitrifying bacteria (Hartop, 2014). In plants, roots exhibit strong mitochondrial capacity to reduce nitrite to NO even in the absence of NR, whereas leaf mitochondria lack this ability, confirming that NR activity is the primary source of NO production in leaves. Overall, mitochondrial nitrite reduction represents a critical survival mechanism that sustains root function under low-oxygen conditions (Gupta et al., 2005). A continuous exchange of nitrogen oxides between the cytoplasm and mitochondria further maintains NO production during hypoxia. Mitochondrially generated NO is converted back to nitrate by class 1 plant hemoglobins, which are strongly induced under oxygen-limiting conditions. The nitrate is subsequently reduced to nitrite by NR, thereby providing mitochondria with a continuous substrate supply for NO synthesis. This mitochondria–cytoplasm NO recycling pathway also contributes to redox homeostasis by lowering the NADH/NAD^+ and NADPH/$NADP^+$ ratios, enabling cells to better adapt to oxygen limitation (Roszer, 2014). Overall, mitochondrial NO production is tightly regulated by metabolic conditions and specific genes. Under hypoxia, NR-dependent nitrite formation fuels mitochondrial NO synthesis, supporting energy balance and stress adaptation. Concurrently, genes such as alternative oxidase prevent excessive NO accumulation by stabilizing electron flow through the ETC. Together, these mechanisms maintain cellular homeostasis and protect plants against environmental stress (Gupta et al., 2022).

2.2.4 By xanthineoxido reductase

Xanthine oxidoreductase (XOR) is a molybdenum-containing oxidoreductase enzyme that exists in homodimeric form and is highly conserved across plants, animals, and microorganisms (Allagulova et al., 2022). Each dimer consists of two identical subunits, and each subunit contains three major functional domains: the 2Fe–S domain (iron–sulfur cluster), the FAD domain, and the Mo–Co domain. These domains are arranged from the N-terminus to the C-terminus in a manner similar to mammalian XOR (Urarte et al., 2014). Under low-oxygen and highly reducing conditions, plant XOR is capable of reducing nitrite to NO by utilizing electrons derived from xanthine or NADH within the peroxisome and cytosol. XOR-dependent NO generation is considered important for conferring resistance against diverse abiotic and biotic stresses (Gupta et al., 2025). In animals, XOR-mediated NO production contributes to key physiological processes, including vasodilation (Berry & Hare, 2004). However, in plants, the role of XOR in NO biosynthesis remains unproven. Although theoretical studies and reviews suggest that plant XOR may reduce nitrite to generate NO, direct experimental evidence in living plants is still very limited (Tewari et al., 2021).

2.3 Nonenzymatic pathways for nitric oxide production

In plants, NO can be generated nonenzymatically through the chemical reduction of nitrite under acidic conditions. This pathway does not require enzymatic activity and instead relies on naturally occurring reductants and pigments (Tewari et al., 2021). Under low-pH conditions, ascorbic acid (vitamin C) can chemically reduce nitrite to NO. Experimental evidence demonstrates that ascorbate promotes NO production from nitrite in acidic environments (Du et al., 2023). In *Oncidium*, flowering is tightly regulated by ascorbic acid, which elevates NO levels. The increased NO, modulated by nitrate and nitrite reductases, represses floral transition by influencing the expression of flowering-related genes (Senthil Kumar et al., 2016). Additionally, carotenoid pigments localized in chloroplasts can also facilitate the nonenzymatic reduction of nitrite to NO. Several studies suggest that when carotenoids are illuminated in the presence of nitrogen dioxide (NO_2), NO is produced (Prochazkova et al., 2014).

2.4 Biosynthesis of nitric oxide in challenging environments

NO is a gaseous, redox-active, and highly RNS synthesized in living cells under both normal and stress conditions (Nabi et al., 2019). Among the most prominent abiotic stresses that negatively affect plant growth and crop productivity are heat, cold, drought, salinity, and HM stress. Under these adverse conditions, NO functions as a critical signaling molecule, contributing to the enhancement of antioxidant defense systems, regulation of osmotic adjustment and water status, maintenance of cellular ion homeostasis, and stabilization of photosynthetic efficiency, thereby supporting overall plant growth and resilience (Allagulova et al., 2023). Although the enzymatic role of NO is well recognized, the detailed molecular mechanisms underlying its

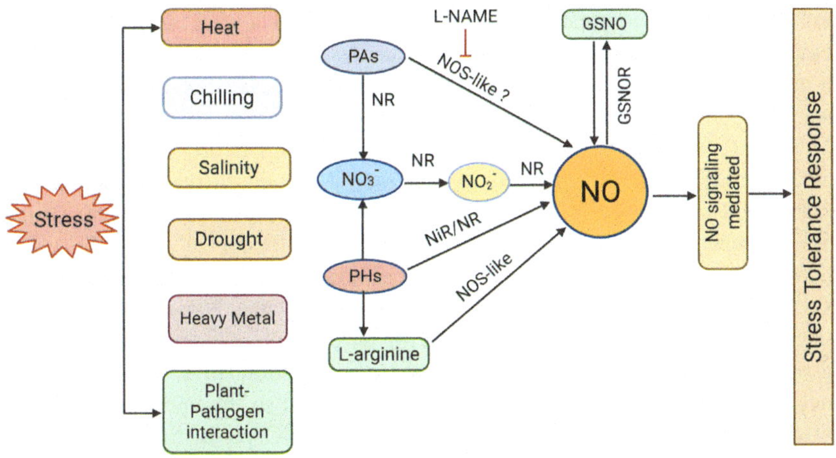

FIGURE 2.2 A diagrammatic illustration of pathways involved in the biosynthesis of NO in plants under challenging environments (NO, nitric oxide).

biosynthesis and its involvement in plant stress responses remain only partially elucidated. Generally, NO in plants is synthesized through two main alternative pathways, oxidative and reductive, via either spontaneous or enzyme-mediated reactions (Allagulova et al., 2023). Notably, NR-dependent NO generation plays a pivotal role in mediating plant defense responses against both biotic and abiotic stress factors (Mohn et al., 2019). NR-derived NO interacts synergistically with PGRs and phytohormones, including IAA, gibberellic acid (GA), brassinosteroids (BRs), salicylic acid, jasmonic acid (JA), and ABA, through complex signaling cascades to maintain ROS homeostasis under stress conditions (Tian et al., 2020). This chapter emphasizes the biosynthesis of NO and its implications in plant stress adaptation during challenging environmental conditions (Fig. 2.2).

2.4.1 Abiotic stress

2.4.1.1 Heat stress

Temperature is an environmental factor that plants cannot actively regulate, and thus, ambient temperature serves as a major determinant of plant growth, productivity, and geographical distribution across climatic zones (Kolupaev et al., 2023). Exposure to temperature extremes, both heat and cold, triggers the biosynthesis of endogenous NO, which subsequently regulates the transcriptional activity of genes encoding transcription factors, antioxidant enzymes, and other stress-responsive components (Karpets et al., 2016). The maintenance of NO homeostasis is therefore essential for ensuring accurate signal transduction during plant acclimation to thermal stress, thereby facilitating tolerance under both high- and low-temperature conditions (Allagulova et al., 2023) . Elevated temperatures have been consistently associated with increased NO accumulation in diverse plant species (Parankusam et al., 2017). Enhanced NO synthesis under heat stress has been documented in tobacco leaves (Xuan et al., 2010) and *A. thaliana* (Gould et al., 2003). Moderately elevated temperatures (+ 30°C, + 38°C, and + 42°C) have been reported to increase NO levels in pea tissues (Corpas et al., 2008), rice leaves (Song et al., 2013), *C. aurantium* (Ziogas et al., 2013), and strawberry plants (Christou et al., 2014; Kolupaev et al., 2023). This thermally induced NO accumulation was almost completely suppressed by the NOS inhibitor L-NAME and partially inhibited by the NR inhibitor sodium tungstate, suggesting the involvement of both oxidative and reductive NO biosynthetic pathways under heat stress. In addition to tomato seedlings, melatonin treatment under heat stress conditions led to the upregulation of PA biosynthetic genes, which subsequently promoted NO production through both NR and NOS activities (Jahan et al., 2019). This finding indicates that PAs play a significant role in NO biosynthesis via the NR-dependent pathway. Similarly, in transgenic *A. thaliana* plants expressing the NO-degrading dioxygenase (NOD) enzyme, NR-mediated NO generation was found to suppress NOD activity, thereby reducing leaf senescence during heat stress (Kolbert et al., 2019). Moreover, the application of exogenous NO_3^- stimulates NR-mediated NO synthesis, which in turn boosts the activities of superoxide dismutase, catalase (CAT), and glutathione peroxidase, thereby helping plants cope with heat stress (Karpets et al., 2016). Notably, NR-derived NO also activates signal transduction pathways linked to exogenous carbon monoxide, regulating ROS levels to further alleviate heat-induced damage (Shkliarevskyi et al., 2021).

2.4.1.2 *Chilling stress*

Low-temperature stress induces NO production via either the nitrite-dependent or arginine-dependent pathway; however, these two routes do not function simultaneously (Puyaubert and baudouin,2014). In *A. thaliana* and *Citrus aurantium*, chilling conditions promote nitrite accumulation and enhance both the enzymatic activity and transcript levels of NR (Ziogas et al., 2013). In contrast, *Capsicum annuum* (bell pepper) exhibits a decline in NO levels under chilling stress (Airaki et al., 2012). The role of nitrite-dependent NO synthesis in *A. thaliana* has been confirmed using *nia1 nia2* double mutants, which display reduced NO accumulation and heightened sensitivity to 4°C treatment compared with wild-type plants (Zhao et al., 2009). NR-derived NO further activates *P5CS1* expression, thereby enhancing proline biosynthesis and improving osmoprotection during cold exposure (Zhao et al., 2009). Additional evidence indicates that NR-mediated NO, together with S-nitrosoglutathione reductase (GSNOR) and MAPK1/2, orchestrates cold-adaptive signaling (Lv et al., 2018). Specifically, NO stimulates MAPK1/2 activity, while GSNOR maintains NO homeostasis to prevent excessive accumulation of RNS. Studies on NO-deficient *A. thaliana* triple mutants (*nia1 nia2 noa1–2*) demonstrate that NO also activates CBF- and ZF/OZF2-dependent signaling cascades, which regulate the expression of key cold-responsive genes (Costa-Broseta et al., 2019). Cold-induced increases in NO synthesis have been observed in both wild-type and *Atnoa1/rif1* mutant lines of *A. thaliana* (Khatoon et al., 2023; Zhao et al., 2009). Research in *Medicago falcata, M. truncatula*, and *nia1* mutants (*nia1-1, nia1–2*) of *M. truncatula* demonstrates that NR-dependent NO is central to cold-regulated glutathione metabolism and redox balance. Notably, *M. falcata*, a cold-tolerant species, exhibits higher glutathione levels, redox capacity, enzyme activities, and transcript abundance compared with *M. truncatula*, underscoring species-specific differences in tolerance (Zhang et al., 2019). Consistently, *nia1 nia2* mutants fail to exhibit elevated NO under cold stress, reinforcing the importance of NR-mediated NO production in *A. thaliana*. Chilling stress also suppresses *NOA1* expression and NOS-like activity, while enhancing *NIA1* expression. Positive associations between NR-derived NO and freezing tolerance have been reported (Khatoon et al., 2023; Sami et al., 2018). Conversely, in pea leaves and *Chorispora bungeana* cell cultures, inhibition of NOS-like activity suppresses cold-induced NO production, suggesting that an oxidative, NOS-dependent pathway contributes to NO formation in these species (Allagulova et al., 2023; Liu et al., 2010). In walnut shoots, NOS-derived NO appears to support chilling tolerance by increasing soluble sugars, proline, phenolics, glutathione, and reduced glutathione/oxidized glutathione (GSH/GSSG) ratios, thereby mitigating membrane injury, oxidative damage, and photosynthetic inhibition (Dong et al., 2018). Under chilling conditions, both NOS activity and NO levels increase; however, NO exerts negative feedback on NOS activity. Specifically, NO scavenging (c-PTIO) enhances NOS activity while reducing NO content, whereas exogenous NO (SNP) suppresses NOS activity despite elevating NO levels. These findings indicate that NO regulates both chilling responses and NOS activity in walnut. In pea, cold-induced NO generation is inhibited by L-NAME, further supporting the involvement of NOS-like pathways (Kolupaev et al., 2023; Puyaubert & Baudouin, 2014). Overall, *nia1 nia2* mutants exhibit minimal NO changes under low-temperature stress, whereas cold exposure substantially induces NR activity and *NIA1* expression in *A. thaliana* and citrus species (Puyaubert & Baudouin, 2014). Thus NR is likely the predominant enzyme driving NO synthesis during hypothermia (Dong et al., 2018). Complementarily, overexpression of JA-biosynthetic genes (*LOX4*, *OPR1*) enhances freezing tolerance even in NO-deficient *A. thaliana* mutants (*nia1–2 noa1–2*), highlighting synergistic regulation between NR-mediated NO signaling and JA pathways in cold stress tolerance (Costa-Broseta et al., 2018; Gupta et al., 2025). Despite extensive research, the primary source of NO under chilling stress remains unresolved, and the mechanisms by which plants coordinate NOS-like and NR-mediated NO production during cold adaptation are still unclear.

2.4.1.3 Drought stress

Under drought stress, nitrate NR-derived NO plays a pivotal role in regulating stomatal behavior, activating antioxidant defenses, and modulating signaling pathways to maintain osmotic balance and minimize water loss (Hou et al., 2024; Khan et al., 2020). During drought stress, plants accumulate ABA, which activates NADPH oxidases RBOHD and RBOHF (respiratory burst oxidase homology D and F), leading to superoxide generation. This redox signal enhances NO production via NR, and the resulting NO activates MAPK signaling, culminating in stomatal closure (Desikan et al., 2002; Khatoon et al., 2023). Ahammed et al. (2021) demonstrated that the transcription factor *SlWRKY81* suppresses NR expression and activity, thereby reducing NO production. Similarly, increasing nitrate supply in sugarcane enhanced NO production and NR activity in roots, thereby mitigating water deficit through the activation of antioxidant enzymes, improved photosynthesis, enhanced stomatal conductance, and stimulation of root development (Pissolato et al., 2020). These findings suggest that NR-mediated NO generation may vary among plant species, enabling fine-tuning of drought-adaptive responses. The beneficial effects of NR-derived NO may also arise from its contribution to improved nitrogen use efficiency under stress, including enhanced inorganic nitrogen uptake and stimulation of lateral root formation (Gupta et al., 2025; Sun et al., 2015). Evidence further supports the involvement of NOS-like pathways in drought-induced NO formation. NOS-generated NO in water-deficient maize leaves, while increased NOS activity was observed in drought-stressed wheat seedlings (Siddiqui et al., 2011). In addition, both NOS activity and NO release rates increased substantially under dehydration. In maize, application of the NOS inhibitor L-NAME blocked drought-induced NO accumulation, indicating that NOS-like activity is a major source of NO underwater deficiency (Hao et al.,2008; Khatoon et al., 2023). Nitrogen form has also been shown to influence the timing of drought-induced NO production. In rice, ammonium (NH_4^+) nutrition stimulated NO accumulation within 3 hours of drought treatment, whereas nitrate (NO_3^-)-fed plants exhibited NO induction only after 6 hours (Cao et al., 2019). This earlier NO burst, likely driven by NOS-like activity in NH_4^+-treated roots, contributes to improved drought tolerance (Lau et al., 2021).

2.4.1.4 Salinity stress

Salinity is the second most widespread abiotic stress after drought and imposes a dual detrimental effect, inducing cellular dehydration and ionic toxicity that ultimately suppresses plant growth and productivity (Allagulova et al., 2023). Numerous studies have demonstrated that salt exposure elevates endogenous NO levels across diverse plant species, where NO functions as a central regulator of adaptive responses to salinity. Enhanced NO production under salt stress has been reported in *A. thaliana* (Qiao et al., 2009), *Helianthus annuus*

(David et al., 2010), *Salvinia auriculata* (Gomes et al., 2017), and *Brassica napus* (Huang et al., 2022), among others. The *A. thaliana* triple mutant *nia1 nia2 noa1* exhibits impaired NO biosynthesis and extreme salt sensitivity, underscoring the involvement of both arginine-dependent and nitrite-dependent pathways in NO generation and highlighting the essential role of NO in salinity tolerance (Xie et al., 2013). Salt stress also induces the transcription of calmodulin genes (*AtCaM1* and *AtCaM4*) in *A. thaliana*. Their protein products interact with and suppress GSNOR, thereby enhancing cellular NO accumulation (Shang et al., 2022). In dune reed callus, NaCl treatment stimulated NOS-like activity and NO release, as confirmed by hemoglobin assays and spectrophotometric analysis of oxyhemoglobin conversion (Allagulova et al., 2023; Zhao et al., 2004). Subsequent studies reported increased NOS activity and NO levels in salt-stressed rice and maize (Kaya & Ashraf, 2021), while NOS-dependent NO production in *A. thaliana* was directly linked to salinity tolerance (Zhao et al., 2007). Furthermore, transgenic rice overexpressing rat nNOS exhibited elevated NOS activity, increased NO accumulation, and enhanced salt resistance (Cai et al., 2015). Similarly, application of 5-aminolevulinic acid promoted NOS-mediated NO synthesis in maize, improving salt tolerance (Kaya & Ashraf, 2021). Collectively, these findings support a positive role for NOS-derived NO in plant salinity resistance (Dong et al., 2018). In parallel, salinity modulates NR activity, and the contribution of NR-derived NO is well established. Salt stress upregulates NR at both transcriptional and posttranscriptional levels, thereby promoting NO production in cucumber roots. Inhibitor-based studies identified glucose-6-phosphate dehydrogenase (G-6-PDH) as a critical regulator of NR-dependent NO synthesis and salt tolerance in red kidney bean roots (Liu et al., 2007; Nabi et al., 2019). In *A. thaliana*, heme oxygenase 1 (HY1) modulates the expression of *NIA/NR* and *NOA1*, fine-tuning NO biosynthesis under salinity (Xie et al., 2013). Salt-stressed plants not only enhance NO production through NR and NOS-like enzymes but also suppress NO degradation, thereby sustaining elevated NO levels (Zhou et al., 2016). When NR activity or endogenous NO is inhibited, plants exhibit increased osmolyte accumulation and antioxidant enzyme activity; however, these responses are insufficient to prevent excessive ROS accumulation, lipid peroxidation, and DNA damage (Gupta & Seth, 2020). In cucumber, salinity induces both cytosolic NR and plasma membrane-associated NR (PM–NR), leading to elevated NO levels primarily associated with a higher NO_2^-/NO_3^- ratio and reduced NiR activity (Reda et al., 2018). NR-derived NO also modulates key hormonal pathways, including ABA, GA, ET, and IAA, and contributes to ion homeostasis by maintaining the Na^+/K^+ balance under salt stress (Kataria et al., 2022). Evidence for NO–ethylene crosstalk includes the downregulation of the ET-responsive transcription factor *EIN3* in *nia1 nia2* mutants and NO-mediated S-nitrosation of ACC synthase and ACC oxidase, which regulates ET biosynthesis under salinity (Li and Pan, 2016). NR-derived NO is also essential for maintaining cellular energy homeostasis during salinity stress. It sustains the ATP/ADP ratio, and *nia1 nia2* mutants display reduced NO levels and a marked decline in mitochondrial abundance under salt stress (Aridhi et al., 2020). Remarkably, exogenous NO_2^- application restores growth and mitochondrial numbers in these mutants, underscoring the critical role of NO in mitochondrial biogenesis and energy metabolism under stress (Gupta et al., 2025; Kumari et al., 2023). In summary, plants regulate NO production under salinity through multiple enzymatic pathways, reflecting long-term adaptive strategies to saline environments. Nevertheless, significant gaps remain in understanding the precise mechanisms and coordination of NR- and NOS-mediated NO synthesis, making NO biosynthesis under salt stress one of the most challenging and unresolved issues in plant stress physiology.

2.4.1.5 *Heavy metal stress*

HM toxicity has emerged as one of the most widespread and alarming abiotic stresses in recent years. Upon entering plant cells, HMs induce ionic imbalances, disrupt nutrient metabolism, and stimulate excessive ROS generation, ultimately impairing cellular function, photosynthesis, and a wide range of physiological processes (Gupta et al., 2025). Increasing evidence indicates that NO acts as a critical signaling molecule in plant adaptation to HM stress (Allagulova et al., 2023; Terrón-Camero et al., 2019). Numerous studies have identified NR as the primary source of NO under HM exposure, as elevated NR activity and gene expression are frequently observed in response to cadmium (Cd), lead (Pb), aluminum (Al), and copper (Cu). Li et al. (2020) reported that magnesium mitigates Cd-induced NO overproduction and Cd accumulation in *Panax notoginseng* by promoting the interaction between 14-3-3 proteins and NR, thereby enhancing NR-dependent NO synthesis. Similarly,(Kan et al., 2016) demonstrated that NR-mediated NO production contributes to Cd retention in *P. notoginseng* roots by modifying cell wall properties. NR-derived NO has also been implicated in Al tolerance in red kidney bean roots (Dong et al., 2018; Wang et al., 2010). Consistently, *Phaseolus vulgaris* and *Hordeum vulgare* exhibit increased NR activity and NO production under Al and Cu toxicity (Hu et al., 2015). Furthermore, calcium- and sulfur-mediated alleviation of Cr toxicity in *Solanum lycopersicum* and *S. melongena* depends on NR-derived NO, which enhances plant growth and photosynthetic performance by strengthening antioxidant defenses, stimulating proline metabolism, and facilitating HM sequestration (Singh et al., 2020). A synergistic interaction between methyl jasmonate and NR-derived NO has been shown to induce the expression of the GSH1 gene, thereby enhancing the accumulation of GSH in both roots and leaves of rice under As stress (Ghorbani et al., 2023). Similarly, exogenous application of SA acts in concert with NR-dependent NO to mitigate oxidative damage in rice exposed to arsenate As(III) stress (Singh et al., 2017). Beyond NR, several studies have highlighted the contribution of L-arginine-dependent NOS-like activity to endogenous NO formation in plants (Kolbert et al., 2019; Gupta et al., 2019). Sun et al. (2018) demonstrated that NOS-like activity significantly contributes to NO generation, thereby enhancing Al tolerance in wheat seedlings. Likewise,(Singh et al., 2024) documented endogenous NO accumulation and in vitro detection of NOS-like activity in tomato seedlings exposed to arsenate As(V), confirming its involvement in NO production during HM stress.

2.4.2 Biotic stress

Similar to abiotic stresses, biotic stresses arising from plant–pathogen interactions also cause substantial reductions in crop productivity. To counter these biological challenges, plants have evolved multilayered defense mechanisms comprising structural barriers, chemical deterrents, and immune responses regulated by diverse signaling molecules, including NO (Gupta et al., 2025; Gupta, 2023). NR-derived NO has been demonstrated to regulate the transcription of defense-associated genes, including those encoding pathogenesis-related proteins and antioxidant enzymes (Deng et al., 2016). Recent evidence also indicates that NR-mediated NO suppresses CAT activity, thereby enhancing H_2O_2 accumulation, which subsequently activates defense signaling against *Xanthomonas axonopodis* in *Manihot esculenta* (Yan et al., 2021). Moloi and Westhuizen (2014) demonstrated that resistant *Triticum aestivum* varieties infested by the Russian wheat aphid (*Diuraphis noxia*) showed early NO accumulation. This response involved NR and NiR in NO production, with NO further regulating defense through the induction of β-1,3-glucanase and peroxidase. Elevated expression of NR genes and increased NO accumulation have been reported in several plant species during pest and pathogen attacks, for instance, in *S. lycopersicum* under root-knot nematode (RKN) infestation (Zhou et al., 2015) and in black pepper challenged by *Phytophthora capsici* (Asha et al., 2024). Studies in *Solanum tuberosum* further confirmed that NO_2^--dependent NO production contributes to adaptive defense against *Phytophthora infestans* (Floryszak-Wieczorek et al., 2016; Gupta et al., 2025). Conversely, silencing NR or pharmacological inhibition of its activity diminishes endogenous NO production and compromises plant resistance, as observed in *Nicotiana benthamiana* (Deng et al., 2016). Zou et al. (2018) demonstrated that NIA1-dependent NO production plays a role in BR-induced viral resistance in *A. thaliana*. Melillo et al. (2011) demonstrated that resistant *S. lycopersicum* varieties generate NO and ROS upon RKN invasion, with rapid NO accumulation primarily mediated by NOS-like activity. Similarly, Leonetti et al. (2011) confirmed that NOS-like activity drives the highest NO production at 12 hours postinoculation hpi) in response to RKN, reinforcing this defense mechanism. Liu et al. (2011) investigated two *Oryza sativa* cultivars with contrasting resistance to the brown plant hopper

(*Nilaparvata lugens*) and demonstrated that NO synthesis in response to infestation is dependent on NOS-like activity, resulting in elevated NO levels. Yu et al. (2012) demonstrated that in *Pinus thunbergii* infected by the pine wood nematode (*Bursaphelenchus xylophilus*), NO cooperates with H_2O_2 in mediating the defense response. Alterations in NO concentration were observed during infection, and treatment with SNP induced NO synthesis through NOS-like activity rather than NR in the infected plants (Graska et al., 2023).

2.5 Conclusions and future directions

In conclusion, NO is a critical signaling molecule in plants that supports growth, development, and stress tolerance. By regulating physiological, biochemical, and epigenetic processes, NO enhances plant health and resilience, underscoring its importance in both fundamental plant biology and agricultural applications. Further, the wide variation observed in endogenous NO levels across studies is largely due to the molecule's reactive physicochemical properties and its highly transient behavior, along with differences in plant species, experimental setups, and stress environments. So far, multiple pathways for NO production have been identified in plants, with NR being the most thoroughly investigated. Despite substantial evidence for arginine-dependent NO synthesis, authentic NOS enzymes have yet to be clearly defined in terrestrial plants. This underscores the need for continued research aimed at accurately pinpointing NO-producing sites within plant cells, tissues, and organs during different developmental phases and under varying environmental conditions. Such progress will significantly deepen our understanding of NO functions and their biosynthetic routes in plants under both normal and stress scenarios. Yet, the intricate molecular processes governing NO formation during stress responses remain inadequately understood, and thus, emphasizing the necessity for further exploration of its regulatory mechanisms and contributions to stress tolerance in plants.

References

Ahammed, G. J., Li, X., Mao, Q., Wan, H., Zhou, G., & Cheng, Y. (2021). The SlWRKY81 transcription factor inhibits stomatal closure by attenuating nitric oxide accumulation in the guard cells of tomato under drought. *Physiologia Plantarum, 172*(2), 885–895. https://doi.org/10.1111/ppl.13243, http://onlinelibrary.wiley.com/journal/10.1111/(ISSN)1399-3054.

Airaki, M., Leterrier, M., Mateos, R. M., Valderrama, R., Chaki, M., Barroso, J. B., del Río, L. A., Palma, J. M., & Corpas, F. J. (2012). Metabolism of reactive oxygen species and reactive nitrogen species in pepper (*Capsicum annuum* L.) plants under low temperature stress. *Plant, Cell and Environment, 35*(2), 281–295. https://doi.org/10.1111/j.1365-3040.2011.02310.x.

Allagulova, C. R., Avalbaev, A. M., Lubyanova, A. R., Lastochkina, O. V., & Shakirova, F. M. (2022). Current concepts of the mechanisms of nitric oxide formation in plants. *Pleiades journals, Russian Federation Russian Journal of Plant Physiology, 69*(4). https://doi.org/10.1134/S1021443722030037, https://rd.springer.com/journal/11183.

Allagulova, C. R., Lubyanova, A. R., & Avalbaev, A. M. (2023). Multiple ways of nitric oxide production in plants and its functional activity under abiotic stress conditions. *International Journal of Molecular Sciences, 24*(14), 11637.

Aridhi, F., Sghaier, H., Gaitanaros, A., Khadri, A., Aschi-Smiti, S., & Brouquisse, R. (2020). Nitric oxide production is involved in maintaining energy state in Alfalfa (*Medicago sativa* L.) nodulated roots under both salinity and flooding. *Planta, 252*(2). https://doi.org/10.1007/s00425-020-03422-1, link.springer.de/link/service/journals/00425/index.htm.

Asha, S., Kattupalli, D., Vijayanathan, M., & Soniya, E. V. (2024). Identification of nitric oxide mediated defense signaling and its microRNA mediated regulation during Phytophthora capsici infection in black pepper. *Physiology and Molecular Biology of Plants, 30*(1), 33–47. https://doi.org/10.1007/s12298-024-01414-z, https://www.springer.com/journal/12298.

Asija, S., Seth, T., Umar, S., & Gupta, R. (2023). Polyamines and their crosstalk with phytohormones in the regulation of plant defense responses. *Journal of Plant Growth Regulation, 42*(8), 5224–5246.

Astier, J., Gross, I., & Durner, J. (2018). Nitric oxide production in plants: an update. *Journal of Experimental Botany, 69*(14), 3401–3411.

Berger, A., Boscari, A., Horta Araújo, N., Maucourt, M., Hanchi, M., Bernillon, S., Rolin, D., Puppo, A., & Brouquisse, R. (2020). Plant nitrate reductases regulate nitric oxide production and nitrogen-fixing metabolism during the *Medicago truncatula–Sinorhizobium meliloti* symbiosis. *Frontiers in Plant Science, 11*. https://doi.org/10.3389/fpls.2020.01313, https://www.frontiersin.org/journals/plant-science.

Berry, C. E., & Hare, J. M. (2004). Xanthine oxidoreductase and cardiovascular disease: Molecular mechanisms and pathophysiological implications. *The Journal of physiology, 555*(3), 589–606.

Cai, W., Liu, W., Wang, W. S., Fu, Z. W., Han, T. T., & Lu, Y. T. (2015). Overexpression of rat neurons nitric oxide synthase in rice enhances drought and salt tolerance. *PLoS ONE, 10*(6). https://doi.org/10.1371/journal.pone.0131599, http://www.plosone.org/article/fetchObject.action?uri=info:doi/10.1371/journal.pone.0131599&representation=PDF.

Cao, X., Zhu, C., Zhong, C., Zhang, J., Wu, L., Jin, Q., & Ma, Q. (2019). Nitric oxide synthase-mediated early nitric oxide burst alleviates water stress-induced oxidative damage in ammonium-supplied rice roots. *BMC Plant Biology, 19*(1). https://doi.org/10.1186/s12870-019-1721-2, http://www.biomedcentral.com/bmcplantbiol/.

Caranto, J. D., & Lancaster, K. M. (2017). Nitric oxide is an obligate bacterial nitrification intermediate produced by hydroxylamine oxidoreductase. *Proceedings of the National Academy of Sciences, 114*(31), 8217–8222. https://doi.org/10.1073/pnas.1704504114.

Chiletti, R., Fincher, S. H., Horton, S. B., Peek, G. J., Checchia, P., & Butt, W. (2025). The role of nitric oxide in the sweep gas for patients receiving extracorporeal membrane oxygenation or cardiopulmonary bypass. *Canadian Journal of Cardiology, 41*(4), 621–629. https://doi.org/10.1016/j.cjca.2024.12.027, https://www.sciencedirect.com/science/journal/0828282X.

Choi, E., Chaudhry, S. I., & Martens-Habbena, W. (2023). Role of nitric oxide in hydroxylamine oxidation by ammonia-oxidizing bacteria. *Applied and Environmental Microbiology, 89*(8), e02173-22.

Christou, A., Filippou, P., Manganaris, G. A., & Fotopoulos, V. (2014). Sodium hydrosulfide induces systemic thermotolerance to strawberry plants through transcriptional regulation of heat shock proteins and aquaporin. *BMC Plant Biology, 14*(1). https://doi.org/10.1186/1471-2229-14-42Cyprus, http://www.biomedcentral.com/1471-2229/14/42.

Corpas, F. J., Barroso, J. B., Carreras, A., Valderrama, R., Palma, J. M., León, A. M., Sandalio, L. M., & Río, L. A. D. (2006). Constitutive arginine-dependent nitric oxide synthase activity in different organs of pea seedlings during plant development. *Planta, 224*(2), 246–254. https://doi.org/10.1007/s00425-005-0205-9.

Corpas, F. J., Chaki, M., Fernandez-Ocana, A., Valderrama, R., Palma, J. M., Carreras, A., Begara-Morales, J. C., Airaki, M., del Rio, L. A., & Barroso, J. B. (2008). Metabolism of reactive nitrogen species in pea plants under abiotic stress conditions. *Plant and Cell Physiology, 49*(11), 1711–1722. https://doi.org/10.1093/pcp/pcn144.

Correa-Aragunde, N., Graziano, M., & Lamattina, L. (2004). Nitric oxide plays a central role in determining lateral root development in tomato. *Planta, 218*(6), 900–905.

Costa-Broseta, Á., Perea-Resa, C., Castillo, M. C., Ruíz, M. F., Salinas, J., & León, J. (2018). Nitric oxide controls constitutive freezing tolerance in arabidopsis by attenuating the levels of osmoprotectants, stress-related hormones and anthocyanins. *Scientific Reports, 8*(1). https://doi.org/10.1038/s41598-018-27668-8, www.nature.com/srep/index.html.

Costa-Broseta, Á., Perea-Resa, C., Castillo, M. C., Ruíz, M. F., Salinas, J., & León, J. (2019). Nitric oxide deficiency decreases C-repeat binding factor-dependent and -independent induction of cold acclimation. *Journal of Experimental Botany, 70*(12), 3283–3296. https://doi.org/10.1093/jxb/erz115, http://jxb.oxfordjournals.org/.

Crawford, N. M. (2006). Mechanisms for nitric oxide synthesis in plants. *Journal of Experimental Botany, 57*(3), 471–478. https://doi.org/10.1093/jxb/erj050.

David, A., Yadav, S., & Bhatla, S. C. (2010). Sodium chloride stress induces nitric oxide accumulation in root tips and oil body surface accompanying slower oleosin degradation in sunflower seedlings. *Physiologia Plantarum, 140*(4), 342–354.

Deng, X. G., Zhu, T., Zou, L. J., Han, X. Y., Zhou, X., Xi, D. H., Zhang, D. W., & Lin, H. H. (2016). Orchestration of hydrogen peroxide and nitric oxide in brassinosteroid-mediated systemic virus resistance in *Nicotiana benthamiana*. *Plant Journal, 85*(4), 478–493. https://doi.org/10.1111/tpj.13120, http://www.blackwellpublishing.com/journals/TPJ.

Desikan, R., Griffiths, R., Hancock, J., & Neill, S. (2002). A new role for an old enzyme: Nitrate reductase-mediated nitric oxide generation is required for abscisic acid-induced stomatal closure in *Arabidopsis thaliana*. *Proceedings of the National Academy of Sciences, 99*(25), 16314–16318. https://doi.org/10.1073/pnas.252461999.

Dong, N., Li, Y., Qi, J., Chen, Y., & Hao, Y. (2018). Nitric oxide synthase-dependent nitric oxide production enhances chilling tolerance of walnut shoots in vitro via involvement chlorophyll fluorescence and other physiological parameter levels. *Scientia Horticulturae, 230*, 68–77. https://doi.org/10.1016/j.scienta.2017.11.016, www.elsevier.com/inca/publications/store/5/0/3/3/1/6.

Du, J., Filipović, M. R., Wagner, B. A., & Buettner, G. R. (2023). Ascorbate mediates the non-enzymatic reduction of nitrite to nitric oxide. *Advances in Redox Research, 9*. https://doi.org/10.1016/j.arres.2023.100079, https://www.journals.elsevier.com/advances-in-redox-research.

Emamverdian, A., Ding, Y., Barker, J., Mokhberdoran, F., Ramakrishnan, M., Liu, G., & Li, Y. (2021). Nitric oxide ameliorates plant metal toxicity by increasing antioxidant capacity and reducing Pb and Cd translocation. *Antioxidants, 10*(12), 1981. https://doi.org/10.3390/antiox10121981.

Ermilova, E. V. (2023). Nitric oxide (II) in the biology of Chlorophyta. *Molecular Biology, 57*(6), 921–928.

Floryszak-Wieczorek, J., Arasimowicz-Jelonek, M., & Izbiańska, K. (2016). The combined nitrate reductase and nitrite-dependent route of NO synthesis in potato immunity to *Phytophthora infestans*. *Plant Physiology and Biochemistry, 108*, 468–477.

Foresi, N., Correa-Aragunde, N., Parisi, G., Caló, G., Salerno, G., & Lamattina, L. (2010). Characterization of a nitric oxide synthase from the plant kingdom: NO generation from the green alga *Ostreococcus tauri* is light irradiance and growth phase dependent. *Plant Cell, 22*(11), 3816–3830. https://doi.org/10.1105/tpc.109.073510, http://www.plantcell.org/cgi/reprint/22/11/3816.pdf.

Ghorbani, A., Pishkar, L., Roodbari, N., Tavakoli, S. A., Jahromi, E. M., & Wu, C. (2023). Nitrate reductase is needed for methyl jasmonate-mediated arsenic toxicity tolerance of rice by modulating the antioxidant defense system, glyoxalase system and arsenic sequestration mechanism. *Journal of Plant Growth Regulation, 42*(2), 1107–1119. https://doi.org/10.1007/s00344-022-10616-2, https://www.springer.com/journal/344.

Gomes, M. A. C., Pestana, I. A., Santa-Catarina, C., Hauser-Davis, R. A., & Suzuki, M. S. (2017). Salinity effects on photosynthetic pigments, proline, biomass and nitric oxide in Salvinia auriculata Aubl. *Acta Limnologica Brasiliensia, 29*(0). https://doi.org/10.1590/s2179-975x4716.

Gould, K. S., Lamotte, O., Klinguer, A., Pugin, A., & Wendehenne, D. (2003). Nitric oxide production in tobacco leaf cells: A generalized stress response? *Plant, Cell & Environment, 26*(11), 1851–1862. https://doi.org/10.1046/j.1365-3040.2003.01101.x.

Graska, J., Fidler, J., Gietler, M., Prabucka, B., Nykiel, M., & Labudda, M. (2023). Nitric oxide in plant functioning: metabolism, signaling, and responses to infestation with ecdysozoa parasites. *Biology, 12*(7), 927. https://doi.org/10.3390/biology12070927.

Gupta, K. J., Fernie, A. R., Kaiser, W. M., & van Dongen, J. T. (2011). On the origins of nitric oxide. *Trends in Plant Science, 16*(3), 160–168. https://doi.org/10.1016/j.tplants.2010.11.007.

Gupta, K. J., Kaladhar, V. C., Fitzpatrick, T. B., Fernie, A. R., Møller, I. M., & Loake, G. J. (2022). Nitric oxide regulation of plant metabolism. *Molecular Plant, 15*(2), 228–242. https://doi.org/10.1016/j.molp.2021.12.012, https://www.cell.com/molecular-plant/home.

Gupta, K. J., Kumari, A., Florez-Sarasa, I., Fernie, A. R., & Igamberdiev, A. U. (2018). Interaction of nitric oxide with the components of the plant mitochondrial electron transport chain. *Journal of Experimental Botany, 69*(14), 3413–3424. https://doi.org/10.1093/jxb/ery119, http://jxb.oxfordjournals.org/.

Gupta, K. J., Stoimenova, M., & Kaiser, W. M. (2005). In higher plants, only root mitochondria, but not leaf mitochondria reduce nitrite to NO, in vitro and in situ. *Journal of Experimental Botany, 56*(420), 2601–2609.

Gupta, K. J., Mur, L. A. J., Wany, A., Kumari, A., Fernie, A. R., & Ratcliffe, R. G. (2019). The role of nitrite and nitric oxide under low oxygen conditions in plants. *New Phytologist, 225*(3), 1143–1151. https://doi.org/10.1111/nph.15969.

Gupta, P., Saxena, G., & Gupta, R. (2025). Nitrate reductase-mediated nitric oxide synthesis in shaping stress resilience in plants. *Journal of Experimental Botany, 76*(22), 6634–6656. https://doi.org/10.1093/jxb/eraf353, http://jxb.oxfordjournals.org/.

Gupta, P., & Seth, C. S. (2020). Interactive role of exogenous 24 Epibrassinolide and endogenous NO in *Brassica juncea* L. under salinity stress: Evidence for NR-dependent NO biosynthesis. *Nitric Oxide, 97*, 33–47. https://doi.org/10.1016/j.niox.2020.01.014.

Gupta, R. (2023). Melatonin: A promising candidate for maintaining food security under the threat of phytopathogens. *Plant Physiology and Biochemistry, 198*, 107691. https://doi.org/10.1016/j.plaphy.2023.107691.

Hao, G. P., Xing, Y., & Zhang, J. H. (2008). Role of nitric oxide dependence on nitric oxide synthase-like activity in the water stress signaling of maize seedling. *Journal of integrative plant biology, 50*(4), 435–442.

Hartop, K. (2014). *The impact of nitrite on aerobic growth of Paracoccusdenitrificans PD1222* (Doctoral dissertation, University of East Anglia).

Hou, X., Dan, Y., Qi, N., Zhang, M., Li, C., Li, Y., Yao, Y., & Liao, W. (2024). Nitric oxide promotes adventitious rooting in cucumber under drought stress through regulating ascorbate glutathione cycle and protein S-nitrosylation. *Journal of Plant Growth Regulation, 43*(8), 2771–2784. https://doi.org/10.1007/s00344-024-11304-z, https://www.springer.com/journal/344.

Hou, X., Shi, M., Zhang, Z., Yao, Y., Li, Y., Li, C., Yu, W., Wang, C., & Liao, W. (2025). DNA demethylation is involved in nitric oxide-induced flowering in tomato. *Journal of Integrative Agriculture, 24*(5), 1769–1785. https://doi.org/10.1016/j.jia.2024.09.037, https://www.keaipublishing.com/en/journals/journal-of-integrative-agriculture/.

Hu, Y., You, J., & Liang, X. (2015). Nitrate reductase-mediated nitric oxide production is involved in copper tolerance in shoots of hulless barley. *Plant Cell Reports, 34*(3), 367–379.

Huang, Q., Farooq, M. A., Hannan, F., Chen, W., Ayyaz, A., Zhang, K., Zhou, W., & Islam, F. (2022). Endogenous nitric oxide contributes to chloride and sulphate salinity tolerance by modulation of ion transporter expression and reestablishment of redox balance in *Brassica napus* cultivars. *Environmental and Experimental Botany, 194*, 104734. https://doi.org/10.1016/j.envexpbot.2021.104734.

Irisa, T., Hira, D., Furukawa, K., & Fujii, T. (2014). Reduction of nitric oxide catalyzed by hydroxylamine oxidoreductase from an anammox bacterium. *Journal of Bioscience and Bioengineering, 118*(6), 616–621. https://doi.org/10.1016/j.jbiosc.2014.05.018, http://www.elsevier.com/wps/find/journaldescription.cws_home/505516/description#description.

Jahan, M. S., Shu, S., Wang, Y., Chen, Z., He, M., Tao, M., Sun, J., & Guo, S. (2019). Melatonin alleviates heat-induced damage of tomato seedlings by balancing redox homeostasis and modulating polyamine and nitric oxide biosynthesis. *BMC Plant Biology, 19*(1). https://doi.org/10.1186/s12870-019-1992-7, http://www.biomedcentral.com/bmcplantbiol/.

Kan, Q., Wu, W., Yu, W., Zhang, J., Xu, J., Rengel, Z., Chen, L., Cui, X., & Chen, Q. (2016). Nitrate reductase-mediated NO production enhances Cd accumulation in Panax notoginseng roots by affecting root cell wall properties. *Journal of Plant Physiology, 193*, 64–70. https://doi.org/10.1016/j.jplph.2016.01.017, www.urbanfischer.de/journals/jpp/p_physio.htm.

Karpets, Y. V., Kolupaev, Y. E., Yastreb, T. O., & Oboznyi, A. I. (2016). Induction of heat resistance in wheat seedlings by exogenous calcium, hydrogen peroxide, and nitric oxide donor: functional interaction of signal mediators. *Russian Journal of Plant Physiology, 63*(4), 490–498. https://doi.org/10.1134/S1021443716040075, http://www.kluweronline.com/issn/1021-4437.

Kataria, S., Anand, A., Kumar Raipuria, R., Kumar, S., Jain, M., Watts, A., & Brestic, M. (2022). Magnetopriming actuates nitric oxide synthesis to regulate phytohormones for improving germination of soybean seeds under salt stress. *Cells, 11*(14), 2174. https://doi.org/10.3390/cells11142174.

Kaya, C., & Ashraf, M. (2021). Nitric oxide is required for aminolevulinic acid-induced salt tolerance by lowering oxidative stress in maize (Zea mays). *Journal of Plant Growth Regulation, 40*(2), 617–627.

Khan, M., Al Azzawi, T. N. I., Ali, S., Yun, B. W., & Mun, B. G. (2023). Nitric oxide, a key modulator in the alleviation of environmental stress-mediated damage in crop plants: A meta-analysis. *Plants, 12*(11), 2121.

Khan, M., Ali, S., Al Azzawi, T. N. I., & Yun, B. W. (2023). Nitric oxide acts as a key signalling molecule in plant development under stressful conditions. *International journal of Molecular Sciences, 24*(5), 4782.

Khan, M. N., AlSolami, M. A., Basahi, R. A., Siddiqui, M. H., Al-Huqail, A. A., Abbas, Z. K., Siddiqui, Z. H., Ali, H. M., & Khan, F. (2020). Nitric oxide is involved in nano-titanium dioxide-induced activation of antioxidant defense system and accumulation of osmolytes under water-deficit stress in *Vicia faba* L. *Ecotoxicology and Environmental Safety, 190*. https://doi.org/10.1016/j.ecoenv.2019.110152, http://www.elsevier.com/inca/publications/store/6/2/2/8/1/9/index.htt.

Khatoon, S., Jain, K., Mahajan, M., & Khan, M. I. R. (2023). Nitric oxide biosynthesis under stressful environments. In *Nitric oxide in developing plant stress resilience*. India: Elsevier, 17–30. https://doi.org/10.1016/B978-0-323-91209-9.00013-0, https://www.sciencedirect.com/book/9780323912099.

Kolbert, Z., Feigl, G., Freschi, L., & Poór, P. (2019). Gasotransmitters in action: nitric oxide-ethylene crosstalk during plant growth and abiotic stress responses. *Antioxidants, 8*(6), 167. https://doi.org/10.3390/antiox8060167.

Kolupaev, Y. E., Yemets, A. I., Yastreb, T. O., & Blume, Y. B. (2023). The role of nitric oxide and hydrogen sulfide in regulation of redox homeostasis at extreme temperatures in plants. *Frontiers in Plant Science, 14*. https://doi.org/10.3389/fpls.2023.1128439, https://www.frontiersin.org/journals/plant-science.

Krasylenko, Y. A., Yemets, A. I., & Blume, Y. B. (2019). Nitric oxide synthase inhibitor L-NAME affects *Arabidopsis* root growth, morphology, and microtubule organization. *Cell Biology International, 43*(9), 1049–1055.

Kumar, D., & Ohri, P. (2023). Say "NO" to plant stresses: Unravelling the role of nitric oxide under abiotic and biotic stress. *Nitric Oxide, 130*, 36–57. https://doi.org/10.1016/j.niox.2022.11.004.

Kumari, A., Kaladhar, V. C., Yadav, N., Singh, P., Reddy, K., & Gupta, K. J. (2023). Nitric oxide regulates mitochondrial biogenesis in plants. *Plant Cell and Environment, 46*(8), 2492–2506. https://doi.org/10.1111/pce.14637, http://onlinelibrary.wiley.com/journal/10.1111/(ISSN)1365-3040.

Lau, S. E., Hamdan, M. F., Pua, T. L., Saidi, N. B., & Tan, B. C. (2021). Plant nitric oxide signaling under drought stress. *Plants, 10*(2), 1–30. https://doi.org/10.3390/plants10020360, https://www.mdpi.com/2223-7747/10/2/360.

Leonetti, P., Melillo, M. T., & Bleve-Zacheo, T. (2011). Nitric oxide and hydrogen peroxide: two players in the defence response of tomato plants to root-knot nematodes. *Communications in Agricultural and Applied Biological Sciences, 76*(3), 371–381.

Li, D., Xiao, S., Ma, W.-N., Peng, Z., Khan, D., Yang, Q., Wang, X., Kong, X., Zhang, B., Yang, E., Rengel, Z., Wang, J., Cui, X., & Chen, Q. (2020). Magnesium reduces cadmium accumulation by decreasing the nitrate reductase-mediated nitric oxide production in Panax notoginseng roots. *Journal of Plant Physiology, 248*, 153131. https://doi.org/10.1016/j.jplph.2020.153131.

Li, X., Pan, Y., Chang, B., Wang, Y., & Tang, Z. (2016). NO promotes seed germination and seedling growth under high salt may depend on EIN3 protein in Arabidopsis. *Frontiers in Plant Science, 6*, 2016. https://doi.org/10.3389/fpls.2015.01203, http://www.journal.frontiersin.org/article/10.3389/fpls.2015.01203/pdf.

Lillo, C., Meyer, C., Lea, U. S., Provan, F., & Oltedal, S. (2004). Mechanism and importance of post-translational regulation of nitrate reductase. *Journal of Experimental Botany, 55*(401), 1275–1282. https://doi.org/10.1093/jxb/erh132.

Liu, J., Xue, T., & Shen, Y. (2019). Effect of nitric oxide on seed germination and dormancy in empress trees. *HortTechnology, 29*(3), 271–275.

Liu, Y., He, J., Jiang, L., Wu, H., Xiao, Y., Liu, Y., Li, G., Du, Y., Liu, C., & Wan, J. (2011). Nitric oxide production is associated with response to brown planthopper infestation in rice. *Journal of Plant Physiology, 168*(8), 739–745. https://doi.org/10.1016/j.jplph.2010.09.018.

Liu, Y., Jiang, H., Zhao, Z., & An, L. (2010). Nitric oxide synthase like activity-dependent nitric oxide production protects against chilling-induced oxidative damage in *Chorispora bungeana* suspension cultured cells. *Plant Physiology and Biochemistry, 48*(12), 936–944. https://doi.org/10.1016/j.plaphy.2010.09.001.

Liu, Y., Wu, R., Wan, Q., Xie, G., & Bi, Y. (2007). Glucose-6-Phosphate dehydrogenase plays a pivotal role in nitric oxide-involved defense against oxidative stress under salt stress in red kidney bean roots. *Plant and Cell Physiology, 48*(3), 511–522. https://doi.org/10.1093/pcp/pcm020.

Lv, X., Li, H., Chen, X., Xiang, X., Guo, Z., Yu, J., & Zhou, Y. (2018). The role of calcium-dependent protein kinase in hydrogen peroxide, nitric oxide and ABA-dependent cold acclimation. *Journal of Experimental Botany, 69*(16), 4127–4139. https://doi.org/10.1093/jxb/ery212, http://jxb.oxfordjournals.org/.

Melillo, M. T., Leonetti, P., Leone, A., Veronico, P., & Bleve-Zacheo, T. (2011). ROS and NO production in compatible and incompatible tomato-*Meloidogyne incognita* interactions. *European Journal of Plant Pathology, 130*(4), 489–502. https://doi.org/10.1007/s10658-011-9768-4.

Meyer, C., Lea, U. S., Provan, F., Kaiser, W. M., & Lillo, C. (2005). Is nitrate reductase a major player in the plant NO (nitric oxide) game? *Photosynthesis Research, 83*(2), 181–189. https://doi.org/10.1007/s11120-004-3548-3, https://link.springer.com/journal/11120.

Mohn, M. A., Thaqi, B., & Fischer-Schrader, K. (2019). Isoform-specific NO synthesis by *Arabidopsis thaliana* nitrate reductase. *Plants, 8*(3), 67.

Moloi, M. J., & Westhuizen, A. J. (2014). Involvement of nitric oxide in the Russian wheat aphid resistance response of wheat. *Cereal Research Communications, 42*(1), 119–125.

Nabi, R. B. S., Tayade, R., Hussain, A., Kulkarni, K. P., Imran, Q. M., Mun, B. G., & Yun, B. W. (2019). Nitric oxide regulates plant responses to drought, salinity, and heavy metal stress. *Environmental and Experimental Botany, 161*, 120–133. https://doi.org/10.1016/j.envexpbot.2019.02.003, www.elsevier.com/inca/publications/store/2/6/7.

Parankusam, S., Adimulam, S. S., Bhatnagar-Mathur, P., & Sharma, K. K. (2017). Nitric oxide (NO) in plant heat stress tolerance: Current knowledge and perspectives. *Frontiers in Plant Science, 8*. https://doi.org/10.3389/fpls.2017.01582, http://journal.frontiersin.org/article/10.3389/fpls.2017.01582/full.

Pissolato, M. D., Silveira, N. M., Prataviera, P. J. C., Machado, E. C., Seabra, A. B., Pelegrino, M. T., et al. (2020). Enhanced nitric oxide synthesis through nitrate supply improves drought tolerance of sugarcane plants. *Frontiers in Plant Science, 11*. https://doi.org/10.3389/fpls.2020.00970.

Prochazkova, D., Haisel, D.& Pavlikova, D. (2014). Nitric oxide biosynthesis in plants-the short overview.

Puyaubert, J., & Baudouin, E. (2014). New clues for a cold case: Nitric oxide response to low temperature. *Plant, Cell & Environment, 37*(12), 2623–2630. https://doi.org/10.1111/pce.12329.

Qiao, W., Xiao, S., Yu, L., & Fan, L. (2009). Expression of a rice gene OsNOA1 re-establishes nitric oxide synthesis and stress-related gene expression for salt tolerance in *Arabidopsis* nitric oxide-associated 1 mutant Atnoa1. *Environmental and Experimental Botany, 65*(1), 90–98. https://doi.org/10.1016/j.envexpbot.2008.06.002.

Reda, M., Golicka, A., Kabała, K., & Janicka, M. (2018). Involvement of NR and PM-NR in NO biosynthesis in cucumber plants subjected to salt stress. *Plant Science, 267*, 55–64. https://doi.org/10.1016/j.plantsci.2017.11.004, http://www.journals.elsevier.com/plant-science/.

Van Ree, K., Gehl, B., Wassim Chehab, E., Tsai, Y. C., & Braam, J. (2011). Nitric oxide accumulation in *Arabidopsis* is independent of NOA1 in the presence of sucrose. *The Plant Journal, 68*(2), 225–233.

Roszer, T. (2014). Biosynthesis of nitric oxide in plants. In *Nitric oxide in plants: metabolism and role in stress physiology*. Springer International Publishing, 17–32. http://doi.org/10.1007/978-3-319-06710-0, https://doi.org/10.1007/978-3-319-06710-0_2.

Rümer, S., Gupta, K. J., & Kaiser, W. M. (2009). Plant cells oxidize hydroxylamines to NO. *Journal of Experimental Botany, 60*(7), 2065–2072.

Sami, F., Faizan, M., Faraz, A., Siddiqui, H., Yusuf, M., & Hayat, S. (2018). Nitric oxide-mediated integrative alterations in plant metabolism to confer abiotic stress tolerance, NO crosstalk with phytohormones and NO-mediated post translational modifications in modulating diverse plant stress. *Nitric Oxide, 73*, 22–38. https://doi.org/10.1016/j.niox.2017.12.005.

Santa-Cruz, D. M., Pacienza, N. A., Polizio, A. H., Balestrasse, K. B., Tomaro, M. L., & Yannarelli, G. G. (2010). Nitric oxide synthase-like dependent NO production enhances heme oxygenase up-regulation in ultraviolet-B-irradiated soybean plants. *Phytochemistry, 71*(14-15), 1700–1707. https://doi.org/10.1016/j.phytochem.2010.07.009.

Sarma, H., Ramchiary, C., Sen, B., Daimary, M., & Prasad, R. (2025). Integration of nitric oxide signaling in plant stress responses: Unveiling its role in enhancing crop tolerance to abiotic stress. *Discover Plants, 2*(1). https://doi.org/10.1007/s44372-025-00102-w.

Šírová, J., Sedlářová, M., Piterková, J., Luhová, L., & Petřivalský, M. (2011). The role of nitric oxide in the germination of plant seeds and pollen. *Plant Science, 181*(5), 560–572. https://doi.org/10.1016/j.plantsci.2011.03.014.

Senthil Kumar, R., Shen, C. H., Wu, P. Y., Suresh Kumar, S., Hua, M. S., & Yeh, K. W. (2016). Nitric oxide participates in plant flowering repression by ascorbate. *Scientific Reports, 6*(1), 35246.

Shang, J.-X., Li, X., Li, C., & Zhao, L. (2022). The role of nitric oxide in plant responses to salt stress. *International Journal of Molecular Sciences, 23*(11), 6167. https://doi.org/10.3390/ijms23116167.

Shkliarevskyi, M. A., Kolupaev, Y. E., Karpets, Y. V., Lugovaya, A. A., & Bessonova, V. P. (2021). Involvement of nitrate reductase and nitric oxide (NO) in implementation of the stress-protective action of a carbon monoxide (CO) donor on wheat seedlings under hyperthermy. *Journal of Plant Physiology, 68*(4), 688–695. https://doi.org/10.1134/S1021443721040166, https://rd.springer.com/journal/11183.

Siddiqui, M. H., Al-Whaibi, M. H., & Basalah, M. O. (2011). Role of nitric oxide in tolerance of plants to abiotic stress. *Protoplasma, 248*(3), 447–455.

Singh, A. P., Dixit, G., Kumar, A., Mishra, S., Kumar, N., Dixit, S., Singh, P. K., Dwivedi, S., Trivedi, P. K., Pandey, V., Dhankher, O. P., Norton, G. J., Chakrabarty, D., & Tripathi, R. D. (2017). A protective role for nitric oxide and salicylic acid for arsenite phytotoxicity in rice (*Oryza sativa* L.). *Plant Physiology and Biochemistry, 115*, 163–173. https://doi.org/10.1016/j.plaphy.2017.02.019, http://www.journals.elsevier.com/plant-physiology-and-biochemistry/.

Singh, P., Jaiswal, S., Tripathi, D. K., & Singh, V. P. (2024). Nitric oxide acts upstream of indole-3-acetic acid in ameliorating arsenate stress in tomato seedlings. *Plant Physiology and Biochemistry, 208*. https://doi.org/10.1016/j.plaphy.2024.108461, https://www.sciencedirect.com/science/journal/09819428.

Singh, P. K., Tiwari, M., Kidwai, M., Srivastava, D., Tripathi, R. D., & Chakrabarty, D. (2020). Role of nitric oxide in overcoming heavy metal stress. In *Protective chemical agents in the amelioration of plant abiotic stress: biochemical and molecular perspectives*. India: wiley, 214–237. https://doi.org/10.1002/9781119552154.ch9, https://onlinelibrary.wiley.com/doi/book/10.1002/9781119552154.

Song, L., Zhao, H., & Hou, M. (2013). Involvement of nitric oxide in acquired thermotolerance of rice seedlings. *Russian Journal of Plant Physiology, 60*(6), 785–790.

Sun, C., Liu, L., Lu, L., Jin, C., & Lin, X. (2018). Nitric oxide acts downstream of hydrogen peroxide in regulating aluminum-induced antioxidant defense that enhances aluminum resistance in wheat seedlings. *Environmental and Experimental Botany, 145*, 95–103. https://doi.org/10.1016/j.envexpbot.2017.10.020, www.elsevier.com/inca/publications/store/2/6/7.

Sun, C., Zhang, Y., Liu, L., Liu, X., Li, B., Jin, C., & Lin, X. (2021). Molecular functions of nitric oxide and its potential applications in horticultural crops. *Horticulture Research, 8*(1). https://doi.org/10.1038/s41438-021-00500-7, www.nature.com/hortres.

Sun, H., Li, J., Song, W., Tao, J., Huang, S., Chen, S., Hou, M., Xu, G., & Zhang, Y. (2015). Nitric oxide generated by nitrate reductase increases nitrogen uptake capacity by inducing lateral root formation and inorganic nitrogen uptake under partial nitrate nutrition in rice. *Journal of Experimental Botany, 66*(9), 2449–2459. https://doi.org/10.1093/jxb/erv030, http://jxb.oxfordjournals.org/.

Talwar, P. S., Gupta, R., Maurya, A. K., & Deswal, R. (2012). *Brassica juncea* nitric oxide synthase like activity is stimulated by PKC activators and calcium suggesting modulation by PKC-like kinase. *Plant physiology and biochemistry, 60*, 157–164.

Terrón-Camero, L. C., Peláez-Vico, M.Á., Del-Val, C., Sandalio, L. M., & Romero-Puertas, M. C. (2019). Role of nitric oxide in plant responses to heavy metal stress: Exogenous application versus endogenous production. *Journal of Experimental Botany, 70*(17), 4477–4488.

Tewari, R. K., Horemans, N., & Watanabe, M. (2021). Evidence for a role of nitric oxide in iron homeostasis in plants. *Journal of Experimental Botany, 72*(4), 990–1006.

Tian, W., Huang, D., Geng, B., Zhang, Q., Feng, J., & Zhu, S. (2020). Regulation of the biosynthesis of endogenous nitric oxide and abscisic acid in stored peaches by exogenous nitric oxide and abscisic acid. *Journal of the Science of Food and Agriculture, 100*(5), 2136–2144. https://doi.org/10.1002/jsfa.10237, http://onlinelibrary.wiley.com/journal/10.1002/(ISSN)1097-0010.

Tun, N. N., Santa-Catarina, C., Begum, T., Silveira, V., Handro, W., Floh, E. I. S., & Scherer, G. F. E. (2006). Polyamines induce rapid biosynthesis of nitric oxide (NO) in *Arabidopsis thaliana* seedlings. *Plant and Cell Physiology, 47*(3), 346–354. https://doi.org/10.1093/pcp/pci252.

Urarte, E., Esteban, R., Moran, J. F., & Bittner, F. (2014). Established and proposed roles of xanthine oxidoreductase in oxidative and reductive pathways in plants. In *Reactive oxygen and nitrogen species signaling and communication in plants*. Cham: Springer International Publishing, 15–42.

Versantvoort, W., Pol, A., Jetten, M. S. M., van Niftrik, L., Reimann, J., Kartal, B., & Op den Camp, H. J. M. (2020). Multiheme hydroxylamine oxidoreductases produce NO during ammonia oxidation in methanotrophs. *Proceedings of the National Academy of Sciences, 117*(39), 24459–24463. https://doi.org/10.1073/pnas.2011299117.

Wang, H. H., Huang, J. J., & Bi, Y. R. (2010). Nitrate reductase-dependent nitric oxide production is involved in aluminum tolerance in red kidney bean roots. *Plant Science, 179*(3), 281–288.

Wilson, I. D., Neill, S. J., & Hancock, J. T. (2008). Nitric oxide synthesis and signalling in plants. *Plant, Cell & Environment, 31*(5), 622–631.

Wimalasekera, R., Tebartz, F., & Scherer, G. F. (2011). Polyamines, polyamine oxidases and nitric oxide in development, abiotic and biotic stresses. *Plant Science, 181*(5), 593–603.

Xie, Y., Mao, Y., Lai, D., Zhang, W., Zheng, T., & Shen, W. (2013). Roles of NIA/NR/NOA1-dependent nitric oxide production and HY1 expression in the modulation of *Arabidopsis* salt tolerance. *Journal of Experimental Botany, 64*(10), 3045–3060. https://doi.org/10.1093/jxb/ert149.

Xuan, Y., Zhou, S., Wang, L., Cheng, Y., & Zhao, L. (2010). Nitric oxide functions as a signal and acts upstream of atCaM3 in thermotolerance in arabidopsis seedlings. *Plant Physiology, 153*(4), 1895–1906. https://doi.org/10.1104/pp.110.160424, http://www.plantphysiol.org/cgi/reprint/153/4/1895.

Yan, Y., Wang, P., Wei, Y., Bai, Y., Lu, Y., Zeng, H., Liu, G., Reiter, R. J., He, C., & Shi, H. (2021). The dual interplay of RAV5 in activating nitrate reductases and repressing catalase activity to improve disease resistance in cassava. *Plant Biotechnology Journal, 19*(4), 785–800. https://doi.org/10.1111/pbi.13505, http://onlinelibrary.wiley.com/journal/10.1111/(ISSN)1467-7652.

Yang, B., Wu, J., Gao, F., Wang, J., & Su, G. (2014). Polyamine-induced nitric oxide generation and its potential requirement for peroxide in suspension cells of soybean cotyledon node callus. *Plant Physiology and Biochemistry, 79*, 41–47. https://doi.org/10.1016/j.plaphy.2014.02.025.

Yu, L. Z., Wu, X. Q., Ye, J. R., Zhang, S. N., & Wang, C. (2012). NOS-like-mediated nitric oxide is involved in *Pinus thunbergii* response to the invasion of *Bursaphelenchus xylophilus*. *Plant Cell Reports, 31*(10), 1813–1821. https://doi.org/10.1007/s00299-012-1294-0.

Zhang, P., Li, S., Guo, Z., & Lu, S. (2019). Nitric oxide regulates glutathione synthesis and cold tolerance in forage legumes. *Environmental and Experimental Botany, 167*, 103851. https://doi.org/10.1016/j.envexpbot.2019.103851.

Zhao, L., Zhang, F., Guo, J., Yang, Y., Li, B., & Zhang, L. (2004). Nitric oxide functions as a signal in salt resistance in the calluses from two ecotypes of reed. *Plant Physiology, 134*(2), 849–857. https://doi.org/10.1104/pp.103.030023, http://www.plantphysiol.org/.

Zhao, M. G., Chen, L., Zhang, L. L., & Zhang, W. H. (2009). Nitric reductase-dependent nitric oxide production is involved in cold acclimation and freezing tolerance in *Arabidopsis*. *Plant Physiology, 151*(2), 755–767. https://doi.org/10.1104/pp.109.140996, http://www.plantphysiol.org/cgi/reprint/151/2/755.

Zhao, M. G., Tian, Q. Y., & Zhang, W. H. (2007). Nitric oxide synthase-dependent nitric oxide production is associated with salt tolerance in *Arabidopsis*. *Plant physiology, 144*(1), 206–217.

Zhou, J., Jia, F., Shao, S., Zhang, H., Li, G., Xia, X., Zhou, Y., Yu, J., & Shi, K. (2015). Involvement of nitric oxide in the jasmonate-dependent basal defense against root-knot nematode in tomato plants. *Frontiers in Plant Science, 6*. https://doi.org/10.3389/fpls.2015.00193, http://journal.frontiersin.org/article/10.3389/fpls.2015.00193/pdf.

Zhou, S., Jia, L., Chu, H., Wu, D., Peng, X., Liu, X., Zhang, J., Zhao, J., Chen, K., Zhao, L., & Pei, Z.-M. (2016). *Arabidopsis* CaM1 and CaM4 promote nitric oxide production and salt resistance by inhibiting S-nitrosoglutathione reductase via direct binding. *PLOS Genetics, 12*(9), e1006255. https://doi.org/10.1371/journal.pgen.1006255.

Zhu, W., Wang, L., Wang, J., Zhan, N., Shi, Z., Sun, Y., Lv, Q., Hu, Y., Bao, F., Li, L., He, Y., & Wang, Y. (2025). Nitric oxide delays floral transition in *Arabidopsis* by inhibiting histone deacetylases HDA5 and HDA6. *The Plant Journal, 123*(2). https://doi.org/10.1111/tpj.70379.

Zhu, Y., Liao, W., Wang, M., Niu, L., Xu, Q., & Jin, X. (2016). Nitric oxide is required for hydrogen gas-induced adventitious root formation in cucumber. *Journal of Plant Physiology, 195*, 50–58. https://doi.org/10.1016/j.jplph.2016.02.018, www.urbanfischer.de/journals/jpp/p_physio.htm.

Ziogas, V., Tanou, G., Filippou, P., Diamantidis, G., Vasilakakis, M., Fotopoulos, V., & Molassiotis, A. (2013). Nitrosative responses in citrus plants exposed to six abiotic stress conditions. *Plant Physiology and Biochemistry, 68*, 118–126. https://doi.org/10.1016/j.plaphy.2013.04.004.

Zou, L. J., Deng, X. G., Zhang, L. E., Zhu, T., Tan, W. R., Muhammad, A., Zhu, L. J., Zhang, C., Zhang, D. W., & Lin, H. H. (2018). Nitric oxide as a signaling molecule in brassinosteroid-mediated virus resistance to cucumber mosaic virus in *Arabidopsis thaliana*. *Physiologia Plantarum, 163*(2), 196–210. https://doi.org/10.1111/ppl.12677, http://onlinelibrary.wiley.com/journal/10.1111/(ISSN)1399-3054.

Chapter 3

Nitric oxide-based posttranslational modifications and their implications in plant stress signaling

Adeela Munir[1], Clément Leonard[1], Cheol Woo Min[1], Gi Hyun Lee[1], Babar Usman[1], Sun Tae Kim[1], and Ravi Gupta[2]

[1]*Department of Plant Bioscience, Pusan National University, Life and Industry Convergence Research Institute, Miryang, Republic of Korea,* [2]*Plant Stress Physiology and Proteomics Laboratory, College of General Education, Kookmin University, Seoul, South Korea*

3.1 Introduction

Nitric oxide (NO) is a gaseous free radical that acts as a fundamental signaling molecule in plants, orchestrating adaptive responses to environmental cues and stress conditions through complex networks of posttranslational modifications (PTMs) (Sedlářová et al., 2025). NO exerts its regulatory influence on plant systems primarily through three major PTMs. The first is metal nitrosylation, a process in which an NO moiety interacts with the transition-metal center, such as iron, zinc, or copper, of metalloproteins through coordination chemistry. This metal nitrosylation of proteins triggers conformational changes that can significantly impact the protein's reactivity or enzymatic activity (Wang et al., 2024). The second modification is protein tyrosine nitration, which involves the covalent addition of a nitro group (-NO_2) to the aromatic ring of a tyrosine residue, typically forming 3-nitrotyrosine (León & Costa-Broseta, 2020). While historically viewed as a permanent marker of nitro-oxidative stress, this modification is now recognized as a selective regulatory mechanism that often results in the inhibition or inactivation of the target protein. The third major PTM is S-nitrosation, which involves the reversible attachment of an NO group to the thiol (-SH) side chain of cysteine residues, resulting in the formation of S-nitrosothiols (SNOs) (Lin et al., 2025; Liu et al., 2024). S-nitrosation is the most extensively studied NO-dependent modification and is vital for orchestrating developmental and stress-response signaling (Lin et al., 2025).

Nitric Oxide in Plant Resilience to Abiotic Stress. DOI: https://doi.org/10.1016/B978-0-443-40497-9.00016-7

In this chapter, we summarize the current understanding of NO-based PTMs and discuss how these modifications regulate protein activity, stability, and signaling networks under both biotic and abiotic stress conditions. Special attention is given to the NO-mediated regulation of diverse proteins associated with phytohormone signaling, antioxidant defense, and other key stress-responsive proteins.

3.2 Nitric oxide-based posttranslational modifications

3.2.1 Metal nitrosylation

NO reacts with metal ions to form metal–nitrosyl complexes (Li & Li, 2016), which facilitate transport and storage of NO inside the biological systems (Ignarro, 2000; Vanin, 2009). The coordination chemistry of metal nitrosyls has grown considerably in the last few decades. Due to the highly reactive nature of NO, it can form stable metal–NO complexes by coordinating with transition metals found in metalloproteins, especially iron, zinc, and copper (Ford, 2010). Depending on the surrounding protein environment, the NO moiety becomes susceptible to nucleophilic and, less commonly, electrophilic attack once it is bound. The possible function of metal–NO complexes in the synthesis of SNOs is supported by this chemical reactivity. Moreover, the reversible character of metal–NO binding due to displacement by oxygen molecules or a change in redox state usually sets conformational changes in the target protein, thus affecting its reactivity or enzymatic activity (Toledo & Augusto, 2012). Several analytical techniques, including infrared spectroscopy, electron paramagnetic resonance, and X-ray crystallography, have been introduced to study metal nitrosylation of proteins (Bellota-Anton et al., 2011).

Hemoglobin is the most well-known protein identified to undergo metal nitrosylation in plants. Three main groups of plant hemoglobins are distinguished by their structural characteristics: class 1 and 2 hemoglobins and class 3 truncated hemoglobin (Gupta et al., 2011; Igamberdiev et al., 2011). High-oxygen-affinity class 1 nonsymbiotic hemoglobins (nsHbs) with low dissociation constants (K_D) make them unlikely to be oxygen carriers (Herold & Puppo, 2005). Studies over the last decade have shown that NO can oxidize oxygenated class 1nsHbs, producing nitrate (Perazzolli et al., 2004). Class 2 symbiotic hemoglobins (sHbs) have also been shown to exhibit similar NO interactions; they can bind and scavenge NO (Sasakura et al., 2006; Mathieu et al., 1998).

3.2.2 Tyrosine nitration

Reactive nitrogen species (RNS) covalently modify tyrosine residues of the proteins, a process that is termed tyrosine nitration (Astier & Lindermayr, 2012; Matamoros & Becana, 2021). During this process, a nitro group (-NO_2) reacts at the ortho position within the phenolic hydroxyl group of the tyrosine residues to form 3-nitrotyrosine. The NO_2 group mainly originates from peroxynitrite ($ONOO^-$), a potent oxidative agent produced as a result of an instant reaction between NO and a superoxide anion (O_2^-). The nitration of tyrosine residues is a critical regulatory mechanism in plant physiology, often serving as a bridge between environmental stress and metabolic adjustment. Tyrosine nitration in plants can be detected using advanced analytical methods such as chromatographic purification in conjunction with mass spectrometry (MS) and immunopurification or western blotting using anti-3-nitrotyrosine antibodies (Berton et al., 2012; Wisastra et al., 2011; Tanou et al., 2012). Moreover, a few studies have also utilized fluorescent tagging of 3-nitrotyrosine residues for the detection of this PTM (Sharov et al., 2010). One study utilized mixed-mode solid-phase extraction coupled with LC-ESI-TOF MS to enrich and detect tyrosine and 3-nitrotyrosine residues with high sensitivity and specificity (Chaki et al., 2018). However, the use of high-resolution tandem MS remains a key method for site-specific identification of nitrated proteins involved in NO signaling pathways (León, 2022).

Currently, the biological role of tyrosine nitration in plants remains poorly defined. It has been reported that this modification may interfere with important signaling processes involving the addition or removal of phosphate groups (phosphorylation/dephosphorylation) at tyrosine residues, especially if both nitration and phosphorylation target the same residue. However, so far, no direct experimental evidence in plants has confirmed this interaction (Kolbert et al., 2017; Mendez et al., 2020). One of the earliest pieces of evidence of plant tyrosine nitration was found in tobacco cells lacking nitrate reductase. It was reported that tobacco lines expressing antisense nitrite-reductase transcripts accumulated nitrite and created high levels of NO, which resulted in elevated peroxynitrite formation and increased protein tyrosine nitration (Morot-Gaudry-Talarmain et al., 2002). A significant consequence of this modification is the loss of protein function, particularly within the photosynthetic apparatus. For instance, nitration of Photosystem II core proteins and Rubisco has been shown to directly inhibit photosynthetic efficiency, as the addition of bulky nitro groups interferes with electron transport and carbon fixation (Galetskiy et al., 2011). At first, it was presumed that tyrosine nitration is an irreversible process, which leads to protein degradation (Radi, 2018; Lee et al., 2020). However, recent research involving animal models indicates that denitrosation can actually occur, though the specific enzymes that facilitate this process are still unknown (León, 2022). A definitive example of enzyme inhibition via this PTM is also observed in nitrogen metabolism. NO inactivates glutamine synthetase in the root nodules of *Medicago truncatula*, thereby redirecting metabolites toward enhanced antioxidant defense within the nodules (Melo et al., 2011).

Notably, tyrosine nitration seems to be crucial for plant defense and nitrosative stress responses. Saito et al. (2006) showed that when tobacco BY-2 cells were treated with an elicitor, they rapidly produced peroxynitrite and tyrosine-nitrated proteins. This suggests that these changes could be a component of early defense signaling. Likewise, sodium nitroprusside (SNP)-induced nitrosative stress in pea plants led to elevated 3-nitrotyrosine synthesis, establishing a connection between protein nitration and the balance of cellular oxidative/nitrosative stress (Valderrama et al., 2007). Romero-Puertas et al. (2007) also discovered that S-nitrosation of peroxiredoxin II E increases its vulnerability to tyrosine nitration mediated by peroxynitrite, suggesting that interactions between various NO-derived modifications may alter protein function under stress. Tyrosine nitration also adversely affects the activity of a number of plant antioxidant enzymes, such as catalase, as reported in pepper (Chaki et al., 2015). Chen et al. (2022) showed that tyrosine nitration partially inhibits mitochondrial and apoplastic manganese superoxide dismutase (Mn-SODs; MSD1 and MSD2), leading to superoxide accumulation, altered redox balance, and modulation of redox-dependent metabolic and stress responses (Chen et al., 2022). Likewise, Mendez et al. (2020) demonstrated that maize cyclin-dependent kinase A;1 (ZmCDKA;1), a key regulator of cell cycle progression, is selectively nitrated at Tyr15 and Tyr19 under nitrosative conditions. These tyrosines are located within the ATP-binding site, and their nitration significantly reduces the affinity of CDKA-1 for ATP, leading to partial inhibition of its kinase activity. These findings establish tyrosine nitration as a regulatory PTM capable of modulating plant cell cycle progression during nitro-oxidative stress (Mendez et al., 2020). In summary, this modification, which was once thought to be only a sign of nitrosative damage, is now being recognized as a selective and possibly regulating PTM that has important effects on redox biology and plant stress signaling (Fig. 3.1).

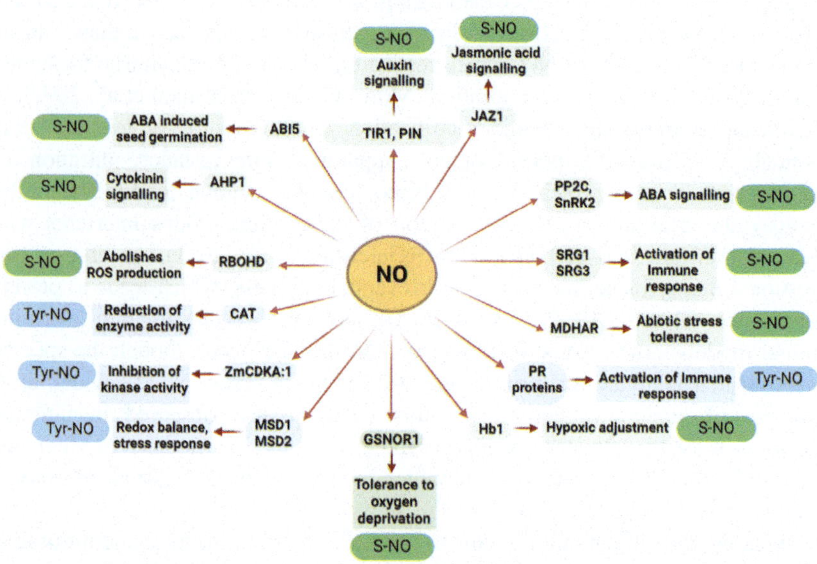

FIGURE 3.1 Regulatory role of nitric oxide (*NO*)-mediated posttranslational modifications (*PTMs*) in plant signaling and stress responses.

3.2.3 Cysteine nitrosation

S-nitrosation is a key NO-dependent protein PTM that has been implicated in a wide range of cellular processes in recent years (Feng et al., 2019). In the literature, the term nitrosation is frequently used to characterize chemical or biochemical reactions involving NO. S-nitrosation creates -SNO by covalently bonding a NO group to the free sulfhydryl group of cysteine residues in the target protein (Hess et al., 2005). This modification is regarded as the most significant NO-based PTM in plants because of its reversible nature, which enables the dynamic control of protein function. Reversibility is a crucial mechanism in redox signaling and stress responses because it is regulated by particular cellular systems, such as thioredoxin and GSNOR (S-nitrosoglutathione reductase), which eliminate the -SNO group and return the protein to its active thiol form (Zaffagnini et al., 2013; Feng et al., 2019). It is directly mediated by NO· or by donors of NO· groups, such as GSNO, SNP, and ONOO-. Depending on the protein and the altered residue, this modification can have profound effects on protein activities (Feng et al., 2019).

3.2.3.1 Role of S-nitrosation in shaping phytohormone signaling

Phytohormones regulate key plant processes and affect nearly every aspect of their life. These molecules regulate growth, development, and stress resilience in very low concentrations. Among the phytohormones, salicylic acid (SA), jasmonic acid (JA), auxin, cytokinin, gibberellins (GA), abscisic acid (ABA), and ethylene are known to work closely with NO (París et al., 2013; Ventimiglia & Mutus, 2020). Together, they form a complex communication network that allows plants to respond precisely to both developmental signals and environmental stresses (Ergashev et al., 2025). Intriguingly, NO is known to regulate some of the key proteins of stress signaling networks, such as the mitogen-activated protein kinase (MAPK) signaling cascade, to shape the stress signaling in plants. For instance, during combined salt stress and S-nitrosoglutathione treatment, core components of the MAPK cascade, specifically MAPK, MAPK Kinase (MAPKK/MKK), and MKK5, are positively regulated by S-nitrosation. This PTM enhances the activation and stability of the cascade, which subsequently triggers downstream stress-responsive genes to improve salt tolerance (Wei et al., 2022).

3.2.3.1.1 Auxin

Auxin is a key phytohormone that primarily shapes root development. NO plays a pivotal role in modulating auxin signaling, especially under stress conditions. It influences root elongation, gravitropic responses, and adaptation to environmental changes by regulating auxin transport and signaling components. Through S-nitrosation of PIN-FORMED (PIN) proteins and auxin response factors (ARFs), NO adjusts the directional flow of auxin toward root tips, optimizing root architecture for improved nutrient uptake and stress resilience (Ni et al., 2017). Interestingly, NO can act downstream of auxin during lateral root and root hair formation, yet in some contexts, elevated NO levels can also increase auxin accumulation, demonstrating a dynamic and context-dependent relationship. For example, in the auxin signaling pathway, Transport Inhibitor Response 1/Auxin Signaling F-Box (TIR1/AFB) proteins function as auxin receptors that promote the degradation of Auxin/Indole-3-Acetic Acid (Aux/IAA) transcriptional repressors (Calderon-Villalobos et al., 2010). This degradation releases ARFs and enables the activation of auxin-responsive genes. NO regulates this pathway through S-nitrosation of TIR1, which enhances its interaction with Aux/IAA proteins, promoting their degradation and thereby modulating auxin signaling during root growth in Arabidopsis seedlings (Terrile et al., 2012).

3.2.3.1.2 Jasmonic acid

JA is central to plant protection against herbivores and pathogens, and NO seems to regulate the activation of JA signaling through S-nitrosation of JAR1 and other signaling components. This interaction leads to increased expression of defense-related genes, including proteinase inhibitors and defensin-like peptides, fortifying the plant against attack. In this way, NO and JA collaborate to activate immune responses and boost pathogen resistance (Pande et al., 2022).

3.2.3.1.3 Cytokinins

Cytokinins are essential plant hormones that primarily promote cell division (cytokinesis) and regulate various developmental processes, including shoot apical meristem maintenance, lateral bud growth, and the delay of leaf senescence (Chen et al., 2021). NO acts as a critical negative regulator of cytokinin signaling by interfering with the phosphorelay activity via S-nitrosation. Specifically, NO-mediated S-nitrosation of the Arabidopsis Histidine Phosphotransfer 1 protein at the Cys-115 residue represses its phosphorylation and hinders the subsequent transfer of the phosphoryl group to the Type-B Response Regulator 1 (ARR1). By physically blocking this relay, NO effectively reduces the plant sensitivity to cytokinins, allowing for a coordinated growth response under conditions of nitro-oxidative stress (Feng et al., 2013).

3.2.3.1.4 Abscisic acid

ABA, the well-known stress hormone, demonstrates one of the most critical partnerships with NO. During stress conditions, ABA triggers stomatal closure, and NO acts as both a mediator and modulator of this process. By S-nitrosating ABA receptors and key signaling proteins such as protein phosphatase 2C and sucrose nonfermenting 1 (SNF1)-related protein kinases (SnRK2s), NO fine-tunes ABA signaling pathways. This regulation enhances stomatal closure, modulates ion channel activity in guard cells, and ultimately improves drought tolerance (Gupta et al., 2025). Similarly, Abscisic Acid Insensitive 5 is S-nitrosated at cysteine 153, which causes its proteasome-mediated degradation, thereby promoting seed germination and seedling establishment (Castillo et al., 2015). NO also modifies pyrabactin resistance1/pyr1-like/regulatory component of ABA receptor proteins through S-nitrosation and tyrosine nitration (Castillo et al., 2015). Furthermore, NO regulates open stomata 1 (ost1)/SNF1-related protein kinase 2.6 (SnRK2.6) through S-nitrosation at cysteine 137 in guard cells, thereby negatively modulating ABA signaling. This PTM impairs ABA-induced stomatal closure during drought stress, as observed in guard cells of NO-overaccumulating *A. thaliana* mutants deficient in *atgsnor1–2* (Wang et al., 2015). Moreover, S-nitrosation of MYB Domain Protein 2 transcription factor at cysteine 53 alters its DNA-binding activity and influences ABA-responsive gene expression in *A. thaliana* during drought (Serpa et al., 2007) (Fig. 3.1).

3.3 S-nitrosation-mediated regulation of key stress-related proteins

3.3.1 Ascorbate peroxidase

A key antioxidant enzyme in plants, ascorbate peroxidase (APX), is essential for detoxifying hydrogen peroxide (H_2O_2) generated during both normal metabolism and environmental stress. It is mostly found in the cytosol and participates in the ascorbate–glutathione cycle, which uses ascorbate as an electron donor to convert H_2O_2 into water (Begara-Morales et al., 2014). During oxidative stress, S-nitrosation of cytosolic ascorbate peroxidase 1 increases its enzymatic activity, strengthening hydrogen peroxide scavenging (Yang et al., 2015). During abiotic stressors such as heat, drought, and salinity, this PTM enables plants to react quickly to oxidative bursts. Apart from its antioxidant function, APX1 also influences the redox environment and signaling pathways associated with ROS and NO crosstalk. As a result, the reversible S-nitrosation of this protein functions as a redox switch, adjusting its enzymatic activity in response to cellular demands. In pea plants (*Pisum sativum*), cytosolic APX is regulated by a dual PTM mechanism by NO. Tyrosine nitration of APX by peroxynitrite at Tyr5 and Tyr235 leads to its irreversible inhibition, while its S-nitrosation by S-nitrosoglutathione (GSNO) at residue Cys32 serves as a reversible activation switch (Begara-Morales et al., 2014). This dual regulation allows the plants to fine-tune their antioxidant defenses; for example, during salinity stress, an increase in S-nitrosated APX improves its enzymatic activity to scavenge H_2O_2, thereby enhancing tolerance to oxidative damage (Begara-Morales et al., 2014; Mata-Pérez et al., 2023). APX is thus regarded as a promising target for enhancing plant stress tolerance through redox-based biotechnological approaches because of its crucial role in redox balance and stress adaptation (Chen et al., 2009).

3.3.2 S-nitrosoglutathione reductase 1

GSNOR1 is an important enzyme that controls the GSNO levels, the primary carrier and reservoir of NO bioactivity in plants (Frungillo et al., 2014). As a master regulator of NO signaling, GSNOR1 indirectly regulates the total amount of protein S-nitrosation by catalyzing the NADH-dependent breakdown of GSNO into glutathione disulfide and ammonia (Liu et al., 2001). Under certain stressors, such as heat and hypoxia, S-nitrosation affects the activity of GSNOR1 itself, which can result in its inactivation and degradation by selective autophagy (Song et al., 2022). During environmental stressors, cells can temporarily raise SNO levels and improve NO-based signaling thanks to this feedback regulation. Furthermore, GSNOR1 integrates redox status with hormonal and stress signaling by influencing hormone signaling pathways, specifically SA and ABA, and regulating the activity of respiratory burst oxidase homologs, which in turn affects ROS production. The broad influence of GSNOR1 on plant physiology and defense is demonstrated by the altered responses to various stresses displayed by Arabidopsis mutants lacking functional *GSNOR1*, including increased susceptibility or tolerance to heat, drought, and oxidative conditions (Jahnová et al., 2019).

3.3.3 Superoxide dismutase

SODs are key antioxidant enzymes that convert superoxide radicals into hydrogen peroxide and oxygen, serving as a first line of defense against oxidative stress in plants. Increasing evidence suggests that NO modulates SOD activity, both by S-nitrosation and tyrosine nitration. According to Begara-Morales et al. (2016), NO can directly alter the activity of antioxidant enzymes, including SOD, with S-nitrosation potentially serving as a reversible regulatory mechanism that fine-tunes redox homeostasis. Arora and Bhatla (2015) showed how NO donors, such as SNP, cause Cu/Zn-SOD and Fe-SOD to be differentially regulated in sunflower seedlings, indicating a tissue-specific and NO concentration-dependent effect under salt stress. Additionally, more general research has demonstrated that NO can indirectly support SOD activity during stress responses by scavenging superoxide radicals and lowering peroxynitrite formation, preventing irreversible inactivation of SOD (Bil et al., 2020). All of these results point to S-nitrosation of SOD as a critical mechanism through which NO joins the redox signaling network and dynamically balances RNS and ROS under abiotic stress. A recent study revealed denitrosylation of SOD in the resistant genotype of rice in response to *Magnaporthe oryzae* infection, leading to a reduction in its activity. The decreased enzyme activity results in enhanced superoxide accumulation, which initiates defense responses against *M. oryzae* (Munir et al., 2025).

3.3.4 Monodehydroascorbate reductase

S-nitrosation of monodehydroascorbate reductase (MDHAR) in tomato under salt stress increased its enzymatic activity, thereby reinforcing ascorbic acid regeneration and improving ROS detoxification (Qi et al., 2020). Taken together, these results confirm that NO shapes redox homeostasis during stress conditions by inducing S-nitrosation of key antioxidant proteins to facilitate stress adaptation in plants.

3.3.5 Other key stress-related proteins

Under low-oxygen conditions, NO levels rise and modify key redox regulators. In *A. thaliana*, S-nitrosation of Arabidopsis Hemoglobin 1 reduces NO emission, contributing to hypoxic adjustment (Perazzolli et al., 2004). S-nitrosation of GSNOR 1 induces selective autophagy, enhancing tolerance to oxygen deprivation (Zhan et al., 2018). In addition, transcription factors such as SENESCENCE-RELATED GENE 1 (SRG1) represent key nodes of NO bioactivity in plant immunity. S-nitrosation of SRG1 at cysteine-87 enhances the transcription of immune repressors, thereby dampening the immune response (Cui et al., 2018). Conversely, S-nitrosation of SRG3 positively regulates plant immune responses, highlighting the dual and context-dependent roles of NO-mediated modifications in plant defense (Cui et al., 2021) (Fig. 3.1).

3.4 Conclusion

NO-based PTMs represent crucial regulatory mechanisms in plant stress biology. Through S-nitrosation, tyrosine nitration, and metal nitrosyl complex formation, NO modulates the function of antioxidant enzymes, transcription factors, hormone receptors, and signaling kinases. These modifications enable rapid and reversible adjustments in protein activity, allowing plants to fine-tune redox homeostasis and coordinate defense and adaptive responses under environmental challenges (Fig. 3.1). Although significant progress has been made in identifying modified proteins and understanding their functional consequences, many aspects, such as the specificity, reversibility, and interplay between different PTMs, remain to be clarified. Continued research into NO-mediated protein modifications will advance our understanding of plant stress signaling networks and may provide novel strategies for improving crop stress tolerance.

Funding

This research was funded by the National Research Foundation of Korea (NRF), funded by the Ministry of Education, Science, and Technology (grant number RS-2026-25492629).

References

Arora, D., & Bhatla, S. C. (2015). Nitric oxide triggers a concentration-dependent differential modulation of superoxide dismutase (FeSOD and Cu/ZnSOD) activity in sunflower seedling roots and cotyledons as an early and long distance signaling response to NaCl stress. *Plant Signaling & Behavior, 10*(10), e1071753. https://doi.org/10.1080/15592324.2015.1071753.

Astier, J., & Lindermayr, C. (2012). Nitric oxide-dependent posttranslational modification in plants: An update. *International Journal of Molecular Sciences, 13*(11), 15193–15208. https://doi.org/10.3390/ijms131115193.

Begara-Morales, J. C., Sánchez-Calvo, B., Chaki, M., Valderrama, R., Mata-Pérez, C., López-Jaramillo, J., Padilla, M. N., Carreras, A., Corpas, F. J., & Barroso, J. B. (2014). Dual regulation of cytosolic ascorbate peroxidase (APX) by tyrosine nitration and S-nitrosylation. *Journal of Experimental Botany, 65*(2), 527–538. https://doi.org/10.1093/jxb/ert396.

Begara-Morales, J. C., Sánchez-Calvo, B., Chaki, M., Valderrama, R., Mata-Pérez, C., Padilla, M. N., & Barroso, J. B. (2016). Antioxidant systems are regulated by nitric oxide-mediated post-translational modifications (NO-PTMs). *Frontiers in Plant Science, 7*, 152.

Bellota-Anton, C., Munnoch, J., Robb, K., Adamczyk, K., Candelaresi, M., Parker, A. W., & Tucker, N. P. (2011). Spectroscopic analysis of protein Fe–NO complexes. *Biochemical Society Transactions, 39*(5), 1293–1298.

Berton, P., Domínguez-Romero, J. C., Wuilloud, R. G., Sánchez-Calvo, B., Chaki, M., Carreras, A., Valderrama, R., Begara-Morales, J. C., Corpas, F. J., Barroso, J. B., Gilbert-López, B., García-Reyes, J. F., & Molina-Díaz, A. (2012). Determination of nitrotyrosine in *Arabidopsis thaliana* cell cultures with a mixed-mode solid-phase extraction cleanup followed by liquid chromatography time-of-flight mass spectrometry. *Analytical and Bioanalytical Chemistry, 404*(5), 1495–1503. https://doi.org/10.1007/s00216-012-6220-3.

Bil, P., Ciesielska, S., Jaksik, R., & Rzeszowska-Wolny, J. (2020). Circuits regulating superoxide and nitric oxide production and neutralization in different cell types: Expression of participating genes and changes induced by ionizing radiation. *Antioxidants, 9*(8), 701. https://doi.org/10.3390/antiox9080701.

Calderon-Villalobos, L. I., Tan, X., Zheng, N., & Estelle, M. (2010). Auxin perception—structural insights. *Cold Spring Harbor Perspectives in Biology, 2*(7), a005546.

Castillo, M. C., Lozano-Juste, J., González-Guzmán, M., Rodriguez, L., Rodriguez, P. L., & León, J. (2015). Inactivation of PYR/PYL/RCAR ABA receptors by tyrosine nitration may enable rapid inhibition of ABA signaling by nitric oxide in plants. *Science Signaling, 8*(392). https://doi.org/10.1126/scisignal.aaa7981, http://stke.sciencemag.org/content/sigtrans/8/392/ra89.full.pdf.

Chaki, M., Álvarez De Morales, P., Ruiz, C., Begara-Morales, J. C., Barroso, J. B., Corpas, F. J., & Palma, J. M. (2015). Ripening of pepper (*Capsicum annuum*) fruit is characterized by an enhancement of protein tyrosine nitration. *Annals of Botany, 116*(4), 637–647. https://doi.org/10.1093/aob/mcv016, http://aob.oxfordjournals.org/.

Chaki, M., Sánchez-Calvo, B., Carreras, A., Valderrama, R., Begara-Morales, J. C., Corpas, F. J., & Barroso, J. B. (2018). Identification of tyrosine and nitrotyrosine with a mixed-mode solid-phase extraction cleanup followed by liquid chromatography–electrospray time-of-flight mass spectrometry in plants. *Methods in Molecular Biology, 1747*, 161–169. https://doi.org/10.1007/978-1-4939-7695-9_13, http://www.springer.com/series/7651.

Chen, H., Lee, J., Lee, J.-M., Han, M., Emonet, A., Lee, J., & Lee, Y. (2022). MSD2, an apoplastic Mn-SOD, contributes to root skotomorphogenic growth by modulating ROS distribution in *Arabidopsis*. *Plant Science, 317*, 111192.

Chen, H., Qin, Y., Pu, J., Hu, J., & Wen, Y. (2021). Phytotoxicity of the chiral herbicide dichlorprop: Cross-talk between nitric oxide, reactive oxygen species and phytohormones. *Science of the Total Environment, 788*, 147866. https://doi.org/10.1016/j.scitotenv.2021.147866.

Chen, R., Sun, S., Wang, C., Li, Y., Liang, Y., An, F., & Zuo, J. (2009). The *Arabidopsis* PARAQUAT RESISTANT2 gene encodes an S-nitrosoglutathione reductase that is a key regulator of cell death. *Cell Research, 19*(12), 1377–1387.

Cui, B., Pan, Q., Clarke, D., Villarreal, M. O., Umbreen, S., Yuan, B., Shan, W., Jiang, J., & Loake, G. J. (2018). S-nitrosylation of the zinc finger protein SRG1 regulates plant immunity. *Nature Communications, 9*(1). https://doi.org/10.1038/s41467-018-06578-3, http://www.nature.com/ncomms/index.html.

Cui, B., Xu, S., Li, Y., Umbreen, S., Frederickson, D., Yuan, B., Jiang, J., Liu, F., Pan, Q., & Loake, G. J. (2021). The Arabidopsis zinc finger proteins SRG2 and SRG3 are positive regulators of plant immunity and are differentially regulated by nitric oxide. *New Phytologist, 230*(1), 259–274. https://doi.org/10.1111/nph.16993, http://onlinelibrary.wiley.com/journal/10.1111/(ISSN)1469-8137.

Ergashev, U., Yang, T., Zhang, Y., Ergashev, I., Luo, L., Yu, M., & Han, Y. (2025). Protein S-nitrosylation: Roles for nitric oxide signaling in the regulation of stress-dependent phytohormones. *Physiologia Plantarum, 177*(2), e70180.

Feng, J., Wang, C., Chen, Q., Chen, H., Ren, B., Li, X., & Zuo, J. (2013). S-nitrosylation of phosphotransfer proteins represses cytokinin signaling. *Nature Communications, 4*(1), 1529.

Feng, J., Chen, L., & Zuo, J. (2019). Protein S -Nitrosylation in plants: Current progresses and challenges. *Journal of Integrative Plant Biology, 61*(12), 1206–1223. https://doi.org/10.1111/jipb.12780.

Ford, P. C. (2010). Reactions of NO and Nitrite with heme models and proteins. *Inorganic Chemistry, 49*(14), 6226–6239. https://doi.org/10.1021/ic902073z.

Frungillo, L., Skelly, M. J., Loake, G. J., Spoel, S. H., & Salgado, I. (2014). S-nitrosothiols regulate nitric oxide production and storage in plants through the nitrogen assimilation pathway. *Nature Communications, 5*. https://doi.org/10.1038/ncomms6401, http://www.nature.com/ncomms/index.html.

Galetskiy, D., Lohscheider, J. N., Kononikhin, A. S., Popov, I. A., Nikolaev, E. N., & Adamska, I. (2011). Phosphorylation and nitration levels of photosynthetic proteins are conversely regulated by light stress. *Plant Molecular Biology, 77*(4-5), 461–473. https://doi.org/10.1007/s11103-011-9824-7.

Gupta, K. J., Hebelstrup, K. H., Mur, L. A. J., & Igamberdiev, A. U. (2011). Plant hemoglobins: Important players at the crossroads between oxygen and nitric oxide. *FEBS Letters, 585*(24), 3843–3849. https://doi.org/10.1016/j.febslet.2011.10.036, http://onlinelibrary.wiley.com/journal/10.1002/(ISSN)1873-3468.

Gupta, P., Saxena, G., & Gupta, R. (2025). Nitrate reductase-mediated nitric oxide synthesis in shaping stress resilience in plants. *Journal of Experimental Botany, 76*(22), 6634–6656.

Herold, S., & Puppo, A. (2005). Oxyleghemoglobin scavenges nitrogen monoxide and peroxynitrite: A possible role in functioning nodules? *JBIC Journal of Biological Inorganic Chemistry, 10*, 935–945.

Hess, D. T., Matsumoto, A., Kim, S. O., Marshall, H. E., & Stamler, J. S. (2005). Protein S-nitrosylation: Purview and parameters. *Nature Reviews. Molecular Cell Biology, 6*(2), 150–166. https://doi.org/10.1038/nrm1569.

Igamberdiev, A. U., Bykova, N. V., & Hill, R. D. (2011). 3 2011/03 IUBMB Life 10.1002/iub.439 15216551 3 146 152 Canada Structural and functional properties of class 1 plant hemoglobins 63

Ignarro, L. J. (2000). *Nitric oxide: Biology and pathobiology*. Academic press.

Jahnová, J., Luhová, L., & Petřivalský, M. (2019). S-nitrosoglutathione reductase—The master regulator of protein S-nitrosation in plant NO signaling. *Plants, 8*(2), 48. https://doi.org/10.3390/plants8020048.

Kolbert, Z., Feigl, G., Bordé, Á., Molnár, Á., & Erdei, L. (2017). Protein tyrosine nitration in plants: Present knowledge, computational prediction and future perspectives. *Plant Physiology and Biochemistry, 113*, 56–63. https://doi.org/10.1016/j.plaphy.2017.01.028.

Lee, C. M., Wilderman, P. R., Park, J. W., Murphy, T. J., & Morgan, E. T. (2020). Tyrosine nitration contributes to nitric oxide-stimulated degradation of CYP2B6. *Molecular Pharmacology, 98*(3), 267–279. https://doi.org/10.1124/molpharm.120.000020, https://molpharm.aspetjournals.org/content/molpharm/98/3/267.full.pdf.

León, J., & Costa-Broseta, Á. (2020). Present knowledge and controversies, deficiencies, and misconceptions on nitric oxide synthesis, sensing, and signaling in plants. *Plant, Cell & Environment, 43*(1), 1–15.

León, J. (2022). Protein tyrosine nitration in plant nitric oxide signaling. *Frontiers in Plant Science, 13*. https://doi.org/10.3389/fpls.2022.859374.

Li, L., & Li, L. (2016). Recent advances in multinuclear metal nitrosyl complexes. *Coordination Chemistry Reviews, 306*, 678–700.

Lin, W., Shang, J. X., Li, X. Y., Zhou, X. F., & Zhao, L. Q. (2025). Nitric oxide regulates multiple signal pathways in plants via protein S-nitrosylation. *Current Issues in Molecular Biology, 47*(6). https://doi.org/10.3390/cimb47060407, https://www.mdpi.com/journal/cimb.

Liu, L., Hausladen, A., Zeng, M., Que, L., Heitman, J., & Stamler, J. S. (2001). A metabolic enzyme for S-nitrosothiol conserved from bacteria to humans. *Nature, 410*(6827), 490–494. https://doi.org/10.1038/35068596.

Liu, Y., Liu, Z., Wu, X., Fang, H., Huang, D., Pan, X., & Liao, W. (2024). Role of protein S-Nitrosation in plant growth and development. *Plant Cell Reports, 43*(8), 204.

Mata-Pérez, C., Sánchez-Vicente, I., Arteaga, N., Gómez-Jiménez, S., Fuentes-Terrón, A., Oulebsir, C. S., & Lorenzo, Ó. (2023). Functions of nitric oxide-mediated post-translational modifications under abiotic stress. *Frontiers in Plant Science, 14*, 1158184.

Matamoros, M. A., & Becana, M. (2021). Molecular responses of legumes to abiotic stress: Post-translational modifications of proteins and redox signaling. *Oxford Journal of Experimental Botany, 72*(16), 5876–5892. https://doi.org/10.1093/jxb/erab008, http://jxb.oxfordjournals.org/.

Mathieu, C., Moreau, S., Frendo, P., Puppo, A., & Davies, M. J. (1998). Direct detection of radicals in intact soybean nodules: Presence of nitric oxide-leghemoglobin complexes. *Free Radical Biology and Medicine, 24*(7-8), 1242–1249. https://doi.org/10.1016/S0891-5849(97)00440-1, www.elsevier.com/locate/freeradbiomed.

Melo, P. M., Silva, L. S., Ribeiro, I., Seabra, A. R., & Carvalho, H. G. (2011). Glutamine synthetase is a molecular target of nitric oxide in root nodules of *Medicago truncatula* and is regulated by tyrosine nitration. *Plant Physiology, 157*(3), 1505–1517. https://doi.org/10.1104/pp.111.186056, http://www.plantphysiol.org/content/157/3/1505.full.pdf+html.

Mendez, A. A., Mangialavori, I. C., Cabrera, A. V., Benavides, M. P., Vazquez-Ramos, J. M., & Gallego, S. M. (2020). Tyr-nitration in maize CDKA; 1 results in lower affinity for ATP binding. *Biochimica et Biophysica Acta (BBA)-Proteins and Proteomics, 1868*(10), 140479.

Morot-Gaudry-Talarmain, Y., Rockel, P., Moureaux, T., Quilleré, I., Leydecker, M., Kaiser, W., & Morot-Gaudry, J. (2002). Nitrite accumulation and nitric oxide emission in relation to cellular signaling in nitrite reductase antisense tobacco. *Planta, 215*(5), 708–715. https://doi.org/10.1007/s00425-002-0816-3.

Munir, A., Min, C. W., Wang, Y., Kim, S. T., & Gupta, R. (2025). S-nitrosoproteome analysis of rice leaves highlights the possible roles of superoxide dismutase in resistance against *Magnaporthe oryzae*. *Physiologia Plantarum, 177*(5), e70515.

Ni, M., Zhang, L., Shi, Y. F., Wang, C., Lu, Y., Pan, J., & Liu, J. Z. (2017). Excessive cellular S-nitrosothiol impairs endocytosis of auxin efflux transporter PIN2. *Frontiers in Plant Science, 8*. https://doi.org/10.3389/fpls.2017.01988, http://journal.frontiersin.org/article/10.3389/fpls.2017.01988/full.

Pande, A., Mun, B. G., Rahim, W., Khan, M., Lee, D. S., Lee, G. M., Al Azzawi, T. N. I., Hussain, A., Kim, C. K., & Yun, B. W. (2022). Phytohormonal regulation through protein s-nitrosylation under stress. *Frontiers in Plant Science, 13*, 865542.

París, R., Iglesias, M. J., Terrile, M. C., & Casalongué, C. A. (2013). Functions of S-nitrosylation in plant hormone networks. *Frontiers in Plant Science, 4*. https://doi.org/10.3389/fpls.2013.00294, http://journal.frontiersin.org/Journal/10.3389/fpls.2013.00294/pdf.

Perazzolli, M., Dominici, P., Romero-Puertas, M. C., Zago, E., Zeier, J., Sonoda, M., Lamb, C., & Delledonne, M. (2004). Arabidopsis nonsymbiotic hemoglobin AHb1 modulates nitric oxide bioactivity. *The Plant Cell, 16*(10), 2785–2794. https://doi.org/10.1105/tpc.104.025379, http://www.plantcell.org/.

Qi, Q., Yanyan, D., Yuanlin, L., Kunzhi, L., Huini, X., & Xudong, S. (2020). Overexpression of SlMDHAR in transgenic tobacco increased salt stress tolerance involving S-nitrosylation regulation. *Plant Science, 299*. https://doi.org/10.1016/j.plantsci.2020.110609, http://wwwjournals.elsevier.com/plant-science/.

Radi, R. (2018). Oxygen radicals, nitric oxide, and peroxynitrite: Redox pathways in molecular medicine. *Proceedings of the National Academy of Sciences, 115*(23), 5839–5848. https://doi.org/10.1073/pnas.1804932115.

Romero-Puertas, M. C., Laxa, M., Mattè, A., Zaninotto, F., Finkemeier, I., Jones, A. M. E., Perazzolli, M., Vandelle, E., Dietz, K. J., & Delledonne, M. (2007). S-nitrosylation of peroxiredoxin II E promotes peroxynitrite-mediated tyrosine nitration. *The Plant Cell, 19*(12), 4120–4130. https://doi.org/10.1105/tpc.107.055061, http://www.plantcell.org/cgi/reprint/19/12/4120.

Saito, S., Yamamoto-Katou, A., Yoshioka, H., Doke, N., & Kawakita, K. (2006). Peroxynitrite generation and tyrosine nitration in defense responses in tobacco by-2 cells. *Plant and Cell Physiology, 47*(6), 689–697. https://doi.org/10.1093/pcp/pcj038.

Sasakura, F., Uchiumi, T., Shimoda, Y., Suzuki, A., Takenouchi, K., Higashi, S., & Abe, M. (2006). A class 1 hemoglobin gene from *Alnus firma* functions in symbiotic and nonsymbiotic tissues to detoxify nitric oxide. *Molecular Plant-Microbe Interactions® 19*(4), 441–450. https://doi.org/10.1094/mpmi-19-0441.

Sedlářová, M., Jedelská, T., Lebeda, A., & Petřivalský, M. (2025). Progress in plant nitric oxide studies: Implications for phytopathology and plant protection. *International Journal of Molecular Sciences, 26*(5), 2087. https://doi.org/10.3390/ijms26052087.

Serpa, V., Vernal, J., Lamattina, L., Grotewold, E., Cassia, R., & Terenzi, H. (2007). Inhibition of AtMYB2 DNA-binding by nitric oxide involves cysteine S-nitrosylation. *Biochemical and Biophysical Research Communications, 361*(4), 1048–1053. https://doi.org/10.1016/j.bbrc.2007.07.133.

Sharov, V. S., Dremina, E. S., Galeva, N. A., Gerstenecker, G. S., Li, X., Dobrowsky, R. T., Stobaugh, J. F., & Schöneich, C. (2010). Fluorogenic tagging of peptide and protein 3-nitrotyrosine with 4-(Aminomethyl)benzenesulfonic acid for quantitative analysis of protein tyrosine nitration. *Chromatographia, 71*(1-2), 37–53. https://doi.org/10.1365/s10337-009-1409-0.

Song, X., Wang, T., Zhang, Y., Yu, J. Q., & Xia, X. J. (2022). S-nitrosoglutathione reductase contributes to thermotolerance by modulating high temperature-induced apoplastic H_2O_2 in *Solanum lycopersicum*. *Frontiers in Plant Science, 13*. https://doi.org/10.3389/fpls.2022.862649, https://www.frontiersin.org/journals/plant-science.

Tanou, G., Filippou, P., Belghazi, M., Job, D., Diamantidis, G., Fotopoulos, V., & Molassiotis, A. (2012). Oxidative and nitrosative-based signaling and associated post-translational modifications orchestrate the acclimation of citrus plants to salinity stress. *The Plant Journal, 72*(4), 585–599. https://doi.org/10.1111/j.1365-313x.2012.05100.x.

Terrile, M. C., París, R., Calderõn-Villalobos, L. I. A., Iglesias, M. J., Lamattina, L., Estelle, M., & Casalongué, C. A. (2012). Nitric oxide influences auxin signaling through S-nitrosylation of the Arabidopsis TRANSPORT INHIBITOR RESPONSE 1 auxin receptor. *Plant Journal, 70*(3), 492–500. https://doi.org/10.1111/j.1365-313X.2011.04885.x.

Toledo, J. C., Jr, & Augusto, O. (2012). Connecting the chemical and biological properties of nitric oxide. *Chemical Research in Toxicology, 25*(5), 975–989.

Valderrama, R., Corpas, F. J., Carreras, A., Fernández-Ocaña, A., Chaki, M., Luque, F., Gómez-Rodríguez, M. V., Colmenero-Varea, P., del Río, L. A., & Barroso, J. B. (2007). Nitrosative stress in plants. *FEBS Letters, 581*(3), 453–461. https://doi.org/10.1016/j.febslet.2007.01.006.

Vanin, A. F. (2009). Dinitrosyl iron complexes with thiolate ligands: Physico-chemistry, biochemistry and physiology. *Nitric Oxide: Biology and Chemistry / Official Journal of the Nitric Oxide Society, 21*(1), 1–13.

Ventimiglia, L., & Mutus, B. (2020). The physiological implications of S-nitrosoglutathione reductase (GSNOR) activity mediating NO signalling in plant root structures. *Antioxidants, 9*(12), 1206. https://doi.org/10.3390/antiox9121206.

Wang, P., Du, Y., Hou, Y. J., Zhao, Y., Hsu, C. C., Yuan, F., & Zhu, J. K. (2015). Nitric oxide negatively regulates abscisic acid signaling in guard cells by S-nitrosylation of OST1. *Proceedings of the National Academy of Sciences, 112*(2), 613–618.

Wang, T., Hou, X., Wei, L., Deng, Y., Zhao, Z., Liang, C., & Liao, W. (2024). Protein S-Nitrosation under abiotic stress: Role and mechanism. *Plant Physiology and Biochemistry, 207*, 108329.

Wei, L., Zhang, J., Wei, S., Hu, D., Liu, Y., Feng, L., Li, C., Qi, N., Wang, C., & Liao, W. (2022). Nitric oxide enhanced salt stress tolerance in tomato seedlings, involving phytohormone equilibrium and photosynthesis. *International Journal of Molecular Sciences, 23*(9), 4539. https://doi.org/10.3390/ijms23094539.

Wisastra, R., Poelstra, K., Bischoff, R., Maarsingh, H., Haisma, H. J., & Dekker, F. J. (2011). Antibody-free detection of protein tyrosine nitration in tissue sections. *Chembiochem: A European Journal of Chemical Biology, 12*(13), 2016–2020. https://doi.org/10.1002/cbic.201100148.

Yang, H., Mu, J., Chen, L., Feng, J., Hu, J., Li, L., Zhou, J.-M., & Zuo, J. (2015). S -nitrosylation positively regulates ascorbate peroxidase activity during plant stress responses. *Plant Physiology, 167*(4), 1604–1615. https://doi.org/10.1104/pp.114.255216.

Zaffagnini, M., Morisse, S., Bedhomme, M., Marchand, C. H., Festa, M., Rouhier, N., Lemaire, S. D., & Trost, P. (2013). Mechanisms of nitrosylation and denitrosylation of cytoplasmic glyceraldehyde-3-phosphate dehydrogenase from *Arabidopsis thaliana*. *Journal of Biological Chemistry, 288*(31), 22777–22789. https://doi.org/10.1074/jbc.M113.475467Italy, http://www.jbc.org/content/288/31/22777.full.pdf+html.

Zhan, N., Wang, C., Chen, L., Yang, H., Feng, J., Gong, X., Ren, B., Wu, R., Mu, J., Li, Y., Liu, Z., Zhou, Y., Peng, J., Wang, K., Huang, X., Xiao, S., & Zuo, J. (2018). S-nitrosylation targets GSNO reductase for selective autophagy during hypoxia responses in plants. *Molecular Cell, 71*(1), 142. https://doi.org/10.1016/j.molcel.2018.05.024.

Chapter 4

Nitric oxide in plant immunity: signaling mechanisms and evolutionary perspectives

Nadra Tabassum[1,2], Qiaona Pan[1], and Beimi Cui[1]
[1]*Institute of Molecular Plant Sciences, School of Biological Sciences, The University of Edinburgh, Edinburgh, United Kingdom,* [2]*Department of Botany, University of Dhaka, Dhaka, Bangladesh*

4.1 Introduction

The identification of nitric oxide (NO) as a crucial component of plant immunity represents a significant advancement in our understanding of plant physiology. Initially, NO was considered a signaling molecule that controlled vascular relaxation and neuronal transmission in mammals (Ignarro, 1990). The involvement of NO in plant defense represented a breakthrough, first reported in late 1998 when rapid NO accumulation was shown to regulate pathogenesis-related responses (Delledonne et al., 1998; Durner & Klessig, 1999; Zaninotto et al., 2006). Since then, the plant science community has uncovered its broad involvement in stress biology, from abiotic tolerance to pathogen resistance (Domingos et al., 2015; Lamattina et al., 2003).

In plant immunity, NO is typically produced within minutes of pathogen perception, acting as both a local signal and a systemic mediator (Delledonne et al., 1998). NO regulates cell-wall fortification, ion fluxes, defense gene expression, and, in many cases, programmed cell death associated with the hypersensitive response (HR) (Hayat et al., 2009; Wang et al., 2013). NO does not act as an isolated molecule; rather, it operates within a highly integrated nitro-oxidative network in concert with reactive oxygen species (ROS) and reactive nitrogenous species (RNS) (Lindermayr & Durner, 2015; Sun et al., 2010; Delledonne et al., 1998). This dynamic interplay of nitrosative and oxidative burst orchestrates a broad spectrum of redox-dependent cellular processes, underscoring their collective role in maintaining physiological homeostasis and mediating pathophysiological responses (Khan et al., 2023; Loake & Grant, 2007).

Nitric Oxide in Plant Resilience to Abiotic Stress. DOI: https://doi.org/10.1016/B978-0-443-40497-9.00011-8

Plant immunity operates through a complex, multilayered defense system conceptualized by the zig-zag model (Jones & Dangl, 2006), which describes the dynamic interplay between plants and pathogens. It comprises two major layers: pattern-triggered immunity (PTI) (Dangl & McDowell, 2006; Yuan et al., 2021), activated by conserved pathogen or damage-associated molecular patterns (P/DAMPs) such as flagellin, chitin, or endogenous pathogens (DAMPs) (Hou et al., 2019), and effector-triggered immunity (ETI) mediated by intracellular resistance (R)-proteins that recognize specific pathogen effectors (Greenberg & Yao, 2004). While PTI provides a broad-spectrum, nonhost-specific defense, some pathogens deploy effectors that suppress PTI, leading to effector-triggered susceptibility (Greenberg & Yao, 2004; Wu et al., 2014). In turn, ETI restores defense through hypersensitive cell death and systemic acquired resistance (SAR), driven by ROS and salicylic acid (SA) accumulation (Fu & Dong, 2013). Within this interconnected framework, NO plays a pivotal role in both PTI and ETI, integrating into the nitro-oxidative network alongside ROS to regulate signaling cascades, transcriptional reprogramming, hormone biosynthesis, and redox homeostasis (Yuan et al., 2021). This functional versatility of NO is rooted in its ability to modify key cellular proteins through redox-based posttranslational mechanisms, thereby linking upstream immune perception to downstream gene expression. The signaling function of NO is primarily executed through posttranslational modifications (PTMs) of target proteins, particularly S-nitrosation, which involves the reversible covalent attachment of an NO group to cysteine thiols (Astier et al., 2012). This mechanism enables NO to modulate the activity, localization, and stability of a wide array of proteins involved in plant defense, including receptors, transcription factors, and antioxidant enzymes (Astier et al., 2012; Lindermayr et al., 2006). NO signaling is intricately interlinked with phytohormonal pathways, notably SA and jasmonic acid (JA), both of which are essential mediators of plant immune responses (Zhang et al., 2025). The homeostatic regulation of cellular S-nitrosothiols (SNOs), the principal reservoir of bioactive NO, is maintained by the enzyme S-nitrosoglutathione reductase (GSNOR), which catalyzes the degradation of S-nitrosoglutathione (GSNO) and thereby fine-tunes cellular redox and signaling balance (Feechan et al., 2005; Salgado et al., 2013).

Recent studies have expanded the exploration of NO-mediated immunity beyond angiosperms to early diverging land plant lineages, including bryophytes (mosses and liverworts) and pteridophytes (ferns), to elucidate the evolutionary origins and conservation of this signaling system. In the liverwort *Marchantia polymorpha*, the *MpGSNOR1* gene was identified as a key regulator of SNO homeostasis and immunity (Tabassum and Goodrich, 2026), supporting the evolutionary conservation of GSNOR-dependent NO signaling. Disruption of *MpGSNOR1* via CRISPR/Cas9 resulted in severe developmental abnormalities and compromised immune responses, confirming that GSNOR-dependent NO signaling was already established in early land plants (Tabassum and Goodrich, 2026). Furthermore, *M. polymorpha* produces a suite of bioactive secondary metabolites, including flavonoids and terpenoids, which likely interact synergistically with NO-dependent pathways to enhance defense capacity (Davies et al., 2020; Kubo et al., 2018; Nowaczyński et al., 2025). Similarly, the moss *Physcomitrium patens* displays highly conserved immune signaling mechanisms, involving SA and JA pathways during pathogen challenge (Castel et al., 2024; Naramoto et al., 2022; Ponce de León, 2011). Infection studies with fungi, such as *Colletotrichum gloeosporioides*, and oomycetes, like *Pythium irregulare*, have revealed a rapid induction of defense-related genes, including pathogenesis-related (PR) proteins, whose overexpression enhances pathogen resistance (Otero-Blanca et al., 2021). While direct S-nitrosation targets in mosses remain to be fully identified, the similarity of defense transcriptional profiles to those of angiosperms strongly suggests that NO and ROS act as evolutionarily conserved mediators of innate immune signaling (Bressendorff et al., 2016). In pteridophytes (ferns), direct molecular evidence linking NO to immune function remains limited; however, the presence of conserved immune receptors and the well-established role of NO in both bryophytes and seed plants strongly supports the notion that NO-mediated signaling represents an ancestral and fundamental component of plant defense (Castel et al., 2025; Han & Tsuda, 2022). Moreover, given the evolution of the vascular system in these early lineages, NO likely plays a pivotal role in systemic defense communication, potentially through the transport of mobile SNOs as long-distance signaling molecules (Besson-Bard et al., 2008; Gurung et al., 2014). Collectively, these findings support the view that NO is an evolutionarily ancient regulator of plant defense, whose core signaling module, S-nitrosation homeostasis controlled by GSNOR, originated in the earliest terrestrial plants. The conservation of this mechanism from bryophytes to angiosperms underscores the fundamental importance of NO in mediating redox balance, defense activation, and interorgan communication throughout plant evolution (Ponce de León, 2011; Salgado et al., 2013; Tabassum and Goodrich, 2026).

4.2 Nitric oxide metabolism in plants: sources, sinks, and homeostasis

4.2.1 Sources of nitric oxide: nitrate reductase and beyond

The origin of endogenous NO is debated because of the complex involvement of both enzymatic and nonenzymatic pathways (Chamizo-Ampudia et al., 2017; Gupta et al., 2025). Collectively, these pathways regulate intracellular NO levels in response to developmental and environmental signals (Gupta et al., 2025). In animals, NO is synthesized by Nitric Oxide Synthases (NOS), which catalyze the oxidation of L-arginine (Domingos et al., 2015; Ignarro, 1990). However, higher plants lack canonical homologs of NOS (Wendehenne et al., 2004), and rely primarily on reductive mechanisms utilizing nitrite (NO_2^-) as a substrate, among which nitrate reductase (NR) is the most well-characterized enzymatic source.

4.2.1.1 *Nitrate reductase-dependent nitric oxide production*

NR, a cytosolic molybdoenzyme central to nitrogen assimilation, catalyzes the reduction of nitrate (NO_3^-) to nitrite (NO_2^-). Under specific physiological conditions, such as hypoxia, low pH, or excess reducing power, NR can further reduce NO_2^- to NO through an NAD(P)H-dependent mechanism (Chamizo-Ampudia et al., 2017; Gupta et al., 2025). Furthermore, NR activity in the *noe1* rice mutant was twice that of the wild type, correlating with higher NO production. This was accompanied by three- and ninefold increases in the expression of *OsNIA1* and *OsNIA2*, respectively, indicating that the excess NO in *noe1* is primarily generated via the NR pathway (Lin et al., 2012). This dual catalytic capacity positions NR as a pivotal enzyme linking nitrogen metabolism with redox signaling. Recent studies have identified a potential NR–NO Forming Nitrite Reductase (NR–NOFNiR) complex, suggesting a dedicated enzymatic system for NO production (Chamizo-Ampudia et al., 2017). The activity of NR is tightly regulated by PTMs, notably phosphorylation, which modulate its activity and enable rapid adjustment of NO output under stress conditions (Kaiser & Huber, 2001). Mutant analyses further demonstrate that NR deficiency correlates with reduced NO accumulation and impaired stress and immune responses (Ahmad et al., 2024; León et al., 2014).

4.2.1.2 *Mitochondrial and alternative reductive pathways*

Beyond the cytosol, organellar compartments also contribute significantly to NO production. The mitochondrial electron transport chain can reduce NO_2^- to NO under low oxygen availability, primarily via cytochrome oxidase and the alternative oxidase, maintaining electron flow and redox balance during hypoxia (Gupta & Igamberdiev, 2011; Gupta Kapuganti et al., 2014). This mitochondrial pathway is particularly relevant during flooding stress or in tissues with fluctuating oxygen concentrations, such as nodules and roots. Additionally, nonenzymatic NO formation can occur when high concentrations of NO_2^- react with reductants such as ascorbate under acidic conditions, particularly in the apoplast and vacuole, offering a rapid and localized NO source during abiotic or biotic stress (Du et al., 2023).

4.2.1.3 *Emerging oxidative and peroxidase-dependent mechanisms*

Although no canonical NOS enzyme has been identified in higher plants, evidence for L-arginine-dependent, NOS-like activity continues to surface, especially in green algae such as *Ostreococcus* (Planchet & Kaiser, 2006; Corpas, 2019). More recently, peroxidase-catalyzed oxidative pathways have been proposed, linking NO production with phytohormone metabolism. In particular, López-Gómez et al. (2024) reported a peroxidase-mediated conversion of oximes, such as indole-3-acetaldoxime, a precursor in auxin biosynthesis, into NO, revealing a novel biochemical bridge between NO signaling and hormonal regulation.

Plant NO biosynthesis represents a multifaceted network where reductive nitrite-dependent pathways, led by NR and mitochondrial reduction systems, serve as principal sources of NO, while oxidative and peroxidase-dependent mechanisms offer additional regulatory flexibility (Yu et al., 2012). This diversity underscores the evolutionary innovation of plant NO metabolism, enabling precise spatial and temporal control over signaling during stress adaptation and development. Thus plant NO metabolism appears to rely on diversified enzymatic and redox routes, with NR as the major player.

4.2.2 Sinks and scavengers: phytoglobins, S-nitrosoglutathione reductase, and S-nitrosothiols buffering

NO plays a central role as a versatile signaling and redox molecule involved in numerous physiological and stress response processes, but it is extremely reactive and short-lived (Domingos et al., 2015; Lamattina et al., 2003). Therefore maintaining its homeostasis is crucial for proper cellular function. The steady state depends on both production and scavenging. The cellular system maintains a tight control over NO levels and its bioactivity through a sophisticated, interconnected homeostatic network (Treffon & Vierling, 2022). This network involves Phytoglobins (Plant hemoglobin/Pgbs), GSNOR, and protein SNOs (Pathak et al., 2024; Stomberski et al., 2019; Wang, 2006). This system ensures NO acts effectively as a signal without accumulating to cytotoxic concentrations and plays complementary roles in maintaining NO and SNO homeostasis (Borrowman et al., 2023). To prevent cellular damage and maintain physiological balance, Pgbs, known as plant heme proteins or nonsymbiotic hemoglobins, serve as a major system that effectively regulates NO concentrations (Seregélyes & Dudits, 2003). Pgbs bind and detoxify NO through a reaction known as NO-dioxygenase activity (Pathak et al., 2024). In this process, oxygenated Class 1 Pgbs rapidly convert NO into the less reactive molecule nitrate (NO_3^-), as shown below:

$$\mathrm{Pgb(Fe^{2+})O_2 + NO \rightarrow Pgb(Fe^{3+}) + NO_3^-}$$

This reaction is the foundation of the Pgbs–NO cycle. Under low-oxygen (hypoxic) conditions, that is, during flooding or seed germination, this cycle plays an important role. It helps to remove excess and simultaneously helps maintain the energy and redox balance of the cell. Pgbs in its reduced form ($PgbFe^{2+}$) is restored through NADH (Nicotinamide adenine dinucleotide). This regeneration ties NO removal to cellular metabolism, ensuring essential processes persist under low oxygen (Zafari, 2022). Research indicates that the expression of Class 1 Pgbs increases markedly under hypoxic conditions, enabling plants to adapt more effectively to environmental stress. For example, the overexpression of Pgbs genes in crops such as barley and rice has been shown to enhance flooding tolerance and improve cellular energy status (Samant et al., 2025; Zhang et al., 2019). Likewise, in legume root nodules, Pgbs serve a protective function by regulating NO concentrations that could otherwise inhibit nitrogenase, the key enzyme responsible for symbiotic nitrogen fixation (Berger et al., 2020; Gupta et al., 2011). Pgbs act as a biochemical safeguard, balancing the dual identity of NO as both a vital signaling molecule and a potential toxin. Through their NO-dioxygenase activity and participation in the Pgbs–NO cycle, these proteins enable plants to fine-tune redox processes, maintain energy homeostasis, and survive environmental stresses where oxygen availability and NO concentrations fluctuate (Seregélyes & Dudits, 2003; Zhang et al., 2025).

4.3 How plants "sense" nitric oxide: molecular targets and second messengers

4.3.1 Protein S-nitrosation as a master switch

It is important to understand how plants perceive and transduce NO signals at the molecular level. The sensing of NO involves diverse molecular targets and secondary messengers, with protein S-nitrosation functioning as a key mechanism that links NO perception to downstream signaling (Umbreen et al., 2018; Wang, 2006). This reversible modification enables precise, spatiotemporal control of protein function in response to developmental and environmental cues, analogous to protein phosphorylation (Aboalroub & Al Azzam, 2024; Yu et al., 2012), to regulate protein functions (Fig. 4.1), including protein–protein interaction (Iglesias et al., 2018), protein PTMs (Cui

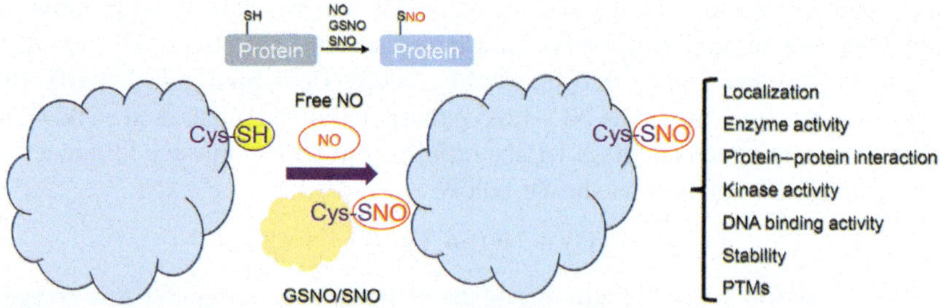

FIGURE 4.1 Mechanism and Functional Outcome of Protein S-Nitrosation. The mechanism depicts the modification of a Cysteine sulfhydryl group (*Cys-SH*) on a target protein (*large blue* structure) by a nitric oxide (*NO*) donor. In this process, the NO group is transferred from a free NO, GSNO, or S-nitrosated protein (Cys-SNO, *yellow* structure) to the Cys-SH of the target protein, resulting in the formation of an S-nitrosated protein (Cys-SNO). This covalent modification is crucial for cell signaling by directly impacting the structure and function of the target proteins, leading to changes in their enzyme activity, protein–protein interactions, cellular localization, stability, or PTM, etc.

et al., 2024; Skelly et al., 2019), protein stability (Zhan et al., 2018), enzyme activity (Yun et al., 2011; Zhang et al., 2025), DNA-binding activity (Cui et al., 2018; Lindermayr et al., 2010; Zhao et al., 2024), and localization (Tada et al., 2008), etc. In plants, this PTM serves as a central integrator of redox, hormonal, and immune signaling networks, coordinating responses to both biotic and abiotic stimuli through rapid, reversible control of protein function (Pande et al., 2022; Zhang & Liao, 2021).

Moreover, it also influences a wide range of cellular processes, including growth, stress adaptation, and defense responses (Astier et al., 2012; Borrowman et al., 2023). Key regulatory proteins such as Nonexpressor Of Pathogenesis-Related Genes 1 (NPR1) (Tada et al., 2008), Respiratory Burst Oxidase Homolog D (RBOHD), TGA1(TGACG-binding factor 1), and BIK1 (Botrytis-induced kinase 1) are well-established S-nitrosation targets that connect NO signaling with immunity (Cui et al., 2024; Kovacs et al., 2015; Lindermayr & Durner, 2015) 4). For instance, S-nitrosation of NPR1 controls its oligomer–monomer transition and nuclear translocation, a critical step in SA-dependent defense activation (Mou et al., 2003; Tada et al., 2008). Similarly, modification of RBOHD modulates ROS generation during pathogen attack, integrating NO and redox signaling pathways (Lee et al., 2020; Zhang et al., 2025).

S-nitrosylation also affects enzymes associated with cellular metabolism and redox regulation, including GSNOR, which maintains the cellular SNO pool, and thioredoxins that mediate denitrosation (Benhar et al., 2009; Chakraborty et al., 2022). The interplay between S-nitrosation and denitrosation thus provides plants with a flexible mechanism to fine-tune redox homeostasis and signal transduction. Dysregulated S-nitrosation has been associated with impaired stress tolerance, growth retardation, and hypersensitivity to pathogens (Benhar et al., 2009).

4.3.2 Second messengers: cGMP, Ca^{2+}, and mitogen-activated protein kinases

In addition to direct modification, NO influences second messenger systems. cGMP (Cyclic guanosine monophosphate) is a pivotal mediator of many NO-induced physiological and defense processes, including stomatal closure, developmental regulation, and even gravitropic responses in certain plant species, including ferns (Gayatri et al., 2013; Salmi et al., 2007; Francis et al., 2010). Pharmacological studies indicate that cGMP contributes to stomatal closure and defense gene activation. NO activates guanylate cyclase (GC) either directly, by modifying the enzyme similarly to the mammalian mechanism, or indirectly through redox signaling. Activated GC catalyzes the conversion of GTP (Guanosine triphosphate) to cGMP, which subsequently binds to and regulates protein kinases and Ca^{2+} channels. Elevated cGMP levels thus serve as a bridge between NO signaling and ion flux regulation, linking redox signaling to physiological and defense outcomes (Gayatri et al., 2013; Salmi et al., 2007).

Calcium functions as a universal second messenger, encoding information in its temporal and spatial "calcium signatures." NO can initiate or modulate these Ca^{2+} transients either directly or via cGMP-dependent pathways (Hetherington & Brownlee, 2004). The NO/cGMP signal often triggers the opening of Ca^{2+} channels in the plasma membrane, endoplasmic reticulum, or vacuole, leading to rapid cytosolic Ca^{2+} increases (Batistič & Kudla, 2012; Hetherington & Brownlee, 2004). These changes are perceived by Ca^{2+} sensors such as Calmodulin and Calcineurin B-like proteins, which activate downstream protein kinases, including Ca^{2+}-dependent protein kinases (CDPKs). This cascade culminates in the activation of defense-related gene expression and stress-adaptive responses (Batistič & Kudla, 2012; Qi et al., 2010).

NO also modulates calcium fluxes, acting in synergy with ROS to activate CDPKs and mitogen-activated protein kinases (MAPKs). These cascades amplify defense signaling, leading to transcriptional reprogramming and antimicrobial metabolite production. Thus NO signaling combines covalent protein modification and classical second-messenger pathways, creating multilayered regulatory control. MAPK cascades represent another major signaling module downstream of NO and Ca^{2+} (Zhang & Klessig, 2001). The NO/cGMP/Ca^{2+} pathways converge to activate MAPK kinase kinases (MAPKKKs), initiating a phosphorylation cascade through MAPKKs to MAPKs (Pagnussat et al., 2004). Activated MAPKs translocate to the nucleus, where they phosphorylate specific transcription factors, leading to the expression of PR-genes and other defense components (Asai et al., 2002). This ensures the amplification of the NO signal into a full transcriptional and metabolic defense response. Thus NO signaling integrates direct covalent modifications with second messenger-mediated amplification involving cGMP, Ca^{2+}, and MAPKs. This multilayered network provides plants with a robust, dynamic, and spatiotemporally controlled regulatory system to fine-tune immune, developmental, and stress-adaptive responses.

4.3.3 Integrative perspective: coupling direct and indirect nitric oxide signaling

Together, S-nitrosation and second messenger signaling represent two interconnected dimensions of NO perception in plants. Through direct covalent modification, S-nitrosation provides rapid, localized control of protein function, while second messengers such as cGMP, Ca^{2+}, and MAPKs amplify and diversify the signal across cellular networks. This integration ensures that NO responses are finely tuned, spatiotemporally coordinated, and context-dependent, linking redox changes to transcriptional reprogramming, metabolic adjustment, and defense activation. Thus plants utilize a hierarchical NO-sensing system that couples biochemical precision with systemic signaling flexibility to maintain redox and physiological homeostasis.

4.4 Early nitric oxide bursts and cell wall defense responses

When plants sense PAMPs (Pathogen-associated molecular pattern), they release NO within minutes (Cui et al., 2024). This rapid burst can be tracked using fluorescent probes like DAF-FM DA (4-Amino-5-methylamino-2′,7′-difluorofluorescein). In Arabidopsis, lipopolysaccharides and flg22 trigger NO accumulation that depends mainly on NR (*NIA1* and *NIA2*) and sometimes on peroxidases (Zeidler et al., 2004). These bursts are usually observed in two phases, an early spike for signaling and a later one associated with HR cell death (Schlicht, 2013).

NO works closely with ROS to regulate plant immunity (Delledonne et al., 1998). The NADPH oxidase RBOHD produces ROS, while NO fine-tunes that response through S-nitrosation. This crosstalk helps direct peroxidase enzymes that harden cell walls by crosslinking proteins and lignin (Almagro et al., 2009; Torres et al., 2002; Yun et al., 2011). In the apoplast, NO and ROS can combine to form peroxynitrite, further modifying wall enzymes and enhancing barrier formation (Haghpanah et al., 2025). For instance, callose deposition over the plant cell walls, in response to pathogen invasion, is dependent on NO. Arabidopsis mutants lacking *GSNOR1* show reduced callose deposition and weaker defense (Feechan et al., 2005; Wang et al., 2021). NO can also influence the activity of enzymes that remodel pectin, including pectin methylesterases (PMEs) and their inhibitors (PMEIs), which shape cell-wall stiffness and disease resistance (Lionetti et al., 2017; Liu et al., 2018). Although direct evidence during pathogen attack is still emerging, studies under stress conditions show that NO modulates pectin demethylesterification, supporting a potential NO-PME/PMEI regulatory link in plant defense (Pan et al., 2017; Xiong et al., 2009).

4.5 Stomatal immunity

Stomata are critical entry points for pathogens, but they also serve as fast-acting defense gates. When plants perceive PAMPs like flg22, they quickly close their stomata to block the infection process known as stomatal immunity. Intriguingly, NO plays a major role here. It is produced in guard cells mainly by NR and possibly by working with ROS NO triggers Ca^{2+} influx, depolarizes the plasma membrane (Desikan et al., 2002; Neill et al., 2002), and activates the anion channel SLAC1, leading to water efflux and closure (Desikan et al., 2002; Lamattina et al., 2003). However, the requirement for NO is not universal. Some species close their stomata in response to ABA even with low NO levels, suggesting that guard-cell regulation varies by species and context (Gayatri et al., 2013; Van Meeteren et al., 2020). Excessive NO can even cause reopening through cyclic GMP signaling or peroxynitrite accumulation (Zhang et al., 2001).

Hormones amplify these effects. ABA enhances NO production, while SA sustains closure, and JA can promote reopening for gas (Szalai et al., 2011; Zhang et al., 2025). Through these interactions, NO balances defense and photosynthetic needs, fine-tuning guard-cell immunity.

4.6 S-nitrosation and plant immunity

As discussed above, NO-dependent PTM, S-nitrosation, a key mechanism that mediates NO bioactivity, plays a pivotal role in modulating disease resistance in *Arabidopsis thaliana* against both compatible (host) and incompatible (nonhost) pathogens, including *Pseudomonas syringae* pv. *tomato* (Pst) DC3000, *Blumeria graminis* f. sp. *tritici* (Bgt), and *Hyaloperonospora parasitica* (Cui et al., 2021; Feechan et al., 2005). Mutants exhibiting elevated global *S*-nitrosation are generally more susceptible to bacterial and fungal infections, whereas plants with enhanced GSNOR1 activity display improved resistance (Feechan et al., 2005). Proteomic analyses have identified *S*-nitrosation across a wide array of proteins (Astier et al., 2012); for instance, infection with both avirulent and virulent strains of *Pst* resulted in *S*-nitrosation of over 100 proteins in *Arabidopsis* (Maldonado-Alconada et al., 2011) and over 500 in rice in response to *Magnaporthe oryze* infection (Munir et al., 2025). Collectively, this evidence underscores that S-nitrosation acts as a vital signaling mechanism in plant immunity, where its tight regulation is essential for an effective defense response.

4.6.1 S-nitrosation in plant immunity

In plants, pathogen perception rapidly triggers a nitrosative burst that elevates NO and drives *S*-nitrosation (SNO) of defense proteins, modulating PTI and ETI (Cui et al., 2024). S-nitrosation exerts a dual regulatory influence on the immune signaling network, both activating and repressing its key components (Yu et al., 2012). For instance, pathogen perception triggers the S-nitrosation of the NADPH oxidase RBOHD at Cys890, which inhibits its function to fine-tune the (ROS burst and programmed cell death (Yun et al., 2011). In contrast, this PTM can activate transcription factors, such as by increasing the DNA-binding affinity of the defense regulator TGA1 (Lindermayr et al., 2010). A clear model of this dual nature is found in a zinc-finger transcription factor family, the SRGs. Research shows that S-nitrosation inhibits SRG1, reducing its ability to bind DNA and control transcription (Cui et al., 2018). Interestingly, its paralog SRG3 is also modified by S-nitrosation, while SRG2 is not (Cui et al., 2021). This specificity is functionally critical, as high SNO levels in a gsnor mutant abolish the disease resistance from overexpressing *SRG3* but do not impact *SRG2*-mediated resistance. These results demonstrate that neofunctionalisation within this transcription factor family has created a differential regulatory circuit sensitive to NO, allowing for precise control of plant immunity.

4.6.2 S-nitrosation in salicylic acid biosynthesis and signaling

SA is a pivotal defense hormone in plants, with its biosynthesis and signaling being crucial for immunity (Spoel & Dong, 2024). Pathogen infection induces SA biosynthesis through transcription factors, such as SARD1 and CBP60g, which in turn activate SA biosynthesis. Accumulation of SA is perceived by the central signaling protein NPR1, which activates defense genes (Zhang & Li, 2019). Research over the past two decades has established that NO and S-nitrosation are integral to the SA biosynthesis pathway, regulating its biosynthesis, perception, and downstream signaling (Cui et al., 2021; Moreau et al., 2010; Skelly et al., 2019; Tada et al., 2008; Wang et al., 2009). The Arabidopsis *gsnor1–3* mutant, which has elevated cellular SNO levels, exhibits a stark reduction in SA accumulation and relative gene expression. Further, this mutant fails to activate SA accumulation and signaling upon *Pseudomonas parasitica* infection, with expression of key SA biosynthesis genes, including *CBP60g, SARD1, ICS1, and PAD4*, being significantly compromised, leading to susceptibility (Cui et al., 2021; Feechan et al., 2005). Unpublished data suggest that SARD1 is S-nitrosated in vitro, which inhibits its DNA-binding activity and could explain the compromised SA biosynthesis in *gsnor1–3*, and this is the direct evidence that S-nitrosation regulates SA biosynthesis. Similarly, as discussed above, the transcription factor SRG1 is subject to a dual-level regulation by NO: its expression is enhanced early in infection, promoting SA biosynthesis and resistance, but the protein itself is later S-nitrosated, potentially blocking its DNA-binding to form a negative feedback loop (Cui et al., 2018). This regulatory mechanism is specific; for instance, the disease resistance from overexpressing the homologous SRG3 is suppressed by high SNO levels, while SRG2 is not affected (Cui et al., 2021). This indicates the SRG family has evolved to be differentially regulated by S-nitrosation, a key redox modification in plant immunity. The evolutionary conservation of the targeted cysteine in SRG1 suggests this may be a widespread regulatory mechanism.

Research has clarified how S-nitrosation regulates SA signaling. A key player is SABP3, a chloroplast-localized carbonic anhydrase (CA) that binds SA and is necessary for the HR (Slaymaker et al., 2002). Mutants lacking SABP3 show enhanced susceptibility to bacterial pathogens. The activity of this protein is directly modulated by S-nitrosation at Cys-280, which occurs during infection and reduces both its SA-binding affinity and enzymatic activity (Wang et al., 2009). Since CA activity is required for resistance, this modification is proposed to create a negative feedback loop to fine-tune the defense response. Recent work suggests that the kinase CPCK2 may promote this inhibitory S-nitrosation of SABP3 via phosphorylation, though in vivo confirmation is needed (Rui et al., 2024). S-nitrosation also targets the central SA regulator NPR1 (Tada et al., 2008). In the resting cells, NPR1 localizes to the cytoplasm in the form of oligomers by intermolecular disulfide bonds. Upon pathogen infection, NPR1 is S-nitrosated at Cys156, predictably close to the multimerization interface of NPR1, and this redox modification can enhance disulfide bond formation. Polymerized NPR1 mainly localizes in the cytosol, inhibiting SA-mediated defense response. This event switches NPR1 from cytosolic, disulfide-bound oligomers to active monomers that are subsequently translocated into the nucleus and interact with the TGA class of basic leucine zipper transcription factors. The result is an enhanced binding activity of TGA1 to the promoter region of PR-genes, stimulating SA-dependent immune defense (Corina Vlot et al., 2009). Upon pathogen attack, SA induces THIOREDOXIN-h5 (TRXh5), which facilitates NPR1 monomerization, nuclear translocation, and activation of PR-genes (Tada et al., 2008). Notably, TGA1 is regulated by S-nitrosation and S-glutathionylation, improving TGA1 binding activity to the PR1 promoter region (Lindermayr et al., 2010). Thus S-nitrosation exerts precise, multilevel control over the SA signaling pathway, both initiating and subsequently dampening the immune response.

4.6.3 S-nitrosation regulates reactive oxygen species homeostasis

ROS are crucial signaling molecules and toxic agents in plant immunity, generated as metabolic byproducts and through dedicated systems (Mittler, 2017). They contribute to defense by directly attacking pathogens, reinforcing the cell wall via callose deposition and glycoprotein crosslinking, and acting as signals to induce stomatal closure and other cellular defenses. A primary source of this orchestrated ROS production is the plasma membrane-localized NADPH oxidase (Torres et al., 2002; Wu et al., 2023). This enzyme, residing in lipid rafts, uses its transmembrane structure to transfer electrons from the cytosol to the apoplast, where oxygen is reduced to superoxide ($O_2^{\bullet-}$). This is then rapidly converted by superoxide dismutase into hydrogen peroxide (H_2O_2), resulting in the defensive oxidative burst (Chu-Puga et al., 2019; Hu et al., 2020). The scale and duration of this ROS burst are precisely regulated by S-nitrosation (Cui et al., 2018, 2024; Yu et al., 2012). Genetic studies on GSNOR1-deficient mutants, *gsnor1–3* and *par2-1*, which accumulate high levels of NO and H_2O_2, demonstrate that balanced S-nitrosation is critical for redox control during immunity (Feechan et al., 2005; Lee et al., 2008). This regulation operates at multiple levels. Early in the immune response, a nitrosative burst leads to the S-nitrosation of the kinase BIK1, promoting its activation and its physical association with RBOHD, which likely promotes ROS production (Cui et al., 2024). Conversely, at later stages, RBOHD itself is S-nitrosated at a specific residue (Cys890), a modification that appears to inhibit its NADPH oxidase activity by potentially disrupting FAD (Flavin adenine dinucleotide) binding. This later suppression of RBOHD serves as a vital negative feedback loop to prevent excessive ROS accumulation and uncontrolled programmed cell death (Yun et al., 2011).

The influence of S-nitrosation extends deeply into the core of the cellular antioxidant system, such as the ascorbate-glutathione cycle (Wang et al., 2015). This regulation provides a direct mechanistic link between NO signaling and the management of ROS. A key target is ascorbate peroxidase (APX), where S-nitrosation can exert isoform-specific effects, either enhancing or inhibiting its H_2O_2-scavenging activity to fine-tune the amplitude and duration of the oxidative burst (Wang et al., 2015). Furthermore, S-nitrosation fine-tunes the ROS landscape by targeting antioxidant enzymes, including APX, catalase (CAT), and SOD (Superoxide dismutase), often enhancing their activity to promote H_2O_2 scavenging and oxidative stress tolerance (Begara-Morales et al., 2014; Fernando et al., 2019; Liu et al., 2022; Ortega-Galisteo et al., 2012). Thus S-nitrosation acts as a central mechanism that dynamically calibrates the ROS level in plant immunity, both by controlling the activity of the RBOHD oxidase and by modulating the capacity of the antioxidant system to ensure an effective yet self-limiting defense response (Fig. 4.2).

4.6.4 Regulation of S-nitrosation

The regulation of cellular SNO level in plants is primarily governed by the GSNOR enzyme, a key player in GSNO catabolism (Feechan et al., 2005; Lee et al., 2008). Initial genetic studies established its central role, demonstrating that the GSNOR1 loss-of-function mutant, *gsnir1–3*, exhibits reduced GSNOR activity and consequently accumulates high levels of SNOs, while *gsnor1-1*, a gain-of-function mutant, shows the opposite phenotype (Feechan et al., 2005). This catabolic pathway is supported by a parallel system involving SNO-CoA reductase, encoded by ALCOHOL DEHYDROGENASE 6 (ADH6), whose mutation similarly leads to SNO accumulation (Anand et al., 2014). Further biochemical studies reveal that *A. thaliana* AKR4C enzymes can degrade SNO-CoA even more efficiently than GSNO, doing so in an NADPH-dependent manner at a three to four-fold higher rate (Treffon et al., 2021), which is similar to the human cell (Hou et al., 2019). Although direct evidence is lacking, the efficient degradation of SNO-CoA by AKRs suggests a potential regulatory role in NO signaling, particularly in the context of plant immune responses.

The activity of GSNOR itself is subject to sophisticated posttranslational control, particularly through S-nitrosation. Research has shown that GSNOR is S-nitrosated on specific, evolutionarily conserved cysteine residues (Cys-10, Cys-271, and Cys-370) in response to NO, leading to its inhibition (Guerra et al., 2016). Since these cysteines are not part of the enzyme's active site, this

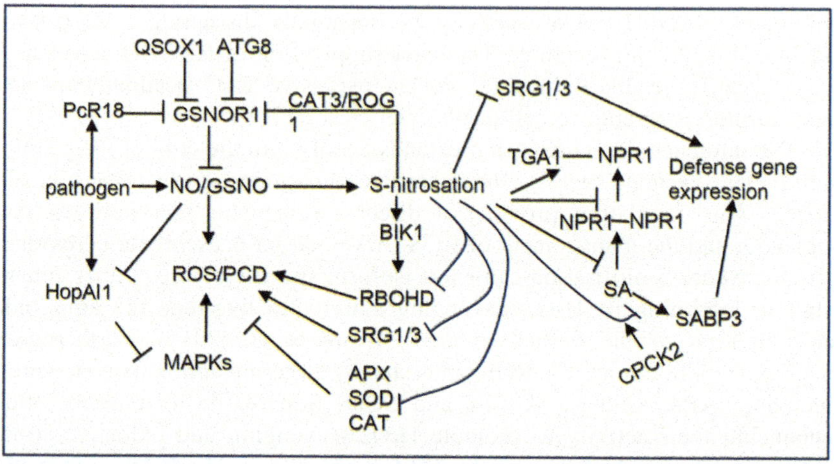

FIGURE 4.2 Integrated model of NO/GSNO-dependent redox regulation of plant immunity. Pathogen perception triggers rapid accumulation of nitric oxide (*NO*) and S-nitrosoglutathione (GSNO), initiating a redox-driven signaling cascade. Elevated NO/GSNO promotes S-nitrosation of key immune regulators, including BIK1, RBOHD, and SRG1/3, thereby modulating apoplastic ROS production and promoting programmed cell death (*PCD*). S-nitrosation also fine-tunes antioxidant capacity through APX, SOD, and CAT, ensuring tight homeostatic control of the oxidative burst. NO/GSNO pathways converge with salicylic acid (*SA*) signaling to regulate NPR1 dynamics and TGA1-dependent transcriptional activation of defense genes. SA perception through receptors and SA-binding proteins such as SABP3, together with CPK2-mediated phosphorylation, enhances NPR1 activity, while NPR1 autoregulatory feedback modulates its own expression. Conversely, the GSNO-reducing enzyme GSNOR1, negatively regulated by QSOX1, ATG8, and PcR18, functions as a central rheostat controlling cellular GSNO levels and, consequently, protein S-nitrosation status. MAPK signaling further integrates into this redox circuit to reinforce ROS/PCD outputs, whereas the pathogen effector HopAI1 suppresses MAPK activity to dampen host defense. Together, these interconnected NO/GSNO, ROS, MAPK, and SA pathways form a multilayered regulatory module that orchestrates robust yet finely tuned immune responses (*OSOX1*, QUIESCIN SULFHYDRYL OXIDASE 1).

regulation is believed to be allosteric in nature. Notably, S-nitrosation at Cys-10 triggers a more drastic fate, inducing a conformational change that exposes a domain recognized by the autophagy-related protein ATG8, thereby targeting GSNOR for autophagic degradation. This mechanism, which is prominent during hypoxic stress, was confirmed by the accumulation of GSNOR in autophagy-deficient mutants but not in the presence of proteasome inhibitors (Zhan et al., 2018). This degradation pathway is further regulated by CATALASE 3 (CAT3), which acts as a transnitrosylase, transferring an NO group from itself to GSNOR at Cys-10. The critical role of CAT3's Cys-343 in this process, independent of its CAT activity, was demonstrated by its ability to rescue the *atgsnor1* mutant phenotype, a function impaired by a C343T mutation (Li et al., 2021). The interaction of CAT3 with cell death regulators like LSD1 suggests this transnitrosylation mechanism may extend to other signaling pathways (Fig. 4.2).

A recent work demonstrated that GSNOR is also controlled by other redox mechanisms. During the immune response, the redox sensor QUIESCIN SULFHYDRYL OXIDASE 1 is activated by the pathogen-induced oxidative burst and directly oxidizes GSNOR, leading to its inactivation and a subsequent increase in protein S-nitrosation (Chae et al., 2021). In addition, an alternative denitrosation pathway exists that is distinct from GSNOR's GSNO-metabolizing activity. TRXh5 can directly remove NO groups from S-nitrosated proteins (Kneeshaw et al., 2014). Both in vitro and in vivo assays have confirmed this denitrosase activity, and the restoration of wild-type SNO levels and disease resistance in the *nox1* mutant by *TRXh5* overexpression underscores its physiological importance in fine-tuning SNO signaling during plant immunity (Kneeshaw et al., 2014).

4.6.5 Effector and S-nitrosoglutathione reductase: a two-way targeting

As established, the S-nitrosation pathway and its master regulator, GSNOR, constitute a central signaling hub in plant immunity, inevitably positioning them as focal points in the coevolutionary arms race between host and pathogen. This conflict is characterized by a sophisticated, bidirectional targeting system wherein both adversaries exploit the S-nitrosation pathway for strategic gain. On the pathogenic front, a key virulence strategy involves the direct disruption of plant NO homeostasis through the targeting of GSNOR. This is a conserved tactic among *Phytophthora* species, as demonstrated by the inhibition of GSNOR1 activity by secreted effectors from *P. parasitica* (Cui et al., 2021). A recent mechanistic breakthrough revealed that the *Phytophthora capsici* RxLR effector PcRD18 executes a precise, two-pronged assault on tomato GSNOR (Li and Kang, 2025). PcRD18 simultaneously inhibits the catalytic function of GSNOR and promotes its autophagic turnover by coopting the host's ATG8c system. This dual interference culminates in a dysregulation of cellular NO and SNO levels, which, in turn, suppresses the pathogen-triggered oxidative burst and potentiates virulence. Critically, the PcRD18-GSNOR protein–protein interaction is the linchpin of this strategy; structure-guided mutations that disrupt this interface render GSNOR resistant to manipulation, thereby restoring plant immunity.

In a striking countermeasure, plants have evolved to weaponize S-nitrosation against the invaders themselves. This host-mediated transnitrosation acts as a molecular defense mechanism to directly neutralize effector proteins. A paradigm for this is the S-nitrosation-dependent inactivation of the bacterial effector HopAI1 (Ling et al., 2017). Upon translocation into the host cell, HopAI1 is selectively S-nitrosated on a critical catalytic cysteine, abolishing its phosphothreonine lyase activity. This inactivation preserves the phosphorylation-based activation of MAP kinase cascades, thereby reinstating immune signaling and curtailing bacterial proliferation. This bidirectional interplay defines a critical redox battleground. Pathogens like *Phytophthora* deploy effectors to subvert GSNOR, inducing a state of immunosuppressive nitrosative stress. Plants, in turn, retaliate through the precision S-nitrosation of key virulence factors. The dynamic control of GSNOR activity, coupled with the competing S-nitrosation of host and pathogen proteomes, underscores the dual nature of this redox modification. Ultimately, the trajectory of infection is frequently determined by which organism gains mastery over this pivotal node, firmly establishing GSNOR not as a mere housekeeping enzyme, but as a central, contested control point in plant immunity (Fig. 4.2).

4.7 Nitric oxide signaling in early land plants: insights from mosses and liverworts

The critical roles of NO in angiosperm physiology, particularly in abscisic acid (ABA)-mediated stomatal regulation, PTI, and redox-based feedback, are instructive for understanding how these signaling elements manifest in the earliest land-plant lineages. While NO itself is evolutionarily ancient, the complexity of its regulatory network in angiosperms likely represents an evolutionary refinement rather than a fully conserved ancestral state. The key question, therefore, is not whether early land plants possessed NO, but whether they shared the multilayered NO–ROS–Ca^{2+}–ion channel integration and the negative regulatory loop observed in seed plants. In higher plants, the canonical pathway, ABA → ROS → NO → cytosolic Ca^{2+} elevation → stomatal closure, is well established (Lamattina et al., 2003; Neill et al., 2002). Importantly, in angiosperms, excess NO and its reactive derivatives ($ONOO^-$ and other RNS) act as a feedback brake, reopening stomata through S-nitrosation of key guard-cell proteins (Garcia-Mata et al., 2003). Whether such a redox-sensitive reopening mechanism also exists in early land plants remains unresolved.

In early land plants, the evidence supports the presence of NO production and some signaling, but the full suite of angiosperm-type regulatory mechanisms remains largely unverified. In the moss *P. patens*, endogenous NO production has been demonstrated and shown to be mediated primarily by NR rather than any canonical NOS-like enzyme (Medina-Andrés et al., 2015). This strongly suggests that the reductive NR-based route to NO was present early in land plants and is likely ancestral (Chamizo-Ampudia et al., 2017; Medina-Andrés et al., 2015). Meanwhile, bryophytes and early vascular plants show inducible defense responses and some stomatal or pore regulation, but often within a predominantly hydraulic or passive framework (Brodribb & McAdam, 2011; Franks, 2013). Specifically for stomata, many pteridophytes/lycophytes retain largely passive water-status-driven aperture changes, and only limited evidence exists for ABA-/NO-mediated active control (Chater et al., 2017; Ruszala et al., 2011).

Critically, the "NO brake" mechanism, the capacity to terminate or reopen closure via NO/RNS-mediated feedback on ion channels, has not been convincingly demonstrated in nonseed land plants. If guard-cell systems in bryophytes or lycophytes lack this redox-sensitive feedback loop, it would suggest that the ancestral role of NO was simpler (e.g., as a stress or defense integrator) and that the advanced regulatory loop seen in angiosperms is a derived feature. Thus the claim that NO-dependent PTI and stomatal regulation are broadly "conserved" must be reframed: the capacity for NO signaling is ancient, but the full fine-tuned regulatory network may have evolved gradually.

Looking forward, comparative electrophysiology in basal lineages (e.g., patch-clamp on K^+_in and K^+_out channels in fern/lycophyte guard cells under NO donor and scavenger treatments) would test the presence of NO-targets and feedback mechanisms. Proteomic and S-nitrosation assays in bryophytes and ferns could reveal whether the same molecular NO-targets as in angiosperms are conserved. Additionally, developmental roles of NO in bryophyte sporophyte stomata, perhaps tied to spore desiccation rather than rapid gas exchange, should be studied (Merced & Renzaglia, 2017). Ultimately, a phylogenetically broad mapping of NO-production routes, guard cell ion channel architecture, and feedback loops will allow us to chart how NO signaling evolved in the conquest of land. Altogether, NO production via NR appears to be ancestral in land plants; active NO-mediated stomatal and defense responses are present in part in early lineages, but the most sophisticated regulatory features (such as NO/RNS-mediated closure termination) remain unproven outside angiosperms.

4.7.1 Conserved pattern-triggered immunity in early land plants

Emerging genomic and molecular evidence indicates that the core PTI machinery is conserved across land-plant lineages, including nonvascular species. For instance, *M. polymorpha* possesses LysM-domain receptor-like kinases (MpLYK1 and MpLYR), which recognize chitin and peptidoglycan fragments, elicitors known to activate PTI in angiosperms (Yotsui et al., 2023). Exposure to these PAMPs triggers transcriptional reprogramming, ROS production, and defense gene activation, confirming that canonical PRR–PTI (Pattern-recognition receptor) signaling modules were already functional in the earliest land plants (Singh et al., 2024; Yu et al., 2024). Comparative transcriptomics reveal that exposure to *Phytophthora palmivora* elicits conserved immune pathways shared with angiosperms, including WRKY transcription factors and ROS/NO-responsive genes (Carella et al., 2018, 2019). A recent study by Tabassum and Goodrich, 2026 presents preliminary evidence for S-nitrosation dynamics in *M. polymorpha* primarily during developmental transitions and in a set of immune-challenge assays analogous to those described by Feechan et al. (2005). These findings tentatively indicate that NO-dependent post-translational regulation was already operational in basal land plants, linking developmental redox homeostasis with incipient immune signaling mechanisms. In *P. patens*, chitin and flg22 similarly induce callose deposition and the expression of defense-related genes such as peroxidases and phenylalanine ammonia-lyase (Bressendorff et al., 2016). These observations demonstrate that the fundamental architecture of PTI perception, signal transduction, and defense gene activation arose before the divergence of bryophytes and vascular plants. While direct measurements of NO accumulation during PTI in mosses remain limited, the presence of NR-linked NO metabolism (Medina-Andrés et al., 2015) and the activation of redox-sensitive pathways imply that NO is likely integrated into these early immune responses.

4.7.2 Nitric oxide metabolism and defense in bryophytes

Bryophytes such as *P. patens* utilize the NR pathway for NO biosynthesis rather than a canonical NOS enzyme (Chamizo-Ampudia et al., 2017; Medina-Andrés et al., 2015). This reductive route supports stress- and pathogen-induced NO production, which cooperates with ROS to regulate defense and metabolic reprogramming (Ponce de León & Montesano, 2013). When NR activity is chemically suppressed, both NO accumulation and expression of defense-related genes decline, underscoring NO's integral role in immune activation. Functional analyses in *P. patens* further highlight NO as a key coordinator of transcriptional and oxidative defenses. The ethylene-response factor PpERF24 integrates NO/ROS bursts with phenylpropanoid and oxylipin (JA-like) biosynthetic pathways, thereby linking redox signaling to antimicrobial metabolite production (Reboledo et al., 2022). Studies (Lehtonen et al., 2009) identified a rapidly secreted class III peroxidase (Prx34) that is indispensable for defense against fungal pathogens, and *Prx34*-deficient mutants exhibit hypersensitivity to fungal invasion. This finding provides molecular evidence that NO- and ROS-dependent peroxidative activity constitutes a frontline defense mechanism in bryophytes, mechanistically paralleling oxidative bursts in seed-plant immunity.

4.7.3 Evolutionary implications

The intersection of PTI signaling, NR-derived NO metabolism, and S-nitrosation in mosses and liverworts provides compelling evidence that nitro-oxidative immunity is an ancestral land-plant trait. Early plants lacked vascular tissues and systemic defense but compensated through cell-autonomous fortification, ROS/NO bursts, and secondary-metabolite production (Carella et al., 2019; Castel et al., 2024; Delaux et al., 2019). Liverwort oil-body terpenoids and phenolics (Bowman et al., 2022) are induced in parallel with NO-linked redox signaling, implying that NO already functions as a bridge between metabolic and immune responses.

From an evolutionary perspective, the transition from these localized, NO-regulated defenses to the complex systemic immune networks of tracheophytes likely involved the gradual elaboration of NO feedback loops and crosstalk with hormones such as ABA and jasmonate. As Poveda (2020) argued, NO was not an innovation of higher plants but a foundational signal coopted and refined throughout land-plant diversification. The demonstration of S-nitrosation-mediated immune regulation in *M. polymorpha* (Tabassum and Goodrich, 2026) now bridges a critical evolutionary gap, revealing mechanistic continuity between bryophyte and angiosperm immune signaling.

4.8 Future perspective

The foundational understanding of NO as a central redox hub in plant immunity, conserved from bryophytes to angiosperms, now propels the field toward a new era of mechanistic decoding and translational application. The paramount challenge is to resolve the spatiotemporal specificity of NO signaling, understanding how a simple molecule achieves precise, context-dependent commands. The integration of genetically encoded biosensors for real-time, subcellular monitoring with single-cell omics will map the NO landscape and its functional consequences at unprecedented resolution. Beyond cataloging S-nitrosation targets, the future lies in dissecting the dynamic crosstalk between NO-mediated PTMs and other regulatory layers, such as phosphorylation, revealing an integrated "redox code" that governs immunity. Evolutionarily, comparative systems biology across land plants will define the minimal NO regulon and identify key innovations that shaped this network. Ultimately, this deep mechanistic knowledge must be translated into sustainable agricultural practice. The emerging field of "Nitro-Agronomy" holds immense promise, focusing on precision breeding of NO signaling components (e.g., GSNOR, NR), developing NO-based priming agents for enhanced resistance, and optimizing the nexus between nitrogen-use efficiency and immunity. By harnessing the power of this ancient signaling module, we can pioneer innovative strategies to engineer future-proof crops, bolstering plant health and productivity in a changing climate.

Acknowledgments

The Prime Minister Fellowship for PhD 2019 (Government of Bangladesh) to NT, the National Natural Science Foundation of China (32170290) to QP, Wellcome-University of Edinburgh ISSF3 award to BC.

Competing interests

The authors declare that there are no competing interests.

AI Disclosure

During the preparation of this work, the author(s) used Grammarly, ChatGPT, and Gemini in order to English Correction. After using this tool/service, the author(s) reviewed and edited the content as needed and take(s) full responsibility for the content of the publication.

References

Aboalroub, A. A., & Al Azzam, K. M. (2024). Protein S-nitrosylation: A chemical modification with ubiquitous biological activities. *Protein Journal, 43*(4), 639–655. https://doi.org/10.1007/s10930-024-10223-y, https://www.springer.com/journal/10930.

Ahmad, B., Mukarram, M., Choudhary, S., Petrík, P., Ahmad Dar, T., & Khan, M. M. A. (2024). Adaptive responses of nitric oxide (NO) and its intricate dialogue with phytohormones during salinity stress. *Plant Physiology and Biochemistry, 208*, 108504. https://doi.org/10.1016/j.plaphy.2024.108504.

Almagro, L., Gómez Ros, L. V., Belchi-Navarro, S., Bru, R., Ros Barceló, A., & Pedreño, M. A. (2009). Class III peroxidases in plant defence reactions. *Journal of Experimental Botany, 60*(2), 377–390. https://doi.org/10.1093/jxb/ern277.

Anand, P., Hausladen, A., Wang, Y. J., Zhang, G. F., Stomberski, C., Brunengraber, H., Hess, D. T., & Stamler, J. S. (2014). Identification of S-nitroso-CoA reductases that regulate protein S-nitrosylation. *Proceedings of the National Academy of Sciences of the United States of America. 10916490. 111*(52), 18572–18577. https://doi.org/10.1073/pnas.1417816112, http://www.pnas.org/content/111/52/18572.full.pdf+html.

Asai, T., Tena, G., Plotnikova, J., Willmann, M. R., Chiu, W.-L., Gomez-Gomez, L., Boller, T., Ausubel, F. M., & Sheen, J. (2002). MAP kinase signalling cascade in *Arabidopsis* innate immunity. *Nature, 415*(6875), 977–983. https://doi.org/10.1038/415977a.

Astier, J., Kulik, A., Koen, E., Besson-Bard, A., Bourque, S., Jeandroz, S., Lamotte, O., & Wendehenne, D. (2012). Protein S-nitrosylation: What's going on in plants? *Free Radical Biology and Medicine, 53*(5), 1101–1110. https://doi.org/10.1016/j.freeradbiomed.2012.06.032.

Batistič, O., & Kudla, J. (2012). Analysis of calcium signaling pathways in plants. *Biochimica et Biophysica Acta - General Subjects, 1820*(8), 1283–1293. https://doi.org/10.1016/j.bbagen.2011.10.012.

Begara-Morales, J. C., Sánchez-Calvo, B., Chaki, M., Valderrama, R., Mata-Pérez, C., López-Jaramillo, J., Padilla, M. N., Carreras, A., Corpas, F. J., & Barroso, J. B. (2014). Dual regulation of cytosolic ascorbate peroxidase (APX) by tyrosine nitration and S-nitrosylation. *Journal of Experimental Botany, 65*(2), 527–538. https://doi.org/10.1093/jxb/ert396.

Benhar, M., Forrester, M. T., & Stamler, J. S. (2009). Protein denitrosylation: Enzymatic mechanisms and cellular functions. *Nature Reviews. Molecular Cell Biology, 10*(10), 721–732. https://doi.org/10.1038/nrm2764.

Berger, A., Guinand, S., Boscari, A., Puppo, A., & Brouquisse, R. (2020). Medicago truncatula phytoglobin 1.1 controls symbiotic nodulation and nitrogen fixation via the regulation of nitric oxide concentration. *New Phytologist, 227*(1), 84–98. https://doi.org/10.1111/nph.16462.

Besson-Bard, A., Pugin, A., & Wendehenne, D. (2008). New insights into nitric oxide signaling in plants. *Annual Review of Plant Biology, 59*, 21–39. https://doi.org/10.1146/annurev.arplant.59.032607.092830.

Borrowman, S., Kapuganti, J. G., & Loake, G. J. (2023). Expanding roles for S-nitrosylation in the regulation of plant immunity. *Free Radical Biology and Medicine, 194*, 357–368. https://doi.org/10.1016/j.freeradbiomed.2022.12.009, www.elsevier.com/locate/freeradbiomed.

Bowman, J. L., Arteaga-Vazquez, M., Berger, F., Briginshaw, L. N., Carella, P., Aguilar-Cruz, A., Davies, K. M., Dierschke, T., Dolan, L., Dorantes-Acosta, A. E., Fisher, T. J., Flores-Sandoval, E., Futagami, K., Ishizaki, K., Jibran, R., Kanazawa, T., Kato, H., Kohchi, T., Levins, J., ... Zachgo, S. (2022). The renaissance and enlightenment of Marchantia as a model system. *The Plant Cell, 34*(10), 3512–3542. https://doi.org/10.1093/plcell/koac219, https://academic.oup.com/plcell/issue.

Bressendorff, S., Azevedo, R., Kenchappa, C. S., Ponce de León, I., Olsen, J. V., Rasmussen, M. W., Erbs, G., Newman, M.-A., Petersen, M., & Mundy, J. (2016). An innate immunity pathway in the moss *Physcomitrella patens*. *The Plant Cell, 28*(6), 1328–1342. https://doi.org/10.1105/tpc.15.00774.

Brodribb, T. J., & McAdam, S. A. M. (2011). Passive origins of stomatal control in vascular plants. *Science (New York, N.Y.), 331*(6017), 582–585. https://doi.org/10.1126/science.1197985Australia, http://www.sciencemag.org/content/331/6017/582.full.pdf.

Carella, P., Gogleva, A., Tomaselli, M., Alfs, C., & Schornack, S. (2018). Phytophthora palmivora establishes tissue-specific intracellular infection structures in the earliest divergent land plant lineage. *Proceedings of the National Academy of Sciences, 115*(16). https://doi.org/10.1073/pnas.1717900115.

Carella, P., Gogleva, A., Hoey, D. J., Bridgen, A. J., Stolze, S. C., Nakagami, H., & Schornack, S. (2019). Conserved biochemical defenses underpin host responses to oomycete infection in an early-divergent land plant lineage. *Current Biology, 29*(14), 2282. https://doi.org/10.1016/j.cub.2019.05.078.

Castel, B., Baker, M., Keller, J., Martinez, Y., Bonhomme, M., Delaux, P. M., & Jacquet, C. (2025). Exploring fern pathosystems and immune receptors to bridge gaps in plant immunity. *BMC Biology, 23*(1). https://doi.org/10.1186/s12915-025-02413-6, https://bmcbiol.biomedcentral.com/.

Castel, B., El Mahboubi, K., Jacquet, C., & Delaux, P.-M. (2024). Immunobiodiversity: Conserved and specific immunity across land plants and beyond. *Molecular Plant, 17*(1), 92–111. https://doi.org/10.1016/j.molp.2023.12.005.

Chae, H. B., Kim, M. G., Kang, C. H., Park, J. H., Lee, E. S., Lee, S. U., Chi, Y. H., Paeng, S. K., Bae, S. B., Wi, S. D., Yun, B. W., Kim, W. Y., Yun, D. J., Mackey, D., & Lee, S. Y. (2021). Redox sensor QSOX1 regulates plant immunity by targeting GSNOR to modulate ROS generation. *Molecular Plant, 14*(8), 1312–1327. https://doi.org/10.1016/j.molp.2021.05.004, https://www.cell.com/molecular-plant/home.

Chakraborty, S., Sircar, E., Bhattacharyya, C., Choudhuri, A., Mishra, A., Dutta, S., Bhatta, S., Sachin, K., & Sengupta, R. (2022). S-Denitrosylation: A crosstalk between glutathione and redoxin systems. *Antioxidants, 11*(10). https://doi.org/10.3390/antiox11101921, www.mdpi.com/journal/antioxidants.

Chamizo-Ampudia, A., Sanz-Luque, E., Llamas, A., Galvan, A., & Fernandez, E. (2017). Nitrate reductase regulates plant nitric oxide homeostasis. *Trends in Plant Science, 22*(2), 163–174. https://doi.org/10.1016/j.tplants.2016.12.001.

Chater, C. C. C., Caine, R. S., Fleming, A. J., & Gray, J. E. (2017). Origins and evolution of stomatal development. *Plant Physiology, 174*(2), 624–638. https://doi.org/10.1104/pp.17.00183, http://www.plantphysiol.org/content/174/2/624.full.pdf.

Chu-Puga, Á., González-Gordo, S., Rodríguez-Ruiz, M., Palma, J. M., & Corpas, F. J. (2019). NADPH oxidase (Rboh) activity is up regulated during sweet pepper (*Capsicum annuum* L.) fruit ripening. *Antioxidants, 8*(1), 9. https://doi.org/10.3390/antiox8010009.

Corina Vlot, A., Dempsey, D. A., & Klessig, D. F. (2009). Salicylic acid, a multifaceted hormone to combat disease. *Annual Review of Phytopathology, 47*, 177–206. https://doi.org/10.1146/annurev.phyto.050908.135202Germany, http://arjournals.annualreviews.org/doi/pdf/10.1146/annurev.phyto.050908.135202.

Corpas, F. J. (2019). Nitric oxide and hydrogen sulfide in higher plants under physiological and stress conditions. *Antioxidants, 8*(10), 457. https://doi.org/10.3390/antiox8100457.

Cui, B., Ma, X., Li, Y., Zhou, Y., Ju, X., Hussain, A., Umbreen, S., Yuan, B., Tabassum, A., Lubega, J., Shan, W., Loake, G. J., & Pan, Q. (2021). Perturbations in nitric oxide homeostasis promote *Arabidopsis* disease susceptibility towards Phytophthora parasitica. *Molecular Plant Pathology, 22*(9), 1134–1148. https://doi.org/10.1111/mpp.13102.

Cui, B., Pan, Q., Cui, W., Wang, Y., Loake, V. I. P., Yuan, S., Liu, F., & Loake, G. J. (2024). S -nitrosylation of a receptor-like cytoplasmic kinase regulates plant immunity. *Science Advances, 10*(11). https://doi.org/10.1126/sciadv.adk3126.

Cui, B., Pan, Q., Clarke, D., Villarreal, M. O., Umbreen, S., Yuan, B., Shan, W., Jiang, J., & Loake, G. J. (2018). S-nitrosylation of the zinc finger protein SRG1 regulates plant immunity. *Nature Communications, 9*(1). https://doi.org/10.1038/s41467-018-06578-3.

Dangl, J. L., & McDowell, J. M. (2006). Two modes of pathogen recognition by plants. *Proceedings of the National Academy of Sciences, 103*(23), 8575–8576. https://doi.org/10.1073/pnas.0603183103.

Davies, K. M., Jibran, R., Zhou, Y., Albert, N. W., Brummell, D. A., Jordan, B. R., Bowman, J. L., & Schwinn, K. E. (2020). The evolution of flavonoid biosynthesis: A bryophyte perspective. *Frontiers in Plant Science, 11*. https://doi.org/10.3389/fpls.2020.00007, https://www.frontiersin.org/journals/plant-science.

Delaux, P. M., Hetherington, A. J., Coudert, Y., Delwiche, C., Dunand, C., Gould, S., Kenrick, P., Li, F. W., Philippe, H., Rensing, S. A., Rich, M., Strullu-Derrien, C., & de Vries, J. (2019). Reconstructing trait evolution in plant evo–devo studies. *Current Biology, 29*(21), R1110. https://doi.org/10.1016/j.cub.2019.09.044, http://www.elsevier.com/journals/current-biology/0960–9822.

Delledonne, M., Xia, Y., Dixon, R. A., & Lamb, C. (1998). Nitric oxide functions as a signal in plant disease resistance. *Nature, 394*(6693), 585–588. https://doi.org/10.1038/29087.

Desikan, R., Griffiths, R., Hancock, J., & Neill, S. (2002). A new role for an old enzyme: Nitrate reductase-mediated nitric oxide generation is required for abscisic acid-induced stomatal closure in *Arabidopsis thaliana*. *Proceedings of the National Academy of Sciences, 99*(25), 16314–16318. https://doi.org/10.1073/pnas.252461999.

Domingos, P., Prado, A. M., Wong, A., Gehring, C., & Feijo, J. A. (2015). Nitric oxide: A multitasked signaling gas in plants. *Molecular Plant, 8*(4), 506–520. https://doi.org/10.1016/j.molp.2014.12.010, https://www.cell.com/molecular-plant/home.

Du, J., Filipović, M. R., Wagner, B. A., & Buettner, G. R. (2023). Ascorbate mediates the non-enzymatic reduction of nitrite to nitric oxide. *Advances in Redox Research, 9*, 100079. https://doi.org/10.1016/j.arres.2023.100079.

Durner, J., & Klessig, D. F. (1999). Nitric oxide as a signal in plants. *Current Opinion in Plant Biology, 2*(5), 369–374. https://doi.org/10.1016/S1369–5266(99)00007-2, www.elsevier.com/inca/publications/store/6/0/1/3/1/4/index.htt.

Feechan, A., Kwon, E., Yun, B.-W., Wang, Y., Pallas, J. A., & Loake, G. J. (2005). A central role for S -nitrosothiols in plant disease resistance. *Proceedings of the National Academy of Sciences, 102*(22), 8054–8059. https://doi.org/10.1073/pnas.0501456102.

Fernando, V., Zheng, X., Walia, Y., Sharma, V., Letson, J., & Furuta, S. (2019). S-Nitrosylation: An emerging paradigm of redox signaling. *Antioxidants, 8*(9), 404. https://doi.org/10.3390/antiox8090404.

Francis, S. H., Busch, J. L., & Corbin, J. D. (2010). cGMP-dependent protein kinases and cGMP phosphodiesterases in nitric oxide and cGMP action. *Pharmacological Reviews, 62*(3), 525–563. https://doi.org/10.1124/pr.110.002907, http://pharmrev.aspetjournals.org/content/62/3/525.full.pdf+html.

Franks, P. J. (2013). Passive and active stomatal control: Either or both? *New Phytologist, 198*(2), 325–327. https://doi.org/10.1111/nph.12228.

Fu, Z. Q., & Dong, X. (2013). Systemic acquired resistance: Turning local infection into global defense. *Annual Review of Plant Biology, 64*, 839–863. https://doi.org/10.1146/annurev-arplant-042811–105606.

Garcia-Mata, C., Gay, R., Sokolovski, S., Hills, A., Lamattina, L., & Blatt, M. R. (2003). Nitric oxide regulates K^+ and Cl - channels in guard cells through a subset of abscisic acid-evoked signaling pathways. *Proceedings of the National Academy of Sciences, 100*(19), 11116–11121. https://doi.org/10.1073/pnas.1434381100.

Gayatri, G., Agurla, S., & Raghavendra, A. S. (2013). Nitric oxide in guard cells as an important secondary messenger during stomatal closure. *Frontiers in Plant Science, 4*. https://doi.org/10.3389/fpls.2013.00425, http://journal.frontiersin.org/Journal/10.3389/fpls.2013.00425/pdf.

Greenberg, J. T., & Yao, N. (2004). The role of regulation of programmed cell death in plant-pathogen interactions. *Cellular Microbiology, 6*(3), 201–211. https://doi.org/10.1111/j.1462-5822.2004.00361.x.

Guerra, D., Ballard, K., Truebridge, I., & Vierling, E. (2016). S-nitrosation of conserved cysteines modulates activity and stability of S -nitrosoglutathione reductase (GSNOR). *Biochemistry, 55*(17), 2452–2464. https://doi.org/10.1021/acs.biochem.5b01373.

Gupta, K. J., Hebelstrup, K. H., Mur, L. A. J., & Igamberdiev, A. U. (2011). Plant hemoglobins: Important players at the crossroads between oxygen and nitric oxide. *FEBS Letters, 585*(24), 3843–3849. https://doi.org/10.1016/j.febslet.2011.10.036, http://onlinelibrary.wiley.com/journal/10.1002/(ISSN)1873-3468.

Gupta, K. J., & Igamberdiev, A. U. (2011). The anoxic plant mitochondrion as a nitrite: NO reductase. *Mitochondrion, 11*(4), 537–543. https://doi.org/10.1016/j.mito.2011.03.005.

Gupta, P., Saxena, G., & Gupta, R. (2025). Nitrate reductase-mediated nitric oxide synthesis in shaping stress resilience in plants. *Journal of Experimental Botany, 76*(22), 6634–6656. https://doi.org/10.1093/jxb/eraf353, http://jxb.oxfordjournals.org/.

Gupta Kapuganti, J., Igamberdiev, A. U., & Kaiser, W. M. (2014). New insights into the mitochondrial nitric oxide production pathways. *Plant Signaling & Behavior, 5*(8), 999–1001. https://doi.org/10.4161/psb.5.8.12229.

Gurung, S., Cohen, M. F., & Yamasaki, H. (2014). Azide-dependent nitric oxide emission from the water fern *Azolla pinnata*. *Russian Journal of Plant Physiology, 61*(4), 543–547. https://doi.org/10.1134/s1021443714040086.

Haghpanah, M., Namdari, A., Kaleji, M. K., Nikbakht-dehkordi, A., Arzani, A., & Araniti, F. (2025). Interplay between ROS and hormones in plant defense against pathogens. *Plants, 14*(9), 1297. https://doi.org/10.3390/plants14091297.

Han, X., & Tsuda, K. (2022). Evolutionary footprint of plant immunity. *Current Opinion in Plant Biology, 67*, 102209. https://doi.org/10.1016/j.pbi.2022.102209.

Hayat, S., Mori, M., Pichtel, J., & Ahmad, A. (2009). Interplay between nitric oxide and other signals involved in plant resistance to pathogens. In *Nitric oxide in plant physiology*. Wiley, 139–159. http://doi.org/10.1002/9783527629138.ch10.

Hetherington, A. M., & Brownlee, C. (2004). The generation of Ca^{2+} signals in plants. *Annual Review of Plant Biology, 55*, 401–427. https://doi.org/10.1146/annurev.arplant.55.031903.141624.

Hou, S., Liu, Z., Shen, H., & Wu, D. (2019). Damage-associated molecular pattern-triggered immunity in plants. *Frontiers in Plant Science, 10*. https://doi.org/10.3389/fpls.2019.00646.

Hu, C.-H., Wang, P.-Q., Zhang, P.-P., Nie, X.-M., Li, B.-B., Tai, L., Liu, W.-T., Li, W.-Q., & Chen, K.-M. (2020). NADPH oxidases: The vital performers and center hubs during plant growth and signaling. *Cells, 9*(2), 437. https://doi.org/10.3390/cells9020437.

Iglesias, M. J., Terrile, M. C., Correa-Aragunde, N., Colman, S. L., Izquierdo-Álvarez, A., Fiol, D. F., et al. (2018). Regulation of SCFTIR1/AFBs E3 ligase assembly by S-nitrosylation of Arabidopsis SKP1-like1 impacts on auxin signaling. *Redox Biology, 18*, 200–210. https://doi.org/10.1016/j.redox.2018.07.003.

Ignarro, L. J. (1990). Nitric oxide. A novel signal transduction mechanism for transcellular communication. *Hypertension, 16*(5), 477–483. https://doi.org/10.1161/01.hyp.16.5.477.

Jones, J. D. G., & Dangl, J. L. (2006). The plant immune system. *Nature, 444*(7117), 323–329. https://doi.org/10.1038/nature05286.

Kaiser, W. M., & Huber, S. C. (2001). Post-translational regulation of nitrate reductase: Mechanism, physiological relevance and environmental triggers. *Journal of Experimental Botany, 52*(363), 1981–1989. https://doi.org/10.1093/jexbot/52.363.1981, http://jxb.oxfordjournals.org/.

Khan, M., Ali, S., Al Azzawi, T. N. I., Saqib, S., Ullah, F., Ayaz, A., & Zaman, W. (2023). The key roles of ROS and RNS as a signaling molecule in plant–microbe interactions. *Antioxidants, 12*(2). https://doi.org/10.3390/antiox12020268, www.mdpi.com/journal/antioxidants.

Kneeshaw, S., Gelineau, S., Tada, Y., Loake, G. J., & Spoel, S. H. (2014). Selective protein denitrosylation activity of thioredoxin-h5 modulates plant immunity. *Molecular Cell, 56*(1), 153–162. https://doi.org/10.1016/j.molcel.2014.08.003.

Kovacs, I., Durner, J., & Lindermayr, C. (2015). Crosstalk between nitric oxide and glutathione is required for nonexpressor of pathogenesis-related genes 1 (NPR1)-dependent defense signaling in *Arabidopsis thaliana*. *New Phytologist, 208*(3), 860–872. https://doi.org/10.1111/nph.13502Germany, http://onlinelibrary.wiley.com/journal/10.1111/(ISSN)1469-8137.

Kubo, H., Nozawa, S., Hiwatashi, T., Kondou, Y., Nakabayashi, R., Mori, T., Saito, K., Takanashi, K., Kohchi, T., & Ishizaki, K. (2018). Biosynthesis of riccionidins and marchantins is regulated by R2R3-MYB transcription factors in *Marchantia polymorpha*. *Journal of Plant Research, 131*(5), 849–864. https://doi.org/10.1007/s10265-018-1044-7.

Lamattina, L., García-Mata, C., Graziano, M., & Pagnussat, G. (2003). Nitric oxide: The versatility of an extensive signal molecule. *Annual Review of Plant Biology, 54*(1), 109–136. https://doi.org/10.1146/annurev.arplant.54.031902.134752.

Lee, D. H., Lal, N. K., Lin, Z.-J. D., Ma, S., Liu, J., Castro, B., Toruño, T., Dinesh-Kumar, S. P., & Coaker, G. (2020). Regulation of reactive oxygen species during plant immunity through phosphorylation and ubiquitination of RBOHD. *Nature Communications, 11*(1). https://doi.org/10.1038/s41467-020-15601-5.

Lee, U., Wie, C., Fernandez, B. O., Feelisch, M., & Vierling, E. (2008). Modulation of nitrosative stress by S -nitrosoglutathione reductase is critical for thermotolerance and plant growth in *Arabidopsis*. *The Plant Cell, 20*(3), 786–802. https://doi.org/10.1105/tpc.107.052647.

Lehtonen, M. T., Akita, M., Kalkkinen, N., Ahola-Iivarinen, E., Rönnholm, G., & Somervuo, P. (2009). Quickly-released peroxidase of moss in defense against fungal invaders. *New Phytologist, 183*(2), 432–443. https://doi.org/10.1111/j.1469-8137.2009.02864.x.

León, J., Castillo, M. C., Coego, A., Lozano-Juste, J., & Mir, R. (2014). Diverse functional interactions between nitric oxide and abscisic acid in plant development and responses to stress. *Journal of Experimental Botany, 65*(4), 907–921. https://doi.org/10.1093/jxb/ert454.

Li, Z. C., Ren, Q. W., Guo, Y., Ran, J., Ren, X. T., Wu, N. N., Xu, H. Y., Liu, X., & Liu, J. Z. (2021). Dual roles of GSNOR1 in cell death and immunity in tetraploid *Nicotiana tabacum*. *Frontiers in Plant Science, 12*. https://doi.org/10.3389/fpls.2021.596234, https://www.frontiersin.org/journals/plant-science.

Li, T., Kang, J., Zhang, H., Wang, L., Lu, M., Cai, L., Li, J., Joosten, M. H. A. J., & Du, Y. (2025). Phytophthora disrupts plant immunity by manipulating nitric oxide homeostasis through GSNOR inhibition. *Advanced Science, 12*(33), 451–464. https://doi.org/10.1002/advs.202503633.

Lin, A., Wang, Y., Tang, J., Xue, P., Li, C., Liu, L., et al. (2012). Nitric oxide and protein S-nytrosylation are integral to hydrogen peroxide-induced leaf cell death in rice. *Plant Physiology, 158*(1), 451–464. https://doi.org/10.1104/pp.111.184531.

Lindermayr, C., Sell, S., Müller, B., Leister, D., & Durner, J. (2010). Redox regulation of the NPR1-TGA1 system of *Arabidopsis thaliana* by nitric oxide. *The Plant Cell, 22*(8), 2894–2907. https://doi.org/10.1105/tpc.109.066464.

Lindermayr, C., & Durner, J. (2015). Interplay of reactive oxygen species and nitric oxide: Nitric oxide coordinates reactive oxygen species homeostasis. *Plant Physiology, 167*(4), 1209–1210. https://doi.org/10.1104/pp.15.00293, http://www.plantphysiol.org/content/167/4/1209.full.pdf.

Lindermayr, C., Saalbach, G., Bahnweg, G., & Durner, J. (2006). Differential inhibition of *Arabidopsis* methionine adenosyltransferases by protein s-nitrosylation. *Journal of Biological Chemistry, 281*(7), 4285–4291. https://doi.org/10.1074/jbc.m511635200.

Ling, T., Bellin, D., Vandelle, E., Imanifard, Z., & Delledonne, M. (2017). Host-mediated s -nitrosylation disarms the bacterial effector hopAI1 to reestablish immunity. *The Plant Cell, 29*(11), 2871–2881. https://doi.org/10.1105/tpc.16.00557.

Lionetti, V., Fabri, E., De Caroli, M., Hansen, A. R., Willats, W. G. T., Piro, G., & Bellincampi, D. (2017). Three pectin methylesterase inhibitors protect cell wall integrity for *Arabidopsis* immunity to botrytis. *Plant Physiology, 173*(3), 1844–1863. https://doi.org/10.1104/pp.16.01185.

Liu, N., Sun, Y., Pei, Y., Zhang, X., Wang, P., Li, X., Li, F., & Hou, Y. (2018). A pectin methylesterase inhibitor enhances resistance to verticillium wilt. *Plant Physiology, 176*(3), 2202–2220. https://doi.org/10.1104/pp.17.01399.

Liu, R., Zhu, T., Chen, X., Wang, Z., Yang, Z., Ren, A., Shi, L., Yu, H., & Zhao, M. (2022). GSNOR regulates ganoderic acid content in Ganoderma lucidum under heat stress through S-nitrosylation of catalase. *Communications Biology, 5*(1). https://doi.org/10.1038/s42003-021-02988-0.

Loake, G., & Grant, M. (2007). Salicylic acid in plant defence-the players and protagonists. *Current Opinion in Plant Biology, 10*(5), 466–472. https://doi.org/10.1016/j.pbi.2007.08.008.

López-Gómez, P., Buezo, J., Urra, M., Cornejo, A., Esteban, R., Fernández de los Reyes, J., Urarte, E., Rodríguez-Dobreva, E., Chamizo-Ampudia, A., Eguaras, A., Wolf, S., Marino, D., Martínez-Merino, V., & Moran, J. F. (2024). A new oxidative pathway of nitric oxide production from oximes in plants. *Molecular Plant, 17*(1), 178–198. https://doi.org/10.1016/j.molp.2023.12.009.

Maldonado-Alconada, A. M., Echevarría-Zomeño, S., Lindermayr, C., Redondo-López, I., Durner, J., & Jorrín-Novo, J. V. (2011). Proteomic analysis of *Arabidopsis* protein S-nitrosylation in response to inoculation with *Pseudomonas syringae*. *Acta Physiologiae Plantarum, 33*(4), 1493–1514. https://doi.org/10.1007/s11738-010-0688-2.

Medina-Andrés, R., Solano-Peralta, A., Saucedo-Vázquez, J. P., Napsucialy-Mendivil, S., Pimentel-Cabrera, J. A., Sosa-Torres, M. E., Dubrovsky, J. G., Lira-Ruan, V., & Scavone, C. (2015). The nitric oxide production in the moss physcomitrella patens is mediated by nitrate reductase. *PLoS One, 10*(3), e0119400. https://doi.org/10.1371/journal.pone.0119400.

Merced, A., & Renzaglia, K. S. (2017). "Structure, function and evolution of stomata from a bryological perspective. *Bryophyte Diversity and Evolution, 39*(1), 7. https://doi.org/10.11646/bde.39.1.4 Available at.

Mittler, R. (2017). ROS Are Good. *Trends in Plant Science, 22*(1), 11–19. https://doi.org/10.1016/j.tplants.2016.08.002.

Moreau, M., Lindermayr, C., Durner, J., & Klessig, D. F. (2010). NO synthesis and signaling in plants – where do we stand? *Physiologia Plantarum, 138*(4), 372–383. https://doi.org/10.1111/j.1399-3054.2009.01308.x.

Mou, Z., Fan, W., & Dong, X. (2003). Inducers of plant systemic acquired resistance Regulate NPR1 function through redox changes. *Cell, 113*(7), 935–944. https://doi.org/10.1016/S0092-8674(03)00429-X, https://www.sciencedirect.com/journal/cell.

Munir, A., Min, C. W., Wang, Y., Kim, S. T., & Gupta, R. (2025). S-nitrosoproteome analysis of rice leaves highlights the possible roles of superoxide dismutase in resistance against *Magnaporthe Oryzae*. *Physiologia Plantarum, 177*(5), e70515.

Naramoto, S., Hata, Y., Fujita, T., & Kyozuka, J. (2022). The bryophytes *Physcomitrium patens* and *Marchantia polymorpha* as model systems for studying evolutionary cell and developmental biology in plants. *The Plant Cell, 34*(1), 228–246. https://doi.org/10.1093/plcell/koab218.

Neill, S. J., Desikan, R., Clarke, A., & Hancock, J. T. (2002). Nitric oxide is a novel component of abscisic acid signaling in stomatal guard cells. *Plant Physiology, 128*(1), 13–16. https://doi.org/10.1104/pp.010707, http://www.plantphysiol.org/.

Nowaczyński, F., Nicoletti, R., Zimowska, B., & Ludwiczuk, A. (2025). Marchantia polymorpha as a source of biologically active compounds. *Molecules (Basel, Switzerland), 30*(3), 558. https://doi.org/10.3390/molecules30030558.

Ortega-Galisteo, A. P., Rodríguez-Serrano, M., Pazmiño, D. M., Gupta, D. K., Sandalio, L. M., & Romero-Puertas, M. C. (2012). S-Nitrosylated proteins in pea (*Pisum sativum* L.) leaf peroxisomes: Changes under abiotic stress. *Journal of Experimental Botany, 63*(5), 2089–2103. https://doi.org/10.1093/jxb/err414.

Otero-Blanca, A., Pérez-Llano, Y., Reboledo-Blanco, G., Lira-Ruan, V., Padilla-Chacon, D., Folch-Mallol, J. L., Sánchez-Carbente, M.del R., Ponce De León, I., & Batista-García, R. A. (2021). Physcomitrium patens infection by colletotrichum gloeosporioides: Understanding the fungal–bryophyte interaction by microscopy, phenomics and RNA sequencing. *Journal of Fungi, 7*(8), 677. https://doi.org/10.3390/jof7080677.

Pagnussat, G. C., Lanteri, M. L., Lombardo, M. C., & Lamattina, L. (2004). Nitric oxide mediates the indole acetic acid induction activation of a mitogen-activated protein kinase cascade involved in adventitious root development. *Plant Physiology, 135*(1), 279–286. https://doi.org/10.1104/pp.103.038554, http://www.plantphysiol.org/.

Pan, C. L., Yao, S. C., Xiong, W. J., Luo, S. Z., Wang, Y. L., Wang, A. Q., Xiao, D., Zhan, J., & He, L. F. (2017). Nitric oxide inhibits al-induced programmed cell death in root tips of peanut (*Arachis hypogaea* L.) by affecting physiological properties of antioxidants systems and cell wall. *Frontiers in Physiology, 8*. https://doi.org/10.3389/fphys.2017.01037, https://www.frontiersin.org/articles/10.3389/fphys.2017.01037/full.

Pande, A., Mun, B. G., Rahim, W., Khan, M., Lee, D. S., Lee, G. M., Al Azzawi, T. N. I., Hussain, A., Kim, C. K., & Yun, B. W. (2022). Phytohormonal regulation through protein S-nitrosylation under stress. *Frontiers in Plant Science, 13*. https://doi.org/10.3389/fpls.2022.865542.

Pathak, P. K., Yadav, N., Kaladhar, V. C., Jaiswal, R., Kumari, A., Igamberdiev, A. U., Loake, G. J., & Gupta, K. J. (2024). The emerging roles of nitric oxide and its associated scavengers-phytoglobins-in plant symbiotic interactions. *Journal of Experimental Botany, 75*(2), 563–577. https://doi.org/10.1093/jxb/erad399, http://jxb.oxfordjournals.org/.

Planchet, E., & Kaiser, W. M. (2006). "Nitric oxide production in plants. *Plant Signaling & Behavior, 1*(2), 46–51. https://doi.org/10.4161/psb.1.2.2435.

Ponce de León, I. (2011). The moss *Physcomitrella patens* as a model system to study interactions between plants and phytopathogenic fungi and oomycetes. *Journal of Pathogens, 2011*, 1–6. https://doi.org/10.4061/2011/719873.

Ponce de León, I., & Montesano, M. (2013). Activation of defense mechanisms against pathogens in mosses and flowering plants. *International Journal of Molecular Sciences, 14*(2), 3178–3200. https://doi.org/10.3390/ijms14023178.

Poveda, J. (2020). Marchantia polymorpha as a model plant in the evolutionary study of plant-microorganism interactions. *Current Plant Biology, 23*, 100152. https://doi.org/10.1016/j.cpb.2020.100152.

Qi, Z., Verma, R., Gehring, C., Yamaguchi, Y., Zhao, Y., Ryan, C. A., & Berkowitz, G. A. (2010). Ca^{2+} signaling by plant *Arabidopsis thaliana* Pep peptides depends on AtPepR1, a receptor with guanylyl cyclase activity, and cGMP-activated Ca^{2+} channels. *Proceedings of the National Academy of Sciences, 107*(49), 21193–21198. https://doi.org/10.1073/pnas.1000191107.

Reboledo, G., Agorio, A., Vignale, L., Alvarez, A., & Ponce De León, I. (2022). The moss-specific transcription factor PpERF24 positively modulates immunity against fungal pathogens in *Physcomitrium patens*. *Frontiers in Plant Science, 13*. https://doi.org/10.3389/fpls.2022.908682.

Rui, L., et al. (2024). "The chloroplast-localized casein kinase ESCAPED_CHARACTER_LESS_THANscp>IIESCAPED_CHARACTER_LESS_THAN/scp> α subunit, CPCK2, negatively regulates plant innate immunity through promoting S-nitrosylation of SABP3," *The Plant Journal, 120*(2), 552–568. https://doi.org/10.1111/tpj.17000 Available at.

Ruszala, E. M., Beerling, D. J., Franks, P. J., Chater, C., Casson, S. A., Gray, J. E., & Hetherington, A. M. (2011). Land plants acquired active stomatal control early in their evolutionary history. *Current Biology, 21*(12), 1030–1035. https://doi.org/10.1016/j.cub.2011.04.044.

Salgado, I., Carmen Martínez, M., Oliveira, H. C., & Frungillo, L. (2013). Nitric oxide signaling and homeostasis in plants: A focus on nitrate reductase and S-nitrosoglutathione reductase in stress-related responses. *Revista Brasileira de Botanica, 36*(2), 89–98. https://doi.org/10.1007/s40415-013-0013-6.

Salmi, M. L., Morris, K. E., Roux, S. J., & Porterfield, D. M. (2007). Nitric oxide and cGMP signaling in calcium-dependent development of cell polarity in Ceratopteris richardii. *Plant Physiology, 144*(1), 94–104. https://doi.org/10.1104/pp.107.096131, http://www.plantphysiol.org/cgi/reprint/144/1/94.pdf.

Samant, S. B., Swain, J., Yadav, N., Yadav, R., Singh, P., Rai, P., Sheri, V., Sreeman, S., Subramanyam, R., Pareek, A., & Gupta, K. J. (2025). Overexpression of phytoglobin1 in rice leads to enhanced nitrogen use efficiency via modulation of nitric oxide. *Plant, Cell & Environment, 48*(4), 2755–2768. https://doi.org/10.1111/pce.15289.

Schlicht, M. (2013). The role of nitric oxide in the interaction of *Arabidopsis thaliana* with the biotrophic fungi, *Golovinomyces orontii* and *Erysiphe pisi*. *Frontiers in Plant Science, 4*. https://doi.org/10.3389/fpls.2013.00351.

Seregélyes, C., & Dudits, D. (2003). Phytoglobins and nitric oxide: New partners in an old signalling system in plants. *Acta Biologica Hungarica, 54*(1), 15–25. https://doi.org/10.1556/abiol.54.2003.1.2.

Singh, V. P., Jaiswal, S., Wang, Y., Feng, S., Tripathi, D. K., Singh, S., Gupta, R., Xue, D., Xu, S., & Chen, Z. H. (2024). Evolution of reactive oxygen species cellular targets for plant development. *Trends in Plant Science, 29*(8), 865–877. https://doi.org/10.1016/j.tplants.2024.03.005, https://www.sciencedirect.com/science/journal/13601385.

Skelly, M. J., Malik, S. I., Le Bihan, T., Bo, Y., Jiang, J., Spoel, S. H., & Loake, G. J. (2019). A role for S-nitrosylation of the SUMO-conjugating enzyme SCE1 in plant immunity. *Proceedings of the National Academy of Sciences, 116*(34), 17090–17095. https://doi.org/10.1073/pnas.1900052116.

Slaymaker, D. H., Navarre, D. A., Clark, D., del Pozo, O., Martin, G. B., & Klessig, D. F. (2002). The tobacco salicylic acid-binding protein 3 (SABP3) is the chloroplast carbonic anhydrase, which exhibits antioxidant activity and plays a role in the hypersensitive defense response. *Proceedings of the National Academy of Sciences, 99*(18), 11640–11645. https://doi.org/10.1073/pnas.182427699.

Spoel, S. H., & Dong, X. (2024). Salicylic acid in plant immunity and beyond. *The Plant Cell, 36*(5), 1451–1464. https://doi.org/10.1093/plcell/koad329.

Stomberski, C. T., Hess, D. T., & Stamler, J. S. (2019). Protein S-nitrosylation: Determinants of specificity and enzymatic regulation of s-nitrosothiol-based signaling. *Antioxidants and Redox Signaling, 30*(10), 1331–1351. https://doi.org/10.1089/ars.2017.7403, www.liebertonline.com/ars.

Sun, J., Druhan, L. J., & Zweier, J. L. (2010). Reactive oxygen and nitrogen species regulate inducible nitric oxide synthase function shifting the balance of nitric oxide and superoxide production. *Archives of Biochemistry and Biophysics, 494*(2), 130–137. https://doi.org/10.1016/j.abb.2009.11.019.

Szalai, G., Pál, M., &Janda, T. (2011). Abscisic acid may alter the salicylic acid-related abiotic stress response in maize Materials and Methods Results Chlorophyll content and electrolyte leakage. http://www.sci.u-szeged.hu/ABSARTICLE.

Tabassum, N., Goodrich, J., & Loake, G. J. (2026). A key role for s-nitrosylation in immune regulation and development in the liverwort marchantia polymorpha. *Journal of Experimental Botany, 77*(14), 4561–4578. https://doi.org/10.1093/jxb/erag171.

Tada, Y., Spoel, S. H., Pajerowska-Mukhtar, K., Mou, Z., Song, J., Wang, C., Zuo, J., & Dong, X. (2008). Plant immunity requires conformational charges of NPR1 via S-nitrosylation and thioredoxins. *Science (New York, N.Y.), 321*(5891), 952–956. https://doi.org/10.1126/science.1156970.

Torres, M. A., Dangl, J. L., & Jones, J. D. G. (2002). Arabidopsis gp91 phox homologues AtrbohD and AtrbohF are required for accumulation of reactive oxygen intermediates in the plant defense response. *Proceedings of the National Academy of Sciences, 99*(1), 517–522. https://doi.org/10.1073/pnas.012452499.

Treffon, P., & Vierling, E. (2022). Focus on nitric oxide homeostasis: Direct and indirect enzymatic regulation of protein denitrosation reactions in plants. *Antioxidants, 11*(7). https://doi.org/10.3390/antiox11071411, www.mdpi.com/journal/antioxidants.

Treffon, P., Rossi, J., Gabellini, G., Trost, P., Zaffagnini, M., & Vierling, E. (2021). Quantitative proteome profiling of a s-nitrosoglutathione reductase (GSNOR) null mutant reveals a new class of enzymes involved in nitric oxide homeostasis in plants. *Frontiers in Plant Science, 12*. https://doi.org/10.3389/fpls.2021.787435.

Umbreen, S., Lubega, J., Cui, B., Pan, Q., Jiang, J., & Loake, G. J. (2018). Specificity in nitric oxide signalling. *Journal of Experimental Botany, 69*(14), 3439–3448. https://doi.org/10.1093/jxb/ery184.

Van Meeteren, U., Kaiser, E., Matamoros, P. M., Verdonk, J. C., Aliniaeifard, S., & Lawson, T. (2020). Is nitric oxide a critical key factor in ABA-induced stomatal closure? *Journal of Experimental Botany, 71*(1), 399–410. https://doi.org/10.1093/jxb/erz437.

Wang, P., Du, Y., Hou, Y.-J., Zhao, Y., Hsu, C.-C., Yuan, F., Zhu, X., Tao, W. A., Song, C.-P., & Zhu, J.-K. (2015). Nitric oxide negatively regulates abscisic acid signaling in guard cells by S-nitrosylation of OST1. *Proceedings of the National Academy of Sciences, 112*(2), 613–618. https://doi.org/10.1073/pnas.1423481112.

Wang, Y. (2006). S-Nitrosylation: An emerging redox-based post-translational modification in plants. *Journal of Experimental Botany, 57*(8), 1777–1784. https://doi.org/10.1093/jxb/erj211.

Wang, Y., Loake, G. J., & Chu, C. (2013). Cross-talk of nitric oxide and reactive oxygen species in plant programed cell death. *Frontiers in Plant Science, 4*. https://doi.org/10.3389/fpls.2013.00314, http://journal.frontiersin.org/Journal/10.3389/fpls.2013.00314/pdf.

Wang, Y., Li, X., Fan, B., Zhu, C., & Chen, Z. (2021). Regulation and function of defense-related callose deposition in plants. *International Journal of Molecular Sciences, 22*(5), 2393. https://doi.org/10.3390/ijms22052393.

Wang, Y. Q., Feechan, A., Yun, B. W., Shafiei, R., Hofmann, A., Taylor, P., Xue, P., Yang, F. Q., Xie, Z. S., Pallas, J. A., Chu, C. C., & Loake, G. J. (2009). S-nitrosylation of AtSABP3 antagonizes the expression of plant immunity. *Journal of Biological Chemistry, 284*(4), 2131–2137. https://doi.org/10.1074/jbc.M806782200, http://www.jbc.org/cgi/reprint/284/4/2131.

Wendehenne, D., Durner, J., & Klessig, D. F. (2004). Nitric oxide: A new player in plant signalling and defence responses. *Current Opinion in Plant Biology, 7*(4), 449–455. https://doi.org/10.1016/j.pbi.2004.04.002.

Wu, B., Qi, F., & Liang, Y. (2023). Fuels for ROS signaling in plant immunity. *Trends in Plant Science, 28*(10), 1124–1131. https://doi.org/10.1016/j.tplants.2023.04.007.

Wu, L., Chen, H., Curtis, C., & Fu, Z. Q. (2014). Go in for the kill: how plants deploy effector-triggered immunity to combat pathogens. *Virulence, 5*(7), 710–721. https://doi.org/10.4161/viru.29755.

Xiong, J., An, L., Lu, H., & Zhu, C. (2009). Exogenous nitric oxide enhances cadmium tolerance of rice by increasing pectin and hemicellulose contents in root cell wall. *Planta, 230*(4), 755–765. https://doi.org/10.1007/s00425-009-0984-5.

Yotsui, I., Matsui, H., Miyauchi, S., Iwakawa, H., Melkonian, K., Schlüter, T., Michavila, S., Kanazawa, T., Nomura, Y., Stolze, S. C., Jeon, H.-W., Yan, Y., Harzen, A., Sugano, S. S., Shirakawa, M., Nishihama, R., Ichihashi, Y., Ibanez, S. G., Shirasu, K., ... Nakagami, H. (2023). LysM-mediated signaling in *Marchantia polymorpha* highlights the conservation of pattern-triggered immunity in land plants. *Current Biology, 33*(17), 3732. https://doi.org/10.1016/j.cub.2023.07.068.

Yu, M., Yun, B.-W., Spoel, S. H., & Loake, G. J. (2012). A sleigh ride through the SNO: regulation of plant immune function by protein S-nitrosylation. *Current Opinion in Plant Biology, 15*(4), 424–430. https://doi.org/10.1016/j.pbi.2012.03.005.

Yu, X.-Q., Niu, H.-Q., Liu, C., Wang, H.-L., Yin, W., & Xia, X. (2024). PTI-ETI synergistic signal mechanisms in plant immunity. *Plant Biotechnology Journal, 22*(8), 2113–2128. https://doi.org/10.1111/pbi.14332.

Yuan, M., Ngou, B. P. M., Ding, P., & Xin, X. F. (2021). PTI-ETI crosstalk: An integrative view of plant immunity. *Current Opinion in Plant Biology, 62*. https://doi.org/10.1016/j.pbi.2021.102030, www.elsevier.com/inca/publications/store/6/0/1/3/1/4/index.htt.

Yun, B. W., Feechan, A., Yin, M., Saidi, N. B. B., Le Bihan, T., Yu, M., Moore, J. W., Kang, J. G., Kwon, E., Spoel, S. H., Pallas, J. A., & Loake, G. J. (2011). S-nitrosylation of NADPH oxidase regulates cell death in plant immunity. *Nature, 478*(7368), 264–268. https://doi.org/10.1038/nature10427.

Zafari, S., Vanlerberghe, G. C., & Igamberdiev, A. U. (2022). Nitric oxide turnover under hypoxia results in the rapid increased expression of the plastid-localized phosphorylated pathway of serine biosynthesis. *Frontiers in Plant Science, 12*. https://doi.org/10.3389/fpls.2021.780842.

Zaninotto, F., Camera, S. L., Polverari, A., & Delledonne, M. (2006). Cross talk between reactive nitrogen and oxygen species during the hypersensitive disease resistance response. *Plant Physiology, 141*(2), 379–383. https://doi.org/10.1104/pp.106.078857.

Zeidler, D., Zähringer, U., Gerber, I., Dubery, I., Hartung, T., Bors, W., Hutzler, P., & Durner, J. (2004). Innate immunity in *Arabidopsis thaliana* lipopolysaccharides activate nitric oxide synthase (NOS) and induce defense genes. *Proceedings of the National Academy of Sciences, 101*(44), 15811–15816. https://doi.org/10.1073/pnas.0404536101.

Zhan, N., Wang, C., Chen, L., Yang, H., Feng, J., Gong, X., Ren, B., Wu, R., Mu, J., Li, Y., Liu, Z., Zhou, Y., Peng, J., Wang, K., Huang, X., Xiao, S., & Zuo, J. (2018). S-Nitrosylation targets gsno reductase for selective autophagy during hypoxia responses in plants. *Molecular Cell, 71*(1), 142. https://doi.org/10.1016/j.molcel.2018.05.024.

Zhang, J., & Liao, W. (2021). Roles of S-nitrosylation in abiotic stress tolerance in plants. In *Nitric oxide in plant biology: an ancient molecule with emerging roles*. China: Elsevier, 453–475. http://doi.org/10.1016/B978-0-12-818797-5.00015-7, https://www.sciencedirect.com/book/9780128187975.

Zhang, J., Buegger, F., Albert, A., Ghirardo, A., Winkler, B., Schnitzler, J.-P., Henrik Hebelstrup, K., & Durner, J. (2019). Christian lindermayr, christine foyer, phytoglobin overexpression promotes barley growth in the presence of enhanced level of atmospheric nitric oxide. *Journal of Experimental Botany, 70*(17), 4521–4537. https://doi.org/10.1093/jxb/erz249.

Zhang, P., Jackson, E., Li, X., & Zhang, Y. (2025). Salicylic acid and jasmonic acid in plant immunity. *Horticulture Research, 12*(7). https://doi.org/10.1093/hr/uhaf082.

Zhang, S., & Klessig, D. F. (2001). MAPK cascades in plant defense signaling. *Trends in Plant Science, 6*(11), 520–527. https://doi.org/10.1016/S1360-1385(01)02103-3.

Zhang, X., Zhang, L., Dong, F., Gao, J., Galbraith, D. W., & Song, C.-P. (2001). Hydrogen Peroxide Is Involved in Abscisic Acid-Induced Stomatal Closure in *Vicia faba*. *Plant Physiology, 126*(4), 1438–1448. https://doi.org/10.1104/pp.126.4.1438.

Zhang, Y., & Li, X. (2019). Salicylic acid: Biosynthesis, perception, and contributions to plant immunity. *Current Opinion in Plant Biology, 50*, 29–36. https://doi.org/10.1016/j.pbi.2019.02.004, www.elsevier.com/inca/publications/store/6/0/1/3/1/4/index.htt.

Zhao, H., et al. (2024). Correction to: S-nitrosylation of the transcription factor MYB30 facilitates nitric oxide–promoted seed germination in Arabidopsis. *The Plant Cell, 36*(3), 794. https://doi.org/10.1093/plcell/koad298.

Chapter 5

Methods for profiling nitric oxide-induced S-nitrosation in plants

Adeela Munir[1], Cheol Woo Min[1], Sun Tae Kim[1], and Ravi Gupta[2]
[1]*Department of Plant Bioscience, Pusan National University, Life and Industry Convergence Research Institute, Miryang, Republic of Korea,* [2]*Plant Stress Physiology and Proteomics Laboratory, College of General Education, Kookmin University, Seoul, South Korea*

5.1 Introduction

Nitric oxide (NO) is a gaseous molecule available in all living organisms, from prokaryotes to eukaryotes, ubiquitously. It performs a wide range of biological functions, such as growth and development, and regulation of stress responses (Hancock, 2020). Extensive research on NO as a gaseous signaling molecule has established its critical role in mammals, vertebrates, invertebrates, plants, and eukaryotic organisms like protozoan, yeast, and bacteria (Torreilles, 2001). The physiological significance of NO was first reported in animals in the late 20th century. In 1978, Gruetter and his colleagues demonstrated that NO acts as a signaling molecule in animal tissues by activating the enzyme soluble guanylate cyclase. Notably, this activation was shown to require the presence of heme, indicating a highly specific mode of biochemical interaction. This discovery provided a basis for the role of NO in vascular relaxation and cellular signaling, thus opening a new field in redox biology (Gruetter et al., 1979). In parallel, the study of NO in plants began to gain momentum when Klepper et al. reported the emission of NO and NO_2 gases in the surrounding environment from herbicide-treated soybean plants. This work marked the first documentation of NO as a naturally occurring molecule in plant systems, suggesting a potential role in plant-environment interactions (Klepper, 1979).

In recent years, NO has emerged as a fundamental signaling molecule in living organisms, with a role in various physiological responses and developmental stages (Khan et al., 2023). These include the breaking of seed dormancy, initiation of germination, root system architecture development, such as lateral and adventitious root formation, primary root elongation, and

Nitric Oxide in Plant Resilience to Abiotic Stress. DOI: https://doi.org/10.1016/B978-0-443-40497-9.00003-9

root hair differentiation, alongside chlorophyll biosynthesis, vegetative growth, vascular tissue differentiation, symbiotic nodule formation, and stomatal movement. NO is also involved in iron homeostasis, regulation of leaf senescence, flowering initiation, and fruit ripening (Simontacchi et al., 2015). In addition to its developmental functions, NO serves as a critical signaling molecule in plant defense, protecting against a wide range of environmental stresses. These include mechanical wounding, pathogen attack, hypoxic conditions, ultraviolet (UV) radiation, drought, salinity, temperature extremes, and heavy metal toxicity (Khan et al., 2023; Rahim et al., 2022). In particular, the functional role of NO in plant immunity was first established in 1998 when the Lamb group presented the evidence that NO plays a critical role in the induction of hypersensitive response (HR) in *Arabidopsis thaliana* leaves in response to the pathogenic bacterium *Pseudomonas syringae*. They found that NO, along with reactive oxygen species (ROS), mediates signaling processes essential for localized cell death and resistance, thereby highlighting NO as a key player in plant defense mechanisms (Delledonne et al., 1998). However, the detection of S-nitrosated proteins in plants is relatively challenging due to their low stoichiometry and light sensitivity; thus information on such proteins is highly limited, especially in plants. In this chapter, we discuss various methods developed so far for the enrichment of S-nitrosated proteins/peptides, along with some of the key S-nitrosated proteins to present the biological relevance of this modification in plants.

5.2 Nitric oxide biosynthesis and its induced posttranslational modifications

Different organisms exhibit distinct NO synthesis pathways. From the early 1990's, it is known that the source of NO production in animals is the oxidation of arginine, which is catalyzed by NO synthase (NOS) (Marietta, 1993; Moncada, 1993; Palmer, 1993). However, NO biosynthesis in plants is considerably more complex, involving both oxidative and reductive pathways (Gupta et al., 2022). Oxidation pathways utilize the NOS-like enzyme activity to catalyze the NO production from L-arginine, whereas reductive pathways majorly involve the conversion of nitrites to NO with the help of nitrate reductase (NR) (Astier et al., 2018) (Fig. 5.1).

NO-driven posttranslational modifications (PTMs) of proteins are key components of NO signaling pathways. All major NO-based PTMs, including S-nitrosation (-SNO), tyrosine nitration, and metal nitrosylation, are driven by reactive nitrogen species (Fig. 5.1). For instance, S-nitrosation is mediated by nitrosating agents, including N_2O_3 or NO^+, tyrosine nitration occurs via peroxynitrite ($ONOO^-$) or nitrogen dioxide radical (NO_2), and metal nitrosylation involves direct coordination of NO or its derivatives with transition metal centers in metalloproteins (Corpas et al., 2008). These PTMs can be reversible or irreversible, depending on their type. Reversible

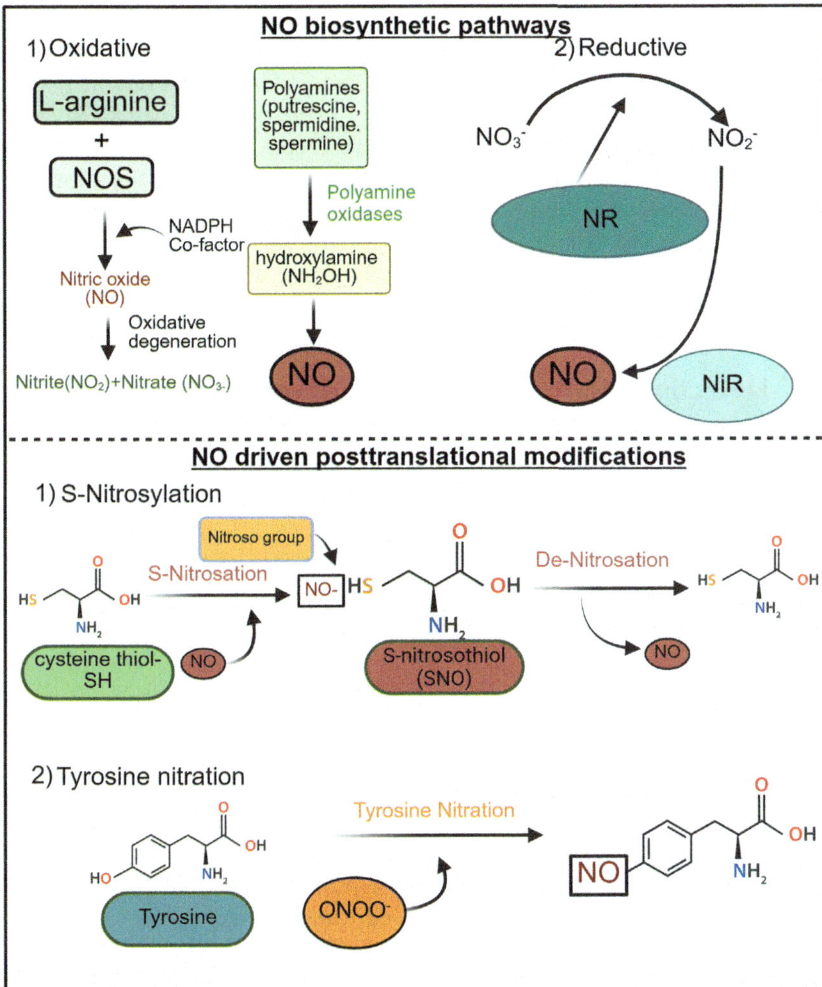

FIGURE 5.1 Nitric oxide (*NO*) homeostasis and major NO-driven posttranslational modifications (*PTMs*). The upper panel shows the dual pathways of NO biosynthesis in plants: (1) Oxidative pathways, involving L-arginine-dependent NOS-like activity and polyamine-mediated hydroxylamine oxidation; and (2) Reductive pathways, primarily driven by nitrate reductase (*NR*) and nitrite reductase (*NiR*) activity. The lower panel illustrates the primary NO-dependent PTMs: (1) S-Nitrosylation/nitrosation, the reversible covalent attachment of a nitroso group to a cysteine thiol-SH to form an S-nitrosothiol (*SNO*); and (2) Tyrosine nitration, the irreversible modification of tyrosine residues by peroxynitrite ($ONOO^-$) to form 3-nitrotyrosine.

modification implies that the chemical change done due to PTM can be undone, allowing the protein to return to its original state, whereas irreversible PTM means that the modification cannot be undone and leads to a permanent alteration in the protein's chemistry. Among the three primary NO-based PTMs, S-nitrosation is a reversible PTM that takes place after the attachment of the nitrosothiol group (-SNO) on the cysteine residue of the protein of interest. In contrast, the tyrosine nitration, which involves covalent attachment of NO to the tyrosine residues, was initially considered irreversible; however, recent studies have highlighted the ability of animal cells to denitrate tyrosine-nitrated proteins. Both S-Nitrosation and tyrosine nitration are recognized as the most critical modifications involved in mediating biological responses, particularly under stress conditions (Corpas et al., 2021; Liu et al., 2021) (Fig. 5.1).

5.3 Detection methods of nitric oxide-based S-nitrosation

Essentially, posttranslational S-nitrosation is the process by which a NO molecule covalently binds to the cysteine thiol side chain to form an SNO (Qiu et al., 2019). Because SNO bonds are sensitive to light/reducing agents and are low-abundance, detecting protein S-nitrosation is technically difficult (Lin et al., 2025). In order to identify S-nitrosated proteins, indirect "switch" techniques have been developed, which first transform SNO groups into more stable labels. The Biotin Switch Technique (BST) is the most traditional and popular approach (Feng et al., 2013; Jaffrey & Snyder, 2001). After blocking free thiols, S-nitrosated cysteines are selectively reduced to thiols, for example, by ascorbate, and labeled with a biotin label for detection (Huang & Chen, 2010). For more than ten years, BST remained the main and only assay to detect S-nitrosated proteins in plants. Lately, newer mass spectrometry (MS)-based techniques evolved as a result of its known limitations in sensitivity and specificity. For instance, an Iodoacetyl Tandem Mass Tag (iodoTMT) labeling method was introduced, which became a potent method for quantitatively capturing S-nitrosated sites with increased throughput over the last 5 years (Lin et al., 2025; Wei et al., 2023; Yu et al., 2022; Chen et al., 2023). In this chapter, we discuss and compare these two techniques, highlighting their principles, merits, and demerits.

5.3.1 Biotin switch technique

5.3.1.1 Principle

The biotin switch method, initially presented by Jaffrey and Snyder (2001), involves switching NO groups to biotin-labeled thiols to detect S-nitrosated proteins indirectly. There are four primary components to the protocol, of which the first step is blocking, which involves permanently blocking all free cysteine thiols in the protein sample by utilizing N-ethylmaleimide, iodoacetamide, or methyl methanethiosulfonate (MMTS) so that they cannot react in later stages. The second step includes the selective reduction of S-nitrosothiols to free thiols by ascorbate, usually sodium ascorbate, so that they can be labeled or tagged at the next step. To prevent nonspecific thiol generation, this step is carried out under controlled conditions and in the dark because light can break SNO bonds. Notably, the effectiveness of ascorbate is critical and can be a limiting factor in this step. The third step is biotin labeling, which involves reacting these newly formed free thiols that were previously S-nitrosated cysteines with a thiol-specific biotinylating reagent, typically biotin-HPDP (N-[6-(biotinamido)hexyl]-3′-(2′-pyridyldithio)propionamide). These tags add a biotin tag to the free thiols through a disulfide bond. To derivatize the released thiols right away and stop them from reoxidizing. Notably, steps two and three are frequently carried out concurrently. Following this switch, S-nitrosated proteins that are now biotin-labeled can be purified using biotin-affinity methods such as streptavidin beads (Foster & Stamler, 2004; Kovacs & Lindermayr, 2013; Lin et al., 2025; Qin et al., 2013) (Fig. 5.2).

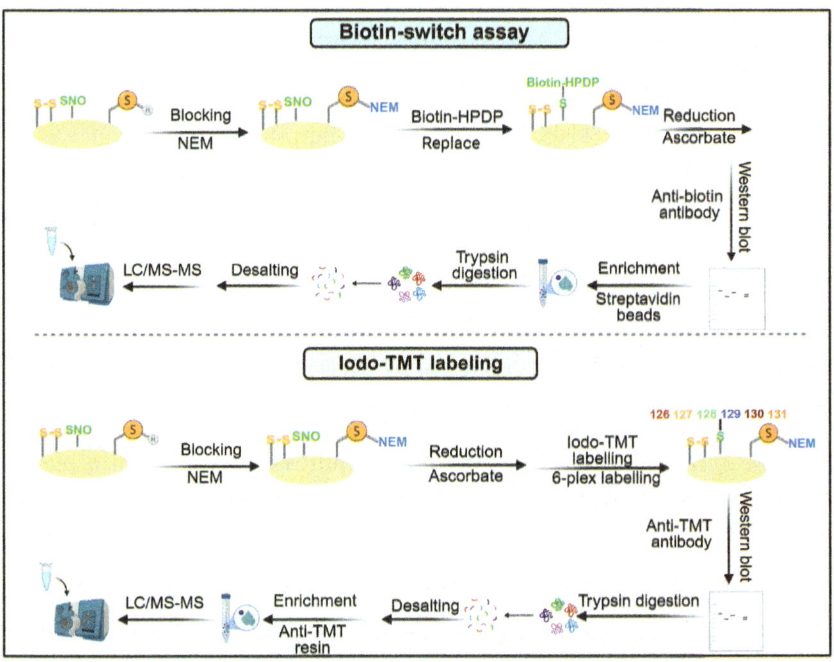

FIGURE 5.2 Methodological workflows for the proteomic profiling of S-nitrosated proteins. This schematic illustrates the two primary strategies for identifying the plant S-nitrosoproteome: the Biotin-switch assay, which replaces labile S-nitroso groups with Biotin-HPDP for streptavidin-based enrichment and LC-MS/MS identification, and Iodo-TMT labeling, a multiplexed approach that utilizes isobaric tags for the simultaneous quantification and site-specific identification of S-nitrosation across multiple experimental conditions via anti-TMT resin enrichment (*HPDP*, N-[6-(biotinamido)hexyl]-3′-(2′-pyridyldithio)propionamide).

5.3.1.2 Detection and use in plants

For the detection of S-nitrosated (-SNO) proteins, both qualitative and quantitative assays can be performed. While the quantitative assays involve detection and quantification of -SNO site intensities by MS, qualitative assays require the probing of biotin-labeled streptavidin-purified proteins on immunoblots. However, cleaving the biotin-HPDP disulfide with the labeled proteins is a prerequisite prior to Liquid Chromatography-Tandem Mass Spectrometry (LC-MS/MS) analysis of the S-nitrosated proteins and is usually carried out with the help of reducing agents such as DTT. In plant systems, BST has been extensively utilized to map the S-nitroproteome during normal as well as challenging environmental conditions. For instance, early BST-based studies reported S-nitrosation of approximately 80 proteins in potato leaves and 13 peroxisomal proteins in pea under different experimental treatments (Ortega-Galisteo et al., 2012; Kato et al., 2013). Under cadmium and 2,4-D treatments, the S-nitrosation levels of catalase and glycolate oxidase were decreased, indicating that S-nitrosation of these proteins may control the amount of H_2O_2 produced under abiotic stress (Ortega-Galisteo et al., 2012). Using a modified BST method coupled with MS, over 80 S-nitrosated proteins were identified in potato leaves and tubers (Kato et al., 2013). Moreover, a modified BST-based proteomic workflow was used in Arabidopsis by Hu et al. (2015), leading to the identification of 1195 endogenously S-nitrosated peptides that corresponded to 926 proteins in wild-type vs. *gsnor1* mutant seedlings. To summarize, BST has been a reliable technique for cataloging S-nitrosated proteins in plants.

5.3.1.3 Merits of biotin switch technique

Like all biochemical detection methods, the biotin switch technique presents distinct advantages that have supported its widespread adoption, particularly in plant nitrosoproteomics.

(a) Widely used and accessible: BST is a tried-and-true method that was once the "primary method" for detecting S-nitrosated proteins in plants. Its protocol is widely accessible as it makes use of standard biochemical reagents and equipment.

5.3.1.4 Limitations of biotin switch technique

Despite its utility, the biotin-switch assay has significant drawbacks that can affect data quality.

1. **False positives/negatives:** BST is a multistep, indirect method that is prone to errors. False positive results may result from biotin labeling of non-SNO cysteines due to incomplete blocking of free thiols in step 1 or adventitious thiol generation. Due to MMTS instability or inadequate blocking reagent concentration, even very low levels of unblocked thiols

may be mistakenly identified as S-nitrosated sites. On the other hand, since ascorbate reduction efficiency varies and certain SNO sites may be resistant or not accessible, ascorbate treatment is not able to reduce S-nitrosothiols in certain cases, leading to false negatives result (Liang et al., 2024). This unpredictability makes biological interpretation more difficult, and significant optimization (changing blocking and reduction conditions) is often required.

2. **Sensitivity and coverage:** For low-abundance S-nitrosation, BST typically has limited sensitivity. For example, according to a summary, BST-based techniques usually only find dozens to hundreds of SNO proteins or peptides in plant tissues, which is quite low. The need to enrich and recover biotinylated proteins contributes to this comparatively low yield; the multistep procedure may result in incomplete recovery and sample loss. Furthermore, SNO modifications are low-stoichiometry and transient; thus many modified sites remain undetected in the absence of highly sensitive downstream detection. This restriction prompted researchers to look for more effective tagging agents in order to increase the depth of the SNO proteome. Chen et al. (2010) made two improvements to the BST method to increase the recognition and identification efficiency of SNO-modified proteins. In order to prevent the Cys residue-blocking failure brought on by the high-temperature instability of MMTS reagents, MMTS was first substituted with iodoacetamide. S-alkylation reagents were utilized to specifically and irreversibly bind SNO bonds, create S-alkylated biotin proteins, and improve the recognition efficiency of SNO-modified proteins because the reduction efficiency of ascorbic acid is unstable due to oxidation in light and air.
3. **Interference of biotin tags in MS:** Although biotin tags are cleaved using DTT prior to MS analysis, insufficient cleavage can lead to the residual biotin-HPDP or associated biotin tags, which hinder subsequent MS analysis. Because of this, some biotinylated peptides may not ionize or fragment well, resulting in lower than desired peptide coverage in BST-MS investigations. Actually, sample loss and insufficient peptide recovery may result from the necessity to elute biotinylated proteins from streptavidin resin under frequently harsh conditions (Chen et al., 2010). The comparatively small number of SNO sites found by BST proteomics is a result of these factors.
4. **Laborious protocol:** BST is labor-intensive and prone to sample loss or variability because it requires several incubation and precipitation steps. To preserve SNO signals, it can be regarded as a "cumbersome" process that needs careful control. Despite its popularity, this method is relatively low-throughput, and it is difficult to locate the modification sites. Additionally, the biotin affinity enrichment may introduce false positive signals due to the existence of endogenous biotin-like molecules, particularly in plant tissues. In conclusion, although BST has been a vital tool for identifying protein S-nitrosylation in plants, its indirect nature and technical difficulties

frequently result in lower-coverage, semiquantitative data that may contain artifactual signals. These drawbacks have spurred the creation of better techniques.

5.3.2 Iodoacetyl tandem mass tag-based detection

5.3.2.1 *Principle*

S-nitrosated cysteines are labeled using iodoacetyl-functionalized Tandem Mass Tags in the iodoTMT assay, a contemporary quantitative variant of the switch technique. Each of the isobaric mass tags (often referred to as a sixplex) in Thermo Fisher's iodoTMT reagents has an iodoacetyl group that reacts selectively and irreversibly with sulfhydryl (-SH) groups. The first steps of the process are similar to those of BST SNO groups are blocked to protect unmodified cysteines and then reduced to thiols, often with ascorbate. At this stage, the sample is treated with the iodoTMT reagents rather than their biotinylation. The TMT label is applied to previously S-nitrosated cysteines by the iodoacetyl group, which alkylates the newly formed thiols. Sample multiplexing is made possible by the fact that each tag in the set has the same mass but carries a distinct isotope distribution. Samples can be combined (if multiplexed) and digested to peptides following tagging and desalting. An anti-TMT resin that selectively binds to TMT-labeled peptides is then used to enrich the tagged peptides. Using the reporter ions released from the tags to measure relative amounts from each sample condition, LC-MS/MS is used to identify and quantify the enriched S-nitrosated peptides. The iodoTMT switch assay, sometimes shortened to ISA, essentially "locks in" S-nitrosation sites with a permanent mass tag, allowing for multiplexed quantitation and highly sensitive detection (Murray et al., 2012; Qu et al., 2014; Yin et al., 2017) (Fig. 5.2).

5.3.2.2 *Detection and use in plants*

Recently, the iodoTMT method has been used to profile the nitrosoproteomes of plants. One noteworthy example is a study by Qiu et al. (2019) in tea leaves (*Camellia sinensis*), which was the first to report on the analysis of plant S-nitrosoproteome using iodoTMT labeling. They used iodoTMT sixplex reagents to label SNO-derived thiols after blocking thiols in tea leaf proteins, ascorbate reduction, anti-TMT affinity enrichment, and MS identification. As a result, 191 proteins were detected in tea leaves containing 228 S-nitrosation sites. The authors pointed out that site-specific techniques, such as iodoTMT, significantly improved detection, while previous plant studies employing biotin-switch "shotgun" methods generally detected far fewer SNO proteins. IodoTMT or comparable tags have also been used in other plant studies. To identify S-nitrosated proteins regulated by GSNOR (S-nitrosoglutathione reductase) under stress, for example, a 2019 study in tomatoes (*Solanum lycopersicum*) used a proteomic approach; iodoTMT or similar chemical labeling were used to improve coverage (Gong & Shi, 2019). These studies show that hundreds of S-nitrosated proteins can be found in plants under a variety of conditions using iodoTMT-based SNO detection. More recently, the iodoTMT method was utilized to profile the changes in rice leaf S-nitrosoproteome in response to *Magnaporthe oryzae* infection, leading to the identification of 511 S-nitrosated peptides corresponding to 335 proteins (Munir et al., 2025). This demonstrates how iodoTMT and its derivatives are advancing the field of plant S-nitrosoproteomics (Qin et al., 2023).

5.3.2.3 Merits of iodoacetyl tandem mass tag-based detection

1. **Quantitative and multiplexed:** IodoTMT is a quantitative method by nature and allows multiplexing of up to six samples at present. Therefore, multiple samples or treatments can be compared in a single MS run. This allows for the high-throughput measurement of variations in S-nitrosation levels between different sample conditions, such as control versus stress or wild-type versus mutant. Moreover, compared to BST or other classical workflows where only one sample can be analyzed at a time by MS, iodoTMT also offers a significant benefit in terms of sample running cost for MS. This is specifically a major benefit for the labs that outsource the MS analyses.
2. **Higher sensitivity:** Cysteine thiols and iodoTMT tags create an irreversible covalent bond. There are two advantages to this: (1) Unlike certain biotin linkers, once a formerly S-nitrosated cysteine is tagged, the information is "locked in" and will not dissociate during MS analysis. Therefore, low-abundance SNO signals are preserved during the analytical procedure. (2) The tags increase the sensitivity of detection by carrying enrichment handles and reporter ions, leading to the detection of more SNO sites from the same sample than with conventional BST. Accordingly, iodoTMT enhances the signal-to-noise ratio for SNO peptides, frequently resulting in a higher number of sites detected (Chung et al., 2015).
3. **Improved reproducibility:** Due to the multiplexing feature offered by iodoTMT, multiple samples can be analyzed in a single MS run, which avoids variations that arise due to separate MS runs. Furthermore, compared to BST, which occasionally pulls down nonspecifically biotinylated proteins, the results are repeatable and have a lower random background due to the higher specificity of iodoTMT labels towards anti-TMT antibodies used for affinity -based enrichment of TMT-labeled peptides (Wolhuter et al., 2018).

5.3.2.4 Limitations of Iodoacetyl Tandem Mass Tag-based detection

1. **Expensive:** One of the major limitations of the iodoTMT workflow is the requirement of costly reagents such as the iodoTMT sixplex kit and anti-TMT resin. Therefore, despite its strength, the higher cost of this workflow may limit its usage.

2. **Complexity and potential artifacts:** The iodoTMT protocol is not without artifact risks because it still uses the ascorbate switch step and alkylation chemistry that is comparable to BST. False positive results may arise from the iodoTMT tag labeling free thiols in the same way that biotin would if the blocking of free cysteines is not complete. To detect any peptides labeled nonspecifically, strict blocking and appropriate controls, such as a sample not treated with ascorbate, are required. Furthermore, studies have shown that different cysteine-reactive labels have varying preferences, which means that some SNO-sites may be labelled inefficiently with iodoacetyl tags. This phenomenon is known as labeling bias. Another factor to take into account is that ratio compression, or the interference between coeluting tagged peptides, can affect isobaric tag quantification in complex samples. To ensure precise quantification of SNO site occupancy, care must be taken in both experimental design and data analysis.

5.3.3 Recent developments beyond biotin switch technique and iodoacetyl tandem mass tag

In order to address the low throughput of BST, some MS-based techniques, such as SNO-Cys site identification, S-nitrosothiol resin-assisted capture (SNO-RAC), S-alkylating labeling strategy, and cysTMT switch assay, have been developed by altering the BST protocol for specifically evaluating SNO-Cys modification status and boosting the throughput of detection (Chen et al., 2010; Forrester et al., 2009; Hao et al., 2006). A straightforward, reliable workflow for SNO-protein identification; an irreversible thiol linkage for efficient peptide digestion and clear SNO-Cys mapping, sample multiplexing for higher throughput; and relative quantification of SNO-proteins under varied conditions are all features that are necessary for thorough characterisation of SNO-Cys modification, but none of these modified BST methods can accomplish them all.

While utilizing the iodoTMT method, Chung et al. (2015) demonstrated that a subset of SNO sites may avoid detection by one tag but be captured by another. This suggests that not all SNO proteins may be captured by iodoTMT and that, in order to conduct thorough profiling, complementary techniques or sequential tagging with different reagents may be required. Compared to more recent tags (such as fluorous "FAT" tags), iodoTMT may still overlook peptides with low stoichiometry or extreme hydrophobicity. Sensitivity was further enhanced by the recent development of the FAT-switch method (fluorinated alkyl tag coupled with TMT labeling) in Arabidopsis, which identified approximately 2500 SNO peptides in a single experiment, significantly more than previous BST or iodoTMT results. This demonstrates how iodoTMT and its derivatives are advancing the field of plant S-nitrosoproteomics (Qin et al., 2023).

Conflict of interest

No competing interests exist.

Funding

This work was supported by a grant from the National Research Foundation of Korea (NRF), funded by the Ministry of Education, Science, and Technology (grant no. RS-2026–25492629).

AI Disclosure

During the preparation of this work, the author(s) used ChatGPT in order to For language improvement. After using this tool/service, the author(s) reviewed and edited the content as needed and take(s) full responsibility for the content of the publication.

References

Astier, J., Gross, I., & Durner, J. (2018). Nitric oxide production in plants: An update. *Journal of Experimental Botany, 69*(14), 3401–3411. https://doi.org/10.1093/jxb/erx420.

Delledonne, M, Xia, Y, Dixon, R. A, & Lamb, C (1998). Nitric oxide functions as a signal in plant disease resistance. *Nature, 394*(6693), 585–588. https://doi.org/10.1038/29087.

Chen, X., Xu, Q., Yue, Y., Duan, Y., Liu, H., Chen, X., & Zheng, L. (2023). Comparative oxidation proteomics analyses suggest redox regulation of cytosolic translation in rice leaves upon *Magnaporthe oryzae* infection. *Plant Communications, 4*(3).

Chen, Y. J., Ku, W. C., Lin, P. Y., Chou, H. C., Khoo, K. H., & Chen, Y. J. (2010). S-alkylating labeling strategy for site-specific identification of the S-nitrosoproteome. *Journal of Proteome Research, 9*(12), 6417–6439. https://doi.org/10.1021/pr100680a.

Chung, H. S., Murray, C. I., Venkatraman, V., Crowgey, E. L., Rainer, P. P., Cole, R. N., Bomgarden, R. D., Rogers, J. C., Balkan, W., Hare, J. M., Kass, D. A., & Van Eyk, J. E. (2015). Dual labeling biotin switch assay to reduce bias derived from different cysteine subpopulations: A method to maximize s-nitrosylation detection. *Circulation Research, 117*(10), 846–857. https://doi.org/10.1161/CIRCRESAHA.115.307336, http://circres.ahajournals.org.

Corpas, F. J., Del Río, L. A., & Barroso, J. B. (2008). Post-translational modifications mediated by reactive nitrogen species: Nitrosative stress responses or components of signal transduction pathways? *Plant Signaling and Behavior, 3*(5), 301–303. https://doi.org/10.4161/psb.3.5.5277, http://www.tandfonline.com/loi/kpsb20.

Corpas, F. J., González-Gordo, S., & Palma, J. M. (2021). Protein nitration: A connecting bridge between nitric oxide (NO) and plant stress. *Plant Stress, 2*. https://doi.org/10.1016/j.stress.2021.100026, www.journals.elsevier.com/plant-stress.

Feng, J., Wang, C., Chen, Q., Chen, H., Ren, B., Li, X., & Zuo, J. (2013). S-nitrosylation of phosphotransfer proteins represses cytokinin signaling. *Nature Communications, 4*(1). https://doi.org/10.1038/ncomms2541.

Forrester, M. T., Thompson, J. W., Foster, M. W., Nogueira, L., Moseley, M. A., & Stamler, J. S. (2009). Proteomic analysis of S-nitrosylation and denitrosylation by resin-assisted capture. *Nature Biotechnology, 27*(6), 557–559. https://doi.org/10.1038/nbt.1545.

Foster, M. W., & Stamler, J. S. (2004). New insights into protein S-nitrosylation: Mitochondria as a model system. *Journal of Biological Chemistry, 279*(24), 25891–25897. https://doi.org/10.1074/jbc.M313853200.

Gong, B., & Shi, Q. (2019). Identifying S-nitrosylated proteins and unraveling S-nitrosoglutathione reductase-modulated sodic alkaline stress tolerance in *Solanum lycopersicum* L. *Plant Physiology and Biochemistry, 142*, 84–93. https://doi.org/10.1016/j.plaphy.2019.06.020.

Gruetter, C. A., Barry, B. K., McNamara, D. B., Gruetter, D. Y., Kadowitz, P. J., & Ignarro, L. (1979). Relaxation of bovine coronary artery and activation of coronary arterial guanylate cyclase by nitric oxide, nitroprusside and a carcinogenic nitrosoamine. *Journal of Cyclic Nucleotide Research, 5*(3), 211–224.

Gupta, K. J., Kaladhar, V. C., Fitzpatrick, T. B., Fernie, A. R., Møller, I. M., & Loake, G. J. (2022). Nitric oxide regulation of plant metabolism. *Molecular Plant, 15*(2), 228–242. https://doi.org/10.1016/j.molp.2021.12.012, https://www.cell.com/molecular-plant/home.

Hancock, J. T. (2020). Nitric oxide signaling in plants. *Plants, 9*(11), 1–5. https://doi.org/10.3390/plants9111550, https://www.mdpi.com/2223-7747/9/11/1550/pdf.

Hao, G., Derakhshan, B., Shi, L., Campagne, F., & Gross, S. S. (2006). SNOSID, a proteomic method for identification of cysteine S-nitrosylation sites in complex protein mixtures. *Proceedings of the National Academy of Sciences, 103*(4), 1012–1017. https://doi.org/10.1073/pnas.0508412103.

Hu, J., Huang, X., Chen, L., Sun, X., Lu, C., Zhang, L., Wang, Y., & Zuo, J. (2015). Site-specific nitrosoproteomic identification of endogenously S -nitrosylated proteins in *Arabidopsis*. *Plant Physiology, 167*(4), 1731–1746. https://doi.org/10.1104/pp.15.00026.

Huang, B., & Chen, C. (2010). Detection of protein S-nitrosation using irreversible biotinylation procedures (IBP). *Free Radical Biology and Medicine, 49*(3), 447–456.

Jaffrey, S. R., & Snyder, S. H. (2001). The biotin switch method for the detection of S-nitrosylated proteins. *Science Signaling, 2001*(86), pl1. https://doi.org/10.1126/scisignal.862001pl1.

Kato, H., Takemoto, D., & Kawakita, K. (2013). Proteomic analysis of S-nitrosylated proteins in potato plant. *Physiologia Plantarum, 148*(3), 371–386. https://doi.org/10.1111/j.1399-3054.2012.01684.x.

Khan, M., Ali, S., Al Azzawi, T. N. I., Saqib, S., Ullah, F., Ayaz, A., & Zaman, W. (2023). The key roles of ROS and RNS as a signaling molecule in plant–microbe interactions. *Antioxidants, 12*(2). https://doi.org/10.3390/antiox12020268, www.mdpi.com/journal/antioxidants.

Klepper, L. (1979). Nitric oxide (NO) and nitrogen dioxide (NO_2) emissions from herbicide-treated soybean plants. *Atmospheric Environment (1967), 13*(4), 537–542. https://doi.org/10.1016/0004-6981(79)90148-3.

Kovacs, I., & Lindermayr, C. (2013). Nitric oxide-based protein modification: Formation and site-specificity of protein S-nitrosylation. *Frontiers in Plant Science, 4*(MAY). https://doi.org/10.3389/fpls.2013.00137.

Liang, F., Wang, M., Li, J., & Guo, J. (2024). The evolution of S-nitrosylation detection methodology and the role of protein S-nitrosylation in various cancers. *Cancer Cell International, 24*(1). https://doi.org/10.1186/s12935-024-03568-y.

Lin, W., Shang, J. X., Li, X. Y., Zhou, X. F., & Zhao, L. Q. (2025). Nitric oxide regulates multiple signal pathways in plants via protein S-nitrosylation. *Current Issues in Molecular Biology, 47*(6). https://doi.org/10.3390/cimb47060407, https://www.mdpi.com/journal/cimb.

Liu, R., Zhu, T., Yang, T., Yang, Z., Ren, A., Shi, L., Zhu, J., Yu, H., & Zhao, M. (2021). Nitric oxide regulates ganoderic acid biosynthesis by the S-nitrosylation of aconitase under heat stress in *Ganoderma lucidum*. *Environmental Microbiology, 23*(2), 682–695. https://doi.org/10.1111/1462-2920.15109.

Marietta, M. A. (1993). Nitric oxide synthase: Function and mechanism. *Chemistry and Biology of Pteridines and Folates,* 281–284.

Moncada, S. (1993). The L-arginine: Nitric oxide pathway, cellular transduction and immunological roles. *Advances in Second Messenger and Phosphoprotein Research, 28,* 97–99.

Munir, A., Min, C. W., Wang, Y., Kim, S. T., & Gupta, R. (2025). S-nitrosoproteome analysis of rice leaves highlights the possible roles of superoxide dismutase in resistance against *Magnaporthe oryzae*. *Physiologia Plantarum, 177*(5), e70515.

Murray, C. I., Uhrigshardt, H., O'Meally, R. N., Cole, R. N., & Van Eyk, J. E. (2012). Identification and quantification of S-nitrosylation by cysteine reactive tandem mass tag switch assay. *Molecular and Cellular Proteomics, 11*(2). https://doi.org/10.1074/mcp.M111.013441, http://www.mcponline.org/content/11/2/M111.013441.full.pdf+html.

Ortega-Galisteo, A. P., Rodríguez-Serrano, M., Pazmiño, D. M., Gupta, D. K., Sandalio, L. M., & Romero-Puertas, M. C. (2012). S-Nitrosylated proteins in pea (*Pisum sativum* L.) leaf peroxisomes: Changes under abiotic stress. *Journal of Experimental Botany, 63*(5), 2089–2103. https://doi.org/10.1093/jxb/err414.

Palmer, R. M. (1993). The L-arginine: Nitric oxide pathway. *Current Opinion in Nephrology and Hypertension, 2*(1), 122–128. https://doi.org/10.1097/00041552-199301000-00018.

Qin, G., Qu, M., Jia, B., Wang, W., Luo, Z., Song, C. P., Tao, W. A., & Wang, P. (2023). FAT-switch-based quantitative S-nitrosoproteomics reveals a key role of GSNOR1 in regulating ER functions. *Nature Communications, 14*(1). https://doi.org/10.1038/s41467-023-39078-0, https://www.nature.com/ncomms/.

Qin, Y., Dey, A., & Daaka, Y. (2013). Protein S-nitrosylation measurement. *Methods in Enzymology, 522*, 409–425. https://doi.org/10.1016/B978-0-12-407865-9.00019-4.

Qiu, C., Sun, J., Wang, Y., Sun, L., Xie, H., Ding, Y., Qian, W., & Ding, Z. (2019). First nitrosoproteomic profiling deciphers the cysteine S-nitrosylation involved in multiple metabolic pathways of tea leaves. *Scientific Reports, 9*(1). https://doi.org/10.1038/s41598-019-54077-2.

Qu, Z., Meng, F., Bomgarden, R. D., Viner, R. I., Li, J., Rogers, J. C., Cheng, J., Greenlief, C. M., Cui, J., Lubahn, D. B., Sun, G. Y., & Gu, Z. (2014). Proteomic quantification and site-mapping of S -nitrosylated proteins using isobaric iodoTMT reagents. *Journal of Proteome Research, 13*(7), 3200–3211. https://doi.org/10.1021/pr401179v, http://pubs.acs.org/journal/jprobs.

Rahim, W., Khan, M., Al Azzawi, T. N. I., Pande, A., Methela, N. J., Ali, S., Imran, M., Lee, D. S., Lee, G. M., Mun, B. G., Moon, Y. S., Lee, I. J., & Yun, B. W. (2022). Exogenously applied sodium nitroprusside mitigates lead toxicity in rice by regulating antioxidants and metal stress-related transcripts. *International Journal of Molecular Sciences, 23*(17). https://doi.org/10.3390/ijms23179729, http://www.mdpi.com/journal/ijms.

Simontacchi, M., Galatro, A., Ramos-Artuso, F., & Santa-María, G. E. (2015). Plant survival in a changing environment: The role of nitric oxide in plant responses to abiotic stress. *Frontiers in Plant Science, 6*(NOVEMBER). https://doi.org/10.3389/fpls.2015.00977, http://journal.frontiersin.org/article/10.3389/fpls.2015.00977/pdf.

Torreilles, J. (2001). Table of contetns. *Frontiers in Bioscience, 6*, d1161–1172.

Wei, G., Wang, C., Lei, X., Gao, X., Li, J., Zhang, S., & Guo, J. (2023). IodoTMT-labeled redox proteomics reveals the involvement of oxidative post-translational modification in response to para-hydroxybenzoic acid and hydrogen peroxide stresses in poplar. *Ecotoxicology and Environmental Safety, 259*, 115033. https://doi.org/10.1016/j.ecoenv.2023.115033.

Wolhuter, K., Whitwell, H. J., Switzer, C. H., Burgoyne, J. R., Timms, J. F., & Eaton, P. (2018). Evidence against stable protein s-nitrosylation as a widespread mechanism of post-translational regulation. *Molecular Cell, 69*(3), 438. https://doi.org/10.1016/j.molcel.2017.12.019, www.molecule.org.

Yin, Z., Balmant, K., Geng, S., Zhu, N., Zhang, T., Dufresne, C., Dai, S., & Chen, S. (2017). Bicarbonate induced redox proteome changes in *Arabidopsis* suspension cells. *Frontiers in Plant Science, 8*. https://doi.org/10.3389/fpls.2017.00058.

Yu, L., Dai, Z., Zhang, Y., Iqbal, S., Lu, S., Guo, L., & Yao, X. (2022). Proteome-wide identification of S-sulfenylated cysteines reveals metabolic response to freezing stress after cold acclimation in *Brassica napus*. *Frontiers in Plant Science, 13*. https://doi.org/10.3389/fpls.2022.1014295.

Section II

NO in abiotic stress resilience

Chapter 6

Nitric oxide-induced signaling and downstream targets in cold stress

Ankita Sehrawat[1], Yaiphabi Sougrakpam[2], and Renu Deswal[3]
[1]*Department of Botany, Maitreyi College, University of Delhi, New Delhi, Delhi, India,* [2]*Division of Plant Physiology, Indian Agricultural Research Institute, New Delhi, Delhi, India,* [3]*Department of Botany, University of Delhi, New Delhi, Delhi, India*

6.1 Introduction

Plants are challenged by multiple environmental stresses such as temperature extremes (low and high), drought, salinity, flooding, UV radiation, and pathogen attack (Satyakam et al., 2022). Among these stresses, cold stress affects membrane fluidity, causes ion leakage, increases reactive oxygen species/reactive nitrogen species (ROS/RNS) production, decreases photosynthetic efficiency, which further leads to cellular and physiological changes, causing reduced growth, development, and productivity. Plants have remarkably adapted to survive in almost every ecological niche on Earth, irrespective of the variations in the temperature (−80°C to 70°C). The ability of plants to survive in such a wide range of climatic conditions shows their developmental plasticity and adaptability (both active and passive). Active adaptation mechanisms support plants in adjusting to sudden changes in temperature and seasonal change. In contrast, passive mechanisms involve changes at the genomic and proteome levels. According to FAO, India had the largest cropland area (≈ 170 Mha), followed by the United States (158 Mha), China, and the Russian Federation (≈ 120 Mha each). Due to the continuous increase in the global population (doubled between 1961 and 2016), there is greater demand for food. Land, being a limited resource, is strained as the cropland area per capita decreased from 0.45 to 0.21 hectares from 1961 to 2016 globally (https://www.fao.org/sustainability/news/detail/en/c/1274219/). Therefore, understanding the plant response is essential for developing highly productive cold/freeze-resilient crops, which could be introduced in non-cultivable areas, as a key step for global food security.

Nitric Oxide in Plant Resilience to Abiotic Stress. DOI: https://doi.org/10.1016/B978-0-443-40497-9.00008-8

In the past decades, remarkable progress has been made in understanding plant's response to low temperature (LT) stress (Abdullah et al., 2022; Hwarari et al., 2022; Rivero et al., 2022). In natural habitats, plants are exposed to mainly two types of LT stress: cold/chilling stress in the temperature range between 0°C and 15°C, while temperatures below 0°C cause freezing stress. Plants follow both avoidance and tolerance strategies to alleviate cold stress (Satyakam et al., 2022). Avoidance mechanism is mainly associated with structural aspects and primarily protects the cellular structure, for example, inhibiting the ice crystal formation inside the cells. However, cold tolerance mechanisms involve alteration in the gene expression as well as biochemical and physiological changes (Ritonga et al., 2021). For instance, during cold stress, plasma membrane rigidification is followed by metabolite transport through the apoplast, endomembrane, and organelles. Furthermore, changes in the protein subcellular localization take place, as they move from one location to another to regulate signaling, growth, and programmed cell death. All these changes lead to the initiation of downstream cold stress signaling pathways.

The ICE-CBF-COR regulatory module is an important signaling pathway involved in regulating cold responses (Chinnusamy et al., 2003). Plants use complex biochemical and molecular mechanisms involving transcriptional and post-translational modifications (PTMs) to adapt to cold stress. These complex mechanisms are predominantly divided into Abscisic acid (ABA)-dependent and ABA-independent pathways. ABA regulates the expression of Cold-Responsive (COR) genes by transcription factors (TFs), including AREB/ABF (ABA-responsive element binding protein/ABA-binding factor), myelocytomatosis oncogene/myeloblastosis oncogene (MYC), bZIP, homobox, HOS members, H4, and Zn finger factors. In the ABA-independent pathway, C-repeat binding factors/dehydration responsive element binding (CBFs/DREBs), transcriptional activators control the expression of COR genes. CBFs are controlled by ICE [Inducer of CBF expression 1, a MYC-like basic helix loop helix transcription factor]. Further, cold-inducible genes provide cold tolerance by regulating cold-responsive proteins [including antifreeze proteins (AFPs), cold shock proteins, dehydrins, heat shock proteins (HSPs), late embryogenesis abundant (LEA) proteins] and by synthesizing proline, soluble sugars, xanthophylls, and carotenoids (Satyakam et al., 2022). AFPs bind to tiny ice-crystals (formed during freezing stress) and inhibit the growth of intracellular ice, providing cold tolerance. Recently, "epigenetic regulators" such as histone modifications, DNA methylation, and microRNA biogenesis have been shown to play regulatory roles during cold stress (Hereme et al., 2021). These epigenetic modifications, being stable in the genome, are passed on as epigenetic stress memory to the next generation.

In this chapter, various aspects of cold stress response are discussed, specifically, Nitric oxide (NO) as a key signaling molecule, as it participates in counteracting the effects of ROS in abiotic and biotic stress. Key regulatory mechanisms and proteins (including TFs) that impart cold/freezing tolerance are reviewed, as these components are potential targets for genetic engineering and their analysis is important for crop improvement.

6.2 Nitric oxide and cold stress signaling

NO is a gaseous signaling molecule that plays an important role in providing abiotic stress tolerance (Kolupaev et al., 2023; Liu et al., 2021; Meng et al., 2022). Till 2000, the majority of research groups were using the animal system to decipher the role of NO in the biological world and its application in the clinical and medical fields. However, only a few reports on NO in plants were published. A few exciting questions to be addressed about NO signaling were:

1. Whether NO is produced in response to cold stress or not? If yes, how?
2. What is the mechanism of action of NO in cold stress? And how to unfold the NO and cold stress signaling network?

The basic approaches employed include NO measurement, purification of subcellular proteome/organelle, S-nitrosation analysis using proteomics technologies (Biotin switch technique, 2DGE, shotgun proteomics and mass spectrometry [MS]).

6.2.1 Cold induces nitric oxide biosynthesis by both enzymatic and non-enzymatic mechanisms

NO synthesis in plants can be either through, reductive, or oxidative pathways (Allagulova et al., 2023). Reductive pathway (non-enzymatic or enzymatic process) involves the reduction of nitrites to NO, while in the oxidative pathway, oxidation of L-arginine and polyamines produces NO (Astier et al., 2018). We have earlier shown that cold stress induces NO production by both oxidative (nitric oxide synthase [NOS]-like enzymatic activity) as well as reductive pathways (nitrate reductase, NR) (Sehrawat et al., 2013). However, this was suggested on the basis of studies performed with total tissue, as at that time, NO production was not assessed at the subcellular level in cold stress. Therefore NO was quantified in the apoplast and nuclei [using NO measuring system (inNO, Innovative instruments)], and showed that cold (4°C) induces NO evolution by nonenzymatic nitrite reduction and by NR in apoplast and nuclei, respectively (Sehrawat & Deswal, 2014a; Sehrawat et al., 2019). Nitrite reduction-mediated NO production in cold stress was reported for the first time. Moreover, cold stress induced a temporal increase in the NO production as an early response (2 h) in the apoplast, and comparatively late (6 h) in the nucleus. Although BjNOS activity was detected, NR was the major contributor to NO evolution in cold stress in Brassica *juncea* (Sehrawat & Deswal, 2014a; Sehrawat et al., 2013; Sehrawat et al., 2019; Talwar et al., 2012). All these studies were performed under in vitro conditions (i.e., in the protein extract) and should be performed in vivo. In addition to NR, other enzymes reported to be involved in nitrite-dependent NO production included Amidoxime

Reducing Component, sulfite oxidase, xanthine oxidoreductase, and aldehyde oxidoreductase (Chamizo-Ampudia et al., 2016). However, cold-induced NO production by any of these enzymes has yet to be reported. In a recent report, exogenous application of putrescine (Put, a polyamine) led to the accumulation of endogenous NO *via* Put oxidation and NOS-dependent pathway, when chilling stress was given to *Anthurium andraeanum* seedlings, indicating synergistic interaction between NO and Put (Sun et al., 2023).

6.2.2 Nitric oxide reacts with protein and nonprotein thiols to activate a nitric oxide-mediated, cold-signaling network

It is very well established that abiotic stress leads to increased production of ROS and RNS. Thiols form disulfide bonds in the tertiary and quaternary structure of the proteins and maintain it. Thiols can be classified as protein-based thiols (TPTs) and nonprotein thiols low-molecular-weight thiols (LMTs) (Sehrawat & Deswal, 2014a; Sehrawat et al., 2013). Thiols play a fundamental role in contributing to plant stress tolerance, as these interact with RNS and ROS, regulating enzymatic activity and maintaining the protein structure and folding through disulfide bonds in proteins. Cold stress increases the available protein thiols and decreases the nonprotein thiols (mainly reduced glutathione (GSH) and free cysteines) in the crude and Ribulose-1,5-bisphosphate carboxylase/oxygenase (RuBisCO) depleted fractions of *B. juncea* seedlings (Sehrawat et al., 2013). In contrast, *B. juncea* apoplastic and nuclear fractions showed cold-induced increase in both the proteins as well as the nonprotein thiols. This was an important observation, as increased LMT by cold suggests their role in maintaining protein structure and in protecting the proteins from oxidative damage by providing a reducing environment. Early increase (2 h, 4°C) in LMTs indicates the immediate reaction between NO and LMT (GSH) for the S-nitrosoglutathione (GSNO) production, which could release NO or acts as a transnitrosylating agent. Interestingly, in the apoplast, cold stress increased total thiols (2–48 h, 4°C), followed by a decrease from 72 to 96 h of cold treatment. These results correlated with the fact that, as the apoplast is the initial sensing site for stress, primarily, there was an increase in the thiol pool to counteract the negative effect of oxidative stress, while during prolonged stress, the thiol pool decreased due to its utilization for maintaining the redox homeostasis. An increase in TPT indicated a reaction with NO to produce S-nitrosoproteins (SNO-protein) by S-nitrosation. Cold stress also increased SNOs in total extract, apoplast, and nuclei (Sehrawat & Deswal, 2014a; Sehrawat et al., 2013; Sehrawat et al., 2019). The enrichment of both protein (including SNOs) and nonprotein thiols is reflected as a common response to stress.

6.2.3 Cold-induced nitric oxide bioactivity by S-nitrosation

Extensive research has been done in recent decades to understand the role of NO as a signaling messenger. NO carries its bioactivity by interacting with target proteins that undergo NO-dependent PTMs, namely S-nitrosation, metal S-nitrosylation, and tyrosine nitration (Sun et al., 2021). S-nitrosation is the covalent and reversible attachment of NO to a reactive thiol group of cysteine in a protein, forming SNO. Analysis of S-nitrosation in a CAM (crassulacean acid metabolism) plant, *Kalanchoe pinnata*, by the Biotin switch technique and neutravidin affinity chromatography led to the identification of many novel S-nitrosated targets, including kinesin-like protein, glycolate oxidase, putative uridine diphosphate glucose 4-epimerase, and putative DNA topoisomerase II (Abat et al., 2008). Cold stress-induced S-nitrosation analysis of *B. juncea* revealed S-nitrosation-mediated RuBisCO inactivation (40%), which could be the reason for cold stress-induced photosynthetic inhibition. It was observed that a major drawback of total S-nitrosoproteome analysis is the masking of the low-abundant stress-responsive regulatory targets by the abundant ones, such as RuBisCO (50% of the total identified targets in cold) during 2-DGE and MS. Masking hinders MS identification and quantification (Anderson & Anderson, 2002). Therefore removal of these highly abundant proteins was essential for analyzing NO and cold stress signaling. To overcome this problem, two strategies were followed 1) analysis of S-nitrosation in the subcellular system of apoplast (or extracellular space) and nuclei and 2) selective depletion of RuBisCO (the most abundant plant protein) by immunoaffinity purification. To strengthen RNS-mediated cold-stress signaling in subcellular system, S-nitrosation analysis in apoplastic fractions of cold treated *B. juncea* seedlings (6 h, 4°C), identified 52% novel targets (by nLC–MS/MS), and highlighted the role of subcellular analysis for identifying low abundant targets (Sehrawat & Deswal, 2014a). Interestingly, 80% of the targets (65% increased S-nitrosation and 35% decreased S-nitrosation) showed differential S-nitrosation in response to cold. A very significant finding was increased S-nitrosation of nonsecretory proteins (leaderless secretory proteins, 62%) in the apoplast in cold. As nonsecretory proteins do not have a signal peptide for the apoplast (i.e., nonresident of apoplast), this indicates that cold stress may act as a stimulus and directs relocalization of the non-classical and nonsecretory stress-responsive multi-localized proteins (MLPs) to the apoplast for neutralizing stress-induced damage. This study highlighted S-nitrosation-mediated translocation of MLPs, namely Cu/Zn superoxide dismutase (SOD), cysteine proteases, dehydroascorbate reductase (DHAR), and acyl-binding/lipid-transfer protein. As the apoplast serves as a first line of defense against stress, it is reasonable that these stress-responsive MLPs are released into the apoplastic space to counteract the stress-induced damage.

The nucleus being the "brain of cell" or CPU (central processing unit), is involved in transcriptional modulation of several genes (de Melo et al., 2022). To get insight into the regulatory function of NO in the nucleus, the first repertoire of cold-modulated S-nitrosated proteins in the nuclei showed increased S-nitrosation of redox-(Cu/Zn SOD and L-ascorbate oxidase) and metabolism-[fructose bisphosphate aldolase (FBA2) and malate dehydrogenase 1 (MDH1)] related pathways

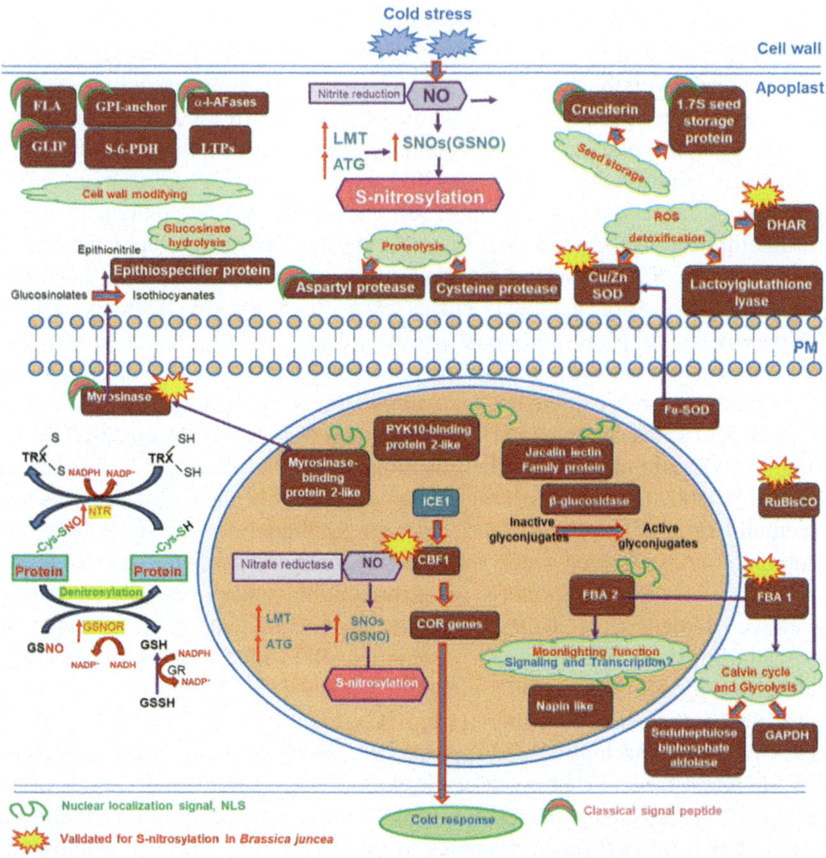

FIGURE 6.1 A putative nitric oxide (*NO*)- cold-stress signaling network in *Brassica juncea*. When a plant senses cold stress, NO production increases by both enzymatic (NOS-like enzyme and nitrate reductase) and nonenzymatic (nitrite reduction) pathways. Increased NO reacts with low-molecular-weight thiols (LMTs/GSH) to form S-nitrosoglutathione (GSNO). Cold-induced SNOs and available thiol groups (ATGs) lead to S-nitrosation of proteins. Increased LMT also helps in maintaining a reduced environment and protect proteins from oxidation. Dehydroascorbate reductase (DHAR), GST, SOD, and myrosinase validated for S-nitrosation in cold and showed that NO positively regulates their activity, therefore contributes to ROS detoxification, regulates glucosinolate hydrolysis system, and provides defense. Novel S-nitrosation of Brassicaceae-specific proteins, cruciferin precursor (apoplast, with classical peptide) and napin (nucleus, NLS) seed storage protein, suggest subcellular regulation by NO. S-nitrosation of FLAs, GPI-anchor, α-1-AFases, S-6PDH, and LTP suggests the regulation of cell-wall-modifying proteins by NO. NO mediated regulations of aspartyl- and cysteine proteases indicate the role of S-nitrosation in regulating proteolysis in cold. S-nitrosation of nonnuclear targets such as myrosinase and beta-glucosidase, along with their associated protein (with NLS) in the nucleus, indicates regulation of glycosyl hydrolase family proteins in cold or NO-mediated nuclear trafficking. RuBisCO activity is negatively regulated, while FBA is positively regulated by S-nitrosation, therefore reflecting the physiological relevance of NO in regulating Calvin cycle and glycolysis. Increased S-nitrosoglutathione reductase (GSNOR) and NADPH-dependent thioredoxin

(*continued*)

in the nucleus. A significant finding of this study was to show S-nitrosation-mediated nuclear trafficking of nonnuclear targets (without nuclear localization signal [NLS], such as beta-d-xylosidase, daikon cysteine protease, and beta-glucosidase) to the nucleus in cold stress. These results were further supported by the previous prediction using bioinformatic tools, which indicated that the majority of the cold-responsive Arabidopsis nuclear proteins were potential candidates for S-nitrosation (Sehrawat & Deswal, 2015). Another interesting observation was that about 40% of the cold-responsive S-nitrosated targets in *B. juncea* were common with the S-nitrosproteome of a model plant (*A. thaliana*), indicating the similarity in the S-nitrosation profile in these two Brassicaceae members (Sehrawat & Deswal, 2014b).

Functional categorization of the cold-responsive S-nitrosated RuBisCO-depleted fractions displayed a shift from the majority of the photosynthetic targets to redox, stress, signaling, and metabolic proteins in the RuBisCO depletome in comparison with the total S-nitrosoproteome, indicating RuBisCO depletion to be a fruitful strategy in enriching the inventory of low-abundant S-nitrosated targets (Sehrawat et al., 2013). A bird's-eye view of cold-responsive S-nitrosoproteome is proposed (Fig. 6.1).

6.2.3.1 Nitric oxide enhances cellular detoxification by activating dehydroascorbate reductase, superoxide dismutase, and GSH S-transferase in cold stress

Almost 59% redox-related cold-responsive proteins are modulated by NO-based PTMs (Sehrawat et al., 2013). One of the noteworthy aspects was S-nitrosation of SOD, DHAR, GSH S-transferase (GST), and glyoxalase I in cold stress. CysNO (a NO donor), along with cold treatment, increased DHAR (52.8%), GST (200%), and SOD (50%) activity, while dithiothreitol (DTT) (a thiol-specific reductant) brought down the activity after CysNO and cold treatment, indicating S-nitrosation-mediated regulation of these during cold stress (Sehrawat & Deswal, 2014a). Interestingly, increased GSH, which is a reductant, further supported cold-mediated increased DHAR activity.

reductase (NTR) enzymes are involved in maintaining NO/SNO homeostasis by regulating denitrosylation by reducing S-nitrosoglutathione (GSNO) and thioredoxins, respectively. FLAs, Fasciclin-Like-arabinogalactan; GPI-anchor, glycosylphos-phatidylinositol (GPI)-anchor; α-1-Afases, alpha-L-arabinofuranosidases; S-6-PDH, sorbitol-6-phosphate dehydrogenase; GLIP, GDSL LIPASE; LTP, Lipid-transfer protein; SOD, superoxide dismutase; DHAR, dehydroascorbate reductase; SNO, S-nitrosothiols; GR, GSH reductase; GSNO, S-nitrosoglutathione; LMTs-Low molecular weight thiols; ATGs-available thiol groups; ABA, abscisic acid; CBFs/DREBs, C-repeat/ dehydration responsive elements (CRT/DRE)- binding factors; ICE1, Inducer of CBF Expression 1; COR, Cold-Regulated genes; NOS-like enzyme, nitric oxide synthase; NR, nitrate reductase; RuBisCO, Ribulose-1,5- bisphosphate carboxylase/Oxygenase; FBA, Fructose bisphosphate aldolase; NADPH-dependent thioredoxin reductase (*NTR*) (*NLS*, Nuclear localization signal; *NOS*, nitric oxide synthase).

Moreover, DHAR, SOD, and GST do not have classical signal peptides (for apoplastic localization), or NLS; therefore their S-nitrosation in the apoplast or nucleus, indicates their probable cold stress-induced trans-localization for maintaining redox status.

6.2.3.2 S-nitrosation-mediated regulation of Brassicaceae-specific glucosinolate hydrolysis pathway in cold stress

Glucosinolates are sulfur-containing secondary metabolites and are a part of the Myrosinase-glucosinolate system or "mustard oil bomb," a Brasssicaceae-specific defense mechanism in which myrosinase hydrolyzes glucosinolates (S-glucosides) to produce hydrolytic compounds (isothiocyanates, thiocyanates, nitriles, epithionitriles, and oxazolidine-2-thione) and discourages feeding by herbivores (Fu et al., 2016; Tang et al., 2023). Notably, cold stress also induces S-nitrosation of enzymes of the glucosinolate hydrolysis pathway, including myrosinase (Sehrawat & Deswal, 2014a), myrosinase-associated protein (in apoplast), myrosinase-binding protein (in nucleus), and epithiospecifier protein (in RuBisCO-depleted fractions). Furthermore, S-nitrosation of myrosinase in the nucleus was validated by western blotting using antibiotin antibody and antimyrosinase antibody. To confirm whether the active form of myrosinase is present, its activity was measured in crude-(1.45 μmoles/min/mL) and nuclear fractions (0.18 μmoles/min/mL). Moreover, myrosinase in the nucleus, crude and apoplast samples, was detected at a similar position on the immunoblot (60–68 kDa), and phylogenetic analysis showed it to belong to the MB gene family. Cold stress decreased the S-nitrosation of myrosinase in the apoplast (Sehrawat & Deswal, 2014a), while increasing it in the nucleus (Sehrawat et al., 2019), indicating differential regulation of myrosinase in subcellular compartments by NO. Interestingly, myrosinase is a classical secretory protein (with signal peptide), and it is getting S-nitrosated in the nucleus, while myrosinase binding protein 2-like (another nuclear S-nitrosated target) possesses an NLS and a jacalin-type lectin domain, which probably hints at its association with myrosinase for supporting translocation of myrosinase to the nucleus. Myrosinase was also identified as a ROS-sensitive target after hydrogen peroxide (H_2O_2) treatment in *B. juncea* seedlings (Chaurasia & Deswal, 2017). These results support the role of NO signaling in regulating the glucosinate hydrolysis pathway to reduce the damaging effects of cold stress. Further studies are needed to understand the conditions that signal myrosinase and its associated proteins to translocate within subcellular systems during stress.

6.2.3.3 Role of nitric oxide in regulating cell wall-modifying enzymes and carbohydrate-binding proteins

Multiple studies have reported that cold stress induces S-nitrosation of many apoplastic cell wall-modifying enzymes involved in cell adhesion [fasciclin-like arabinogalactan proteins, glycosylphos-phatidylinositol-anchor, and alpha-L-arabinofuranosidases], sorbitol synthesis and osmoticum maintenance (NADP-dependent sorbitol-6-phosphate dehydrogenase), cell wall patterning [GDSL LIPASE (GLIP)] and fatty-acid dynamics (lipid-transfer protein), indicating the role of NO in regulating the reorganization of cell wall components (Sehrawat & Deswal, 2014a). Interestingly, in the nucleus, many lectins (carbohydrate-binding proteins), including jacalin-related lectin 34 and PYK10-binding protein 1 with NLS, were identified (Sehrawat et al., 2019). Moreover, beta-glucosidase 23-like protein without NLS was also identified as a cold-responsive S-nitrosated target in the nucleus, involved in cell wall remodeling and activation of phytohormones. It showed 90% amino acid sequence similarity with PYK 10. As jacalin-related lectin 34 and PYK10-binding protein have NLS, S-nitrosation probably mediates the cold-induced transfer of beta-glucosidase to the nucleus on binding with these lectin-binding proteins.

6.2.3.4 *Nitric oxide S-nitrosate seed storage and proteolytic enzymes in apoplast*

Cold stress also induces S-nitrosation of two major seed storage proteins of Brassicaceae, that is, cruciferins (in apoplast) and napins (in nucleus). S-nitrosated seed storage proteins followed both classical (with signal peptide, 1.7S seed storage protein, cruciferin precursor and germin-like protein) and nonclassical secretory pathways (cruciferin) to the apoplast, while napin with an NLS and a glutamine-rich domain (characteristic of few TFs) was present in the nucleus, indicating differential/spatial targeting by NO. Interestingly, in the apoplast, proteolytic enzymes such as aspartyl/aspartic proteases and cysteine proteases showed S-nitrosation, which indicates that these NO-sensitive proteases probably initiate proteolysis of proteins in the apoplast during cold stress.

6.2.3.5 *Nitric oxide regulates primary metabolic pathways (Glycolysis and Calvin cycle)*

Interestingly, enzymes of both light and dark reactions (Calvin cycle) of photosynthesis, including β-carbonic anhydrase, oxygen-evolving enhancer protein 2, RuBisCO (both large and small subunit), FBA, sedoheptulose-1,7-bisphosphatase, and GAPDH, are S-nitrosated in cold stress in *B. juncea* seedlings (Abat & Deswal, 2009; Sehrawat & Deswal, 2014a; Sehrawat et al., 2013; Sehrawat et al., 2019). Notably, GSNO treatment increases FBA activity in a dose-dependent manner, while DTT treatment shows reversal, therefore authenticating its positive regulation by NO (Sehrawat et al., 2013). Cold stress also increases S-nitrosation of FBA2 and MDH1 in the nucleus, suggesting the role of NO in modulating their moonlighting function (primary enzymatic role with acquired secondary nonenzymatic role). Inhibition of RuBisCO activity by S-nitrosation probably saves energy by diverting the glucose equivalents to the pentose phosphate pathway and stores NADPH for maintaining redox status (Abat et al., 2008).

6.2.3.6 Cold and nitric oxide signaling networks are connected by C-repeat binding factor S-nitrosation

Three cold-responsive CBFs, CBF1/DREB1b, CBF2/DREB1c, and CBF3/DREB1a, play an important role in regulating the ABA-independent cold stress signaling pathway. In 2011, two reports were published simultaneously, which indicated a link between NO and CBF. First was by Cantrel et al. (2011); which showed that NO positively regulates CBF1 and CBF3 transcripts in Arabidopsis (Cantrel et al., 2011). Zhao et al. (2011) reported that when tomato (*Solanum lycopersicum*) fruits are treated with a NO donor (SNP, sodium nitroprusside) before storage (2°C), it enhances the relative expression of LeCBF1 and protects tomatoes from cold injury (Zhao et al., 2011). These findings were important as they showed a link between NO and the CBF-dependent cold stress signaling pathway. However, the mechanism of regulation of CBF by NO was unknown. In the same year, 2011, another report also showed that cold treatment (4°C, 3 h) along with SNP (1.5 mM) elevated the expression of LeCBF1 by 4.5fold (350%) in comparison with cold treatment only. In contrast, cPTIO (a NO scavenger) treatment decreases the expression (37% decrease) to the basal level, which confirms that the elevated expression of LeCBF1 was due to NO. Further, recombinant LeCBF1 was also shown to be endogenously S-nitrosated as well as in response to GSNO (a NO donor) treatment. Cold-induced endogenous S-nitrosation of LeCBF1 was confirmed (Kashyap et al., 2015). To further explore downstream targets of NO through which NO is contributing to cold acclimation, proline content was measured in cold-treated seedlings in the presence of SNP and cPTIO, which showed an increase in 22% in cold stress (+SNP) and only 7% increase in cold stress alone, and decrease (61%) with cPTIO, indicating endogenous NO via S-nitrosation helps in proline production.

In 2016, Liu et al. showed upregulation of CBF3 in chilled *CsNOA1* (*Cucumis sativus NITRIC OXIDE ASSOCIATED 1*) overexpressing lines in comparison with the WT plants (Liu et al., 2016). This further supported a putative network indicating *NOA*-dependent NO accumulation in cold to regulate CBFs at transcriptional and post-transcriptional levels as a cold tolerance mechanism. However, any report of S-nitrosation or any other NO-based modification of CBF3 or CBF2 is not yet available in the literature, and should be taken up to get a complete understanding of CBF and NO crosstalk in cold stress signaling.

To sum, the above-mentioned studies have provided evidence of differential S-nitrosation of proteins in cold stress, as SOD, GST, DHAR, FBA, myrosinase, and CBF1 showed positive regulation by NO, while RuBisCO activity was inhibited in the presence of NO. Moreover, S-nitrosation analysis in subcellular systems (apoplast and nucleus) and RuBisCO depletome showed spatial regulation, as in the apoplast, S-nitrosation exclusively regulates cell wall modifying and proteolytic events, while in the nucleus, it regulates carbohydrate-binding proteins and seed storage proteins. In addition, the total proteome and RuBisCO depletome, S-nitrosation regulates mainly photosynthetic and metabolic pathways. Identification of 47 novel S-nitrosated targets in the *B. juncea* indicated the relevance of S-nitrosoproteome analysis in the subcellular system (Abat et al., 2008; Sehrawat & Deswal, 2014a; Sehrawat et al., 2013; Sehrawat et al., 2019).

Tyrosine nitration is another NO-based PTM, mediated by peroxynitrite ($ONOO^-$, an NO-derived species), and forms a 3-nitrotyrosine residue. However, a comprehensive list of tyrosine-nitrated proteins in cold stress is still missing. Nevertheless, NADP-malic enzyme (NADP-ME, a NADPH-generating enzyme) was shown to be nitrated at Tyr-73 by $ONOO^-$ and cold stress negatively modulates cytosolic NADP-ME activity through tyrosine nitration (Begara-morales et al., 2019). To the best of our knowledge, it is the first study validating nitration of a protein in cold stress. To completely understand the NO signaling and its physiological relevance in cold stress, both tyrosine nitration and S-nitrosation could be analyzed in the same tissue together.

6.2.4 Denitrosation in cold stress

It is worth mentioning that till 2015, NO signaling in plants was mainly focused on identifying S-nitrosation targets and understanding their physiological relevance in the stress signaling mechanisms (Kolbert et al., 2019; Kolbert et al., 2021). However, the information on denitrosation was limited till late 2000 (Benhar et al., 2009). Denitrosation is the removal of NO groups, primarily from Cys thiol side chains in proteins, and it plays an important role in ameliorating the negative effects of nitrosative stress (Chakraborty et al., 2022). Analyzing denitrosation will give a more comprehensive understanding of the molecular mechanism of NO/SNO homeostasis. S-nitrosation is regulated by both enzymatic [by NADH-dependent S-nitrosoglutathione reductase (GSNOR) and thioredoxin (TRX)-NADPH-dependent thioredoxin reductase (NTR)] and nonenzymatic denitrosation [by light, reducing agents (glutathione and ascorbate), metal ions, and nucleophilic compounds] pathways (Jedelská et al., 2020; Kalinina & Novichkova, 2021). GSNOR denitrosates GSNO, while NTR catalyzes the conversion of oxidized TRX to reduced TRX. Reduced TRX further participates in the transnitrosation reaction. To unfold the denitrosation pathways in cold stress, a comparative analysis of both of these enzymes was carried out. Stress-responsive elements, including low-temperature responsiveness and drought inducibility (DRE and MBS), were detected in the *BjNTR* and *BjGSNOR* promoters (Babuta et al., 2023). Both NTR and GSNOR (*BjNTRA-3, BjNTRB-4, BjNTRC-4*, and

BjGSNOR-3) showed a low-temperature regulatory element. Dehydration-responsive element (DRE/DRE-core) specific to *BjGSNOR-1, BjNTRA-1*, and *BjNTRB-4* was observed. GSNOR and NTR showed increased activity by 14.3% and 36.4%, respectively, after cold stress treatment (4°C, 48 h) in *B. juncea* seedlings, hinting at the role of these two enzymes in denitrosation in cold stress. To sum up, denitrosation and nitrosation work in coordination during cold stress. However, the mechanism of denitrosation at the spatial and temporal levels in providing cold tolerance is not yet dissected and needs further work.

6.3 Antifreeze proteins in responses to cold and freezing stress

As a result of climate change, plants are continuously subjected to sudden episodes of frosting. Due to this, some plants experience lethal freezing injury, while others avoid freezing or limit the damage caused by the formation of large ice crystals in their tissues by producing antifreeze compounds. AFPs (ice-binding proteins or IRI proteins) are known as "biomolecular gifts of nature," as these improve plants' antifreeze ability and promote the sustainability of life in extremely cold environments (Ekpo et al., 2022). Interestingly, AFPs are homologous to PR proteins (involved in providing protection from pathogens), indicating a "dual cold/LT survival strategy" to the crops, first by playing the role of an AFP and preventing the rupture of the plasma membrane and secondly, by providing protection against pathogen attack. AFPs prevent the merging of minute ice crystals into large crystals in a noncolligative manner, therefore prevent water from freezing and support the organisms in surviving in subzero temperatures. Thermal hysteresis (TH) and ice recrystallization inhibition (IRI) are the two characteristic properties of AFPs. TH involves the separation of the melting and freezing points, while IRI is the capacity to inhibit the growth of ice crystals. In 1969, DeVries reported AFPs in Antarctic marine fishes as a protective mechanism from freezing. In plants, AFPs were first discovered in winter rye in 1992 (DeVries & Wohlschlag, 1969).

6.4 Nitric oxide, antifreeze proteins, and cold stress – is there any connection?

The role of NO as a signaling molecule in cold acclimation is now described in many research publications. However, direct regulation of AFP by NO has not been investigated in detail. Therefore it will be interesting to explore the link between NO and AFPs in the cold signaling network. Interestingly, chitinases, which also function as AFPs, are S-nitrosated during cold stress, indicating a strong link between NO-mediated regulation of AFPs (including chitanases) *via the CBF-dependent* signaling pathway during cold stress. In a recent study, the effect of exogenous application of AFP III (from A/F Protein, commercially available) on the transcript levels of *Nia1* and *Nia2* genes (involved in

NR-dependent NO production) was analyzed in *Hosta capitata* shoot meristems (Pe et al., 2021). As expected, cold (4°C, 6 h and 24 h) treatment induced *Nia1* and *Nia2* genes, leading to increased NO production. In contrast, meristems treated with AFP III along with cold treatment (6 h and 24 h) did not show enhanced expression of *Nia1* and *Nia2*, probably due to the cryoprotective effect of AFP III in cold, which did not trigger expression of *Nia1* and *Nia2*, indicating lower sensitivity to cold stress. This could probably be because, after the treatment of AFP, the plant did not immediately require signaling for higher gene expression for the NO production. However, this study was performed on shoot meristems growing in MS medium under in vitro laboratory conditions, and to understand the link between AFP and NO cold stress signaling mechanism, similar studies should be performed involving foliar application of AFPs on plants growing naturally in LT conditions, followed by analysis of NO signaling at both transcriptional and post-transcriptional levels.

6.5 Conclusions

Cold stress signaling mechanisms involves perception of stress, which triggers multiple responses such as (1) changes in proteins and lipids in cellular membrane, membrane rigidification and cytoskeletal rearrangement, (2) activation of calcium channels and Ca^{2+} accumulation in cytosol, (3) Calcium-binding proteins activates mitogen-activated protein kinase dependent cascade, causing phosphorylation interacts with ICE1 and controls the expression of CBF genes, which further activates COR genes, (4) COR genes activates cold induced transcription factors, encodes proteins for osmolytes, cryoprotectants, hormones, lipid, and stress-responsive proteins (AFPs, HSPs, LEA and dehydrins) that provide cold tolerance, (5) AFPs (in apoplast and cuticle) bind to ice surfaces and reduce or inhibit the melting of minute ice crystals into large crystals in a noncolligative manner, and support the organisms in surviving in subzero temperatures, (6) Low temperature also induces redox imbalances, including increased production of ROS and oxidative damage, (7) Increased NO production may be mediated by nitrate reductase, further regulating cold acclimation response by activating or inhibiting proteins involved in redox, proteolysis, metabolism, and photosynthetic pathways, (8) NO is also involved in the CBF CBF-dependent cold stress signaling pathway via regulating CBF1 at the transcription level and by posttranslational modification via S-nitrosation, (9) NO also plays a significant role in maintaining redox homeostasis in Dioscorea alata tuber and supports germination by increasing the activity of the ascorbate glutathione cycle. Nanoparticles (NPs) are reported to affect plant growth and yield in a dose-dependent manner. NPs, once they enter a plant cell, may induce changes in transcriptome, proteome, and metabolome and mitigate the effect of stress. In contrast, at higher concentrations, NP may induce cytotoxic effects and

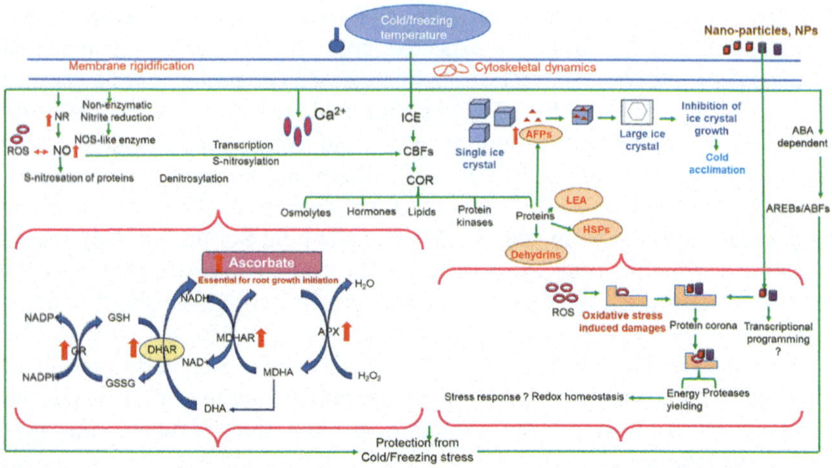

FIGURE 6.2 Putative cold/freezing stress signaling pathway in plants.

therefore reduce the growth and development of the plant and further decrease the yield Fig. 6.2.

Acknowledgments

This work was partially supported by an FRP grant provided by the Institute of Eminence, University of Delhi [IOE/2023-24/12/FRP]. AS thanks Maitreyi College, University of Delhi, for supporting her research. YS acknowledges DBT [DBT-RA/2024-25] for support.

Conflict of interest

The authors have no relevant financial or nonfinancial interests to disclose.

Author contribution statement

All authors read and approved the final manuscript.

References

Abat, J. K., & Deswal, R. (2009). Differential modulation of S-nitrosoproteome of *Brassica juncea* by low temperature: Change in S-nitrosylation of Rubisco is responsible for the inactivation of its carboxylase activity. *Proteomics, 9*(18), 4368–4380. https://doi.org/10.1002/pmic.200800985, http://www3.interscience.wiley.com/cgi-bin/fulltext/122534833/PDFSTART.

Abat, J. K., Mattoo, A. K., & Deswal, R. (2008). S-nitrosylated proteins of a medicinal CAM plant *Kalanchoe pinnata* - Ribulose-1,5-bisphosphate carboxylase/oxygenase activity targeted for inhibition. *FEBS Journal, 275*(11), 2862–2872. https://doi.org/10.1111/j.1742-4658.2008.06425.x.

Abdullah, S. N. A., Azzeme, A. M., & Yousefi, K. (2022). Fine-tuning cold stress response through regulated cellular abundance and mechanistic actions of transcription factors. *Frontiers in Plant Science, 13*. https://doi.org/10.3389/fpls.2022.850216, https://www.frontiersin.org/journals/plant-science.

Allagulova, C. R., Lubyanova, A. R., & Avalbaev, A. M. (2023). Multiple ways of nitric oxide production in plants and its functional activity under abiotic stress conditions. *Federation International Journal of Molecular Sciences, 24*(14). https://doi.org/10.3390/ijms241411637, http://www.mdpi.com/journal/ijms.

Anderson, N. L., & Anderson, N. G. (2002). The human plasma proteome: History, character, and diagnostic prospects. *Molecular & Cellular Proteomics MCP, 1*(11), 845–867. https://doi.org/10.1074/mcp.R200007-MCP200.

Astier, J., Gross, I., & Durner, J. (2018). Nitric oxide production in plants: An update. *Journal of Experimental Botany, 69*(14), 3401–3411. https://doi.org/10.1093/jxb/erx420.

Babuta, P., Paritosh, K., & Deswal, R. (2023). Genome-wide identification and functional analysis of denitrosylases (S-nitrosoglutathione reductases and NADPH-dependent thioredoxin reductases) in *Brassica juncea*. *Plant Biotechnology Reports, 17*(4), 541–559. https://doi.org/10.1007/s11816-023-00831-y.

Begara-morales, J. C., Sánchez-calvo, B., Gómez-rodríguez, M. V., Chaki, M., Valderrama, R., Mata-pérez, C., López-jaramillo, J., Corpas, F. J., & Barroso, J. B. (2019). Short-term low temperature induces nitro-oxidative stress that deregulates the nadp-malic enzyme function by tyrosine nitration in *Arabidopsis thaliana*. *Antioxidants, 8*(10). https://doi.org/10.3390/antiox8100448, https://www.mdpi.com/2076-3921/8/10/448/pdf.

Benhar, M., Forrester, M. T., & Stamler, J. S. (2009). Protein denitrosylation: Enzymatic mechanisms and cellular functions. *Nature Reviews. Molecular Cell Biology, 10*(10), 721–732. https://doi.org/10.1038/nrm2764.

Cantrel, C., Vazquez, T., Puyaubert, J., Rezé, N., Lesch, M., Kaiser, W. M., Dutilleul, C., Guillas, I., Zachowski, A., & Baudouin, E. (2011). Nitric oxide participates in cold-responsive phosphosphingolipid formation and gene expression in *Arabidopsis thaliana*. *New Phytologist, 189*(2), 415–427. https://doi.org/10.1111/j.1469-8137.2010.03500.x, http://onlinelibrary.wiley.com/journal/10.1111/(ISSN)1469-8137.

Chakraborty, S., Sircar, E., Bhattacharyya, C., Choudhuri, A., Mishra, A., Dutta, S., Bhatta, S., Sachin, K., & Sengupta, R. (2022). S-denitrosylation: A crosstalk between glutathione and redoxin systems. *Antioxidants, 11*(10), 1921. https://doi.org/10.3390/antiox11101921.

Chamizo-Ampudia, A., Sanz-Luque, E., Llamas, Á., Ocaña-Calahorro, F., Mariscal, V., Carreras, A., Barroso, J. B., Galván, A., & Fernández, E. (2016). A dual system formed by the ARC and NR molybdoenzymes mediates nitrite-dependent NO production in *Chlamydomonas*. *Plant, Cell & Environment, 39*(10), 2097–2107. https://doi.org/10.1111/pce.12739.

Chaurasia, S. P., & Deswal, R. (2017). Identification and In silico analysis of major redox modulated proteins from brassica juncea seedlings using 2D Redox SDS PAGE (2-dimensional diagonal redox sodium dodecyl sulfate polyacrylamide gel electrophoresis). *Protein Journal, 36*(1), 64–76. https://doi.org/10.1007/s10930-017-9698-x, http://www.kluweronline.com/issn/1572-3887.

Chinnusamy, V., Ohta, M., Kanrar, S., Lee, Bh, Hong, X., Agarwal, M., & Zhu, J. K. (2003). ICE1: A regulator of cold-induced transcriptome and freezing tolerance in *Arabidopsis*. *Genes and Development, 17*(8), 1043–1054. https://doi.org/10.1101/gad.1077503.

de Melo, B. P., Carpinetti, Pd. A., Fraga, O. T., Rodrigues-Silva, P. L., Fioresi, V. S., de Camargos, L. F., & da Silva Ferreira, M. F. (2022). Abiotic stresses in plants and their markers: A practice view of plant stress responses and programmed cell death mechanisms.

Plants, 11(9). https://doi.org/10.3390/plants11091100, https://www.mdpi.com/2223-7747/11/9/1100/pdf?version=1650334801.

DeVries, A. L., & Wohlschlag, D. E. (1969). Freezing resistance in some antarctic fishes. *Science (New York, N.Y.), 163*(3871), 1073–1075. https://doi.org/10.1126/science.163.3871.1073.

Ekpo, M. D., Xie, J., Hu, Y., Liu, X., Liu, F., Xiang, J., Zhao, R., Wang, B., & Tan, S. (2022). Antifreeze proteins: Novel applications and navigation towards their clinical application in cryobanking. *International Journal of Molecular Sciences, 23*(5), 2639. https://doi.org/10.3390/ijms23052639.

Fu, L., Wang, M., Han, B., Tan, D., Sun, X., & Zhang, J. (2016). Arabidopsis myrosinase genes AtTGG4 and AtTGG5 are root-tip specific and contribute to auxin biosynthesis and root-growth regulation. *International Journal of Molecular Sciences, 17*(6), 892. https://doi.org/10.3390/ijms17060892.

Hereme, R., Galleguillos, C., Morales-Navarro, S., & Molina-Montenegro, M. A. (2021). What if the cold days return? Epigenetic mechanisms in plants to cold tolerance. *Planta, 254*(3). https://doi.org/10.1007/s00425-021-03694-1, link.springer.de/link/service/journals/00425/index.htm.

Hwarari, D., Guan, Y., Li, R., Movahedi, A., Chen, J., & Yang, L. (2022). Comprehensive bioinformatics and expression analysis of TCP transcription factors in liriodendron chinense reveals putative abiotic stress regulatory roles. *Forests, 13*(9), 1401. https://doi.org/10.3390/f13091401.

Jedelská, T., Luhová, L., & Petřivalský, M. (2020). Thioredoxins: Emerging players in the regulation of protein S-nitrosation in plants. *Plants, 9*(11), 1426. https://doi.org/10.3390/plants9111426.

Kalinina, E., & Novichkova, M. (2021). Glutathione in protein redox modulation through S-glutathionylation and S-nitrosylation. *Molecules (Basel, Switzerland), 26*(2), 435. https://doi.org/10.3390/molecules26020435.

Kashyap, P., Sehrawat, A., & Deswal, R. (2015). Nitric oxide modulates *Lycopersicon esculentum* C-repeat binding factor 1 (LeCBF1) transcriptionally as well as post-translationally by nitrosylation. *Plant Physiology and Biochemistry, 96*, 115–123. https://doi.org/10.1016/j.plaphy.2015.07.032.

Kolbert, Z., Lindermayr, C., & Loake, G. J. (2021). The role of nitric oxide in plant biology: Current insights and future perspectives. *Journal of Experimental Botany, 72*(3), 777–780. https://doi.org/10.1093/jxb/erab013, http://jxb.oxfordjournals.org/.

Kolbert, Z., Feigl, G., Freschi, L., & Poór, P. (2019). Gasotransmitters in action: Nitric oxide-ethylene crosstalk during plant growth and abiotic stress responses. *Antioxidants, 8*(6), 167. https://doi.org/10.3390/antiox8060167.

Kolupaev, Y. E., Yemets, A. I., Yastreb, T. O., & Blume, Y. B. (2023). The role of nitric oxide and hydrogen sulfide in regulation of redox homeostasis at extreme temperatures in plants. *Frontiers in Plant Science, 14*. https://doi.org/10.3389/fpls.2023.1128439, https://www.frontiersin.org/journals/plant-science.

Liu, L., Huang, L., Sun, C., Wang, L., Jin, C., & Lin, X. (2021). Cross-talk between hydrogen peroxide and nitric oxide during plant development and responses to stress. *Journal of Agricultural and Food Chemistry, 69*(33), 9485–9497. https://doi.org/10.1021/acs.jafc.1c01605.

Liu, X., Liu, B., Xue, S., Cai, Y., Qi, W., Jian, C., Xu, S., Wang, T., & Ren, H. (2016). Cucumber (*Cucumis sativus* L.) nitric oxide synthase associated gene1 (CsNOA1) plays a role in chilling stress. *Frontiers in Plant Science, 7*, 2016. https://doi.org/10.3389/fpls.2016.01652.

Meng, Y., Jing, H., Huang, J., Shen, R., & Zhu, X. (2022). The role of nitric oxide signaling in plant responses to cadmium stress. *International Journal of Molecular Sciences, 23*(13), 6901. https://doi.org/10.3390/ijms23136901.

Pe, P. P. W., Naing, A. H., Kim, C. K., & Park, K. I. (2021). Antifreeze protein improves the cryopreservation efficiency of hosta capitata by regulating the genes involved in the low-temperature tolerance mechanism. *Horticulturae, 7*(4). https://doi.org/10.3390/horticulturae7040082, https://www.mdpi.com/2311-7524/7/4/82/pdf.

Ritonga, F. N., Ngatia, J. N., Wang, Y., Khoso, M. A., Farooq, U., & Chen, S. (2021). AP2/ERF, an important cold stress-related transcription factor family in plants: A review. *Physiology and Molecular Biology of Plants, 27*(9), 1953–1968. https://doi.org/10.1007/s12298-021-01061-8, https://www.springer.com/journal/12298.

Rivero, R., Sønsteby, A., Heide, O. M., Solhaug, K. A., & Remberg, S. F. (2022). Growth analysis of the everbearing strawberry 'Delizzimo' under controlled temperature and photoperiod conditions. *CABI Agriculture and Bioscience, 3*(1). https://doi.org/10.1186/s43170-022-00110-w.

Satyakam, G., Zinta, R. K., Singh, R., & Kumar (2022). Cold adaptation strategies in plants—An emerging role of epigenetics and antifreeze proteins to engineer cold resilient plants. *Frontiers in Genetics, 13*. https://doi.org/10.3389/fgene.2022.909007, https://www.frontiersin.org/journals/genetics#.

Sehrawat, A., & Deswal, R. (2015). Dissecting nitric oxide signaling in nucleus: Role of S-nitrosylation in regulating nuclear proteins. In K. Gupta, & A. Igamberdiev (Vol. Eds.), *Reactive oxygen and nitrogen species signaling and communication in plants. Signaling and communication in plants: vol 23*, Springer. https://doi.org/10.1007/978-3-319-10079-1_12.

Sehrawat, A., Abat, J. K., & Deswal, R. (2013). RuBisCO depletion improved proteome coverage of cold responsive S-nitrosylated targets in *Brassica juncea. Frontiers in Plant Science, 4*. https://doi.org/10.3389/fpls.2013.00342, http://journal.frontiersin.org/Journal/10.3389/fpls.2013.00342/pdf.

Sehrawat, A., & Deswal, R. (2014a). Sub-proteome S-nitrosylation analysis in *Brassica juncea* hints at the regulation of Brassicaceae specific as well as other vital metabolic pathway(s) by nitric oxide and suggests post-translational modifications cross-talk. *Nitric Oxide Biology and Chemistry / Official Journal of the Nitric Oxide Society, 43*, 97–111. https://doi.org/10.1016/j.niox.2014.08.010.

Sehrawat, A., & Deswal, R. (2014b). S-Nitrosylation analysis in *Brassica juncea* apoplast highlights the importance of nitric oxide in cold-stress signaling. *Journal of Proteome Research, 13*(5), 2599–2619. https://doi.org/10.1021/pr500082u.

Sehrawat, A., Gupta, R., & Deswal, R. (2013). Nitric oxide-cold stress signalling cross-talk, evolution of a novel regulatory mechanism. *Proteomics, 13*(12-13), 1816–1835. https://doi.org/10.1002/pmic.201200445.

Sehrawat, A., Sougrakpam, Y., & Deswal, R. (2019). Cold modulated nuclear S-nitrosoproteome analysis indicates redox modulation of novel Brassicaceae specific, myrosinase and napin in *Brassica juncea. Environmental and Experimental Botany, 161*, 312–333. https://doi.org/10.1016/j.envexpbot.2018.10.010.

Sun, C., Zhang, Y., Liu, L., Liu, X., Li, B., Jin, C., & Lin, X. (2021). Molecular functions of nitric oxide and its potential applications in horticultural crops. *Horticulture Research, 8*(1). https://doi.org/10.1038/s41438-021-00500-7.

Sun, X., Yuan, Z., Wang, B., Zheng, L., & Tan, J. (2023). Nitric oxide synthase (NOS)-dependent nitric oxide generation is involved in putrescine-induced chilling tolerance of *Anthurium andraeanum. Scientia Horticulturae, 312*, 111849. https://doi.org/10.1016/j.scienta.2023.111849.

Talwar, P. S., Gupta, R., Maurya, A. K., & Deswal, R. (2012). *Brassica juncea* nitric oxide synthase like activity is stimulated by PKC activators and calcium suggesting modulation by PKC-like kinase. *Plant Physiology and Biochemistry, 60*, 157–164. https://doi.org/10.1016/j.plaphy.2012.08.005.

Tang, Y., Zhang, G., Jiang, X., Shen, S., Guan, M., Tang, Y., Sun, F., Hu, R., Chen, S., Zhao, H., Li, J., Lu, K., Yin, N., & Qu, C. (2023). Genome-wide association study of glucosinolate metabolites (mGWAS) in *Brassica napus* L. *Plants, 12*(3), 639. https://doi.org/10.3390/plants12030639.

Zhao, R., Sheng, J., Lv, S., Zheng, Y., Zhang, J., Yu, M., & Shen, L. (2011). Nitric oxide participates in the regulation of LeCBF1 gene expression and improves cold tolerance in harvested tomato fruit. *Postharvest Biology and Technology, 62*(2), 121–126. https://doi.org/10.1016/j.postharvbio.2011.05.013.

Chapter 7

Role of nitric oxide in the mitigation of heat stress

Arun Kumar Maurya[1], Yogendra Prasad Saksena[1], and Dwaipayan Sinha[2]
[1]*Department of Botany, Multanimal Modi College, Ghaziabad, Uttar Pradesh, India,*
[2]*Government General Degree College, Paschim Medinipur, West Bengal, India*

7.1 Introduction

The continuous rise in human population and increased anthropogenic activities have led to the overexploitation of natural resources, especially fossil fuels, resulting in a doubling of atmospheric CO_2 compared to the preindustrial period. This has contributed to global climate change, characterized by an average temperature change of 2°C–4°C as reported by Assessment Report-6 of the International Panels on Climate Change (Ming et al., 2021). In all scenarios, the average atmospheric temperature is likely to increase by 1.5°C by 2050. It is also expected that if greenhouse gases emissions are not significantly reduced, a 2.1°C increase in global surface temperatures will occur by 2100 compared to 1850–1900 (Ming et al., 2021). In this scenario, the increased temperature is expected to induce heat stress (HS) and negatively affect plant growth (Balla et al., 2009). This results in the burning of leaves, branches, and delicate stems, causing leaf senescence and abscission, as well as fruit discoloration and damage, leading to reduced yield (Guilioni et al., 1997; Ismail & Hall, 1999; Vollenweider & Günthardt-Goerg, 2005). Plants are sessile and have no choice but to endure the effects of high temperatures by evolving innovations or activating morphological and physiological-biochemical defense mechanisms. Accordingly, the heat-tolerant plants have evolved a thermo-perception and response mechanism for optimizing their growth under elevated temperatures.

HS impacts key physiological processes such as membrane stability by inducing lipid peroxidation (Dhanda & Munjal, 2009), denaturation and misfolding of membrane-associated proteins (Kan et al., 2023), and pigment bleaching (Bertrand & Schoefs, 1999). In particular, HS-induced enzyme inactivation causes the generation of reactive oxygen or nitrogen species

Nitric Oxide in Plant Resilience to Abiotic Stress. DOI: https://doi.org/10.1016/B978-0-443-40497-9.00014-3

(ROS/RNS), including singlet oxygen ($^1O_2$), superoxide ($O_2^{\cdot-}$), hydrogen peroxide (H_2O_2), hydroxyl radical ($OH^{\cdot}$), nitric oxide (NO), and peroxynitrate ($OONO^-$)(Maurya, 2020). These ROS/RNS cause thylakoid membrane reorganization (Karim et al., 1997), stroma lamellae swelling, destabilization of photosystem II in the antennas, vacuolar content clumping, and loss of cristae integrity in mitochondria, leading to reduced photosynthesis and respiration (Zhang et al., 2005). HS effects are also seen in the cytoskeleton, which serves as a structural support system, negatively affecting plant cell differentiation, elongation, and growth, and ultimately altering the expression of HS-responsive genes (Haider et al., 2022).

NO, a small, diatomic, highly diffusible and reactive, gaseous signaling molecule, is now well recognized in the plant system (Domingos et al., 2015). NO is synthesized in plants by enzymatic and nonenzymatic pathways (Sedlářová et al., 2025). It interacts with other signaling molecules, such as H_2S, H_2O_2, carbon monoxide (CO), salicylic acid (SA), and cytosolic Ca^{2+} to regulate the expression of various genes (Lamotte et al., 2005; Neill et al., 2003; Sana et al., 2025; Wendehenne et al., 2006). Literature survey revealed that NO is deeply involved in stress signaling associated with various abiotic (salinity, drought, heavy metal, UV radiation, and temperature stress) and biotic stresses (Fancy et al., 2017; Sharma et al., 2020; Siddiqui et al., 2011). NO performs these functions by modulating antioxidant enzymes, osmolyte synthesis, and the reduction of ROS-induced lipid peroxidation (Corpas et al., 2011; Khan et al., 2023; Siddiqui et al., 2011). However, the effects of NO can be both cytoprotective and cytotoxic, depending on its concentration and the permutations and combinations of stress (Belenichev et al., 2024). This chapter attempts to connect the role of NO to high-temperature stress in plants.

7.2 Nitric oxide: synthesis and regulation

NO, a key reactive nitrogen species, acts as a signaling molecule that regulates diverse physiological and stress responses in plants and animals (Domingos et al., 2015). Three NO synthase isoforms (eNOS, nNOS, iNOS) oxidize L-arginine to NO in animals (Alderton et al., 2001). In the case of plants, NO is synthesized via both oxidative and reductive pathways. The oxidative routes include NO synthesis from L-arginine, polyamines, and hydroxylamine, while the reductive route majorly involves conversion of nitrate to nitrite, which is then reduced to NO, mainly via nitrate reductase and the mitochondrial electron transport chain (Allagulova et al., 2022) (Fig. 7.1). Nonenzymatic nitrite reduction at acidic pH can also contribute to NO formation (Bethke et al., 2004a). Although strong evidence supports NOS-like activities in higher plants, a canonical NOS gene has not yet been identified (Astier et al., 2017); instead, proteins such as *Arabidopsis thaliana* NO-associated protein 1 (Moreau et al., 2008) and algal NOS (Foresi et al., 2010), as well as reactions involving oximes, peroxidases, and flavins, are proposed to contribute to NO

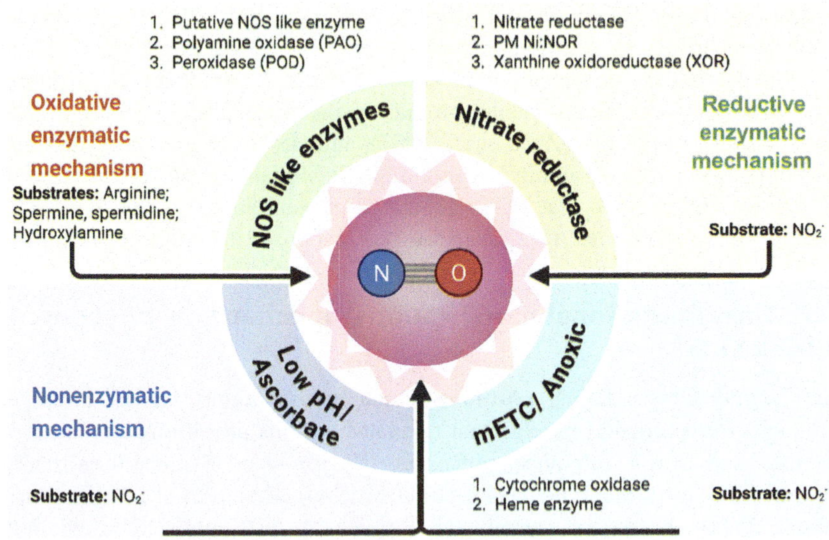

FIGURE 7.1 Different mechanisms of nitric oxide (NO) synthesis in plants.

production (López-Gómez et al., 2024). Although NO generation is well reported in HS, limited studies and mechanisms are available in plants (Bouchard & Yamasaki, 2008; Parankusam et al., 2017; Rai et al., 2020). NO generation has been reported in different cellular components, such as guard cells, subsidiary cells, and long and short trichomes, in leaf epidermal peels and in a mesophyll suspension culture of *Nicotiana tabacum* cv. Xanth under high temperatures (Gould et al., 2003), *Brassica juncea* (Talwar et al., 2012), and *Pisum sativum* (Corpas et al., 2009). Two possible interrelated NO-mediated mechanisms have been proposed that reduce stress severity in plants: either NO functions as an antioxidant (Wink et al., 1999; Wink et al., 2001) or NO can react rapidly with superoxide radicals to form peroxynitrite ($ONOO^-$) (Sedlářová et al., 2025). Although peroxynitrite and its protonated form, peroxynitrous acid (ONOOH, pKa 6.8), are themselves harmful, they are considered to be less toxic than peroxides and may therefore limit cell damage (Wink et al., 1999). The generated ONOOH is subsequently converted into less toxic $OH^•$ and nitric oxide ($NO^•$) (Prolo et al., 2024). It has been found that formation of $ONOO^-$ is associated with a form of cell death showing apoptotic features and in the regulation of the levels of proteins involved in the response to HS in tobacco cultured cells (Malerba & Cerana, 2018). Different types of NO donors and application methods, such as seed priming, foliar application, or irrigation, have been found very effective for managing HS-induced damage in many crops (Savvides et al., 2016).

HS can cause both an increase and a decrease in NO emissions depending upon plant species (Leshem, 2000; Yu et al., 2014), and the severity and duration of stress (Bouchard & Yamasaki, 2008; Yu et al., 2014), for example, a decrease in NO burst was observed at 35°C compared to cells at 55°C in *Nicotiana tabacum* suspension cells (Locato et al., 2008). It has been shown that wild-type *A. thaliana* plants produce a basal endogenous level of NO under control conditions (Zhao et al., 2007), and HS leads to an NO burst that participates in stress signaling (He et al., 2022; Mohanty et al., 2025). Thus it is now well established that NO is generated in plant species under normal as well as stress conditions to regulate various responses according to need.

7.3 Posttranslational modifications: mechanisms of action of nitric oxide

NO functions by inducing posttranslational modifications (PTMs) of target proteins, a fundamental process that regulates various physiological, developmental, and stress tolerance processes. NO induces S-nitrosation under temperature stress (heat, cold, and chilling) (Sougrakpam et al., 2023; Wang et al., 2024). There are various PTM processes, including SUMOylation, phosphorylation, persulfidation, and acetylation (Gupta et al., 2020). During the HS, NO activates the cellular responses through the *S*-nitrosation of GT-element binding transcription factor1, which further binds to Heat Shock Transcription Factor A2 (HsfA2), responsible for thermotolerance (He et al., 2022). It has been found that the level of S-nitrosothiols (SNOs) increases during high-temperature stress (Abat & Deswal, 2009). The accumulation of total SNO pool, including S-nitrosoglutathione (GSNO), was also shown to be increased in sunflower (*Helianthus annuus* L.) seedlings under HT, along with tyrosine nitration (Chaki et al., 2011).

7.4 Nitric oxide and thermomemory

Rising global temperatures and consequent climate instability drive plants to evolve adaptive strategies (Chaudhry & Sidhu, 2022). Thermomemory, the ability to retain information about past events, such as HS, helps modulate counteractive responses to enhance thermotolerance and survival when the same phenomenon recurs. Instructions for such modulation are stored somatically without becoming part of genetic change. This enhanced thermotolerance mediated by thermomemory is commonly known as thermally acquired tolerance (Naaz et al., 2025; Nishad & Nandi, 2021). Thermomemory-based regulation, also known as epigenetic regulation, involves a network of molecular pathways and mechanisms established within plants. Epigenetic regulation (at the transcriptional level) of memory can be classified into two subgroups: temporary (short-term) and long-term exposure-based regulation (Rehman et al., 2023). Temporary regulation develops thermotolerance in plants after a short HS exposure (Nishad & Nandi, 2021). This short-term exposure is called priming. It helps plants withstand HS levels similar to or greater than those in the future, leading to "acclimatization" and improved tolerance and survival. Plants primed exhibit thermally acquired tolerance (TAT), which enhances their growth and survival (Chopra et al., 2025; Nishad & Nandi, 2021).

Plants are a sessile group of organisms and, therefore have evolved a robust, flexible, and multichannel mechanism for sensing and responding to temperature changes (Peng et al., 2025). Temperature stress, including HS, affects the structure and function of the plasma membrane of cells and organelles, altering its fluidity, folding, mobility, and the activity of membrane-associated proteins. The loss of membrane integrity leads to the loss of electrolytes, such as Ca^{2+}, through changes in the 3D structure of Ca^{2+}-transporting proteins, initiating a cascade of downstream responses, including enhanced ROS/RNS (NO) generation, which contributes to thermal perception (Staacke et al., 2025).

A recent report revealed that NO, or a reserve source of NO, GSNO, acts as a signaling molecule, acting by swiftly activating HSFA2 in the *Arabidopsis* shoot apex, followed by delayed activation in other tissues. HSFA2 is a transcription factor, induced by heat stress and involved in the transcriptional regulation of HsP genes during prolonged heat stress or in recovery after HS, suggesting their critical role in the development of acquired thermotolerance. NO causes S-nitrosation of the transcription factor GT-1, which interacts with the HSFA2 promoter in an HS-dependent manner, enhances its binding to NO-responsive elements, and induces HSFA2 for thermotolerance. Such a response was found absent in NO-deficient mutants (He et al., 2022; Olas et al., 2021). In addition, other crucial heat perception mechanisms have been recognized in plants, such as Phytochrome B (Jungjh & Klose, 2016; Legris et al., 2016), PHYTOCHROME-INTERACTING FACTOR 4 and PIF7 (Chung et al., 2020; Koini et al., 2009; Legris et al., 2016), histone H2A.Z-containing nucleosome (Kumar & Wigge, 2010); EARLY FLOWERING 3 (Box et al., 2015), THERMO-WITH ABA-RESPONSE 1 in *Arabidopsis* (Bohn et al., 2024). These mechanisms are also shown to crosstalk with NO or to regulate HS tolerance.

High temperature is found to suppress seed germination and trigger protein S-nitrosation of LONG HYPOCOTYL IN FAR-RED 1 (HFR1) at C164, resulting in its degradation (Ying et al., 2022). HFR1 is a basic helix-loop-helix protein involved in the mediation of R/FR-induced germination. The presence of light causes HFR1 accumulation, but dark conditions promote its degradation by the E3 ubiquitin ligase CONSTITUTIVE PHOTOMORPHOGENESIS 1 (Fairchild et al., 2000; Jang et al., 2005; Yang et al., 2005). The degradation of HFR1 releases PIF1, therefore antagonizing PIF1 transcriptional activity by forming a heterodimer complex to prevent the binding of PIF1 to its target DNA. On the other hand, light induces seed germination through activating the photoreceptor phytochrome B (phyB). This results in stabilization of the HFR1, which in turn sequesters the suppressor PIF1 (Ying et al., 2022). The activation of PIF1-targeted SOMNUS (SOM) expression to alter gibberellin (GA) and abscisic acid (ABA) metabolism under heat stress. The active phyB antagonizes HFR1 S-nitrosation and degradation by increasing S-nitrosoglutathione reductase (GSNOR) activity during seed germination (Lee et al., 2008). To further ascertain the role, the substitution of cysteine-164 of HFR1 with serine abolishes the S-nitrosation. It decreases the HT-induced degradation of HFR1, suggesting that HT and phyB antagonistically modulate the S-nitrosation level of HFR1 to coordinate seed germination and open the window to enhance seed thermotolerance through gene-editing of HFR1 (Lee et al., 2008; Ying et al., 2022).

NO is reported to regulate the DNA methylation pattern during HS acclimatization (Rai et al., 2020) and to be involved in thermomemory formation by interacting with other players in this process (Naaz et al., 2025). NO represents a critical signaling molecule in abiotic stress responses, including priming during HS (Chopra et al., 2025). Well-known heat-inducible transactivators that sustain heat shock protein (*Hsp)* gene expression and extend the duration of thermotolerance include HSFA2 and HSFA3, as well as histone modifiers such as relative of early flowering 6 and Jumonji (JMJ) family members, which play essential roles in the regulation of memory processes (Naaz et al., 2025). The demethylation of H3K27me3 (a distinct group of chromatin modifiers or histone demethylases known as JMJ family proteins) is responsible for establishing the memory phase, and heat shock proteins (HSPs), such as HSP21, a plastidial small heat shock protein, contribute to long-term protection (Qian et al., 2015). NO induces the HS transcription factor to express and accumulate the heat shock protein 18.2 (HSP18.2) via the calcium-binding protein *A. thaliana* calmodulin isoform 3 (AtCaM3), which regulates thermotolerance (Kang et al., 2024; Xuan et al., 2010). The NO is involved in building thermomemory by influencing gene repression and translation via PTMs, such as stimulating the expression of HSFs and HSPs (Xuan et al., 2010). The NO is also involved in posttranslational modifications and the ubiquitination of various target proteins (Pande et al., 2022) (Table 7.1).

7.5 Nitric oxide and thermotolerance

Exceeding a plant's optimal temperature tolerance range is considered to be a form of stress. High temperatures can induce HS and affect various plant processes in plants, ranging from seed germination to the fruiting stage. HS can also affect the sensitivity of different crop species (Kaushal et al., 2016). Plants respond to counteract the HS by upregulating TFs known as HSF, which are responsible for the synthesis of HSPs, also known as molecular chaperones (Li et al., 2025). These proteins help to maintain membrane integrity and scavenge ROS. They also produce antioxidants and osmolytes, a process known as acquired thermotolerance (Kerbler & Wigge, 2023). It was found that endogenous NO and ROS are involved in the accumulation of Hsp70

TABLE 7.1 Effects of NO donors, scavengers and NOS inhibitors on different temperature as well as plant species.

S. no.	Plant species	Temperature	NO donor/ scavenger	Effects	References
1.	Lettuce (*Lactuca sativa*)	27oC–35oC	0.1–05 mM SNP; 0.1 mM SNAP, 1.0 mM Fe(III)CN	NO promotes seed germination and de-etiolation, and inhibits hypocotyl and internode elongation	Beligni and Lamattina (2000)
2.	Rice (Zhong you No.9801) (*Oryza sativa* L.)	40, 45 and 50°C, separately, for 4 h	10 μM SNP, 10 μM GSNO	Elevate the net photosynthesis rate	Song et al. (2013)
3.	Tomato (*Solanum lycopersicon*)	44°C	0.1 mM SNP	SNP, along with Ca^{+2}, increased chlorophyll content	Siddiqui et al. (2017)
4.	Fescue (*Festuca arundinacea*)	44°C	0.1 mM SNP	Upregulates the transcription of the PSII core proteins, Psb complex subunits, and photosynthesis recovery	Chen et al. (2013)
5.	Reed (*Phragmites communis*)	45°C	0.05–0.5 mM SNP and SNAP	Manage oxidative stress by reducing MDA, H_2O_2, and activating antioxidant enzyme activity	Song et al. (2006)
6.	Arabidopsis (*Arabidopsis thaliana*)	32°C	0.5 mM cPTIO	High endogenous NO inhibits seed germination	Hossain et al. (2010)
7.	Arabidopsis(*Arabidopsis thaliana*)	45°C	SNP S-nitroso-N-acetylpenicillamine	NO stimulates DNA-binding activity of heat shock TFs and the accumulation of HSPs by acting upstream of *AtCaM3*	Xuan et al. (2010)

(*Continued*)

TABLE 7.1 (Continued)

S. no.	Plant species	Temperature	NO donor/ scavenger	Effects	References
8.	Rice (*Oryza sativa*)	50°C	0.001–0.02 mM SNP	NO activates antioxidant enzyme, induces the expression of HSPs, and enhances photosynthesis efficiency	Song et al. (2013), Uchida et al. (2002)
9.	Mung bean (*Phaseolus radiatus*)	45°C	0.15 mM SNP	Enhanced photosynthesis and membrane integrity, reduced MDA content, and increased antioxidant enzymes	Yang et al. (2006)
10.	Maize (*Zea mays*)	48°C for 18 h	0.05–0.4 mM SNP	Manages ion leakage, lipid peroxidation, and cell viability	Li et al. (2013)
11.	Wheat (*Triticum aestivum L.*)	38°C for 24 h and 48 h	0.5 mM SNP	Reduced lipid peroxidation, H_2O_2 content; increased content of Chl, AsA, GSH, GSH/GSSG ratio, and upregulation of APX, MDHAR, DHAR, GR, GST, CAT, and Gly I	Hasanuzzaman et al. (2012)
12.	Lentil (*Lens culinaris* Medik.)	(32/20°C)	1 mM SNP	Increased photosynthesis, and increased activity of RuBisCo, sucrose synthase, and vascular acid invertase, and reduced sugar content; accumulation of osmolyte in leaves and anthers	Sita et al. (2021)
13.	Hyacinth bean (*Lablab purpureus* L.)	(40°C–42°C)	SA (0.50 and 1.0 mM) and SNP (0.35 mM)	SNP, along with SA, regulates epigenetic genes and stress-responsive genes	Rai et al. (2018)

14.	Chrysanthemum (*Chrysanthemum morifolium*)	45°C (3, 6, 12 or 24 h)	200 μM SNP	increased photosynthetic pigments and antioxidant enzyme activity	Yang, He, et al. (2011)
15.	Wheat Raj3765 as thermotolerant and HD2932 as thermossusceptible (*Triticum aestivum L.*)	HS (T1 30°C, 1 h; T2 - 38 °C, 1 h)	SNP 150 μM	Expression of HSPs and upregulation of antioxidant enzymes, along with the accumulation of proline and free amino acids	Kumar et al. (2019)
16.	Wheat (*Triticum aestivum* L.)	40°C for 6 h every day for 15 days	100 μM	Reduced ROS accumulation, increased antioxidant enzyme activity, and promoted osmolyte accumulation.	Iqbal et al. (2022)
17.	Wonder green bell pepper (*Capsicum annuum L.*)	25°C (control), 35°C, and 40°C for 24 hours	foliar application of 0.0 μM, 50 μM, or 100 μM melatonin and SNP	Increased leaf N content and the accumulation of Cu and Fe simultaneously enhanced the activities of SOD, APx, and POD	Zargar et al. (2025)
18.	Strawberry (*Fragaria × ananassa*)	25, 35, and 40°C) for 24 h	Pretreatments of 0, 50, and 100 μM SNP	Mitigated oxidative damage, enhanced antioxidant and heat-stress-associated gene expression.	Manafi et al. (2021)

proteins in tomato leaves, which was induced by abiotic (heat) stress stimuli to leaf disks, leading to the detection of two Hsp70 family proteins (heat-inducible 72 and constitutive 75 kDa). Heat stress influenced Hsp72 production, and treatment with a NO donor and a NO scavenger significantly modulated Hsp70 levels in a genotype-dependent manner, and their accumulation correlated with endogenous NO level in *Solanum lycopersicum* and ROS levels in *S. chmielewskii* (Piterková et al., 2013).

7.6 Exogenous nitric oxide, inhibitors, mutants, and heat stress

The role of endogenous NO or exogenously supplied NO from NO donors such as sodium nitroprusside (SNP), *S*-nitroso-*N*-acetylpenicillamine, is well reported in augmenting HS to various species, such as rice (Song et al., 2013), Arabidopsis (*A. thaliana*), faba bean (*Vicia faba* L.) (Alamri et al., 2019), spring maize (*Zea mays* L.) (Kaur & Kaur, 2018), strawberry (Manafi et al., 2021,), and wheat (*Triticum aestivum* L.) (Iqbal et al., 2022). Exogenous NO application enhanced thermotolerance in rice (*Oryza sativa* L.) seedlings at 50°C by lowering leaf H_2O_2 and malondialdehyde (MDA) levels and reducing ion leakage, thereby improving survival. This protective effect was markedly suppressed when seedlings were cotreated with the NO scavenger 2-(4-carboxyphenyl)-4,4,5,5-tetramethylimidazoline-1-oxyl-3-oxide (cPTIO) (Song et al., 2013). Similar responses were observed in embryogenic calluses derived from mature seeds of two reed ecotypes (Phragmites *communis* Trin.; DR and SR), conferring thermotolerance (Song et al., 2006). The effects were significantly improved by SNP application, as indicated by the overall behavior of PSII, as probed by chlorophyll a fluorescence transients (activated RC, improved ETC, oxygen-evolving complex (OEC) to D1 protein) under HS (Chen et al., 2013). These functions were disrupted due to the application of NO scavenger (cPTIO) and putative NOS (N^G-nitro-l-arginine-methyl ester (L-NAME, NO synthase inhibitor) in tall fescue leaves (Chen et al., 2013). Further, it was found that the presence of H_2O_2 and NO induced the expression of *psbA,psbB*, and *psbC*, which encode subunits of the PSII core reaction center (Psb) complex. These data showed that exogenous NO could protect plants by increasing the amount of activated reactive centers (RCs), improving electron transport from the OEC to the D1 protein, and altering relevant gene expression (Chen et al., 2013).

Faba bean (*Vicia faba* L.) plants exposed to 48 h of heat stress showed increased oxidative stress markers [ROS, including H_2O_2, O_2- $^-$ and NO], accumulation of proline and MDA contents, greater electrolyte leakage, enhanced DNA damage, chlorophyll (Chl) degradation, and reduced leaf relative water content and Chl content. As heat stress intensity rose from 32°C to 37°C and 42°C, NO accumulation further increased, accompanied by higher proline levels, upregulation of proline-metabolizing enzymes (Δ^1-pyrroline-5-carboxylate synthetase, ornithine-δ-aminotransferase and proline dehydrogenase), and enhanced expression of heat shock proteins Hsp17.6, Hsp70, Hsp90-1 and Hsp101 (Alamri et al., 2019). Such changes led to enhanced antioxidant enzyme activities of SOD, CAT, GR, and APX, along with increased ascorbate and oxidized glutathione (GSSG). These effects were reversed by NO scavenger cPTIO, suggesting that NO acts as both a signal molecule and a promoter of proline biosynthesis, thereby enhancing HS tolerance (Alamri et al., 2019). Another vital food crop, wheat (*Triticum aestivum* L.) plants, when exposed to 40°C for 6 h every day for consecutive 15 days with NO donor (100 μM SNP), led to alleviation of HS effects on photosynthesis and growth through reduction in H_2O_2 and thiobarbituric acid reactive substances content, maximized osmolytes production, and the antioxidant enzyme expression. NO scavenger application reversed these enhanced responses (Iqbal et al., 2022).

NO–dependent HS tolerance in Arabidopsis was analyzed using two NO-deficient mutants (*noa1/rif1 and nia1nia2*) and the wild type. Both mutants, which generate less NO, showed reduced survival under HS compared with wild-type plants. NO-donor treatments revealed that endogenous NO modulates heat sensitivity in a concentration-dependent manner and, under HS, enhances thermotolerance by regulating *AtCaM3* mRNA levels, increasing HS transcription factor DNA-binding activity, and promoting HSP accumulation. Calmodulin 3 (AtCaM3), a key HS-signaling component, operates downstream of NO: NO donors improved survival in wild-type and *noa1(rif1)* seedlings but not in *cam3* or *cam3noa1(rif1)* double mutants, indicating that AtCaM3 is required for NO signal transduction. This was confirmed using AtCaM3-overexpressing lines in a *noa1(rif1)* background, where electrophoretic mobility-shift and western-blot analyses showed that, after HS, NO stimulated HS transcription factor binding and HSP18.2 accumulation via AtCaM3. (Xuan et al., 2010). Additionally, NO pretreatment was found to enhance the activities of ROS-scavenging enzymes and also upregulate the expression of stress-responsive genes, including those encoding sucrose-phosphate synthase, Δ′-pyrroline-5-carboxylate synthase, and small heat shock protein 26 (Uchida et al., 2002). *HOT5* is a gene of *A. thaliana* that encodes a protein, GSNOR1, responsible for NO homeostasis by SNOs. Under heat stress, two missense alleles and two T-DNA insertion, protein-null alleles were characterized in *A. thaliana*, which failed to exhibit thermotolerance (Lee et al., 2008).

When tobacco Bright Yellow-2 suspension cells were exposed for 10 min at 55°C and then 27°C, the production of NO increased compared to cells exposed to 35°C. NO production was well correlated with cell viability and DNA integrity, as well as with the presence of oxidative stress markers (Locato et al., 2008). Similarly, pretreatment with a NO donor enhanced the heat tolerance of maize seedlings, and this was associated with increased activity of L-cystine desulfhydrase (L-CDs), leading to the accumulation of endogenous H_2S. The application of H_2S donors, such as NaHS and GYY4137, elevated endogenous H_2S levels, thereby reducing electrolyte leakage and improving seedling survival under heat stress (Li et al., 2013). NO also interacts with

H_2S, along with H_2O_2. The accumulation of NO and L-CDs activity gets increased in the presence of exogenous H_2O_2, but in the presence of NO scavenger, and consequent to this, the concentration of internal H_2S gets reduced. The crosstalk among these is required for thermotolerance in maize seedlings (*Zea mays* L.). NO, along with salicylic acid (SA), initiates the heat stress response (HSR) under HS, activating many genes involved in proline, photosynthetic pigments, and antioxidants synthesis (Rai et al., 2020).

NO is essential for auxin-regulated gene expression and cell fate determination in tree tobacco guard cell protoplasts (GCPs) (Beard et al., 2012). The NO shows crosstalk with other signaling molecules, such as H_2O_2 and Ca^{2+}, which are involved in growth and development, and in cellular and physiological responses to abiotic stresses (Niu & Liao, 2016). Endogenous NO, acting as a signal molecule, regulates the *AtCaM3* under HS (Zhang et al., 2009). Calmodulin 3 (CaM3) is a key component of HS signaling in Arabidopsis (Xuan et al., 2010). NO also acts together with H_2O_2 under HS to confer thermotolerance by increasing HS factor DNA-binding activity and HS protein accumulation (Wang et al., 2014). Treatment of SNP under hardening prevents heat damage in wheat seedlings, whereas seedlings fail to protect themselves from HS (Zhang et al., 2023). The interaction between NO and H_2O_2 is essential for providing resistance against HS in 3-day-old seedlings of winter soft wheat (*Triticum aestivum* L., cv. Elegiya) (Karpets et al., 2015). Cross-talk between NO and Ca^{2+} also plays a significant role in abiotic stress (Lang et al., 2014). NO, along with other signaling molecules such as abscisic acid, H_2O_2, Ca^{2+}, and CaM, plays a critical role in HSR (Kaur et al., 2019). Under heat stress, NO, along with these signaling molecules, increases the DNA-binding activity of HSFs and the accumulation of HSPs, leading to resistance to heat stress (Nasir Khan et al., 2015). Nitric oxide, along with ethylene (ET), regulates adventitious root development in cucumber (*Cucumis sativus* L.) explants (Xu et al., 2017), fruit ripening (Mukherjee, 2019), and protects photosynthetic metabolism (Gautam et al., 2022), regulation of Mg^{2+} (Liu et al., 2017), along with Fe and P (García et al., 2021) in plants.

It is found that NO delays natural senescence by reducing ET emission. Under short-term HS, a sudden and substantial increase in ET reduces the endogenous NO (Leshem & Wills, 1998). NO from NO donor (SNP), along with hydrosulfide (NaHS, as an H_2S donor) and other signaling molecules under HS, reduced the suppression of photosynthetic ability by reducing oxidative stress and strengthening the antioxidant system by synthesis of proline, glycine betaine, calcium (Ca^{2+}), and HSP (Iqbal et al., 2021a,2021b). It was found that Ca^{2+} activates NO generation and signal transduction, thereby enhancing heat resistance (Karpets et al., 2016). The NO donor is found to stimulate a temporary increase in cytosolic Ca^{2+} ($[Ca^{2+}]_{cyt}$) in *Nicotiana plumbaginifolia* cells (Lamotte et al., 2006). The CNGC6 is a Ca^{2+}-permeable channel that operates under HS and relays HS signaling via NO.

The NO level increased with Ca^{2+} treatments in Arabidopsis seedlings under HS conditions, but decreased with the Ca^{2+} chelator Ethylene glycol-bis(β-aminoethyl ether)-*N,N,N′,N′*-tetraacetic acid (EGTA) (Peng et al., 2019). Thermotolerance involves Calmodulin isoform 3 (CaM3), a key player in signal transduction during HS in Arabidopsis, where NO acts as an upstream signal to AtCaM3. The binding of AtCaM3 to GSNOR increased GSNOR activity by reducing excess NO accumulation and increasing thermotolerance via a feedback inhibition mechanism between AtCaM3 and NO. This conclusion was drawn from cam3 mutant seedlings, which exhibited a high internal NO accumulation under HS, and a *gsnor* deficiency decreased the survival rate of *cam3* plants, accompanied by an increased internal NO level. However, treatment with NO scavenger enhanced the survival ratio of *cam3* mutant seedlings under HS conditions (Zhang et al., 2020).

7.7 Nitric oxide mitigates heat stress-induced oxidative damage

Among the effects of HS, oxidative damage is the primary concern in plants in the current scenario, driven by the rise in global mean surface air temperature. This is expected to reach 1.5°C by 2050 and continue to rise (Ruane, 2024). The rising HS induces oxidative damage by generating ROS, including lipid peroxides, singlet oxygen ($^1O_2$), $O_2^{\cdot-}$, H_2O_2, and $OH^{\cdot}$ as well as the generation of cellular RNS, such as NO and peroxynitrite (Puglia, 2024). These ROS/RNS molecules induce membrane injury, resulting in electrolyte leak, bleaching of photosynthetic pigments, or damage to photosynthetic complexes, associated proteins, common soluble and membrane proteins, and nucleic acids. These effects ultimately lead to reduced plant growth and development (Puglia, 2024).

Although NO is classified as an RNS, it is also considered a gaseous signaling molecule with several regulatory roles in plant growth (Khan et al., 2023), showing dual roles as an antioxidant and cytoprotective agent, as well as a cytotoxic agent (Belenichev et al., 2024). As discussed above, NO is generated endogenously or supplied exogenously through NO donors, which is found to help manage oxidative damage by activating antioxidant enzymes (CAT, SOD, POD) as observed in coleoptiles of 4-day-old etiolated seedlings of wheat (*Triticum aestivum* L., cv. Elegiya) (Karpets et al., 2011), confer DR callus thermotolerance (Song et al., 2006), and upregulate both antioxidant defense and methylglyoxal (MG) detoxification system in wheat seedlings (Hasanuzzaman et al., 2012). Apart from these actions, NO also alleviated HS-induced oxidative damage by decreasing H_2O_2 and $O_2^{\cdot-}$ contents by acting as an antioxidant molecule that helps enhance leaf photochemical activity and maintain cell membrane integrity in mung bean leaf disks (Yang et al., 2006). NO, along with Ca^{2+}, is also found to impart strength against HS by reducing the accumulation of ROS and MDA, along with the biosynthesis of compatible solutes, such as proline and glycine betaine, total soluble carbohydrates, and photosynthetic pigments (chlorophyll a and b). It increased the activities of key enzymes, including RUBISCO, carbonic anhydrase (CA), nitrate reductase (NR), and antioxidant enzymatic systems (Siddiqui et al., 2017), and managed oxidative damage in leaf and anther in lentil (*Lens culinaris* Medik.) plants (Sita et al., 2021), wheat (Bavita et al., 2012; El-Beltagi et al., 2016), and modulated the mRNA level of certain enzymes (Rai et al., 2018).

7.8 Nitric oxide manages heat stress-induced osmotic imbalance

To counter the deleterious effects of HS, several protective low-MW organic molecules and compounds are synthesized and accumulated in plants, referred to as compatible solutes or osmolytes. These osmolytes include proline, sugars, tertiary sulphoniums, quaternary ammoniums, glycine betaine, and sugar alcohols (Zulfiqar et al., 2020). Nitric oxide is found to increase the osmolyte accumulation in leaves and anthers of lentil (*Lens culinaris* Medik.) (Sita et al., 2021), and faba bean (*Vicia faba* L.)(Alamri et al., 2019). The crass talk between NO and ABA, or Ca^{2+}, involves lowering HS stress by increasing osmolyte accumulation of compatible solutes such as proline and glycine betaine (Iqbal et al., 2022; Siddiqui et al., 2017). NO, along with ROS such as H_2O_2, helps regulate Hsp70 generation and accumulation under several abiotic and biotic stresses (Piterková et al., 2013). In addition, H_2O_2 also regulates NO synthesis, which is essential for regulating HS factor DNA-binding activity and HS protein accumulation (Wang et al., 2014). Therefore it seems that NO, crosstalk with other signaling molecules, helps augmentation and tolerance of osmotic balance by increasing omolytes accumulation, Hsp synthesis, and HS protein accumulation.

7.9 Nitric oxide restores photosynthesis under heat stress

Current adverse global climatic conditions, especially rising temperatures associated with global warming, pose significant threats and limitations to crop productivity (Júnior et al., 2025; Ming et al., 2021). The impact of increased temperature is primarily manifested as HS, leading to excessive ROS generation and accumulation in plants. Such cytotoxic species disrupt the normal function of biomolecules, including lipids, proteins, and nucleic acids, through processes of peroxidation and oxidation. This damages the plasma membrane, making it leaky and injured, and can cause proteins or enzymes that support normal cellular functions, such as photosynthesis and respiration, to malfunction (Abdul Wahid, 2007). HS impairs photosynthesis by reducing the capture of light energy by PSI and PSII, photosynthetic electron transport, and Calvin cycle activities (Stasik & Jones, 2007). The process of capturing light and electron flow by influencing the photochemical reaction, D1 and D2 protein turnover present in PSII, and photosynthetic pigments biosynthesis, such as chlorophyll, carotenoids, and xanthophylls. The photosynthetic machinery, mainly photosystem II (PSII) with its OEC, the adenosine triphosphate (ATP)-generating, and the carbon assimilation processes, are the most stress-sensitive components (Aro et al., 1993; Bukhov & Carpentier, 2000; Murata et al., 2007). Heat stress also affects photosynthesis by increasing the leaf temperature; plants' net photosynthesis (Pn) decreased in pea (*Pisum sativum* L) (Haldimann & Feller, 2005).

The photosynthetic machinery and function of heat-sensitive proteins, many of which are involved in the Calvin cycle, are inhibited by HS, including RUBISCO activase, which causes redox imbalance and ultimately reduces photosynthetic efficiency (Zahra et al., 2023). The moderate HS reduces the photosynthesis process by compromising the reactivation and regeneration of RUBISCO in cotton (Salvucci & Crafts-Brandner, 2004). Moderate HS affects the thylakoid reaction, along with the deactivation of RUBISCO and increased thylakoid leakiness (Sharkey, 2005). The HS also increases photorespiration, which thereby negatively impacts net photosynthesis (P_n). Under various abiotic stresses, NO regulates many morphological, physiological, biochemical, and developmental processes, such as seed germination, plant root growth, respiration, photosynthesis, and stomatal regulation (Besson-Bard et al., 2008; Borisjuk et al., 2007; Moreau et al., 2010). NO is reported to influence photosynthetic performance under heat stress by augmenting and strengthening the photosynthetic machinery, enhancing chlorophyll synthesis, anthocyanins, and flavonoids (Lamattina et al., 2003; Uchida et al., 2002); by affecting the light reaction by binding to PSII, and critical binding sites for NO within the PSII complex are the nonheme iron between the Q_A and Q_B binding sites (Diner & Petrouleas, 1990), interacting with the $Y_D^{\cdot}$ Tyr residue of the D2 protein (Sanakis et al., 1997); binding with the manganese (Mn) cluster of the oxygen-evolving enzyme of photosystem II (Schansker et al., 2002); Rubisco formation, and plastid development in rice under HS through OsNOA1 (Yang, He, et al., 2011). On the other hand, NO also regulates photosynthesis by mitigating heat-induced oxidative stress by promoting the synthesis of carotenoids and zeaxanthin, which help scavenge ROS, and by promoting the dissipation of excess energy as heat (Song et al., 2013), and a decline in photosynthetic pigment content and net photosynthetic rate, reducing the increase in nonphotochemical quenching of fluorescence and malondialdehyde content, and maintaining higher antioxidant enzyme activities, such as SOD, CAT, POD, and APX, in *Chrysanthemum morifolium* (Yang et al., 2011). Exogenous NO application showed improved photosynthetic properties by increasing the number of activated reaction centers (RC) and enhancing electron transport from the OEC to the D1 protein and induced the expression of the *psbA*, *psbB*, and *psbC* genes, responsible for encoding subunits of the PSII core reaction center (Psb) complex in the presence of H_2O_2 (Chen et al., 2013) and helped maintain higher photosynthetic-NUE and photosynthetic-

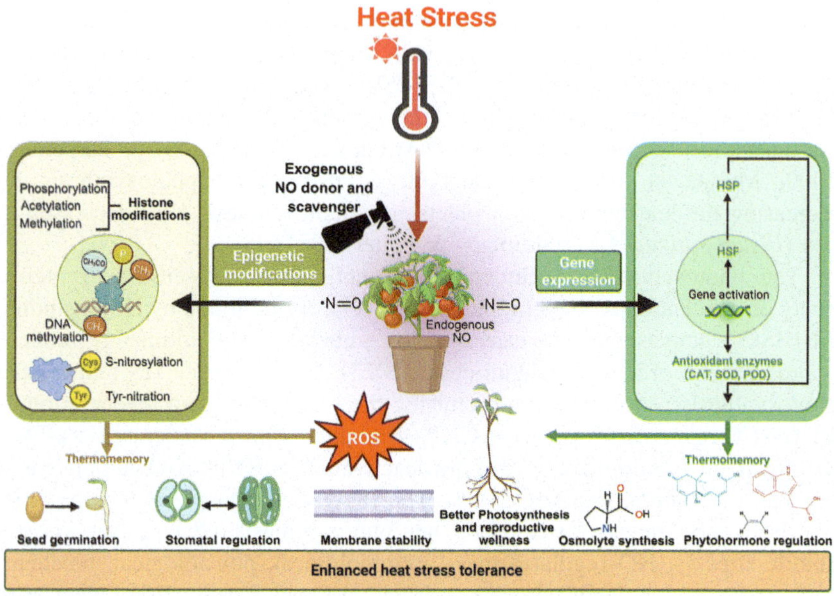

FIGURE 7.2 Effects of heat stress and tolerance mechanism in plants.
(Created in BioRender. Maurya, A. (2026). https://BioRender.com/4a75hhp).

SUE, activity of RUBISCO, and synthesis of reduced GSH under HS in rice (Gautam et al., 2021). The combinational supplementation of HS and NO improved RUBISCO activity and increased sucrose concentration by enhancing carbon assimilation in leaves and anthers (Sita et al., 2021). Some previous studies have reported that NO reduces net photosynthetic rates in oat (*Avena sativa*) and alfalfa (*Medicago sativa*) leaves (Hill & Bennett, 1970), and the amount of glucose, a product of photosynthesis, in mung bean leaves (Lum et al., 2005). Therefore, it can be seen that overall NO is a helpful bioactive molecule that helps improve photosynthesis under HS conditions (Fig. 7.2).

7.10 Nitric oxide, seed germination, and heat stress

Seed germination is the first crucial step for the establishment of a plant (Gavazzi et al., 2008). Heat stress, in general, has a strong negative potential to halt seed germination by affecting membrane stability, mRNA, protein-enzyme denaturation, and hormonal inactivation present in the seeds or synthesized during the early phases of seed germination (Kumar et al., 2023). NO is found to be effective in coping with the adverse process and helps to release from seed dormancy as observed in *A. thaliana*, barley grains (Bethke et al., 2004b), and lettuce (*Lactuca sativa* L. cv. Grand Rapids), and helps to overcome thermo-inhibition in Lettuce (*L. sativa* L. "Jianye Xianfeng

No. 1") (Deng & Song, 2012). These germination processes were reversed by the application of the NO scavenger cPTIO (Hossain et al., 2010), suggesting its role during the seed germination process.

7.11 Nitric oxide ensures reproductive wellness under heat stress

Reproductive wellness in plants includes the proper development, maintenance, and function of reproductive organs and structures and is negatively affected by heat stress (Begcy, Mendes, and De Storme, 2024). NO has been shown to improve pollen germination, viability, stigma receptivity, and ovular viability under HS (Sita et al., 2021). Climate change induced HS induced reproductive growth, flower, and pod losses in lentil (*Lens culinaris* Medik.), a crop susceptible to supra-optimal temperatures (30°C). This loss was recovered through exogenous foliar NO (1 mM SNP) application twice during heat stress (32/20°C) in lentil genotypes differing in heat sensitivity, showing promising results by reducing oxidative stress and increasing pod and seed numbers (Sita et al., 2021).

7.12 Nitric oxide, stomatal regulation, and heat stress

Both in natural and agricultural conditions, plants are frequently exposed to stress and evolve several flexible strategies to adapt to dynamic changes in the surrounding environment (Maurya & Rani, 2017). Stomatal development and its regulation, though primarily aimed at water use and gaseous exchange, have been assigned several secondary functions over time, such as to cope with environmental stress, including heat stress, to prevent water loss (Casson & Gray, 2008; Schulze et al., 1973). NO is well reported to be generated in guard cells of stomata due to several external and internal cues such as ABA, methyl jasmonate, and chitosan, which generate ROS and RNS (including NO) in the Arabidopsis, faba bean, and pea through enzymatic means marked as putative NOS-like enzyme and NR (Beard et al., 2012; Gayatri et al., 2013). Their NO-generative functions were diminished by application of NO inhibitors or NO scavengers (Gayatri et al., 2013), validating their presence and roles. Apart from stomatal regulation, stomatal development was also found to be linked to NO, which promotes stomatal development, leading to increased stomatal index and the relative number of meristemoids and guard mother cells compared to wild-type Arabidopsis. These roles of NO were further verified by using the *nox1* mutant (exhibiting higher NO levels) and the *noa1* mutant (exhibiting low NO accumulation). Further peeping into the mechanism, it was found that NO induced the expression of the master regulator genes involved in the initiation of stomatal development, such as *SPCH*, *MUTE*, and *SCRM2*, and downregulated the expression of the *MPK6* gene (Fu et al., 2016).

Heterotrimeric G proteins consisting of Gα, Gβ, and Gγ subunits act as downstream effectors to regulate multiple functions, including abiotic stress tolerance. The involvement of the pea (*Pisum sativum*) Gβ subunit (PsGβ) in HS was tested in transgenic tobacco plants overexpressing *PsGβ*, which performed better under heat stress than control plants. It was found to be associated with NO induction in guard cells of transgenic plants. Heat stress induced NO-mediated stomatal closure in the leaves of transgenic tobacco plants overexpressing *PsGβ* (Bhardwaj et al., 2020). ABA-induced stomatal closure involves NO as a signaling component, but no report is available that shows NO's effect on the inhibition of stomatal opening. The reports show that a critical Ca^{2+}-dependent NO increase is required for ABA-mediated inhibition of stomatal opening in faba bean (*Vicia faba*) guard cells, which was reversed by a NO scavenger. Further, exogenous NO application inhibited a light-induced stomatal opening in *V. faba* epidermal strips. To further explore the signaling cascade, the effects of Ca^{2+} concentration on ABA-mediated inhibition of stomatal opening were examined. ABA and Ca^{2+} actions were blocked, but not NO-mediated stomatal opening inhibition, revealing that NOS-like activity was required for Ca^{2+} signaling. Further use of the fluorescent probe indicated that Ca^{2+} induces an increase of endogenous NO, suggesting that the light signal-ABA-Ca^{2+}-NO-response pathway regulates stomata (Garcia-Mata & Lamattina, 2007).

Similarly, ABA is also connected with enhanced NO generation in mesophyll cells of maize grown at 22°C–28°C, and such generation was diminished by application of NO scavenger or NOS inhibitor, but didn't affect the ABA-induced H_2O_2 production. The reduced NO level partially blocked the activation of MAPK and other antioxidant genes as induced by ABA. This inhibition and activation of MAPK was overcome by exogenous application of exogenous NO through donor but conversely did not induce the production of H_2O_2 suggesting that ABA-induced H_2O_2 production upstream to the NO generation mediates downstream actions such as activation of MAPK and the upregulation in the expression and the activities of antioxidant enzymes through involvement of the ABA–H_2O_2–NO–cellular response signal transduction chain (Zhang et al., 2007).

The ABA-NO linkage reported in Reed (*Phragmites communis* Trin.) plants, a hydrophyte that grows in the fresh- and brackish-water areas of swamps, riverbanks, and lake shores, and sometimes in terrestrial habitats. Several ecotypes exhibit differential resistance to abiotic stresses, including drought, salinity, and temperature (Chen et al., 2003). It was found that calluses from two ecotypes of reed plant (dune reed [DR, growing in a drought habitat] and swamp reed [SR, growing in an aquatic habitat] display different sensitivity to heat stress (Song et al., 2012). Exogenous NO and endogenous NO depletion in the two calli were used to study plant acclimation to thermotolerance. The DR callus showed that NO production increased significantly under heat stress, whereas it remained stable in the SR callus, thereby augmenting thermotolerance. DR callus maintained a higher relative growth rate, cell viability, and lower ion leakage than SR callus under heat stress. Furthermore, decreased H_2O_2 and MDA contents, enhanced activities of antioxidant enzymes (SOD, CAT, APX, and POD) in both calli in the presence of NO donors under heat stress. The physiological scenario changed upon the application of a NO scavenger, which significantly enhanced oxidative stress in DR callus (Song et al., 2006). The induction of thermotolerance in DR was associated with both ABA and NO, as HS-induced ion leakage and MDA accumulation in two calli were significantly alleviated by exogenous ABA and NO. This report revealed that endogenous NO and ABA contents and NOS activity increased markedly under heat stress in DR callus, while remaining stable in SR callus. Such enhancement was diminished by applications of NO scavenger, NOS inhibitor, and ABA synthesis inhibitor, which had no influence on the protective effect of exogenous NO, but these activities were reversed by exogenous ABA treatment in DR callus and NO scavenger and inhibitor had no impact on ABA, suggesting an intermediary role of NO in ABA-induced responses such as thermotolerance (Song et al., 2008). Similar responses were also observed in *Stylosanthes guianensis*, where tolerance has been associated with endogenous NO synthesis (Song et al., 2006; Zhou et al., 2005).

A Solanaceae member known as tree tobacco (*Nicotiana glauca*) shows a high basal thermotolerance. To see the effect of heat on cultured GCPs on fate-determining auxin-cytokinin hormonal signaling involving NO. It was found that auxin and cytokinin caused GCPs to expand 20- to 30-fold, regenerate cell walls, dedifferentiate, re-enter the cell cycle, and divide at lower temperature (32°C or less), but decline in expansion and complete inhibition in regeneration, dedifferentiation, cell cycle entry, and division were observed at HS conditions (34°C) and complete inhibition at 38°C. This inhibition is associated with NO, as HS inhibits NO accumulation by GCPs. The inhibitor used for NOS mimics the effects of heat by limiting cell expansion, preventing cell wall regeneration, inhibiting cell cycle re-entry, promoting dedifferentiation, and suppressing activation of the *BA* auxin-responsive promoter. This suggests that reduced NO levels suppress auxin signaling in heat-stressed guard cells of thermotolerant plants (Beard et al., 2012).

7.13 Nitric oxide coordinates with phytohormones for heat stress tolerance

Phytohormones play a vital role not only in growth and development but also in stress mitigation (Bhatt et al., 2020). Emerging gaseous plant growth regulators, such as NO, play a similar critical role in diverse plant taxa, identical to that of other conventional phytohormones, including abscisic acid, the jasmonates, and ethylene (Leshem & Kuiper, 1996). The role of exogenously supplemented NO and SA in the presence of SO_4^{2-} was found to impart HS tolerance (supplied at 40°C for 6 h daily for 15 days) in wheat

(*Triticum aestivum* L. cv. HD-3226) among six cultivars assessed. Such thermotolerance is characterized by high photosynthetic sulfur use efficiency, reduced oxidative stress (lower ROS levels, including H_2O_2 content and lipid peroxidation), enhanced antioxidant enzyme activities, and enhanced photosynthesis and growth compared with wild plants. These effects were reversed by application of the NO scavenger and the SA inhibitor, suggesting that the mitigating effects of NO and SA are involved in the induction of sulfur assimilation (Rasheed et al., 2022).

Melatonin (*N*-acetyl-5-methoxytryptamine; MT), another emerging plant growth regulator (Cai et al., 2025), is reported to induce stomatal closure via a dedicated receptor known as phytomelatonin receptor1 (Cai et al., 2025), is reported to induce stomatal closure via a dedicated receptor, the phytomelatonin receptor 1 (PMTR1), which further activates ROS generation (Wei et al., 2018). Generated ROS and NO interaction was observed, as was MT-induced stomatal closure in *Arabidopsis*, and this interaction was diminished by a NO scavenger and by the use of the mutant lines nitrate reductase *1* and *2* (*nia1nia2*) and *NO-associated 1* (*noa1*). However, the presence or absence of MT did not affect H_2O_2-induced NO production or stomatal closure, but MT-induced NO production was impaired in the *rhohC* and *rbohD/F* (NADPH oxidase respiratory burst oxidase homologs) mutant plants. The ROS levels in *nia1nia2* and *noa1* did not differ significantly from the wild-type plants, indicating that NO is a downstream component in the melatonin-induced ROS production. To understand the cascade of generation, exogenous MT did not induce NO or ROS production in the guard cells of *pmtr1* mutant lines, suggesting that NO generation occurs downstream of ROS in PMTR1-mediated stomatal closure in Arabidopsis. It was also observed that reduced NO levels suppressed auxin signaling in heat-stressed guard cells of a thermotolerant tree tobacco (*N. glauca*) plant, where NO was found essential for auxin-regulated gene expression and cell fate determination in GCPs cells (Beard et al., 2012). Auxin and cytokinin caused GCPs to expand 20- to 30-fold, regenerate cell walls, dedifferentiate, re-enter the cell cycle, and divide at lower temperatures (32°C or less), but decline in expansion and complete inhibition in regeneration, dedifferentiation, cell cycle entry, and division was observed at HS conations (34°C) and complete inhibition at 38°C. The NOS inhibitor application mimicked the effects of heat by limiting cell expansion, preventing cell wall regeneration, inhibiting cell cycle re-entry, dedifferentiation, and cell division, and suppressing activation of the *BA* auxin-responsive promoter. These findings connect NO crosstalk with other signaling molecules, such as H_2O_2 and Ca^{2+} (Niu & Liao, 2016).

7.14 Conclusion

It is now well established that NO plays a multifaceted and critical role in mitigating HS by augmenting physio-biochemical protection through several molecular mechanisms, either alone generated in vivo or applied exogenously through NO donors or in combination with other phyto-regulatory molecules. The ability of NO to activate antioxidant enzyme systems through posttranslational mechanisms, regulate osmolyte synthesis, and induce heat shock proteins provides plants with a robust defense system against heat stress. Apart from its direct interventions in HS tolerance, NO plays a pivotal role in basal and stress signaling, coordinating crosstalk with other molecules, such as phytohormones, another gasotransmitter, H_2S, and redox molecules, to optimize cellular responses. These actions collectively help maintain cellular metabolic stability, shield photosynthetic machinery, and support plant growth and development under HS conditions. Further, a deeper understanding involving various species can offer promising avenues for developing HS-resilient crops through exogenous NO donors, integrated breeding strategies, or biotechnological interventions.

AI Disclosure

During the preparation of this work, the author(s) used Grammarly in order to For the purpose of English improvement and amendments accordingly. After using this tool/service, the author(s) reviewed and edited the content as needed and take(s) full responsibility for the content of the publication.

References

Abat, J. K., & Deswal, R. (2009). Differential modulation of S-nitrosoproteome of *Brassica juncea* by low temperature: Change in S-nitrosylation of Rubisco is responsible for the inactivation of its carboxylase activity. *Proteomics, 9*(18), 4368–4380. https://doi.org/10.1002/pmic.200800985, http://www3.interscience.wiley.com/cgi-bin/fulltext/122534833/PDFSTART.

Abdul Wahid, A. W. (2007). Physiological implications of metabolite biosynthesis for net assimilation and heat-stress tolerance of sugarcane (*Saccharum officinarum*) sprouts.

Alamri, S. A., Siddiqui, M. H., Al-Khaishany, M. Y., Khan, M. N., Ali, H. M., & Alakeel, K. A. (2019). Nitric oxide-mediated cross-talk of proline and heat shock proteins induce thermotolerance in *Vicia faba* L. *Environmental and Experimental Botany, 161*, 290–302. https://doi.org/10.1016/j.envexpbot.2018.06.012, www.elsevier.com/inca/publications/store/2/6/7.

Alderton, W. K., Cooper, C. E., & Knowles, R. G. (2001). Nitric oxide synthases: Structure, function and inhibition. *Biochemical Journal, 357*(3), 593–615. https://doi.org/10.1042/0264-6021:3570593.

Allagulova, C. R., Avalbaev, A. M., Lubyanova, A. R., Lastochkina, O. V., & Shakirova, F. M. (2022). Current concepts of the mechanisms of nitric oxide formation in plants. *Federation Russian Journal of Plant Physiology, 69*(4). https://doi.org/10.1134/S1021443722030037, https://rd.springer.com/journal/11183.

Aro, E. M., McCaffery, S., & Anderson, J. M. (1993). Photoinhibition and D1 protein degradation in peas acclimated to different growth irradiances. *Plant Physiology, 103*(3), 835–843. http://www.plantphysiol.org/.

Astier, J., Gross, I., & Durner, J. (2017). Nitric oxide production in plants: An update. *Journal of Experimental Botany, 69*(14), 3401–3411. https://doi.org/10.1093/jxb/erx420.

Balla, K., Bencze, S., Janda, T., & Veisz, O. (2009). Analysis of heat stress tolerance in winter wheat. *Acta Agronomica Hungarica, 57*(4), 437–444. https://doi.org/10.1556/aagr.57.2009.4.6.

Bavita, A., Shashi, B., & Navtej, S. B. (2012). Nitric oxide alleviates oxidative damage induced by high temperature stress in wheat. *Indian Journal of Experimental Biology, 50*(5), 372–378. http://nopr.niscair.res.in/bitstream/123456789/14132/3/IJEB%2050%285%29%20372-378.pdf.

Beard, R. A., Anderson, D. J., Bufford, J. L., & Tallman, G. (2012). Heat reduces nitric oxide production required for auxin-mediated gene expression and fate determination in tree tobacco guard cell protoplasts. *Plant Physiology, 159*(4), 1608–1623. https://doi.org/10.1104/pp.112.200089, http://www.plantphysiol.org/content/159/4/1608.full.pdf+html.

Begcy, K., Mendes, M. A., & De Storme, N. (2024). Plant reproduction under environmental stress. *Frontiers in Plant Science, 15*, 1369070.

Belenichev, I., Popazova, O., Bukhtiyarova, N., Savchenko, D., Oksenych, V., & Kamyshnyi, O. (2024). Modulating nitric oxide: Implications for cytotoxicity and cytoprotection. *Antioxidants, 13*(5), 504. https://doi.org/10.3390/antiox13050504.

Beligni, M. V., & Lamattina, L. (2000). Nitric oxide stimulates seed germination and de-etiolation, and inhibits hypocotyl elongation, three light-inducible responses in plants. *Argentina Planta, 210*(2), 215–221. https://doi.org/10.1007/PL00008128, link.springer.de/link/service/journals/00425/index.htm.

Bertrand, M., & Schoefs, B. (1999). Photosynthetic pigment metabolism in plants during stress. *Handbook of Plant and Crop Stress,* 527–543.

Besson-Bard, A., Courtois, C., Gauthier, A., Dahan, J., Dobrowolska, G., Jeandroz, S., Pugin, A., & Wendehenne, D. (2008). Nitric oxide in plants: Production and cross-talk with Ca^{2+} Signaling. *Molecular Plant, 1*(2), 218–228. https://doi.org/10.1093/mp/ssm016.

Bethke, P. C., Gubler, F., Jacobsen, J. V., & Jones, R. L. (2004a). Dormancy of *Arabidopsis* seeds and barley grains can be broken by nitric oxide. *Planta, 219*(5), 847–855. https://doi.org/10.1007/s00425-004-1282-x.

Bethke, P. C., Badger, M. R., & Jones, R. L. (2004b). Apoplastic synthesis of nitric oxide by plant tissues. *The Plant Cell, 16*(2), 332–341. https://doi.org/10.1105/tpc.017822, http://www.plantcell.org/.

Bhardwaj, D., Sahoo, R. K., Naqvi, A. R., Lakhanpaul, S., & Tuteja, N. (2020). Pea Gβ subunit of G proteins has a role in nitric oxide-induced stomatal closure in response to heat and drought stress. *Protoplasma, 257*(6), 1639–1654. https://doi.org/10.1007/s00709-020-01529-6, www.springer.co.at/springer.py?Page=40&Key=456&cat=9&id_journal=4.

Bhatt, D., Nath, M., Sharma, M., Bhatt, M. D., Bisht, D. S., & Butani, N. V. (2020). Role of growth regulators and phytohormones in overcoming environmental stress. In *Protective chemical agents in the amelioration of plant abiotic stress: Biochemical and molecular perspectives*. India: wiley, 254–279. https://onlinelibrary.wiley.com/doi/book/10.1002/9781119552154, 10.1002/9781119552154.ch11.

Bohn, L., Huang, J., Weidig, S., Yang, Z., Heidersberger, C., Genty, B., Falter-Braun, P., Christmann, A., & Grill, E. (2024). The temperature sensor TWA1 is required for thermotolerance in *Arabidopsis*. *Nature, 629*(8014), 1126–1132. https://doi.org/10.1038/s41586-024-07424-x.

Borisjuk, L., Macherel, D., Benamar, A., Wobus, U., & Rolletschek, H. (2007). Low oxygen sensing and balancing in plant seeds: A role for nitric oxide. *New Phytologist, 176*(4), 813–823. https://doi.org/10.1111/j.1469-8137.2007.02226.x.

Bouchard, J. N., & Yamasaki, H. (2008). Heat stress stimulates nitric oxide production in symbiodinium microadriaticum: A possible linkage between nitric oxide and the coral bleaching phenomenon. *Plant and Cell Physiology, 49*(4), 641–652. https://doi.org/10.1093/pcp/pcn037.

Box, M. S., Huang, B. E., Domijan, M., Jaeger, K. E., Khattak, A. K., Yoo, S. J., Sedivy, E. L., Jones, D. M., Hearn, T. J., & Webb, A. A. (2015). ELF3 controls thermoresponsive growth in *Arabidopsis*. *Current Biology, 25*(2), 194–199.

Bukhov, N. G., & Carpentier, R. (2000). Heterogeneity of photosystem II reaction centers as influenced by heat treatment of barley leaves. *Physiologia Plantarum, 110*(2), 279–285.

Cai, H., Li, J., Li, J., & Teng, H. (2025). Melatonin—Angel of plant growth regulation and protection. *Advanced Agrochem.*

Casson, S., & Gray, J. E. (2008). Influence of environmental factors on stomatal development. *New Phytologist, 178*(1), 9–23. https://doi.org/10.1111/j.1469-8137.2007.02351.x.

Chaki, M., Valderrama, R., Fernández-Ocaña, A. M., Carreras, A., Gómez-Rodríguez, M. V., Pedrajas, J. R., Begara-Morales, J. C., Sánchez-Calvo, B., Luque, F., Leterrier, M., Corpas, F. J., & Barroso, J. B. (2011). Mechanical wounding induces a nitrosative stress by down-regulation of GSNO reductase and an increase in S-nitrosothiols in sunflower (*Helianthus annuus*) seedlings. *Journal of Experimental Botany, 62*(6), 1803–1813. https://doi.org/10.1093/jxb/erq358.

Chaudhry, S., & Sidhu, G. P. S. (2022). Climate change regulated abiotic stress mechanisms in plants: A comprehensive review. *Plant Cell Reports, 41*(1), 1–31. https://doi.org/10.1007/s00299-021-02759-5.

Chen, K., Chen, L., Fan, J., & Fu, J. (2013). Alleviation of heat damage to photosystem II by nitric oxide in tall fescue. *Photosynthesis Research, 116*(1), 21–31. https://doi.org/10.1007/s11120-013-9883-5.

Chen, K.-M., Gong, H.-J., Chen, G.-C., Wang, S.-M., & Zhang, C.-L. (2003). Up-regulation of glutathione metabolism and changes in redox status involved in adaptation of reed (*Phragmites communis)* ecotypes to drought-prone and saline habitats. *Journal of Plant Physiology, 160*(3), 293–301. https://doi.org/10.1078/0176-1617-00927.

Chopra, P., Sapia, N., Karami, O., Kumar, P., Honys, D., Colombo, L., Mendes, M., Benhamed, M., Fotopoulos, V., Lieberman-Lazarovich, M., Mueller-Roeber, B., Kaiserli, E., Hafidh, S., & Fragkostefanakis, S. (2025). Priming thermotolerance: Unlocking heat resilience for climate-smart crops. *Philosophical Transactions of the Royal Society B: Biological Sciences, 380*(1927). https://doi.org/10.1098/rstb.2024.0234.

Chung, B. Y., Balcerowicz, M., Di Antonio, M., Jaeger, K. E., Geng, F., Franaszek, K., Marriott, P., Brierley, I., Firth, A. E., & Wigge, P. A. (2020). An RNA thermoswitch regulates daytime growth in *Arabidopsis*. *Nature Plants, 6*(5), 522–532.

Corpas, F. J., Hayashi, M., Mano, S., Nishimura, M., & Barroso, J. B. (2009). Peroxisomes are required for in vivo nitric oxide accumulation in the cytosol following salinity stress of *Arabidopsis* plants. *Plant Physiology, 151*(4), 2083–2094.

Corpas, F. J., Leterrier, M., Valderrama, R., Airaki, M., Chaki, M., Palma, J. M., & Barroso, J. B. (2011). Nitric oxide imbalance provokes a nitrosative response in plants under abiotic stress. *Plant Science, 181*(5), 604–611.

Deng, Z., & Song, S. (2012). Sodium nitroprusside, ferricyanide, nitrite and nitrate decrease the thermo-dormancy of lettuce seed germination in a nitric oxide-dependent manner in light. *South African Journal of Botany, 78*, 139–146.

Dhanda, S., & Munjal, R. (2009). Cell membrane stability: Combining ability and gene effects under heat stress conditions. *Cereal Research Communications, 37*(3), 409–417.

Diner, B. A., & Petrouleas, V. (1990). Formation by NO of nitrosyl adducts of redox components of the photosystem II reaction center. II. Evidence that HCO^{-3}/CO_2 binds to the acceptor-side non-heme iron. *Biochimica et Biophysica Acta (BBA)-Bioenergetics, 1015*(1), 141–149.

Domingos, P., Prado, A. M., Wong, A., Gehring, C., & Feijo, J. A. (2015). Nitric oxide: A multitasked signaling gas in plants. *Molecular Plant, 8*(4), 506–520.

El-Beltagi, H. S., AHMED, O. K., & HEGAZY, A. E. (2016). Protective effect of nitric oxide on high temperature induced oxidative stress in wheat (*Triticum aestivum*) callus culture. *Notulae Scientia Biologicae, 8*(2), 192–198.

Fairchild, C. D., Schumaker, M. A., & Quail, P. H. (2000). HFR1 encodes an atypical bHLH protein that acts in phytochrome A signal transduction. *Genes & Development, 14*(18), 2377–2391.

Fancy, N. N., Bahlmann, A. K., & Loake, G. J. (2017). Nitric oxide function in plant abiotic stress. *Plant, Cell & Environment, 40*(4), 462–472.

Foresi, N., Correa-Aragunde, N., Parisi, G., Calo, G., Salerno, G., & Lamattina, L. (2010). Characterization of a nitric oxide synthase from the plant kingdom: NO generation from the green alga *Ostreococcus tauri* is light irradiance and growth phase dependent. *The Plant Cell, 22*(11), 3816–3830.

Fu, Z.-W., Wang, Y.-L., Lu, Y.-T., & Yuan, T.-T. (2016). Nitric oxide is involved in stomatal development by modulating the expression of stomatal regulator genes in *Arabidopsis*. *Plant Science, 252*, 282–289. https://doi.org/10.1016/j.plantsci.2016.08.005.

García, M. J., Lucena, C., & Romera, F. J. (2021). Ethylene and nitric oxide involvement in the regulation of Fe and P deficiency responses in dicotyledonous plants. *International Journal of Molecular Sciences, 22*(9), 4904.

Garcia-Mata, C., & Lamattina, L. (2007). Abscisic acid (ABA) inhibits light-induced stomatal opening through calcium- and nitric oxide-mediated signaling pathways. *Nitric Oxide Biology and Chemistry / Official Journal of the Nitric Oxide Society, 17*(3), 143–151. https://doi.org/10.1016/j.niox.2007.08.001.

Gautam, H., Fatma, M., Sehar, Z., Mir, I. R., & Khan, N. A. (2022). Hydrogen sulfide, ethylene, and nitric oxide regulate redox homeostasis and protect photosynthetic metabolism under high temperature stress in rice plants. *Antioxidants, 11*(8), 1478.

Gautam, H., Sehar, Z., Rehman, M. T., Hussain, A., AlAjmi, M. F., & Khan, N. A. (2021). Nitric oxide enhances photosynthetic nitrogen and sulfur-use efficiency and activity of ascorbate-glutathione cycle to reduce high temperature stress-induced oxidative stress in rice (*Oryza sativa* L.) plants. *Biomolecules, 11*(2), 305.

Gavazzi, G., Cocucci, M., Consonni, G., Lucchin, M., Masin, R., Negrini, N., & Zanin, G. (2008). Germination: A crucial step in plant growth. In *Advances in Plant Physiology* (Vol. 10, pp. 245–272). Hemantaranjan A.

Gayatri, G., Agurla, S., & Raghavendra, A. S. (2013). Nitric oxide in guard cells as an important secondary messenger during stomatal closure. *Frontiers in Plant Science, 4*, 425.

Gould, K., Lamotte, O., Klinguer, A., Pugin, A., & Wendehenne, D. (2003). Nitric oxide production in tobacco leaf cells: A generalized stress response? *Plant, Cell & Environment, 26*(11), 1851–1862.

Guilioni, L., Wery, J., & Tardieu, F. (1997). Heat stress-induced abortion of buds and flowers in pea: Is sensitivity linked to organ age or to relations between reproductive organs? *Annals of Botany, 80*(2), 159–168.

Gupta, K. J., Kolbert, Z., Durner, J., Lindermayr, C., Corpas, F. J., Brouquisse, R., Barroso, J. B., Umbreen, S., Palma, J. M., & Hancock, J. T. (2020). Regulating the regulator: Nitric oxide control of post-translational modifications. *New Phytologist, 227*(5), 1319–1325.

Haider, S., Iqbal, J., Naseer, S., Shaukat, M., Abbasi, B. A., Yaseen, T., Zahra, S. A., & Mahmood, T. (2022). Unfolding molecular switches in plant heat stress resistance: A comprehensive review. *Plant Cell Reports, 41*(3), 775–798.

Haldimann, P., & Feller, U. (2005). Growth at moderately elevated temperature alters the physiological response of the photosynthetic apparatus to heat stress in pea (*Pisum sativum* L.) leaves. *Plant, Cell & Environment, 28*(3), 302–317.

Hasanuzzaman, M., Nahar, K., Alam, M. M., & Fujita, M. (2012). Exogenous nitric oxide alleviates high temperature induced oxidative stress in wheat ('*Triticum aestivum*'L.) seedlings by modulating the antioxidant defense and glyoxalase system. *Australian Journal of Crop Science, 6*(8), 1314–1323.

He, N.-Y., Chen, L.-S., Sun, A.-Z., Zhao, Y., Yin, S.-N., & Guo, F.-Q. (2022). A nitric oxide burst at the shoot apex triggers a heat-responsive pathway in *Arabidopsis*. *Nature Plants, 8*(4), 434–450. https://doi.org/10.1038/s41477-022-01135-9.

Hill, A. C., & Bennett, J. (1970). Inhibition of apparent photosynthesis by nitrogen oxides. *Atmospheric Environment (1967), 4*(4), 341–348.

Hossain, K., Itoh, R., Yoshimura, G., Tokuda, G., Oku, H., Cohen, M., & Yamasaki, H. (2010). Effects of nitric oxide scavengers on thermoinhibition of seed germination in *Arabidopsis thaliana*. *Russian Journal of Plant Physiology, 57*(2), 222–232.

Iqbal, N., Sehar, Z., Fatma, M., Umar, S., Sofo, A., & Khan, N. A. (2022). Nitric oxide and abscisic acid mediate heat stress tolerance through regulation of osmolytes and antioxidants to protect photosynthesis and growth in wheat plants. *Antioxidants, 11*(2), 372.

Iqbal, N., Umar, S., Khan, N. A., & Corpas, F. J. (2021a). Crosstalk between abscisic acid and nitric oxide under heat stress: Exploring new vantage points. *Plant Cell Reports, 40*(8), 1429–1450.

Iqbal, N., Umar, S., Khan, N. A., & Corpas, F. J. (2021b). Nitric oxide and hydrogen sulfide coordinately reduce glucose sensitivity and decrease oxidative stress via ascorbate-glutathione cycle in heat-stressed wheat (*Triticum aestivum* L.) plants. *Antioxidants, 10*(1), 108.

Ismail, A. M., & Hall, A. E. (1999). Reproductive-stage heat tolerance, leaf membrane thermostability and plant morphology in cowpea. *Crop Science, 39*(6), 1762–1768.

Jang, I. C., Yang, J. Y., Seo, H. S., & Chua, N. H. (2005). HFR1 is targeted by COP1 E3 ligase for post-translational proteolysis during phytochrome A signaling. *Genes & Development, 19*(5), 593–602.

Jungjh, D. M., & Klose, C. (2016). Phytochromes functionas thermosensorsin *Arabidopsis*. *Science (New York, N.Y.), 354*(6314), 886–889.

Júnior, R.d S. N., Asseng, S., Müller, C., Deswarte, J.-C., Cohan, J.-P., & Martre, P. (2025). Negative impacts of climate change on crop yields are underestimated. *Trends in Plant Science*.

Kan, Y., Mu, X.-R., Gao, J., Lin, H.-X., & Lin, Y. (2023). The molecular basis of heat stress responses in plants. *Molecular Plant, 16*(10), 1612–1634.

Kang, X., Zhao, L., & Liu, X. (2024). Calcium signaling and the response to heat shock in crop plants. *International Journal of Molecular Sciences, 25*(1), 324.

Karim, M., Fracheboud, Y., & Stamp, P. (1997). Heat tolerance of maize with reference to some physiological characteristics.

Karpets, Y. V., Kolupaev, Y. E., & Vayner, A. A. (2015). Functional interaction between nitric oxide and hydrogen peroxide during formation of wheat seedling induced heat resistance. *Russian Journal of Plant Physiology, 62*(1), 65–70. https://doi.org/10.1134/s1021443714060090.

Karpets, Y. V., Kolupaev, Y. E., & Yastreb, T. O. (2011). Effect of sodium nitroprusside on heat resistance of wheat coleoptiles: Dependence on the formation and scavenging of reactive oxygen species. *Russian Journal of Plant Physiology, 58*(6), 1027–1033. https://doi.org/10.1134/S1021443711060094.

Karpets, Y. V., Kolupaev, Y. E., Yastreb, T., & Oboznyi, A. (2016). Induction of heat resistance in wheat seedlings by exogenous calcium, hydrogen peroxide, and nitric oxide donor: Functional interaction of signal mediators. *Russian Journal of Plant Physiology, 63*(4), 490–498.

Kaur, K., & Kaur, K. (2018). Nitric oxide improves thermotolerance in spring maize by inducing varied genotypic defense mechanisms. *Acta Physiologiae Plantarum, 40*(3), 55. https://doi.org/10.1007/s11738-018-2632-9.

Kaur, R., Sinha, K., & Bhunia, R. K. (2019). Can wheat survive in heat? Assembling tools towards successful development of heat stress tolerance in *Triticum aestivum* L. *Molecular Biology Reports, 46*(2), 2577–2593.

Kaushal, N., Bhandari, K., Siddique, K. H. M., & Nayyar, H. (2016). Food crops face rising temperatures: An overview of responses, adaptive mechanisms, and approaches to improve heat tolerance. *Cogent Food & Agriculture, 2*(1), 1134380. https://doi.org/10.1080/23311932.2015.1134380.

Kerbler, S. M., & Wigge, P. A. (2023). Temperature sensing in plants. *Annual Review of Plant Biology, 74*(1), 341–366.

Khan, M., Ali, S., Al Azzawi, T. N. I., & Yun, B.-W. (2023). Nitric oxide acts as a key signaling molecule in plant development under stressful conditions. *International Journal of Molecular Sciences, 24*(5), 4782.

Koini, M. A., Alvey, L., Allen, T., Tilley, C. A., Harberd, N. P., Whitelam, G. C., & Franklin, K. A. (2009). High temperature-mediated adaptations in plant architecture require the bHLH transcription factor PIF4. *Current biology, 19*(5), 408–413. https://doi.org/10.1016/j.cub.2009.01.046.

Kumar, R., Tasleem, M., Jain, M., Ahuja, S., Goswami, S., Bakshi, S., Jambhulkar, S., Singh, S., Singh, G., & Pathak, H. (2019). Nitric oxide triggered defense network in wheat: Augmenting tolerance and grain-quality related traits under heat-induced oxidative damage. *Environmental and Experimental Botany, 158*, 189–204.

Kumar, S. V., & Wigge, P. A. (2010). H2A.Z-containing nucleosomes mediate the thermosensory response in *Arabidopsis*. *Cell, 140*(1), 136–147. https://doi.org/10.1016/j.cell.2009.11.006.

Kumar, S., Bhushan, B., Wakchaure, G. C., Dutta, R., Jat, B. S., Meena, K. K., Rakshit, S., & Pathak, H. (2023). Unveiling the impact of heat stress on seed biochemical composition of major cereal crops: Implications for crop resilience and nutritional value. *Plant Stress, 9*, 100183. https://doi.org/10.1016/j.stress.2023.100183.

Lamattina, L., García-Mata, C., Graziano, M., & Pagnussat, G. (2003). Nitric oxide: The versatility of an extensive signal molecule. *Annual Review of Plant Biology, 54*(1), 109–136.

Lamotte, O., Courtois, C., Barnavon, L., Pugin, A., & Wendehenne, D. (2005). Nitric oxide in plants: The biosynthesis and cell signalling properties of a fascinating molecule. *Planta, 221*(1), 1–4.

Lamotte, O., Courtois, C., Dobrowolska, G., Besson, A., Pugin, A., & Wendehenne, D. (2006). Mechanisms of nitric-oxide-induced increase of free cytosolic Ca^{2+} concentration in Nicotiana plumbaginifolia cells. *Free Radical Biology and Medicine, 40*(8), 1369–1376.

Lang, T., Sun, H., Li, N., Lu, Y., Shen, Z., Jing, X., Xiang, M., Shen, X., & Chen, S. (2014). Multiple signaling networks of extracellular ATP, hydrogen peroxide, calcium, and nitric oxide in the mediation of root ion fluxes in secretor and non-secretor mangroves under salt stress. *Aquatic Botany, 119*, 33–43.

Lee, U., Wie, C., Fernandez, B. O., Feelisch, M., & Vierling, E. (2008). Modulation of nitrosative stress by S-nitrosoglutathione reductase is critical for thermotolerance and plant growth in *Arabidopsis*. *The Plant Cell, 20*(3), 786–802.

Legris, M., Klose, C., Burgie, E. S., Rojas, C. C. R., Neme, M., Hiltbrunner, A., Wigge, P. A., Schäfer, E., Vierstra, R. D., & Casal, J. J. (2016). Phytochrome B integrates light and temperature signals in *Arabidopsis*. *Science (New York, N.Y.), 354*(6314), 897–900.

Leshem, Y.a Y. (2000). Nitric oxide gas as an endogenous regulator of fruit, vegetable and flower maturation and senescence. In Y. A. Y. Leshem (Ed.). *Nitric oxide in plants: occurrence, function and use* (pp. 33–62). Springer Netherlands. https://doi.org/10.1007/978-94-011-4239-7_3.

Leshem, Y. Y., & Wills, R. B. H. (1998). Harnessing senescence delaying gases nitric oxide and nitrous oxide: A novel approach to postharvest control of fresh horticultural produce. *Biologia Plantarum, 41*(1), 1–10. https://doi.org/10.1023/A:1001779227767.

Leshem, Y., & Kuiper, P. (1996). Is there a GAS (general adaptation syndrome) response to various types of environmental stress? *Biologia Plantarum, 38*(1), 1–18.

Li, Y., Gong, K., Wang, X., Sun, Z., & Ding, F. (2025). Heat shock transcription factors as central integrators of plant stress responses: From thermotolerance to multi-stress resilience. *Biology, 14*(12), 1800.

Li, Z. G., Yang, S. Z., Long, W. B., Yang, G. X., & Shen, Z. Z. (2013). Hydrogen sulphide may be a novel downstream signal molecule in nitric oxide-induced heat tolerance of maize (*Zea mays* L.) seedlings. *Plant, Cell & Environment, 36*(8), 1564–1572.

Liu, M., Liu, X. X., He, X. L., Liu, L. J., Wu, H., Tang, C. X., Zhang, Y. S., & Jin, C. W. (2017). Ethylene and nitric oxide interact to regulate the magnesium deficiency-induced root hair development in *Arabidopsis*. *New Phytologist, 213*(3), 1242–1256.

Locato, V., Gadaleta, C., De Gara, L., & De Pinto, M. C. (2008). Production of reactive species and modulation of antioxidant network in response to heat shock: A critical balance for cell fate. *Plant, Cell & Environment, 31*(11), 1606–1619.

López-Gómez, P., Buezo, J., Urra, M., Cornejo, A., Esteban, R., Fernández de los Reyes, J., Urarte, E., Rodríguez-Dobreva, E., Chamizo-Ampudia, A., Eguaras, A., Wolf, S., Marino, D., Martínez-Merino, V., & Moran, J. F. (2024). A new oxidative pathway of nitric oxide production from oximes in plants. *Molecular Plant, 17*(1), 178–198. https://doi.org/10.1016/j.molp.2023.12.009.

Lum, H.-K., Lee, C.-H., Butt, Y. K.-C., & Lo, S. C.-L. (2005). Sodium nitroprusside affects the level of photosynthetic enzymes and glucose metabolism in *Phaseolus aureus* (mung bean). *Nitric Oxide Biology and Chemistry / Official Journal of the Nitric Oxide Society, 12*(4), 220–230. https://doi.org/10.1016/j.niox.2005.03.002.

Malerba, M., & Cerana, R. (2018). Role of peroxynitrite in the responses induced by heat stress in tobacco BY-2 cultured cells. *Protoplasma, 255*(4), 1079–1087. https://doi.org/10.1007/s00709-017-1200-2.

Manafi, H., Baninasab, B., Gholami, M., & Talebi, M. (2021). Nitric oxide induced thermotolerance in strawberry plants by activation of antioxidant systems and transcriptional regulation of heat shock proteins. *The Journal of Horticultural Science and Biotechnology, 96*(6), 783–796. https://doi.org/10.1080/14620316.2021.1927206.

Maurya, A. K. (2020). Oxidative stress in crop plants. In *Agronomic crops: Volume 3: Stress responses and tolerance*. Springer, 349–380.

Maurya, A. K., & Rani, A. (2017). Nitric oxide (NO) and physio-biochemical adaptation in plants against stress. In V. Shukla, S. Kumar, & N. Kumar (Eds.). *Plant adaptation strategies in changing environment* (pp. 347–386). Springer Singapore. https://doi.org/10.1007/978-981-10-6744-0_15.

Ming, A., Rowell, I., Lewin, S., Rouse, R., Aubry, T., & Boland, E. (2021). Key messages from the IPCC AR6 climate science report.

Mohanty, D., Peláez-Vico, M.Á., Myers, R. J., Jr, Sánchez-Vicente, M. I., Lorenzo, O., & Mittler, R. (2025). Aboveground whole-plant live imaging method for nitric oxide (NO) reveals an intricate relationship between NO and H_2O_2. *New Phytologist.* https://doi.org/10.1111/nph.70094.

Moreau, M., Lee, G. I., Wang, Y., Crane, B. R., & Klessig, D. F. (2008). AtNOS/AtNOA1 Is a Functional *Arabidopsis thaliana* cGTPase and Not a Nitric-oxide Synthase* *Journal of Biological Chemistry, 283*(47), 32957–32967. https://doi.org/10.1074/jbc.M804838200.

Moreau, M., Lindermayr, C., Durner, J., & Klessig, D. F. (2010). NO synthesis and signaling in plants–where do we stand? *Physiologia Plantarum, 138*(4), 372–383.

Mukherjee, S. (2019). Recent advancements in the mechanism of nitric oxide signaling associated with hydrogen sulfide and melatonin crosstalk during ethylene-induced fruit ripening in plants. *Nitric Oxide Biology and Chemistry / Official Journal of the Nitric Oxide Society, 82*, 25–34.

Murata, N., Takahashi, S., Nishiyama, Y., & Allakhverdiev, S. I. (2007). Photoinhibition of photosystem II under environmental stress. *Biochimica et Biophysica Acta (BBA)-Bioenergetics, 1767*(6), 414–421.

Naaz, S., Pande, A., & Laxmi, A. (2025). Nitric oxide-mediated thermomemory: A new perspective on plant heat stress resilience [Review]. *Frontiers in Plant Science, Volume 16*, 2025. https://doi.org/10.3389/fpls.2025.1525336.

Nasir Khan, M., Mobin, M., & Khorshid Abbas, Z. (2015). Nitric oxide and high temperature stress: A physiological perspective. In *Nitric oxide action in abiotic stress responses in plants*. Springer, 77–93.

Neill, S. J., Desikan, R., & Hancock, J. T. (2003). Nitric oxide signalling in plants. *New Phytologist, 159*(1), 11–35.

Nishad, A., & Nandi, A. K. (2021). Recent advances in plant thermomemory. *Plant Cell Reports, 40*(1), 19–27.

Niu, L., & Liao, W. (2016). Hydrogen peroxide signaling in plant development and abiotic responses: Crosstalk with nitric oxide and calcium [review]. *Frontiers in Plant Science, Volume 7*, 2016. https://doi.org/10.3389/fpls.2016.00230.

Olas, J. J., Apelt, F., Annunziata, M. G., John, S., Richard, S. I., Gupta, S., Kragler, F., Balazadeh, S., & Mueller-Roeber, B. (2021). Primary carbohydrate metabolism genes participate in heat-stress memory at the shoot apical meristem of *Arabidopsis thaliana*. *Molecular Plant, 14*(9), 1508–1524. https://doi.org/10.1016/j.molp.2021.05.024.

Pande, A., Mun, B.-G., Khan, M., Rahim, W., Lee, D.-S., Lee, G.-M., Al Azawi, T. N. I., Hussain, A., & Yun, B.-W. (2022). Nitric oxide signaling and its association with ubiquitin-mediated proteasomal degradation in plants. *International Journal of Molecular Sciences, 23*(3), 1657.

Parankusam, S., Adimulam, S. S., Bhatnagar-Mathur, P., & Sharma, K. K. (2017). Nitric oxide (NO) in plant heat stress tolerance: current knowledge and perspectives [review]. *Frontiers in Plant Science, Volume 8*, 2017. https://doi.org/10.3389/fpls.2017.01582.

Peng, J., Yu, Y., & Fang, X. (2025). Stress sensing and response through biomolecular condensates in plants. *Plant Communications, 6*(2), 101225. https://doi.org/10.1016/j.xplc.2024.101225.

Peng, X., Zhang, X., Li, B., & Zhao, L. (2019). Cyclic nucleotide-gated ion channel 6 mediates thermotolerance in *Arabidopsis* seedlings by regulating nitric oxide production via cytosolic calcium ions. *BMC Plant Biology, 19*(1), 368. https://doi.org/10.1186/s12870-019-1974-9.

Piterková, J., Luhová, L., Mieslerová, B., Lebeda, A., & Petřivalský, M. (2013). Nitric oxide and reactive oxygen species regulate the accumulation of heat shock proteins in tomato leaves in response to heat shock and pathogen infection. *Plant Science, 207*, 57–65. https://doi.org/10.1016/j.plantsci.2013.02.010.

Prolo, C., Piacenza, L., & Radi, R. (2024). Peroxynitrite: A multifaceted oxidizing and nitrating metabolite. *Current Opinion in Chemical Biology, 80*, 102459. https://doi.org/10.1016/j.cbpa.2024.102459.

Puglia, G. D. (2024). Reactive oxygen and nitrogen species (RONS) signalling in seed dormancy release, perception of environmental cues, and heat stress response. *Plant Growth Regulation, 103*(1), 9–32. https://doi.org/10.1007/s10725-023-01094-x.

Qian, S., Wang, Y., Ma, H., & Zhang, L. (2015). Expansion and functional divergence of Jumonji C-containing histone demethylases: Significance of duplications in ancestral angiosperms and vertebrates. *Plant Physiology, 168*(4), 1321–1337.

Rai, K. K., Pandey, N., & Rai, S. P. (2020). Salicylic acid and nitric oxide signaling in plant heat stress. *Physiologia Plantarum, 168*(2), 241–255. https://doi.org/10.1111/ppl.12958.

Rai, K. K., Rai, N., & Rai, S. P. (2018). Salicylic acid and nitric oxide alleviate high temperature induced oxidative damage in *Lablab purpureus* L plants by regulating bio-physical processes and DNA methylation. *Plant Physiology and Biochemistry, 128*, 72–88.

Rasheed, F., Mir, I. R., Sehar, Z., Fatma, M., Gautam, H., Khan, S., Anjum, N. A., Masood, A., Sofo, A., & Khan, N. A. (2022). Nitric oxide and salicylic acid regulate glutathione and ethylene production to enhance heat stress acclimation in wheat involving sulfur assimilation. *Plants, 11*(22), 3131.

Rehman, S., Ahmad, Z., Ramakrishnan, M., Kalendar, R., & Zhuge, Q. (2023). Regulation of plant epigenetic memory in response to cold and heat stress: Towards climate resilient agriculture. *Functional & Integrative Genomics, 23*(4), 298. https://doi.org/10.1007/s10142-023-01219-5.

Ruane, A.C. (2024). Synthesis report of the IPCC sixth assessment report (AR6).

Salvucci, M. E., & Crafts-Brandner, S. J. (2004). Relationship between the heat tolerance of photosynthesis and the thermal stability of Rubisco activase in plants from contrasting thermal environments. *Plant Physiology, 134*(4), 1460–1470.

Sana, S., Aftab, T., Naeem, M., Jha, P. K., & Prasad, P. (2025). Production of secondary metabolites under challenging environments: Understanding functions and mechanisms of signalling molecules. *Frontiers in Plant Science, 16*, 1569014.

Sanakis, Y., Goussias, C., Mason, R. P., & Petrouleas, V. (1997). NO interacts with the tyrosine radical YD• of photosystem II to form an iminoxyl radical. *Biochemistry, 36*(6), 1411–1417.

Savvides, A., Ali, S., Tester, M., & Fotopoulos, V. (2016). Chemical priming of plants against multiple abiotic stresses: Mission possible? *Trends in Plant Science, 21*(4), 329–340.

Schansker, G., Goussias, C., Petrouleas, V., & Rutherford, A. W. (2002). Reduction of the Mn cluster of the water-oxidizing enzyme by nitric oxide: Formation of an S-2 state. *Biochemistry, 41*(9), 3057–3064.

Schulze, E. D., Lange, O. L., Kappen, L., Buschbom, U., & Evenari, M. (1973). Stomatal responses to changes in temperature at increasing water stress. *Planta, 110*(1), 29–42. https://doi.org/10.1007/BF00386920.

Sedlářová, M., Jedelská, T., Lebeda, A., & Petřivalský, M. (2025). Progress in plant nitric oxide studies: Implications for phytopathology and plant protection. *International Journal of Molecular Sciences, 26*(5), 2087.

Sharkey, T. D. (2005). Effects of moderate heat stress on photosynthesis: Importance of thylakoid reactions, rubisco deactivation, reactive oxygen species, and thermotolerance provided by isoprene. *Plant, Cell & Environment, 28*(3), 269–277.

Sharma, A., Soares, C., Sousa, B., Martins, M., Kumar, V., Shahzad, B., Sidhu, G. P., Bali, A. S., Asgher, M., & Bhardwaj, R. (2020). Nitric oxide-mediated regulation of oxidative stress in plants under metal stress: A review on molecular and biochemical aspects. *Physiologia Plantarum, 168*(2), 318–344.

Siddiqui, M. H., Alamri, S. A., Al-Khaishany, M. Y., Al-Qutami, M. A., Ali, H. M., Al-Rabiah, H., & Kalaji, H. M. (2017). Exogenous application of nitric oxide and spermidine reduces the negative effects of salt stress on tomato. *Horticulture, Environment, and Biotechnology, 58*(6), 537–547. https://doi.org/10.1007/s13580-017-0353-4.

Siddiqui, M. H., Al-Whaibi, M. H., & Basalah, M. O. (2011). Role of nitric oxide in tolerance of plants to abiotic stress. *Protoplasma, 248*(3), 447–455.

Sita, K., Sehgal, A., Bhardwaj, A., Bhandari, K., Kumar, S., Prasad, P. V., Jha, U., Siddique, K. H., & Nayyar, H. (2021). Nitric oxide secures reproductive efficiency in heat-stressed lentil (Lens culinaris Medik.) plants by enhancing the photosynthetic ability to improve yield traits. *Physiology and Molecular Biology of Plants, 27*(11), 2549–2566.

Song, L., Ding, W., Shen, J., Zhang, Z., Bi, Y., & Zhang, L. (2008). Nitric oxide mediates abscisic acid induced thermotolerance in the calluses from two ecotypes of reed under heat stress. *Plant Science, 175*(6), 826–832. https://doi.org/10.1016/j.plantsci.2008.08.005.

Song, L., Ding, W., Zhao, M., Sun, B., & Zhang, L. (2006). Nitric oxide protects against oxidative stress under heat stress in the calluses from two ecotypes of reed. *Plant Science, 171*(4), 449–458. https://doi.org/10.1016/j.plantsci.2006.05.002.

Song, L., Jiang, Y., Zhao, H., & Zhang, Z. (2012). Comparative study on calli from two reed ecotypes under heat stress. *Russian Journal of Plant Physiology, 59*(3), 381–388. https://doi.org/10.1134/S1021443712030168.

Song, L., Zhao, H., & Hou, M. (2013). Involvement of nitric oxide in acquired thermotolerance of rice seedlings. *Russian Journal of Plant Physiology, 60*(6), 785–790. https://doi.org/10.1134/S1021443713060149.

Sougrakpam, Y., Babuta, P., & Deswal, R. (2023). Nitric oxide (NO) modulates low temperature-stress signaling via S-nitrosation, a NO PTM, inducing ethylene biosynthesis inhibition leading to enhanced post-harvest shelf-life of agricultural produce. *Physiology and Molecular Biology of Plants, 29*(12), 2051–2065.

Staacke, T., Mueller-Roeber, B., & Balazadeh, S. (2025). Stress resilience in plants: The complex interplay between heat stress memory and resetting. *New Phytologist, 245*(6), 2402–2421.

Stasik, O., & Jones, H. G. (2007). Response of photosynthetic apparatus to moderate high temperature in contrasting wheat cultivars at different oxygen concentrations. *Journal of Experimental Botany, 58*(8), 2133–2143.

Talwar, P. S., Gupta, R., Maurya, A. K., & Deswal, R. (2012). Brassica juncea nitric oxide synthase like activity is stimulated by PKC activators and calcium suggesting modulation by PKC-like kinase. *Plant Physiology and Biochemistry, 60*, 157–164.

Uchida, A., Jagendorf, A. T., Hibino, T., Takabe, T., & Takabe, T. (2002). Effects of hydrogen peroxide and nitric oxide on both salt and heat stress tolerance in rice. *Plant Science, 163*(3), 515–523.

Vollenweider, P., & Günthardt-Goerg, M. S. (2005). Diagnosis of abiotic and biotic stress factors using the visible symptoms in foliage. *Environmental Pollution, 137*(3), 455–465.

Wang, L., Guo, Y., Jia, L., Chu, H., Zhou, S., Chen, K., Wu, D., & Zhao, L. (2014). Hydrogen peroxide acts upstream of nitric oxide in the heat shock pathway in arabidopsis seedlings. *Plant Physiology, 164*(4), 2184–2196. https://doi.org/10.1104/pp.113.229369.

Wang, T., Hou, X., Wei, L., Deng, Y., Zhao, Z., Liang, C., & Liao, W. (2024). Protein S-nitrosylation under abiotic stress: Role and mechanism. *Plant Physiology and Biochemistry, 207*, 108329.

Wei, J., Li, D.-X., Zhang, J.-R., Shan, C., Rengel, Z., Song, Z.-B., & Chen, Q. (2018). Phytomelatonin receptor PMTR1-mediated signaling regulates stomatal closure in *Arabidopsis thaliana. Journal of Pineal Research, 65*(2), e12500. https://doi.org/10.1111/jpi.12500.

Wendehenne, D., Courtois, C., Besson, A., Gravot, A., Buchwalter, A., Pugin, A., & Lamotte, O. (2006). NO-based signaling in plants. In nitric oxide in plant growth, development and stress physiology (. Springer, 35–51.

Wink, D. A., Miranda, K. M., Espey, M. G., Pluta, R. M., Hewett, S. J., Colton, C., Vitek, M., Feelisch, M., & Grisham, M. B. (2001). Mechanisms of the antioxidant effects of nitric oxide. *Antioxidants and Redox Signaling, 3*(2), 203–213.

Wink, D. A., Vodovotz, Y., Grisham, M. B., DeGraff, W., Cook, J. C., Pacelli, R., Krishna, M., & Mitchell, J. B. (1999). Antioxidant effects of nitric oxide. *Methods in Enzymology, 301*, 413–424. https://doi.org/10.1016/s0076-6879(99)01105-2.

Xu, X.-T., Jin, X., Liao, W.-B., Dawuda, M. M., Li, X.-P., Wang, M., Niu, L.-J., Ren, P.-J., & Zhu, Y.-C. (2017). Nitric oxide is involved in ethylene-induced adventitious root development in cucumber (*Cucumis sativus* L.) explants. *Scientia Horticulturae, 215*, 65–71.

Xuan, Y., Zhou, S., Wang, L., Cheng, Y., & Zhao, L. (2010). Nitric oxide functions as a signal and acts upstream of AtCaM3 in thermotolerance in *Arabidopsis* seedlings. *Plant Physiology, 153*(4), 1895–1906. https://doi.org/10.1104/pp.110.160424.

Yang, J., Lin, R., Sullivan, J., Hoecker, U., Liu, B., Xu, L., & Wang, H. (2005). Light regulates COP1-mediated degradation of HFR1, a transcription factor essential for light signaling in Arabidopsis. *The Plant Cell, 17*(3), 804–821.

Yang, J.-D., Yun, J.-Y., Zhang, T.-H., & Zhao, H.-L. (2006). Presoaking with nitric oxide donor SNP alleviates heat shock damages in mung bean leaf discs. *Bot Studies, 47*, 129–136.

Yang, Q., He, H., Li, H., Tian, H., Zhang, J., Zhai, L., Chen, J., Wu, H., Yi, G., & He, Z.-H. (2011). NOA1 functions in a temperature-dependent manner to regulate chlorophyll biosynthesis and Rubisco formation in rice. *PLoS One, 6*(5), e20015.

Yang, W., Sun, Y., Chen, S., Jiang, J., Chen, F., Fang, W., & Liu, Z. (2011). The effect of exogenously applied nitric oxide on photosynthesis and antioxidant activity in heat stressed chrysanthemum. *Biologia Plantarum, 55*(4), 737–740. https://doi.org/10.1007/s10535-011-0178-4.

Ying, S., Yang, W., Li, P., Hu, Y., Lu, S., Zhou, Y., Huang, J., Hancock, J. T., & Hu, X. (2022). Phytochrome B enhances seed germination tolerance to high temperature by reducing S-nitrosylation of HFR1. *EMBO Reports, 23*(10), e54371. https://doi.org/10.15252/embr.202154371.

Yu, M., Lamattina, L., Spoel, S. H., & Loake, G. J. (2014). Nitric oxide function in plant biology: A redox cue in deconvolution. *New Phytologist, 202*(4), 1142–1156.

Zahra, N., Hafeez, M. B., Kausar, A., Al Zeidi, M., Asekova, S., Siddique, K. H., & Farooq, M. (2023). Plant photosynthetic responses under drought stress: Effects and management. *Journal of Agronomy and Crop Science, 209*(5), 651–672.

Zargar, M., Ramezan, D., Bayat, M., & Soltani Nejad, M. (2025). Alleviation of high temperature stress in bell pepper through foliar application of melatonin and sodium nitroprusside. *Greenhouse Plant Production Journal,* 18–43.

Zhang, A., Jiang, M., Zhang, J., Ding, H., Xu, S., Hu, X., & Tan, M. (2007). Nitric oxide induced by hydrogen peroxide mediates abscisic acid-induced activation of the mitogen-activated protein kinase cascade involved in antioxidant defense in maize leaves. *New Phytologist, 175*(1), 36–50. https://doi.org/10.1111/j.1469-8137.2007.02071.x.

Zhang, J. H., HUANG, W. D., LIU, Y. P., & PAN, Q. H. (2005). Effects of temperature acclimation pretreatment on the ultrastructure of mesophyll cells in young grape plants *(Vitis vinifera* L. cv. Jingxiu) under cross-temperature stresses. *Journal of Integrative Plant Biology, 47*(8), 959–970.

Zhang, W., Zhou, R.-G., Gao, Y.-J., Zheng, S.-Z., Xu, P., Zhang, S.-Q., & Sun, D.-Y. (2009). Molecular and genetic evidence for the key role of AtCaM3 in heat-shock signal transduction in *Arabidopsis*. *Plant Physiology, 149*(4), 1773–1784.

Zhang, X., Ma, M., Wu, C., Huang, S., & Danish, S. (2023). Mitigation of heat stress in wheat (*Triticum aestivum* L.) via regulation of physiological attributes using sodium nitroprusside and gibberellic acid. *BMC Plant Biology, 23*(1), 302. https://doi.org/10.1186/s12870-023-04321-9.

Zhang, X., Wang, W., Kang, X., & Zhao, L. (2020). *Arabidopsis* CaM3 inhibits nitric oxide accumulation and improves thermotolerance by promoting S-nitrosoglutathione reductase via direct binding. *Plant Growth Regulation, 90*(1), 41–50.

Zhao, M.-G., Tian, Q.-Y., & Zhang, W.-H. (2007). Nitric oxide synthase-dependent nitric oxide production is associated with salt tolerance in *Arabidopsis*. *Plant Physiology, 144*(1), 206–217. https://doi.org/10.1104/pp.107.096842.

Zhou, B., Guo, Z., Xing, J., & Huang, B. (2005). Nitric oxide is involved in abscisic acid-induced antioxidant activities in *Stylosanthes guianensis*. *Journal of Experimental Botany, 56*(422), 3223–3228.

Zulfiqar, F., Akram, N. A., & Ashraf, M. (2020). Osmoprotection in plants under abiotic stresses: New insights into a classical phenomenon. *Planta, 251*(1), 3.

Chapter 8

Role of nitric oxide in the mitigation of flooding and hypoxia stress

Sandip Pal[1], Dip Pal[1], Suraj Panja[2], Pratyasha Samanta[1], Retwika Ganguly[3], Nilanjana Das[4], and Narottam Dey[1]

[1]*Department of Biotechnology, Rice Biotechnology Laboratory, Visva-Bharati, Siksha Bhavana, Santiniketan, West Bengal, India,* [2]*Department of Biosciences and Bioengineering, Indian Institute of Technology, Guwahati, Assam, India,* [3]*Department of Biology, Suriano Lab, Montclair State University, Montclair, NJ, United States,* [4]*Department of Biotechnology, Oxidative Stress Biology Laboratory, Visva-Bharati, Siksha Bhavana, Santiniketan, West Bengal, India*

8.1 Introduction

One of the frequent effects of continuous global change is alterations in precipitation patterns. As a result, some areas receive more frequent and severe droughts, while other areas receive more frequent and intense rainfall (Pedersen et al., 2017). Notably, excessive floods brought on either by excessive rainfall due to global climate change or by dam failures seriously harm the environment (Ge et al., 2020; Ge et al., 2021). Moreover, inadequate drainage systems, increased soil degradation, and climate change also contribute to increased floods worldwide. These adverse effects of flood affect plant growth and yield (Khator & Shekhawat, 2020) and can cause loss of up to 80% of crop production (Voesenek & Bailey-Serres, 2015). Therefore, for sustainable agricultural and environmental management, it is crucial to comprehend these issues and create plants that mitigate the consequences of flooding, which also induces hypoxia conditions due to limited diffusion of oxygen to the below-ground plant tissues (Jan et al., 2022; Khan et al. 2023).

To adapt to changing environmental conditions, plants have developed sophisticated defense systems. Stress conditions lead to the overproduction of reactive species in several plant compartments, including mitochondria, peroxisomes, apoplasts, and chloroplasts, which disrupt the normal plant metabolism (Choudhury et al., 2017; Kaur et al., 2019). These reactive species, consisting of reactive oxygen species (ROS), reactive nitrogen species (RNS),

Nitric Oxide in Plant Resilience to Abiotic Stress. DOI: https://doi.org/10.1016/B978-0-443-40497-9.00002-7

reactive sulfur species, and reactive carbon species (Zhou et al., 2021), can have both beneficial and detrimental effects on plants, depending on their concentration. For example, if their threshold level is crossed, they may become toxic to different plant cells and organelles (Khare et al., 2020). Depending on their abundance in plant cells, RNS and ROS can either cause an oxidative burst or function as signaling molecules against various stresses (Patel et al., 2019). To withstand these oxidative bursts, plants have an armor in the form of Catalase (CAT), Superoxide dismutase (SOD), Glutathione S-transferase, Monodehydroascorbate reductase, and Glutathione reductase. In addition to the accumulation of these reactive species, plants also undergo several cellular, physiological, and molecular alterations in response to environmental stressors (Xie et al., 2019). These include modifications in the structure of the cell wall, photosynthetic and nutrient adjustments, and alterations in hormone levels (Manghwar et al., 2024).

RNS is a group of highly instinctive molecules, including nitric oxide (NO), nitrogen dioxide, and S-nitroglutathione, that significantly influence the growth and development of vegetative and reproductive tissues of plants. These RNS affect the seed dormancy and germination, root and shoot growth, nodulation, nutrient uptake, stomatal movement, vascular differentiation, flowering, and fruit ripening (Table 8.1) (Signorelli & Considine, 2018; Rahim et al., 2022; Niyoifasha et al., 2023). Among these, NO has emerged as a free radical gastrotransmitter that functions as a key signaling molecule during the activation of plant defense (Hancock & Veal, 2020). Numerous targets, including proteins, metal complexes, and free radicals, are known to react with NO (Toledo & Augusto, 2012). Furthermore, NO quickly combines with other free radicals, including superoxide anion ($O_2.^-$), to produce strong peroxynitrite, whereas the reaction between glutathione and NO yields immensely substantial product S-nitrosolglutathione, which is considered to be the predominant NO reservoir in plant cells (Khator et al., 2024). Additionally, NO can be oxidized to produce nitrogen dioxide and nitrate ion, and at low concentrations, it can scavenge the hydroxyl radical, which hinders its antioxidant properties (Wink et al., 2001). In this chapter, we discuss the synthesis of NO in plants under flooding and its different signaling roles under hypoxia, leading to different phenotypic effects on plants.

8.2 Synthesis of nitric oxide

In animals, NO is majorly generated by three isoforms of NO synthase (NOS), that is, endothelial, neuronal, and inducible. NOS generates the formation of NO from L-arginine and molecular oxygen (Foresi et al., 2010; Thalineau et al., 2016). In spite of evidence of its viable presence in plants, the metabolic root of NO and its function in plant signaling cascades still remains largely unclear (Berger et al., 2019). Although the pharmacological assays using NOS

TABLE 8.1 **Induced feedback of nitric oxide in various plants under hypoxia.**

Event	Plant species	Function	References
Phytoglobin–NO Cycle	*Oryza sativa* L.	Nitrite promotes the formation of nitrate reductase (NR) and NO in low oxygen environments; phytoglobin (Pgb) controls NO, promoting fermentation and ATP synthesis while lowering ROS and lipid peroxidation.	Kumari et al. (2021)
Alternative energy pathway	*Cicer arietinum* L.	By promoting ATP exchange and fermentation, NO via the Pgb–NO cycle acts as an alternate respiration pathway, assisting in the maintenance of cellular energy under anoxia.	Pandey et al. (2019)
Respiratory regulation	*Cicer arietinum* L.	Anaerobic germination can be facilitated by the ability of NO to reversibly bind cytochrome oxidase in mitochondria, which lowers oxygen consumption and helps prevent internal anoxia.	Pandey et al. (2019)
Fermentation support	*Pisum sativum* L.	NO enhancing activities of key enzymes (e.g., ADH, LDH) via fermentation that are essential for anaerobic ATP production during germination.	Zabalza et al. (2009)
Hormonal and dormancy regulation	*Arabidopsis thaliana* L.	NO promotes α-amylase activity (key for starch mobilization), degrades ABA-related dormancy proteins (such as ABI5), enhances ABA catabolism, and supports GA-mediated responses; however, these effects are time- and dose-dependent, with high concentrations potentially inhibitory.	Timilsina et al. (2022)

inhibitors reveal NOS-like activity in various plant species, the NOS-like enzyme responsible for the generation of NO from L-arginine has not yet been identified in higher plants (Astier et al., 2017). However, various other enzymatic and nonenzymatic mechanisms can produce NO in chloroplasts, mitochondria, peroxisomes, plasma membrane, cytoplasm, and apoplast (Nabi et al., 2019; Sun et al., 2012; Tan et al., 2012). Numerous investigations have identified eight distinct enzymatic (oxidative and reductive) and non-enzymatic processes involved in NO production (Gupta et al., 2011; Crawford & Guo, 2005; Procházková et al., 2014). Non-enzymatic production of NO is the probable outcome of the reduction of nitrite in the acidic cells of plant tissues (León et al., 2021; Kaya et al., 2020). The oxidative pathway produces NO by oxidizing hydroxylamine and L-arginine (Mur et al., 2013). Nitrate reductase (NR) and other reductive enzymes, including cytochrome c reductase and nitrite reductase, which are found in mitochondria, are the main components of the reductive pathway (Roszer et al., 2012; León et al., 2021) (See Fig. 8.1).

Additionally, using reduced cytochrome c as an electron acceptor, NO_2/NO reductase may also convert nitric dioxide into NO (Mohn et al., 2019). During ATP synthesis, nitrogen dioxide is utilized as an alternative electron acceptor in mitochondrial complex III and IV, where the enzymatic reduction of nitrogen dioxide to NO also occurs (Kolbert et al., 2019). This advances to hypoxia in plant cells and results in mitochondrial NO production, leading to suppression of the activity of cytochrome c oxidase, which boosts the energy status of hypoxic cells (Gupta et al., 2016). The reduction of nitrogen dioxide in plant peroxisomes under hypoxic conditions is another mechanism for reductive NO generation (Igamberdiev et al., 2010). Cytosol acidification is

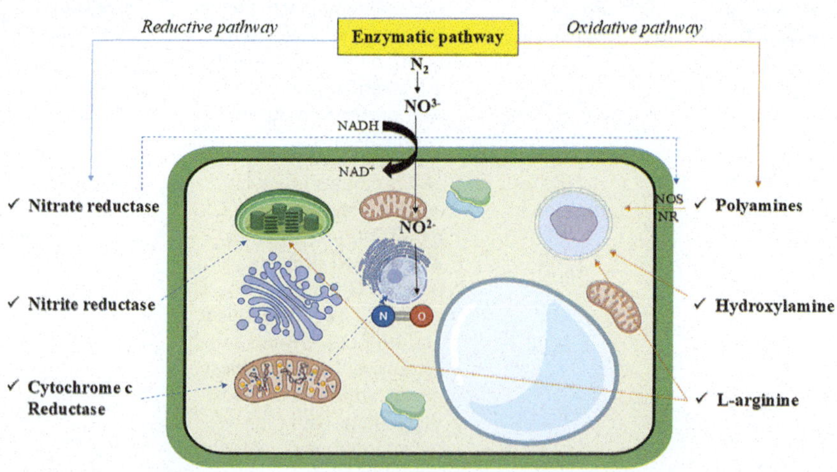

FIGURE 8.1 Synthesis of nitric oxide via oxidative and reductive pathways in plants.

known to enhance NR activity under hypoxia. In addition, low oxygen levels also cause an increased expression of NR encoding genes, *NIA1* and *NIA2* (Morard et al., 2004) to promote NO synthesis.

8.3 Regulation of nitric oxide by phytoglobins

Several factors influence cellular NO levels and, consequently, cell behavior. Among these factors are phytoglobins, which are well-characterized stress-related proteins known for their effective NO-scavenging properties (Hill et al., 2012). It has been suggested that Pgbs may function as universal molecular switches that regulate cell fate through their interaction with NO (Stasolla et al., 2017). One of the primary roles of phytoglobins in plants is thought to be related to NO metabolism (Dordas et al., 2003). Under very low atmospheric oxygen concentrations (below 1%), the phytoglobin/NO cycle converts NO into nitrate (Hebelstrup et al., 2012). This cycle consumes NADH and NADPH while producing NAD^+ and $NADP^+$, which are essential for ATP production through the glycolytic pathway (Igamberdiev et al., 2004). During hypoxia, increased fermentation leads to cytoplasmic acidosis, and nitrite is known to mitigate this process (Libourel et al., 2006). Nitrite enhances the anaerobic germination and growth of deep-water rice seedlings under anoxic conditions. This improvement is achieved through the modulation of fermentation, the phytoglobin-NO (Pgb-NO) cycle, and the reduction of ROS by inducing alternative oxidase (AOX) (Kumari et al., 2021). The cycle begins when nitrite is transported into the mitochondria via a proposed nitrite transporter. Within the mitochondria, nitrite is reduced to NO at various sites, including complex III (bc1), complex IV (cytochrome c oxidase), and the AOX. The produced NO then moves to the cytosol, where it is scavenged by phytoglobin1, resulting in the formation of metphytoglobin and nitrate. This complete cycle ultimately leads to the production of a limited amount of ATP (Kumari et al, 2021; Stoimenova et al., 2007). Further, it was revealed that growing plants under low nitrate levels leads to reduced levels of NO and an increased expression of high-affinity transporters (HATs), such as *NRT2.1*, *NRT2.3*, and *NRT2.4* (Samant et al., 2025). Additionally, phytoglobin over-expression in rice resulted in enhanced *HAT* expression, improved growth, and increased nitrogen accumulation in low nitrate environments. Together, these findings suggest that phytoglobins play an important role in nitrogen uptake and assimilation. Later, it was found that the phytoglobin genes *PGB1* and *PGB2*, which are involved in NO scavenging, were significantly down-regulated during nitrate starvation by approximately 1000-fold and 100,000-fold, respectively, under normal oxygen levels. This downregulation resulted in a 2.5-fold increase in NO concentration (Safavi-Rizi et al., 2024). Additionally, in the leaves, the genes *HRA1*, *SUS4*, and *RAP2.3* were upregulated approximately 20-fold due to hypoxia during nitrate starvation; however, this upregulation was nearly absent when nitrate was present. These findings emphasize the interplay between a plant's responses to nitrate starvation and hypoxia. In 2025, (Nejamkin et al., 2023) conducted a study that highlighted the role of NO, a by-product of nitrate (NO^{3-}) metabolism and a well-characterized signaling molecule in plants, as a key player in the Primary Nitrate Response (PNR). They found that *PGB1*, which is involved in maintaining NO homeostasis, is rapidly induced during the PNR in wild-type plants (wt) but not in mutants lacking the nitrate transceptor *NTR1.1* or the transcription factor nodule inception-like protein 7 (*NLP7*). This indicates that the *NRT1.1-NLP7* pathway regulates *PGB1* gene expression. Biotin switch experiments revealed that *NLP7*, the master regulator of PNR, undergoes S-nitrosation in vitro. Furthermore, the depletion of NO during PNR significantly increases the decline in ROS levels while elevating the CAT and APX activities. Apart from phytoglobins, the NO scavenger system is also linked to processes, including nitrosation of thiols and other substances, lipoamide dehydrogenase involvement, and peroxynitrite synthesis (Igamberdiev et al., 2004; Gupta et al., 2020).

8.4 Effects of nitric oxide in anaerobic germination

In the induction stage, somatic cells are reprogrammed to become capable of forming embryos. As development continues, these cells undergo maturation, during which they lose water and accumulate storage nutrients. These changes prepare the somatic embryo for successful germination at a later stage (Jimenez et al., 2001). Notably, moderate hypoxia also accompanies zygotic embryo development, often favors this process (Rolletschek et al., 2003). Hypoxia is also associated with other conditions of specific development that include seed imbibition and germination. The outermost layers of the seed prevent oxygen diffusion, which causes the interior seed tissues to be anoxic (Borisjuk et al., 2007). According to Igamberdiev and Hill (2004), nitrite reduction to NO aids in the recycling of NAD^+, which is necessary for glycolysis to function. It was also shown that mitochondria can only generate a certain quantity of ATP when oxygen levels are low (Stoimenova et al., 2007). In situations such as hypoxia, nitrite-dependent NO also induces AOX, which can support ROS homeostasis (Vishwakarma et al., 2018). Under these circumstances, anoxia-induced NO synthesis also inhibits ROS formation via the Phytoglobin-NO cycle (Kumari et al., 2021). It was also found that when nitrite was applied to deep-water rice to promote the Phytoglobin-NO cycle, anaerobic germination and seedling development were improved (Gupta et al., 2021; Kumari et al., 2021). The tuning of sugar formed from reserved starch and nitrate increases via fermentative pathways, which maintains the cellular balance, initiates NO synthesis, and activates antioxidant enzymes (Borella et al., 2019) (See Fig. 8.2).

Notably, low oxygen levels limit mitochondrial respiration and promote fermentation to reoxidize NADH produced during glycolysis (Magneschi &

FIGURE 8.2 Role of Nitric oxide in the regulation of fermentation pathway under anaerobic germination in plant seeds.

Perata, 2008). Under such conditions, plants predominantly rely on glycolysis-dependent ATP production. Hypoxia triggers extensive reprogramming of gene expression and metabolic pathways, leading to the induction of fermentative metabolism (Loreti et al., 2016). More precisely, anaerobic energy production yields a net gain of two ATP molecules per glucose, with pyruvate being converted either to lactate by lactate dehydrogenase (LDH) or to ethanol through the sequential action of pyruvate decarboxylase (PDC) and alcohol dehydrogenase (ADH) (Zabalza et al., 2009). Although this net ATP generated is much lower than that produced during aerobic ATP production, anaerobic-tolerant plants can sustain improved fermentation, which can supply the energy needed for a variety of metabolic processes. It was shown that in anoxic germinating seeds, an increased glycolytic flow occurs in contrast to aerobic controls (Kumari et al., 2021). The function of additional pathways that contribute to energy production during anaerobic germination has also been deciphered in recent years. Nitrate reduction and the phytoglobin-NO cycle, for example, may be potential sources of NAD^+, which is necessary for the operation of fermentation under low oxygen conditions (Igamberdiev et al., 2004). However, the role of this pathway in anaerobic germination of rice is not fully understood, though nitrate reduction has been demonstrated to preserve mitochondrial activity throughout the germination process (Gupta et al., 2016; Vartapetian, 2003).

8.5 Role of nitric oxide in the formation of aerenchyma during hypoxia

Aerenchyma are vast intracellular gaps found in plant tissues that form quickly under hypoxic conditions. The growth of schizogenous intercellular spaces (caused by division of cells in the middle lamella during cell development) or a varying degree of lysigeny (caused by death of cortical cells that produces a gaseous space) are the two ways that aerenchyma develops in roots (Pan et al., 2021). These are found in plants such as *Oryza sativa* and *Zea mays*, and ethylene and abscisic acid play an antagonistic role in the formation of lysigenous aerenchyma in response to ROS and RNS signaling (Basu et al., 2021). Though ethylene alone cannot cause aerenchyma formation, it generally requires the assistance of NO in the formation of aerenchyma during hypoxia. Notably, NADPH oxidase (NOXs) play a vital role in the generation of ROS in response to both physical and chemical stimuli (Hong et al., 2020). NO ameliorates the expression of NOX-responsive genes, which assist the formation of aerenchyma during prolonged flooding (Wany et al., 2017). In particular, ERFVII-transcription factors, the key components of ethylene signaling, regulate hypoxia-induced ROS generation by modulating NADPH oxidase activity. Flood-induced calcium influx activates calcium-dependent protein kinases CDPK5 and CDPK13, which phosphorylate NADPH oxidase, leading to enhanced ROS generation. In parallel, NO is produced and mediates post-translational modifications, such as S-nitrosation of the target proteins to further influence ROS signaling. The coordinated action of ROS and NO ultimately promotes aerenchyma formation under hypoxic conditions (Samant et al., 2025).

8.6 Impact of nitric oxide in plants by induction of hypoxia-responsive genes

Plants have distinct survival mechanisms to deal with persistent hypoxic conditions during waterlogging or submergence stress (Wang et al., 2025). For future food security, an in-depth understanding of these adaptation mechanisms is critical to develop climate-resilient, economically important crops in the face of a changing global environment. Hypoxia/anoxia-triggered responses involve multiple adaptive signaling mechanisms involving NO, ROS (Pucciariello & Perata, 2017; Al-Hakeem & Khan, 2025), ethylene (Hartman et al., 2019; Voesenek & Sasidharan, 2013), and cytosolic Ca^{2+}(Igamberdiev & Hill, 2018). Among these, NO signaling is of particular significance during the initial response to low oxygen stress (Timilsina et al., 2022; Wang et al., 2025; Samant et al., 2025). Intriguingly, NO serves as a central regulator for survival under low oxygen conditions by regulating hypoxia-responsive genes (HRGs) (Pucciariello & Perata, 2017). Under induced hypoxia, biosynthesis of NO occurs as a direct consequence of NR activity. In order to maintain redox equilibrium in the presence of limited oxygen, this signaling molecule modifies ROS homeostasis, interacts with stress-responsive hormones, and triggers posttranslational changes in the pathway linked to nitrate metabolism (Wang et al., 2025; Al-Hakeem & Khan, 2025).

Under hypoxic conditions, nitrogen metabolism switches toward NO generation, and the AOX complex stimulates mitochondrial nitrite-dependent NO synthesis even more. NO, together with ROS, further regulates downstream hormonal signaling pathways associated with abscisic acid and ethylene, influencing gene expression responses of these phytohormones (Hartman et al., 2019). These interactions, whether directly mediated by NO or via NO-dependent hormonal control, emphasize the complexities of signaling under low oxygen stress. In particular, the interrelationship of NO, ROS, and oxygen is crucial for controlling the stability and activity of group VII ethylene response factors (ERFs), which are important regulators during low oxygen escape syndrome. However, no study has yet determined the precise NO/ROS ratio that controls the fate of group VII ERFs. While low oxygen levels, NO, and ethylene all influence ERFs, the dynamics of NO production during flooding are still unknown. It has been shown that an increase in NO levels during hypoxia has operational implications for survival (Gupta & Igamberdiev, 2016). NO impacts the stability of the ERF-VII transcription factor, which activates HRGs, by suppressing cysteine oxidase activity (Taylor-Kearney & Flashman, 2021). Previous investigations have repeatedly demonstrated that NO accumulates in numerous plant species during hypoxia/anoxia (Gupta et al., 2020; León et al., 2021). NO, so produced, is implicated in the activation of genes encoding for glycolysis, fermentation, the tricarboxylic acid cycle (TCA cycle), and antioxidant pathways. Interestingly, studies in mutants such as Hb^+ and nia^+ found somewhat increased ROS levels under normoxia than the wild type. However, in hypoxia, all genotypes had higher ROS levels, indicating that NO can alter ROS homeostasis. Furthermore, wild-type roots produced more NO during hypoxia than mutants, suggesting a relationship between NO levels and hypoxia tolerance. During hypoxia, ROS are generated largely in mitochondria and at the plasma membrane via NADPH oxidases, which are also regulated by oxygen levels and NO (Wany & Gupta, 2018). NO also promotes plant resilience in hypoxic conditions by upregulating ROS detoxifying genes such as APX, SOD, and other genes of the ascorbate-glutathione cycle and modifying TCA cycle genes to promote reducing equivalent regeneration (See Fig. 8.3). Hypoxia-induced NO and ROS levels also influence fermentative enzymes such as ADH and PDC. NO-mediated activation of carbohydrate metabolism-related genes enables ATP generation in the absence of oxygen (Sasidharan et al., 2021). NO also plays a significant role in the development of cortical aerenchyma by inducing programmed cell death. In hypoxic conditions, increased NO activates NADPH oxidase gene expression and lipid peroxidation, which leads to the production of lysigenous cortical cells (Wany & Gupta, 2018).

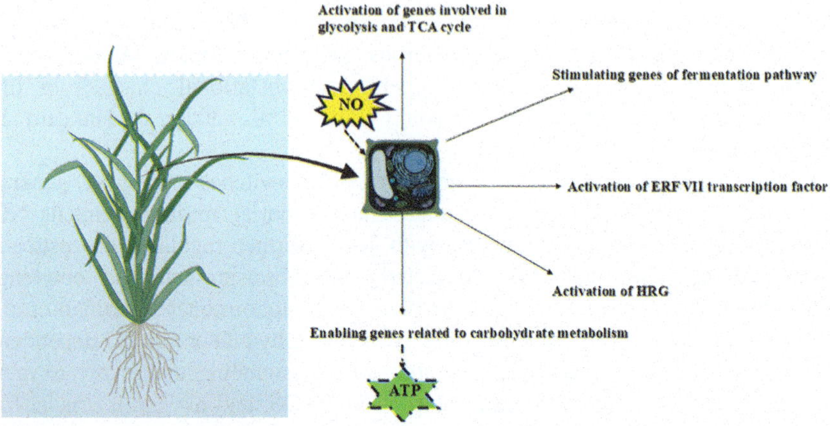

FIGURE 8.3 Multidimensional role of nitric oxide in activating hypoxia-responsive genes.

8.7 Conclusion

NO is a signaling molecule that controls the growth and development of plants, particularly in response to hypoxix conditions by modifying the antioxidant system and activating genes associated with ATP production. NO is generally synthesized by a combination of both oxidative and reductive enzymatic pathways, amidst hypoxia, the reductive pathway remains prominent. Notably, during induced flooding, the NO levels are superiorly regulated by the early detection of low oxygen levels, ATP production, NADP signaling, aerenchyma formation, and the generation of antioxidant defenses. These reactions are the probable outcome of an effective cellular signaling system that is primarily controlled by the tyrosine nitration and S-nitrosation of particular proteins. In a nutshell, NO plays a crucial role in maintaining the adaptability of plants under flood through the regulation of various biochemical as well as genetic mechanisms that assist plants phenotypically and metabolically to survive under prolonged submergence.

Declaration of interest

The authors declare that they have no conflict of interest.

Authors contribution

SP: Writing- original draft, Writing- review editing, Formal analysis; DP: Writing- original draft; SPN: Writing- original draft; PS: Writing- original draft; RG: Writing- review editing; ND: Writing- review editing; ND: Writing- review editing, Conceptualization, Supervision

Funding assistance

No financial assistance was provided for this work.

References

Al-Hakeem, H. F. H., & Khan, M. (2025). The role of statistics in advancing nitric oxide research in plant biology: From data analysis to mechanistic insights. *Frontiers in Plant Science, 16*. https://doi.org/10.3389/fpls.2025.1597030.

Astier, J., Gross, I., & Durner, J. (2017). Nitric oxide production in plants: An update. *Journal of Experimental Botany, 69*(14), 3401–3411. https://doi.org/10.1093/jxb/erx420.

Basu, S., Kumari, S., Kumar, A., Shahid, R., Kumar, S., & Kumar, G. (2021). Nitro-oxidative stress induces the formation of roots' cortical aerenchyma in rice under osmotic stress. *Physiologia Plantarum, 172*(2), 963–975. https://doi.org/10.1111/ppl.13415.

Berger, A., Boscari, A., Frendo, P., Brouquisse, R., & Foyer, C. (2019). Nitric oxide signaling, metabolism and toxicity in nitrogen-fixing symbiosis. *Journal of Experimental Botany, 70*(17), 4505–4520. https://doi.org/10.1093/jxb/erz159.

Borella, J., Becker, R., Lima, M. C., De Oliveira, D. D. S. C., Braga, E. J. B., De Oliveira, A. C. B., & Amarante, L. D. (2019). Nitrogen source influences the antioxidative system of soybean plants under hypoxia and re-oxygenation. *Scientia Agricola, 76*(1), 51–62. Available from https://doi.org/10.1590/1678-992x-2017-0195, http://www.scielo.br/pdf/sa/v76n1/1678-992X-sa-76-01-0051.pdf.

Borisjuk, L., Macherel, D., Benamar, A., Wobus, U., & Rolletschek, H. (2007). Low oxygen sensing and balancing in plant seeds: A role for nitric oxide. *New Phytologist, 176*(4), 813–823. https://doi.org/10.1111/j.1469-8137.2007.02226.x.

Choudhury, F. K., Rivero, R. M., Blumwald, E., & Mittler, R. (2017). Reactive oxygen species, abiotic stress and stress combination. *Plant Journal, 90*(5), 856–867. Available from https://doi.org/10.1111/tpj.13299, http://www.blackwellpublishing.com/journals/TPJ.

Crawford, N., & Guo, F. (2005). New insights into nitric oxide metabolism and regulatory functions. *Trends in Plant Science, 10*(4), 195–200. https://doi.org/10.1016/j.tplants.2005.02.008.

Dordas, C., Hasinoff, B. B., Igamberdiev, A. U., Manac'h, N., Rivoal, J., & Hill, R. D. (2003). Expression of a stress-induced hemoglobin affects NO levels produced by alfalfa root cultures under hypoxic stress. *The Plant Journal, 35*(6), 763–770. 10.1046/j.1365-313x.2003.01846.x.

Foresi, N., Correa-Aragunde, N., Parisi, G., Caló, G., Salerno, G., & Lamattina, L. (2010). Characterization of a nitric oxide synthase from the plant kingdom: NO generation from the green alga *Ostreococcus tauri* is light irradiance and growth phase dependent. *The Plant Cell, 22*(11), 3816–3830. Available from https://doi.org/10.1105/tpc.109.073510, http://www.plantcell.org/cgi/reprint/22/11/3816.pdf.

Ge, W., Wang, X., Li, Z., Zhang, H., Guo, X., Wang, T., Gao, W., Lin, C., & Van Gelder, P. (2021). Interval analysis of the loss of life caused by dam failure. *Journal of Water Resources Planning and Management, 147*(1), Available from https://doi.org/10.1061/(ASCE)WR.1943-5452.0001311, https://ascelibrary.org/journal/jwrmd5.

Ge, W., Sun, H., Zhang, H., Li, Z., Guo, X., Wang, X., Qin, Y., Gao, W., & van Gelder, P. (2020). Economic risk criteria for dams considering the relative level of economy and industrial economic contribution. *Science of the Total Environment, 725*, 138139. https://doi.org/10.1016/j.scitotenv.2020.138139.

Gupta, K. J., & Igamberdiev, A. U. (2016). Reactive nitrogen species in mitochondria and their implications in plant energy status and hypoxic stress tolerance. *Frontiers in Plant Science, 7*, 2016. Available from https://doi.org/10.3389/fpls.2016.00369, http://journal.frontiersin.org/article/10.3389/fpls.2016.00369/pdf.

Gupta, K. J., Igamberdiev, A. U., Manjunatha, G., Segu, S., Moran, J. F., Neelawarne, B., Bauwe, H., & Kaiser, W. M. (2011). The emerging roles of nitric oxide (NO) in plant mitochondria. *Plant Science, 181*(5), 520–526. https://doi.org/10.1016/j.plantsci.2011.03.018.

Gupta, K. J., Lee, C. P., & Ratcliffe, R. G. (2016). Nitrite protects mitochondrial structure and function under hypoxia. *Plant and Cell Physiology,* pcw174. https://doi.org/10.1093/pcp/pcw174.

Gupta, K. J., Mur, L. A. J., Wany, A., Kumari, A., Fernie, A. R., & Ratcliffe, R. G. (2020). The role of nitrite and nitric oxide under low oxygen conditions in plants. *New Phytologist, 225*(3), 1143–1151. Available from https://doi.org/10.1111/nph.15969, http://onlinelibrary.wiley.com/journal/10.1111/(ISSN)1469-8137.

Gupta, K. J., Kaladhar, V. C., Fitzpatrick, T. B., Fernie, A. R., Møller, I. M., & Loake, G. J. (2021). Nitric oxide regulation of plant metabolism. *Molecular Plant, 15*(2), 228–242. https://doi.org/10.1016/j.molp.2021.12.012.

Hancock, J. T., & Veal, D. (2020). Nitric oxide, other reactive signalling compounds, redox, and reductive stress. *Journal of Experimental Botany, 72*(3), 819–829. https://doi.org/10.1093/jxb/eraa331.

Hartman, S., Liu, Z., van Veen, H., Vicente, J., Reinen, E., Martopawiro, S., Zhang, H., van Dongen, N., Bosman, F., Bassel, G. W., Visser, E. J. W., Bailey-Serres, J., Theodoulou, F. L., Hebelstrup, K. H., Gibbs, D. J., Holdsworth, M. J., Sasidharan, R., & Voesenek, L. A. C. J. (2019). Ethylene-mediated nitric oxide depletion pre-adapts plants to hypoxia stress. *Nature Communications, 10*(1), Available from https://doi.org/10.1038/s41467-019-12045-4, http://www.nature.com/ncomms/index.html.

Hebelstrup, Kim H., van Zanten, Martijn, Mandon, Julien, Voesenek, Laurentius A. C. J., Harren, Frans J. M., Cristescu, Simona M., Møller, Ian M., & Mur, Luis A. J. (2012). Haemoglobin modulates NO emission and hyponasty under hypoxia-related stress in Arabidopsis thaliana. *Journal of Experimental Botany, 63*(15), 5581–5591. 10.1093/jxb/ers210.

Hill, Robert D. (2012). Non-symbiotic haemoglobins—What's happening beyond nitric oxide scavenging? *AoB PLANTS, 2012*. 10.1093/aobpla/pls004.

Hong, C.-P., Wang, M.-C., & Yang, C.-Y. (2020). NADPH oxidase RbohD and ethylene signaling are involved in modulating seedling growth and survival under submergence stress. *Plants, 9*(4), 471. https://doi.org/10.3390/plants9040471.

Igamberdiev, A. U., & Hill, R. D. (2018). Elevation of cytosolic Ca^{2+} in response to energy deficiency in plants: The general mechanism of adaptation to low oxygen stress. *Biochemical Journal, 475*(8), 1411–1425. Available from https://doi.org/10.1042/BCJ20180169, http://www.biochemj.org/content/ppbiochemj/475/8/1411.full.pdf.

Igamberdiev, A. U., Bykova, N. V., Shah, J. K., & Hill, R. D. (2010). Anoxic nitric oxide cycling in plants: Participating reactions and possible mechanisms. *Physiologia Plantarum, 138*(4), 393–404. https://doi.org/10.1111/j.1399-3054.2009.01314.x.

Igamberdiev, A. U., Seregèlyes, C., Manac'h, N., & Hill, R. D. (2004). NADH-dependent metabolism of nitric oxide in alfalfa root cultures expressing barley hemoglobin. *Planta, 219*(1), 95–102. https://doi.org/10.1007/s00425-003-1192-3.

Jan, R, Khan, M, Adnan, M, Asaf, S, Asif, S, Kim, KM, & Murad, W (2022). Exogenous Phytohormones and Fertilizers Enhance Jatropha curcas L. Growth through the Regulation of

Physiological, Morphological, and Biochemical Parameters. *Plants (Basel), 11*(24). 10.3390/plants11243584.

JIMÉNEZ, VÍCTOR M. (2001). Regulation of in vitro somatic embryogenesis with emphasis on to the role of endogenous hormones. *Revista Brasileira de Fisiologia Vegetal,* 196–223. 10.1590/s0103-31312001000200008 132.

Kaur, P., Handa, N., Verma, V., Bakshi, P., Kalia, R., Sareen, S., et al. (2019). Cross talk among reactive oxygen, nitrogen and sulfur during abiotic stress in plants. In M. Hasanuzzaman, V. Fotopolous, K. Nahar, & M. Fujita (Eds.). *Reactive oxygen, nitrogen and sulfur species in plants: Production, metabolism, signaling and defense mechanisms* (pp. 857–871). Wiley.

Kaya, C., Ashraf, M., Alyemeni, M. N., & Ahmad, P. (2020). Responses of nitric oxide and hydrogen sulfide in regulating oxidative defence system in wheat plants grown under cadmium stress. *Physiologia Plantarum, 168*(2), 345–360. Available from https://doi.org/10.1111/ppl.13012, http://onlinelibrary.wiley.com/journal/10.1111/(ISSN)1399-3054.

Khan, M., Al Azzawi, T. N. I., Ali, S., Yun, B.-W., & Mun, B.-G. (2023). Nitric Oxide, a Key Modulator in the Alleviation of Environmental Stress-Mediated Damage in Crop Plants: A Meta-Analysis. *Plants, 12*(11), 2121. 10.3390/plants12112121.

Khare, T., Srivastava, A. K., Suprasanna, P., & Kumar, V. (2020). Individual and additive stress impacts of Na^+ and Cl^- on proline metabolism and nitrosative responses in rice. *Plant Physiology and Biochemistry, 152*, 44–52. Available from https://doi.org/10.1016/j.plaphy.2020.04.028, http://www.journals.elsevier.com/plant-physiology-and-biochemistry/.

Khator, K., & Shekhawat, G. S. (2020). Cd- and Cu-induced phytotoxicity on 2–3 leaf stage of *Cyamopsis tetragonoloba* and its regulation by nitrate reductase and ROS quenching enzyme. *Acta Physiologiae Plantarum, 42*(7). https://doi.org/10.1007/s11738-020-03105-0.

Khator, K., Parihar, S., Jasik, J., & Shekhawat, G. S. (2024). Nitric oxide in plants: An insight on redox activity and responses toward abiotic stress signaling. *Plant Signaling & Behavior, 19*(1). https://doi.org/10.1080/15592324.2023.2298053.

Kolbert, Z., Barroso, J. B., Brouquisse, R., Corpas, F. J., Gupta, K. J., Lindermayr, C., Loake, G. J., Palma, J. M., Petřivalský, M., Wendehenne, D., & Hancock, J. T. (2019). A forty year journey: The generation and roles of NO in plants. *Nitric Oxide Biology and Chemistry / Official Journal of the Nitric Oxide Society, 93*, 53–70. https://doi.org/10.1016/j.niox.2019.09.006.

Kumari, A., Singh, P., Kaladhar, V. C., Manbir, N., Paul, D., Pathak, P. K., & Gupta, K. J. (2021). Phytoglobin-NO cycle and AOX pathway play a role in anaerobic germination and growth of deepwater rice. *Plant, Cell & Environment, 45*(1), 178–190. https://doi.org/10.1111/pce.14198.

León, J., Castillo, M. C., & Gayubas, B. (2021). The hypoxia-reoxygenation stress in plants. *Journal of Experimental Botany, 72*(16), 5841–5856. Available from https://doi.org/10.1093/jxb/eraa591, http://jxb.oxfordjournals.org/.

Libourel, I. G. L., van Bodegom, P. M., Fricker, M. D., & Ratcliffe, R. G. (2006). Nitrite Reduces Cytoplasmic Acidosis under Anoxia. *Plant Physiology, 142*(4), 1710–1717. 10.1104/pp.106.088898.

Loreti, E., van Veen, H., & Perata, P. (2016). Plant responses to flooding stress. *Current Opinion in Plant Biology, 33*, 64–71. https://doi.org/10.1016/j.pbi.2016.06.005.

Magneschi, L., & Perata, P. (2008). Rice germination and seedling growth in the absence of oxygen. *Annals of Botany, 103*(2), 181–196. https://doi.org/10.1093/aob/mcn121.

Manghwar, H., Hussain, A., Alam, I., Khoso, M. A., Ali, Q., & Liu, F. (2024). Waterlogging stress in plants: Unraveling the mechanisms and impacts on growth, development, and productivity.

Environmental and Experimental Botany, 224, 105824. https://doi.org/10.1016/j.envexpbot.2024.105824.

Mohn, M. A., Thaqi, B., & Fischer-Schrader, K. (2019). Isoform-specific NO synthesis by *Arabidopsis thaliana* nitrate reductase. *Plants, 8*(3), 67. https://doi.org/10.3390/plants8030067.

Morard, P., Silvestre, J., Lacoste, L., Caumes, E., & Lamaze, T. (2004). Nitrate uptake and nitrite release by tomato roots in response to anoxia. *Journal of Plant Physiology, 161*(7), 855–865. https://doi.org/10.1016/j.jplph.2003.11.003.

Mur, L. A. J., Mandon, J., Persijn, S., Cristescu, S. M., Moshkov, I. E., Novikova, G. V., Hall, M. A., Harren, F. J. M., Hebelstrup, K. H., & Gupta, K. J. (2013). Nitric oxide in plants: An assessment of the current state of knowledge. *AoB Plants, 5*(0), pls052. https://doi.org/10.1093/aobpla/pls052.

Nabi, R. B. S., Tayade, R., Hussain, A., Kulkarni, K. P., Imran, Q. M., Mun, B. G., & Yun, B. W. (2019). Nitric oxide regulates plant responses to drought, salinity, and heavy metal stress. *Environmental and Experimental Botany, 161*, 120–133. Available from https://doi.org/10.1016/j.envexpbot.2019.02.003, www.elsevier.com/inca/publications/store/2/6/7.

Nejamkin, A., Del Castello, F., Lamattina, L., Correa-Aragunde, N., & Foresi, N. (2023). Nitric oxide is required for primary nitrate response in *Arabidopsis*: Evidence for S-nitrosation of NLP7. *Antioxidants and Redox Signaling*. https://doi.org/10.1089/ars.2022.0210.

Niyoifasha, C. J., Borena, B. M., Ukob, I. T., Minh, P. N., Al Azzawi, T. N. I., Imran, M., Ali, S., Inthavong, A., Mun, B. G., Lee, I. J., Khan, M., & Yun, B. W. (2023). Alleviation of Hg-, Cr-, Cu-, and Zn-induced heavy metals stress by exogenous sodium nitroprusside in rice plants. *Plants, 12*(6), Available from https://doi.org/10.3390/plants12061299, www.mdpi.com/journal/plants.

Pan, J., Sharif, R., Xu, X., & Chen, X. (2021). Mechanisms of waterlogging tolerance in plants: Research progress and prospects. *Frontiers in Plant Science, 11*. https://doi.org/10.3389/fpls.2020.627331.

Pandey, S., Kumari, A., Shree, M., Kumar, V., Singh, P., Bharadwaj, C., Loake, G. J., Parida, S. K., Masakapalli, S. K., & Gupta, K. J. (2019). Nitric oxide accelerates germination via the regulation of respiration in chickpea. *Journal of Experimental Botany, 70*(17), 4539–4555. Available from https://doi.org/10.1093/jxb/erz185, http://jxb.oxfordjournals.org/.

Patel, M. K., Pandey, S., Burritt, D. J., & Tran, L. S. P. (2019). Plant responses to low-oxygen stress: Interplay between ROS and NO signaling pathways. *Environmental and Experimental Botany, 161*, 134–142. Available from https://doi.org/10.1016/j.envexpbot.2019.02.013, www.elsevier.com/inca/publications/store/2/6/7.

Pedersen, O., Perata, P., & Voesenek, L. A. C. J. (2017). Flooding and low oxygen responses in plants. *Functional Plant Biology, 44*(9), iii. https://doi.org/10.1071/FPv44n9_FO, : http://www.publish.csiro.au/nid/103.htm.

Procházková, D., Haisel, D., & Pavlíková, D. (2014). Nitric oxide biosynthesis in plants - the short overview. *Plant, Soil and Environment, 60*(3), 129–134. 10.17221/901/2013-pse.

Pucciariello, C., & Perata, P. (2017). New insights into reactive oxygen species and nitric oxide signalling under low oxygen in plants. *Plant, Cell & Environment, 40*(4), 473–482. https://doi.org/10.1111/pce.12715.

Rahim, W., Khan, M., Al Azzawi, T. N. I., Pande, A., Methela, N. J., Ali, S., Imran, M., Lee, D. S., Lee, G. M., Mun, B. G., Moon, Y. S., Lee, I. J., & Yun, B. W. (2022). Exogenously applied sodium nitroprusside mitigates lead toxicity in rice by regulating antioxidants and metal stress-related transcripts. *International Journal of Molecular Sciences, 23*(17), Available from https://doi.org/10.3390/ijms23179729, http://www.mdpi.com/journal/ijms.

Rolletschek, H., Weber, H., & Borisjuk, L. (2003). Energy status and its control on embryogenesis of legumes. Embryo photosynthesis contributes to oxygen supply and is coupled to biosynthetic fluxes. *Plant Physiology, 132*(3), 1196–1206. https://doi.org/10.1104/pp.102.017376.

Rőszer, T. (2012). *Nitric oxide synthesis in leaf peroxisomes and in plant-type mitochondria. Springer eBooks*, 67–80. https://doi.org/10.1007/978-94-007-2819-6_4.

Safavi-Rizi, V., Uhlig, T., Lutter, F., Safavi-Rizi, H., Krajinski-Barth, F., & Sasso, S. (2024). Reciprocal modulation of responses to nitrate starvation and hypoxia in roots and leaves of *Arabidopsis thaliana. Plant Signaling & Behavior, 19*(1). https://doi.org/10.1080/15592324.2023.2300228.

Samant, S. B., Swain, J., Yadav, N., Yadav, R., Singh, P., Rai, P., Sheri, V., Sreeman, S., Subramanyam, R., Pareek, A., & Gupta, K. J. (2025). Overexpression of phytoglobin1 in rice leads to enhanced nitrogen use efficiency via modulation of nitric oxide. *Plant, Cell and Environment, 8*(4), 2755–2768. Available from https://doi.org/10.1111/pce.15289, http://onlinelibrary.wiley.com/journal/10.1111/(ISSN)1365-3040.

Sasidharan, R., Schippers, J. H. M., & Schmidt, R. R. (2021). Redox and low-oxygen stress: Signal integration and interplay. *Plant Physiology, 186*(1), 66–78. Available from https://doi.org/10.1093/plphys/kiaa081, https://academic.oup.com/plphys/article/186/1/66/6070215.

Signorelli, S., & Considine, M. J. (2018). Nitric oxide enables germination by a four-pronged attack on ABA-Induced seed dormancy. *Frontiers in Plant Science, 9*. https://doi.org/10.3389/fpls.2018.00296.

Stasolla, C., & Hill, R. D. (2017). Determining Cellular Responses: Phytoglobins May Direct the Traffic. *Trends in Plant Science, 22*(10), 820–822. 10.1016/j.tplants.2017.08.002.

Stoimenova, M., Igamberdiev, U., Gupta, K. J., & Hill, R. D. (2007). Nitrite-driven anaerobic ATP synthesis in barley and rice root mitochondria. *Planta, 226*(2), 465–474. 10.1007/s00425-007-0496-0.

Sun, Y.-F., Liu, S.-B., Meng, F.-L., Liu, J.-Y., Jin, Z., Kong, L.-T., & Liu, J.-H. (2012). Metal oxide nanostructures and their gas sensing properties: A review. *Sensors, 12*(3), 2610–2631. https://doi.org/10.3390/s120302610.

Tan, D.-X., Hardeland, R., Manchester, L. C., Korkmaz, A., Ma, S., Rosales-Corral, S., & Reiter, R. J. (2012). Functional roles of melatonin in plants, and perspectives in nutritional and agricultural science. *Journal of Experimental Botany, 63*(2), 577–597. https://doi.org/10.1093/jxb/err256.

Taylor-Kearney, L. J., & Flashman, E. (2021). Targeting plant cysteine oxidase activity for improved submergence tolerance. *The Plant Journal, 109*(4), 779–788. https://doi.org/10.1111/tpj.15605.

Thalineau, E., Truong, H.-N., Berger, A., Fournier, C., Boscari, A., Wendehenne, D., & Jeandroz, S. (2016). Cross-regulation between n metabolism and nitric oxide (NO) signaling during plant immunity. *Frontiers in Plant Science, 7*(2016). https://doi.org/10.3389/fpls.2016.00472.

Timilsina, A., Dong, W., Hasanuzzaman, M., Liu, B., & Hu, C. (2022). Nitrate–nitrite–nitric oxide pathway: A mechanism of hypoxia and anoxia tolerance in plants. *International Journal of Molecular Sciences, 23*(19), 11522. https://doi.org/10.3390/ijms231911522.

Toledo, J. C., & Augusto, O. (2012). Connecting the chemical and biological properties of nitric oxide. *Chemical Research in Toxicology, 25*(5), 975–989. https://doi.org/10.1021/tx300042g.

Vartapetian, B. B. (2003). Functional electron microscopy in studies of plant response and adaptation to anaerobic stress. *Annals of Botany, 91*(2), 155–172. https://doi.org/10.1093/aob/mcf244.

Voesenek, L.a C. J., & Bailey-Serres, J. (2015). Flood adaptive traits and processes: An overview. *New Phytologist, 206*(1), 57–73. https://doi.org/10.1111/nph.13209.

Vishwakarma, A., Kumari, A., Mur, L. A. J., & Gupta, K. J. (2018). A discrete role for alternative oxidase under hypoxia to increase nitric oxide and drive energy production. *Free Radical Biology and Medicine, 122*, 40–51. 10.1016/j.freeradbiomed.2018.03.045.

Voesenek, L.a C. J., & Sasidharan, R. (2013). Ethylene – and oxygen signalling – drive plant survival during flooding. *Plant Biology, 15*(3), 426–435. https://doi.org/10.1111/plb.12014.

Wang, L.-N., Wang, W.-C., Liao, K., Xu, L.-J., Xie, D.-X., Xie, R.-H., & Shi, X. (2025). Survival mechanisms of plants under hypoxic stress: Physiological acclimation and molecular regulation. *Journal of Integrative Plant Biology, 67*(3), 440–454. https://doi.org/10.1111/jipb.13880.

Wany, A., & Gupta, K. J. (2018). Reactive oxygen species, nitric oxide production and antioxidant gene expression during development of aerenchyma formation in wheat. *Plant Signaling & Behavior, 13*(2), e1428515. https://doi.org/10.1080/15592324.2018.1428515.

Wany, A., Kumari, A., & Gupta, K. J. (2017). Nitric oxide is essential for the development of aerenchyma in wheat roots under hypoxic stress. *Plant Cell and Environment, 40*(12), 3002–3017. Available from https://doi.org/10.1111/pce.13061, http://www.blackwellpublishing.com/journals/PCE.

Wink, D. A., Miranda, K. M., Espey, M. G., Pluta, R. M., Hewett, S. J., Colton, C., Vitek, M., Feelisch, M., & Grisham, M. B. (2001). Mechanisms of the antioxidant effects of nitric oxide. *Antioxidants and Redox Signaling, 3*(2), 203–213. Available from https://doi.org/10.1089/152308601300185179, www.liebertonline.com/ars.

Xie, X., He, Z., Chen, N., Tang, Z., Wang, Q., & Cai, Y. (2019). The roles of environmental factors in regulation of oxidative stress in plant. *BioMed Research International, 2019*, 1–11. https://doi.org/10.1155/2019/9732325.

Zabalza, Van Dongen, J. T., Froehlich, A., Oliver, S. N., Faix, B., Gupta, K. J., Schmälzlin, E., Igal, M., Orcaray, L., Royuela, M., & Geigenberger, P. (2009). Regulation of respiration and fermentation to control the plant internal oxygen concentration. *Plant Physiology, 149*(2), 1087–1098. Available from https://doi.org/10.1104/pp.108.129288, http://www.plantphysiol.org/cgi/reprint/149/2/1087.

Zhou, X., Joshi, S., Patil, S., Khare, T., & Kumar, V. (2021). Reactive oxygen, nitrogen, carbonyl and sulfur species and their roles in plant abiotic stress responses and tolerance. *Journal of Plant Growth Regulation, 41*(1), 119–142. https://doi.org/10.1007/s00344-020-10294-y.

Chapter 9

Role of nitric oxide in the mitigation of drought stress

Kalpita Singh[1], Akhil Kumar[2], Zahra Tahmasebi[1], Nadeem Iqbal[3], Gábor Kocsy[1], and Riyazuddin Riyazuddin[4]

[1]*HUN-REN Centre for Agricultural Research, Agricultural Institute, Martonvásár, Hungary,* [2]*Hungarian University of Agriculture and Life Sciences, Institute of Genetics and Biotechnology, Gödöllő, Hungary,* [3]*School of Environmental Science and Engineering, Hunan University, Changsha, Hunan, P.R. China,* [4]*Instituto de Bioingeniería, Universidad Miguel Hernández, Elche, Spain*

9.1 Introduction

Water is the essence of life. However, the current scenario of climate change and water scarcity poses a significant threat not only to human life but also to plants. The consequences of drought are multiple and severe, depending on the duration, severity, growth stage of plants, and crop type. The contributing factors for drought conditions include abrupt precipitation, elevated temperatures above normal, insufficient irrigation, and other activities such as degrading soil quality and excessive grazing (Dirie et al., 2024). In addition, below-average rainfall combined with frequent heatwaves intensifies the problem even further. A 29% of the global wheat area was significantly affected due to a lack of water from 1981 to 2016 (Jobbágy et al., 2025). Lack of water affects yield by disrupting plant metabolism, altering cell structure, flower formation, and grain filling, thereby reducing crop yields by 30%–90% (Dirie et al., 2024). Furthermore, drought stress also affects seed germination, seedling establishment, leaf abscission, turgor loss, leaf wilting, damages the plasma membrane, negatively affects photosynthesis and key enzymes such as RuBisCO, causes oxidative stress, and damages cell organelles, including chloroplasts, mitochondria, and the nucleus (Riyazuddin et al., 2023). With limited water resources and rising food demand, researchers are diligently working to develop stress-resilient crop alternatives. Recent research has focused on improving the adaptation of crop plants through signaling molecules such as hydrogen peroxide (H_2O_2) and nitric oxide (NO), which are integral for plant growth and stress response mechanisms.

Nitric Oxide in Plant Resilience to Abiotic Stress. DOI: https://doi.org/10.1016/B978-0-443-40497-9.00001-5

NO is an essential gaseous signaling molecule. It regulates various physiological and metabolic plant processes as well as mediates abiotic and biotic stress responses (Khan et al., 2023). It is a lipophilic free radical and possesses the ability to diffuse through membranes, thus facilitating inter and intracellular signaling. It promotes seed germination, breaks dormancy, enhances seedling vigor, promotes root development, and biomass accumulation under normal and stressful conditions (Fig. 9.1) (Khan et al., 2023). Through S-nitrosation and crosstalk with phytohormones and other signaling molecules such as melatonin and hydrogen sulfide (H_2S), NO can influence respiration, antioxidant mechanisms, nutrient uptake, modulate senescence, fruit ripening, and symbiotic associations (Fig. 9.1). Furthermore, NO can trigger certain defense-related genes via the specific transcription factors. For instance, the exogenous application of NO to cucumber seedlings under

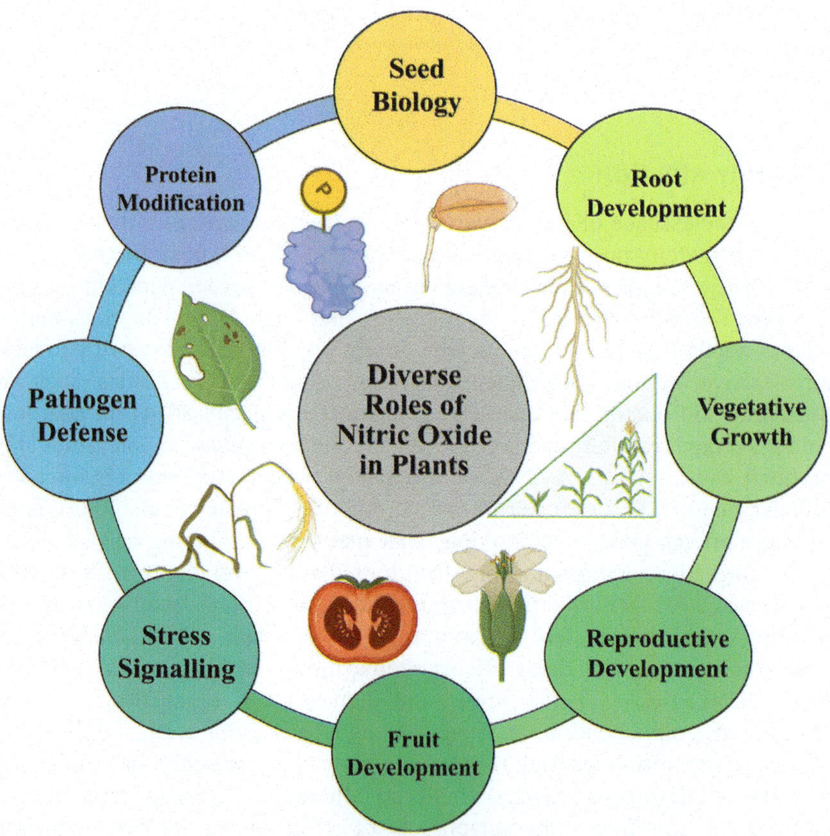

FIGURE 9.1 Various roles of NO in plant growth and development (*NO*, Nitric oxide). *The figure is created with the icons from BioRender (https://BioRender.com).*

drought stress enhanced adventitious root development and improved stress tolerance. The underlying mechanism for the same was that NO application boosted the ascorbate–glutathione antioxidant cycle (Asc-GSH cycle) and enhanced protein S-nitrosation, thus reducing the oxidative damage and elevating root formation under drought stress (Hou et al., 2024).

Interestingly, NO is also involved in mediating cellular redox homeostasis. It exhibits crosstalk with reactive oxygen species (ROS), such as H_2O_2, and modulates growth. For instance, NO harmonizes H_2O_2 levels via the enzymes respiratory burst oxidase homologs and scavenging by activating the Asc-GSH cycle and its related enzymes, namely, ascorbate peroxidase (APX) and catalase (CAT) (Mukherjee & Corpas, 2023). Together, these signaling molecules regulate root growth, root meristem maintenance, and post-translational modifications (PTMs). For instance, H_2O_2-mediated MPK6 activity and NO generation promote lateral root growth in Arabidopsis. Likewise, NO and H_2O_2 crosstalk-mediated adventitious root growth under water scarcity in *Tagetes erecta* (Mukherjee & Corpas, 2023). Despite the available literature, the diverse roles and underlying mechanisms of NO in crop plants are not fully understood, especially under drought stress. There is a need to translate the knowledge obtained from the model plant Arabidopsis to major food crops such as wheat, rice, and maize. The present chapter focuses on the role of NO in mitigating drought stress, with a focus on the exhibited crosstalk with the phytohormones.

9.2 Nitric oxide biosynthesis pathways

NO is a small, redox-active signaling molecule that plays a key role in plant growth, development, and stress response. Its biosynthesis can occur either through a reductive or an oxidative route. The oxidative pathway involves the oxidation of nitrogenous compounds such as L-arginine, polyamines, and hydroxylamine (Fig. 9.2). Unlike animals, the clear enzymatic nitric oxide synthase (NOS) activity in plants is not known; however, L-arginine acts as a substrate for the enzyme exhibiting NOS-like activity. The second pathway is the reductive pathway of NO formation, which mainly involves a sequential nitrite reduction (Fig. 9.2). The precise subcellular localization and dedicated enzymes for oxidative NO synthesis are not fully identified, but the reductive pathway is well documented in organelles such as peroxisomes and chloroplasts. Nitrate is converted to nitrite and then to NO by the dedicated enzymes nitrate reductase (NR) and by mitochondrial or other nitrite-reducing activities (Berger et al., 2020). NR is the most consistent source of NO in plants, and literature documents the dual role of the enzyme in nitrate assimilation and nitrite reduction to NO under various physiological conditions. NR activity directly correlates with cellular nitrite and NO levels, as there are studies that exhibit a change in nitrate assimilation when overexpressing NR (Bovet et al., 2024). Furthermore, under low oxygen (either hypoxia or anoxia), plant

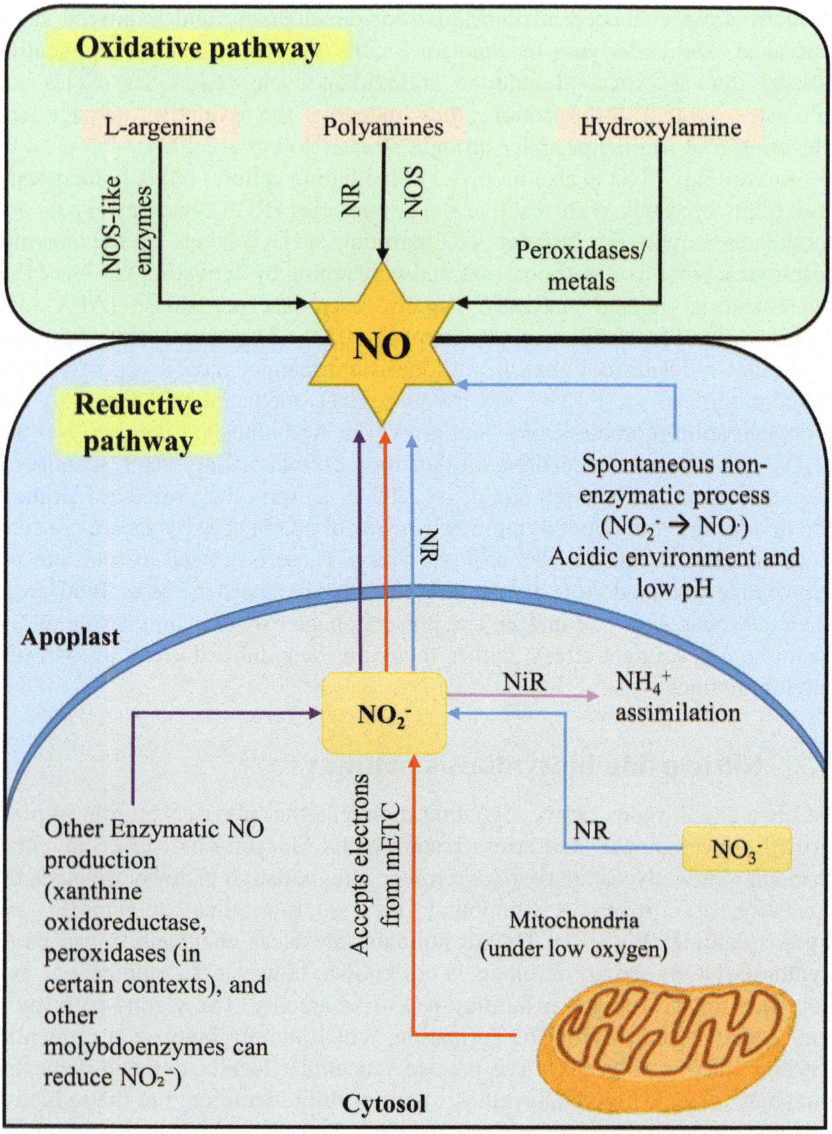

FIGURE 9.2 The representation of NO biosynthesis in plants. L-arginine, polyamines, and hydroxylamines are oxidatively converted to NO. In the reductive pathway, either the spontaneous conversion of NO_2^- happens, leading to NO formation, or NO_2^- may accept an electron under low oxygen levels and form NO (depicted in *red arrow*). The *blue arrow* indicates the sequential conversion of NO_3^- to NO via nitrate reductase (*NR*). In addition, other enzymes also convert NO_2^- to NO (shown in *purple arrow*) (*NO*, Nitric oxide).

mitochondria can reduce nitrite to NO via the electron transport chain (mETC). For instance, isolated mitochondria and in vivo experiments of nitrite feeding with hypoxia treatment exhibited mitochondrial NO production, preservation of ATP production, and lower ROS accumulation, thus exhibiting an adaptive and nitrosative signaling under stress (Gupta & Igamberdiev, 2011). There are some instances of nitrite to NO conversion via membrane-bound activities or peroxidases in the apoplast, where they contribute to a localized wave of NO burst, especially during stress or pathogen attack response. Though the evidence is mixed and context-dependent (Allagulova et al., 2023). Thus, further studies are required for the specific details. It is interesting to note that S-nitrosoglutathione reductase could indirectly control the S-nitrosation pattern and thus NO signaling. It is often considered a major controller of NO bioactivity (Jahnová et al., 2019).

9.3 Role of nitric oxide in modulating drought stress

NO regulates various aspects of plant growth and development. Under unfavorable conditions, it influences stress response and regulates biomass, antioxidant response, anatomy, and gene expression associated with stress signaling as well (Fig. 9.3). Nitrogen is the most abundant in the atmosphere, and it is in close crosstalk with plant development. For instance, a linear relationship was observed between drought and biomass accumulation and nitrogen fixation, with increasing drought severity causing decreased biomass accumulation and nitrogen fixation. However, if nitrogen availability is maintained, it helps in maintaining high yields even under drought stress in peanuts (Pimratch et al., 2008). Similarly, Song et al. (2019) observed that

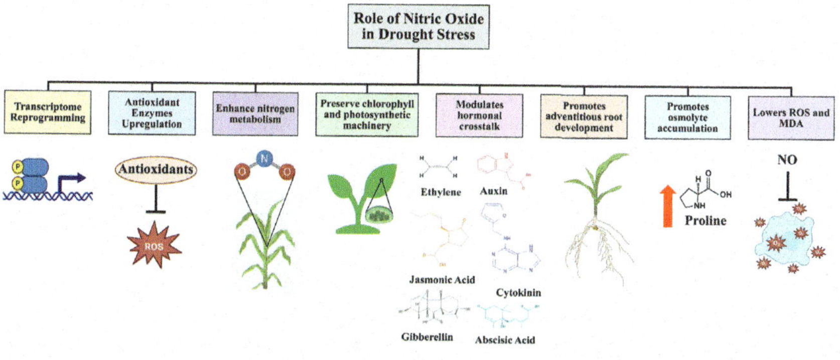

FIGURE 9.3 Role of nitric oxide (*NO*) under drought stress. It mitigates the adverse effects of drought stress by activating the antioxidant system, adjusting osmotic balance, regulating nitrogen metabolism, lowering ROS and membrane injury, reprogramming the stress-responsive gene network, and promoting adaptive developmental responses. These effects are strongly dose-, species-, and nitrogen-dependent (*ROS*, Reactive oxygen species).
The figure is created with the icons from BioRender (https://BioRender.com).

chlorophyll a and b content, photosynthetic machinery, and stomatal conductance, along with transpiration rate, were negatively affected by drought. However, moderate nitrogen levels (7.5 mM) alleviated these declines. Antioxidant enzyme activity, NO_3^- levels, NR activity, and abscisic acid (ABA) levels were increased after moderate nitrogen application. The study concluded that moderate nitrogen levels were beneficial in inducing stress tolerance, but too low or too high nitrogen levels were detrimental and even caused higher drought sensitivity in maize (Song et al., 2019). Likewise, nitrogen availability was identified to modulate biochemical parameters in maize, and optimal levels could lower drought-induced oxidative stress (Ahmadi et al., 2010; Zhang et al., 2007). Another study in soyabean also investigated the role of exogenous NO application in alleviating the detrimental effects of drought stress. Sodium nitroprusside (SNP; NO donor) treatment under PEG-6000 induced osmotic stress caused higher activities of antioxidant enzymes, namely APX, superoxide dismuatse (SOD), and peroxidase (POD) along with an elevation in proline and ABA levels (Fig. 9.3). Thus the study concluded that NO triggers antioxidant enzymes and aids in stress tolerance (Zhou et al., 2023). Drought caused yield losses in mungbean; however, the NO application improved growth, pigment content, and photosynthesis; triggered the accumulation of osmolytes (proline, sugars, and glycine betaine); and activated nitrogen metabolism-related enzymes (NR, GS, and GOGAT) along with the upregulation of *Heat Shock Protein 70* (*HSP70*) and *Late Embryogenesis Abundant* (*LEA*) genes. The underlying mechanism identified was that NO application activated nitrogen metabolism and its related enzymes, which in turn activated the antioxidant system, promoted osmolyte accumulation, lowered oxidative stress markers (MDA and ROS), and upregulated the accumulation of secondary metabolites such as phenols and flavonoids (Lian et al., 2023). Moreover, similar results of NO modulating the antioxidant system and stomatal response were documented under drought stress in *Poncirus trifoliata* as well (Fan & Liu, 2012). Drought stress effects were also investigated in a winter wheat variety to assess the beneficial role of nitrogen fertilization on the postanthesis stage. A medium nitrogen supply could prevent damage to the photosynthetic machinery, activate the antioxidant system, promote growth, and result in a higher yield. The study also concluded that under single stress, either drought or heat, medium nitrogen levels were required and showed the best results; however, under combined stress of both (drought + heat), lower levels of nitrogen were ideal for maintaining the yield at the cost of water loss and poorer water use efficiency (Ru et al., 2023).

The transgenic Arabidopsis plants constitutively expressing the *rat neural NO synthase* (*nNOS*) gene had higher endogenous NO levels. They were assessed for their survival after different durations of drought stress. Transgenic plants had higher NO levels under normal and drought conditions. At physiological levels, transgenic plants exhibited less electrolyte leakage, higher survival rates after rewatering, accumulated higher biomass, and increased plant height under drought stress than in the wild type. In addition, higher osmolyte accumulation and an activated antioxidant system were observed. At the transcriptomic levels, *nNOS* lines exhibited more differentially expressed genes (~447 -drought vs their control) than in the wild type (~154- drought vs control). Interestingly, the *C-repeat binding factor* (*CBF*) gene family, many *zinc-finger transcription factors*, and ABA pathway-related genes were upregulated in *nNOS* lines (Shi et al, 2014). The study stated that elevated NO levels could extensively reprogram the transcriptome, notably involving stress-responsive genes, redox metabolism, hormone signaling (especially ABA), transcription factors (zinc fingers, CBF), and photosynthetic and carbohydrate metabolism genes, thus inducing drought stress tolerance (Shi et al., 2014).

The role of NO in inducing adventitious root formation under drought stress is also well documented. For instance, dose-dependent positive roles of NO were identified in marigolds under drought stress on adventitious root development. NO application could maintain the structural integrity of mesophyll cells in addition to the photosynthetic parameters and mobilization of reserve starch content to soluble sugars to maintain efficient nitrogen metabolism. It was concluded that NO and H_2O_2 crosstalk could alleviate the adverse effects of drought stress on marigold explants by helping preserve cell structure, boosting photosynthetic efficiency, shifting carbohydrate metabolism, and thereby enabling better adventitious root development (Liao et al., 2012). A meta analysis conducted on the available data (48 studies) showed that NO significantly improves growth, photosynthesis, antioxidant status, and osmotic regulation and reduces oxidative damage in drought-stressed plants. The main factors affecting the NO mode of action and its beneficial effects were the method and dose of delivery, species and genotype, along with the stress severity and time of NO application (Lei et al., 2025).

The available literature explores the role of NO in inducing stress tolerance by modulating the antioxidant system, reducing oxidative stress, reprogramming the transcriptome, metabolome, and hormonal network. However, there still lies a gap in our understanding of the exact mechanism and the pathways that are modulated in the crop plants. Thus, the knowledge needs to be translated from the model plant Arabidopsis to food crops and from the lab results to field application.

9.4 Role of nitric oxide in drought tolerance

9.4.1 Abscisic acid with nitric oxide under drought

Drought stress is a major plant stress negatively affecting plant development and productivity (Shabbir et al., 2016). NO and ABA collectively regulate various plant processes, including stomatal functioning, seed dormancy, germination, decrease in photosynthetic rate, activation of stress machinery, ripening, and leaf senescence (Freschi, 2013). Stomatal closure is the primary response that occurs in plants after sensing water shortage. This leads to reduced transpiration rate and ion leakage, resulting in decreased photosynthesis (Silveira et al., 2016). When plants are under drought stress, a significant amount of ABA is produced. Once the ABA molecules reach the guard cells through the ABA transporters, they bind to the pyrabactin resistance1/PYR1-like/regulatory components (PYR/PYL/RCAR) of ABA receptors to trigger a complex signaling network that leads to stomatal closure (Silveira et al., 2016). This cascade involves the participation of various signaling molecules such as Ca^{2+}, NO, and H_2O_2, and protein kinases. Previous reports showed that NO generation is essential for ABA-induced stomatal closure and that ABA can increase NO generation in the guard cells as well. On the other hand, crosstalk between NO and ABA in the ABA-dependent signaling network upregulated the cytosolic Ca^{2+} level to regulate crassulacean acid metabolism (CAM) expression in bromeliads, which significantly improved plant tolerance in a water-limited environment. It seems that ABA-induced NO production is associated with increased NR activity that controls stomatal movements in Arabidopsis and CAM expression in bromeliads. Besides this, ABA-induced ROS induction is regulated by the antioxidant systems such as SOD, CAT, Cu/Zn-SOD, glutathione peroxidase (GPXs), and POD, and also the ascorbate—glutathione cycle, which is modulated by NO-induced PTMs (Desikan et al., 2002; Freschi et al., 2010; Mioto & Mercier, 2013). Previous studies suggest that the exogenous application of NO results in the ABA-induced NO generation in guard cells, which requires NOS and NR. NO triggers the activity of the mitogen-activated protein kinase, and cGMP is required in ABA-dependent stomatal closure to reduce the water loss and activation of the antioxidant defense system to protect plants from oxidative stress (Neill et al., 2008). Interestingly, NO–ABA and nitrogen assimilation together improve the drought tolerance in *B. juncea*. NO and ABA alone and together remove the toxic effect of drought by enhancing the N-metabolism enzymes activities and N-content, along with the other nutrients, and increasing the expression of *Nitrate transporters*, *Ammonium transporter*, *NR*, *Nitrite reductase*, *Glutamine synthetase*, *Glutamate Dehydrogenase*, and *Glutamate synthase* genes (Sahay et al., 2021). In Arabidopsis, NAD Kinase 2 (NADK2) is a vital component of chloroplast protection and chlorophyll synthesis, controlling the function by modulating the ABA signaling by increasing the level of H_2O_2, ROS, Ca^{2+}, and NO in guard cells, which leads to stomatal closure (Sun et al., 2017). Also, the endogenous NO negatively regulates the ABA signaling by constraining the *sucrose nonfermenting 1-related protein kinase 2.6* (*SnRK2.6*) and *open stomata 1* by S-nitrosation during seed germination and seedling development in Arabidopsis (Wang et al., 2015). Besides this, NO production from ABA signaling results in the blockage of kinase activity due to the S-nitrosation of SnRK2.6 at the cysteine residue near the kinase catalytic site (Wang et al.,

2015). However, NO is involved in both promoting and inhibiting the ABA-induced/phospholipid-mediated signaling to trigger stomatal closure by avoiding the hormonal flux (Laxalt et al., 2016).

9.4.2 Ethylene with nitric oxide under drought

Ethylene (ET) is one of the crucial gaseous and long-distance signaling hormones. It is involved in various physiological and biological processes in plants, such as flowering, senescence, fruit ripening, and leaf abscission. Methionine is the precursor of ET; it undergoes various enzymatic reactions to produce ET predominantly through 1-aminocyclopropane-1-carboxylic acid oxidase (ACC oxidase) (Salvi et al., 2021). The crosstalk between NO and ET plays a crucial role in plant adaptation to both abiotic and biotic stresses, as these signaling molecules can influence each other's biosynthesis and concentration. ET and polyamines (PAs) share a common biosynthetic precursor, S-adenosylmethionine (SAM), while NO and PAs are formed from arginine (Kolbert et al., 2019). Under drought stress, NO induces the expression of several stress-responsive genes. For instance, in soybean roots, NO positively regulates the transcription of *cytosolic glucose-6-phosphate dehydrogenase* genes (*Cyt-G6PD, G6PD6, GPD5*, and *G6PD7*), which contribute to drought tolerance (Wang et al., 2020). Moreover, during drought stress, NO modulates nitrogen flux toward polyamine biosynthesis or alternative pathways, such as ET production. This dynamic regulation of nitrogen allocation and signaling pathways contributes to enhanced drought tolerance, as observed in barley (Montilla-Bascón et al., 2017). Similarly, changes in the levels of NO were found to improve genomic stability during drought stress in plants such as sugarcane, broccoli, and soybeans. Remarkably, ET acts in Arabidopsis to antagonize drought as well as ABA-mediated stomatal closure (Tanaka et al., 2005). Besides this, the convergent signaling and response pathways induced by the drought and ozone stress increase the production of H_2O_2 and NO. ABA and ET interact together as well to improve drought and ozone stress, leading to improved stomatal and shoot growth (Wilkinson et al., 2012). Also, it is important to know whether H_2O_2 or NO is involved in the control of growth by ET, potentially indicated by the impact of H_2O_2 or NO signaling in leaves/stomata (Acharya & Assmann, 2009; Moeder et al., 2002). It is suggested that the *Hemoglobins* (*Hb*) gene influences signal transduction pathways, possibly by affecting NO levels, as indicated by the phenotype of normoxic *Hb* transgenic plants. When the expression of the *Hb* gene is lower, the ET level increases, resulting in the formation of root aerenchyma (Igamberdiev et al., 2005). Although drought conditions reduced ET and NO production, waterlogging enhanced ET production and reduced NO. The *Nitrate Reductase 1, 2* (*NIA1 NIA2*) defective mutants of NR and the *Ethylene Resistant 1* (*ETR1*) and *Ethylene Overproducer* (*ETO*) mutants produced less NO than in the wild-type Arabidopsis plants. It is concluded that NO serves as a signaling molecule, and NO emission may be useful in water stress in plants (Magalhaes et al., 2000). In another study, it was found that the salicylic acid (SA) and NO application in wheat helps to protect from heat stress in the presence of sulfur. This is due to the activation of the defense system, ET production, increased glutathione level, enhanced chlorophyll content, and maintaining the optimum ROS level (Rasheed et al., 2022). Exogenous treatment of Arabidopsis plants with ET induced H_2S and an increased level of *L-/D*-cysteine desulfhydrase enzyme activity in the leaves. Whereas the inhibitors of H_2S synthesis are unable to affect the ET-mediated NO accumulation and NR activity, which indicates that H_2S and NO participate in the signal transduction pathway of ET-mediated stomatal closure in Arabidopsis leaves (Liu et al., 2011). Further studies are needed to explore the molecular mechanisms of ET acting in NO production and how ET and NO interact in drought stress conditions.

9.4.3 Brassinosteroids with nitric oxide under drought

Brassinosteroids (BRs) are naturally occurring steroid hormones that play a crucial role in plant growth and development, as well as stress adaptation (Ahammed et al., 2020). It was first reported in the pollen of the *Brassica napus* (Grove et al., 1979). BRs regulate vital plant processes, such as cell elongation and division, senescence, photomorphogenesis, fruit development, and adaptation to various adverse environmental conditions (Clouse, 2011; Xia et al., 2015). BRs are perceived by the plasma membrane-localized receptor *BRASSINOSTEROID INSENSITIVE 1* (*BRI1*) and the coreceptor *BRI1-ASSOCIATED RECEPTOR KINASE 1* (*BAK1/SERK3*), which promotes the reciprocal phosphorylation of *BRI1*. The *BRI1ASSOCIATED RECEPTOR KINASE1* inhibits the activity of the negative regulators GSK3/Shaggy-like kinase (*BRINSENSITIVE2, BIN2*) and enables the degradation of *BIN2* (Peng et al., 2008). Therefore, *BIN2* becomes unable to phosphorylate the downstream transcription factors *BRASSINOAZOLE RESISTANT 1* (*BZR 1*) and *BZI1-EMS-SUPPRESSOR 1* (*BES1*). *PROTEIN PHOSPHATASE 2A* (*PP2A*) is involved in dephosphorylating the phosphorylated *BZR1/BES1*, which regulates the expression of the downstream genes to promote growth and development of the plants (Kono & Yin, 2020; Sun et al., 2010). NO and BR together play a significant role in overcoming the detrimental effects of drought. In a study by Tossi et al. (2015), it was shown that natural and synthetic BRs induced the NO production in tomato cell cultures. Another study showed that BR induced NO production under polyethylene glycol (PEG) treatment. This produced NO, then triggered ABA biosynthesis in maize leaves to enhance the water stress tolerance (Zhang et al., 2011). Also, the combined application of the BR and NO helps in improving the adventitious root development compared to applying BR and NO alone. But little is known about whether BR and NO have a synergistic effect on the plant stress resistance (Li et al., 2020). Similarly, the combined application of 24-epibrassinolide and NO improves drought tolerance by upregulating the expression of the proline biosynthesis gene *Pyrroline-5-carboxylate reductase* (*P5CR1*) and nitrogen metabolism, along with lowering proline degradation to mitigate drought in kiwifruit seedlings (Xia et al., 2022). In addition, it is shown that the application of 24-epibrassinolide increases the endogenous level of NO, which acts as a signaling molecule to upregulate the antioxidant defense system to ameliorate the toxic effect of water stress in pepper plants (Kaya et al., 2019). Evidence showed that the production of NO occurs in a ROS-dependent manner through the NOS-like and nitrate/nitrite enzyme-associated pathway when exogenous 24-epibrassinolide is applied. In response to this, antioxidant gene expression is induced, leading to the elevated activities of the antioxidant enzymes and serving to cope with oxidative stress in cucumber (Cui et al., 2011). In another study, it was found that 24-epibrassinolide induces H_2O_2 and NO in the guard cells and further elevates the activity of L-/D-cysteine desulfhydrase and H_2S, probably participating in a signal transduction pathway that results in stomatal closure of *Vicia faba* L. (Ma et al., 2022).

9.4.4 Jasmonic acid with nitric oxide under drought

Jasmonic acid (JA) is a lipid-based hormone signal that originates from the α-linolenic acid in plants. It plays a key role in controlling various processes of plants, such as seed germination, wounding, senescence, biotic stress, and abiotic stress (Liu et al., 2005). NO acts as a major signal, affecting the JA, SA, and ET signaling not only during biotic stress but also under abiotic stress as well (Zhang et al., 2011). ROS, pH, and Ca^{2+} act as a primary signal with NO, JA, ABA, and ET during stomatal closure (Gayatri et al., 2013). It is known that JA regulates a group of reactions in plants, but only a few of these reactions show a connection with NO. The pretreatment with an NO inhibitor, NG-nitro-L-arginine methyl ester, induced stomatal closure mediated by JA, reflecting the involvement of NO downstream of JA (Liu et al., 2005). In a study, it is shown that the JA could enhance the NO production in the guard cell, and therefore, NO results in the signal transduction of JA-induced stomatal closure, and it can be reversed by the NO scavenger (Liu et al., 2005). The exogenous application of NO induces the expression of JA biosynthesis-related genes such as *oxophytodienoate reductases 1 (OPR1)*, *oxophytodienoate reductases 2 (OPR2)*, and *oxophytodienoate reductases 3 (OPR3)* and *lipoxygenases* (*LOX3*) and a putative *LOX3* protein. Palmieri et al. (2008). Similarly, the exogenous treatment of methyl jasmonate (MeJA) in the *Taxus* cell culture instantly induced the production of NO. On the other hand, other indicators of stress, such as H_2O_2, MDA, and *LOX3*, increase with the MeJA application but are suppressed by the NO donor. This suggested NO acts as a signal molecule to activate MeJA-mediated defense responses and secondary metabolism in the *Taxus* plant cells (Wang & Wu, 2005). In another study, it is concluded that the exogenous application of JA improves the drought tolerance in wheat seedlings through the production of NO and upregulation of the activity of the AsA-GSH cycle (Shan et al., 2015). During water-stress conditions, MeJA pretreatment regulates the NO accumulation and induces the growth, proline accumulation, and water balance regulation in wheat plants (Lubyanova et al., 2022). Similarly, in another study, it was concluded that MeJA application induces NO, which possibly participates in the L-arginine oxidation and nitrate/nitrite reduction, which regulate the stomatal apparatus in Arabidopsis (Yastreb et al., 2018). However, evidence showed that the application of JA increases the production of H_2O_2 and NO levels, along with this, the activities of the enzymes such as glutathione reductase (GR), monodehydroascorbate reductase, dehydroascorbate reductase, L-galactono-1,4-lactone dehydrogenase, and γ-glutamylcysteine synthetase. Moreover, the gene expression of the mentioned enzymes is also enhanced, whereas the ascorbate and glutathione content are reduced with the application of JA. Therefore, it is concluded that during the JA signaling, NO acts downstream of H_2O_2, which results in upregulation of ascorbate and glutathione metabolism in leaves of *Agropyron cristatum* (Shan & Yang, 2017). These are a few cases that demonstrate the crosstalk between NO and JA; however, further investigation is still needed to elucidate the molecular mechanisms of JA and NO signaling in response to drought stress.

9.5 Conclusion and future perspective

NO plays a central and multifaceted role in modulating plant response to drought stress through its complex interactions with other signaling molecules and phytohormones. NO also regulates stomatal closure, antioxidant metabolism, osmolyte accumulation, nitrogen assimilation, and the expression of stress-responsive genes, thereby enhancing drought tolerance. Despite significant progress, several parts of NO biology remain unresolved, including the identity and regulation of its enzymatic synthesis, the spatiotemporal dynamics, and the precise molecular mechanism underlying its crosstalk with other signaling molecules. Moreover, the majority of mechanistic insights have been derived from model species like Arabidopsis, and translating this knowledge to major crops and field conditions remains a critical challenge. Future research should focus on integrating multiomics approaches with advanced imaging and biosensing technologies to unravel the spatial and temporal regulation of NO. Understanding and utilizing the synergistic relationships between NO and phytohormones with redox pathways will be key to designing next-generation, stress-tolerant crops that will be capable of sustaining production and yield under stressful conditions.

Acknowledgment and funding

Gábor Kocsy is funded by the National Research, Development, and Innovation Office (grants K131638 and TKP2021-NKTA-06). Riyazuddin Riyazuddin holds a postdoctoral contract (CIAPOS/2023/138) from the Generalitat Valenciana. Kalpita Singh (SHE-079837-004/2022), Zahra Tahmasebi (SHE-102333-004/2022), and Akhil Kumar (SHE-18483-004/2022 are funded by the Stipendium Hungaricum program of Tempus Public Foundation.

Disclaimer

AI was used to improve the language and readability of the text.

AI disclosure

During the preparation of this work, the author(s) used ChatGPT 4.1 and NotebookLM in order to correct grammar and improve language. After using this tool/service, the author(s) review(s) and edit(s) the content as needed and take(s) full responsibility for the content of the publication.

References

Acharya, B. R., & Assmann, S. M. (2009). Hormone interactions in stomatal function. *Plant Molecular Biology, 69*(4), 451–462. https://doi.org/10.1007/s11103-008-9427-0.

Ahammed, G. J., Li, X., Liu, A., & Chen, S. (2020). Brassinosteroids in plant tolerance to abiotic stress. *Journal of Plant Growth Regulation, 39*(4), 1451–1464. https://doi.org/10.1007/s00344-020-10098-0.

Ahmadi, A., Emam, Y., & Pessarakli, M. (2010). Biochemical changes in maize seedlings exposed to drought stress conditions at different nitrogen levels. *Journal of Plant Nutrition, 33*(4), 541–556. https://doi.org/10.1080/01904160903506274.

Allagulova, C. R., Lubyanova, A. R., & Avalbaev, A. M. (2023). Multiple ways of nitric oxide production in plants and its functional activity under abiotic stress conditions. *International Journal of Molecular Sciences, 24*(14), 11637. https://doi.org/10.3390/ijms241411637.

Berger, A., Boscari, A., Horta Araújo, N., Maucourt, M., Hanchi, M., Bernillon, S., Rolin, D., Puppo, A., & Brouquisse, R. (2020). Plant nitrate reductases regulate nitric oxide production and nitrogen-fixing metabolism during the *Medicago truncatula–Sinorhizobium meliloti* symbiosis. *Frontiers in Plant Science, 11*. https://doi.org/10.3389/fpls.2020.01313.

Bovet, L., Battey, J., Lu, J., Sierro, N., Dewey, R. E., & Goepfert, S. (2024). Nitrate assimilation pathway is impacted in young tobacco plants overexpressing a constitutively active nitrate reductase or displaying a defective CLCNt2. *BMC Plant Biology, 24*(1). https://doi.org/10.1186/s12870-024-05834-7.

Clouse, S. D. (2011). Brassinosteroids. *The Arabidopsis Book, 9*, e0151. https://doi.org/10.1199/tab.0151.

Cui, J. X., Zhou, Y. H., Ding, J. G., Xia, X. J., Shi, K., Chen, S. C., Asami, T., Chen, Z., & Yu, J. Q. (2011). Role of nitric oxide in hydrogen peroxide- dependent induction of abiotic stress tolerance by brassinosteroids in cucumber. *Plant, Cell and Environment, 34*(2), 347–358. https://doi.org/10.1111/j.1365-3040.2010.02248.x.

Desikan, R., Griffiths, R., Hancock, J., & Neill, S. (2002). A new role for an old enzyme: Nitrate reductase-mediated nitric oxide generation is required for abscisic acid-induced stomatal closure in *Arabidopsis thaliana. Proceedings of the National Academy of Sciences, 99*(25), 16314–16318. https://doi.org/10.1073/pnas.252461999.

Dirie, K. A., Maamor, S., & Alam, M. M. (2024). Impacts of climate change in post-conflict Somalia: Is the 2030 Agenda for SDGs endangered? *World Dev. Perspect. 35*, 100598. https://doi.org/10.1016/j.wdp.2024.100598.

Fan, Q. J., & Liu, J. H. (2012). Nitric oxide is involved in dehydration/drought tolerance in *Poncirus trifoliata* seedlings through regulation of antioxidant systems and stomatal response. *Plant Cell Reports, 31*(1), 145–154. https://doi.org/10.1007/s00299-011-1148-1.

Freschi, L. (2013). Nitric oxide and phytohormone interactions: Current status and perspectives. *Frontiers in Plant Science, 4*. https://doi.org/10.3389/fpls.2013.00398.

Freschi, L., Rodrigues, M. A., Domingues, D. S., Purgatto, E., van Sluys, M. A., Magalhaes, J. R., Kaiser, W. M., & Mercier, H. (2010). Nitric oxide mediates the hormonal control of Crassulacean acid metabolism expression in young pineapple plants. *Plant Physiology, 152*(4), 1971–1985. https://doi.org/10.1104/pp.109.151613.

Gayatri, G., Agurla, S., & Raghavendra, A. S. (2013). Nitric oxide in guard cells is an important secondary messenger during stomatal closure. *Frontiers Plant Sciences. 4*, 425. https://doi.org/10.3389/fpls.2013.00425.

Grove, M. D., Spencer, G. F., Rohwedder, W. K., Mandava, N., Worley, J. F., Warthen, J. D., Steffens, G. L., Flippen-Anderson, J. L., & Cook, J. C. (1979). Brassinolide, a plant growth-promoting steroid isolated from *Brassica napus* pollen. *Nature, 281*(5728), 216–217. https://doi.org/10.1038/281216a0.

Gupta, K. J., & Igamberdiev, A. U. (2011). The anoxic plant mitochondrion as a nitrite: NO reductase. *Mitochondrion, 11*(4), 537–543. https://doi.org/10.1016/j.mito.2011.03.005.

Hou, X., Dan, Y., Qi, N., Zhang, M., Li, C., Li, Y., Yao, Y., & Liao, W. (2024). Nitric oxide promotes adventitious rooting in cucumber under drought stress through regulating ascorbate glutathione cycle and protein S-nitrosylation. *Journal of Plant Growth Regulation, 43*(8), 2771–2784. https://doi.org/10.1007/s00344-024-11304-z.

Igamberdiev, A. U., Baron, K., Manac'h-Little, N., Stoimenova, M., & Hill, R. D. (2005). The haemoglobin/nitric oxide cycle: Involvement in flooding stress and effects on hormone signalling. *Annals of Botany, 96*(4), 557–564. https://doi.org/10.1093/aob/mci210.

Jahnová, J., Luhová, L., & Petřivalský, M. (2019). S-nitrosoglutathione reductase—The master regulator of protein S-nitrosation in plant NO signaling. *Plants, 8*(2), 48. https://doi.org/10.3390/plants8020048.

Jobbágy, K., Singh, K., Kulman, K., Szalai, G., Pál, M., Molnár, I., Panda, S. K., Kocsy, G., & Gulyás, Z. (2025). Differences in the metabolic response to osmotic stress between *Triticum aestivum* L. and two ecologically disparate aegilops biuncialis vis. genotypes differing in drought tolerance. *Journal of Agronomy and Crop Science, 211*(5). https://doi.org/10.1111/jac.70110.

Kaya, C., Ashraf, M., Wijaya, L., & Ahmad, P. (2019). The putative role of endogenous nitric oxide in brassinosteroid-induced antioxidant defence system in pepper (*Capsicum annuum* L.) plants under water stress. *Plant Physiology and Biochemistry, 143*, 119–128. https://doi.org/10.1016/j.plaphy.2019.08.024.

Khan, M., Ali, S., Al Azzawi, T. N. I., & Yun, B. W. (2023). Nitric oxide acts as a key signaling molecule in plant development under stressful conditions. *International Journal of Molecular Sciences, 24*(5). https://doi.org/10.3390/ijms24054782.

Kolbert, Z., Feigl, G., Freschi, L., & Poór, P. (2019). Gasotransmitters in action: Nitric oxide-ethylene crosstalk during plant growth and abiotic stress responses. *Antioxidants, 8*(6), 167. https://doi.org/10.3390/antiox8060167.

Kono, A., & Yin, Y. (2020). Updates on BES1/BZR1 regulatory networks coordinating plant growth and stress responses. *Frontiers in Plant Science, 11*. https://doi.org/10.3389/fpls.2020.617162.

Laxalt, A. M., García-Mata, C., & Lamattina, L. (2016). The dual role of nitric oxide in guard cells: Promoting and attenuating the ABA and phospholipid derived signals leading to the stomatal closure. *Frontiers in Plant Science, 7*, 476. https://doi.org/10.3389/fpls.2016.00476.

Lei, Y., Chen, S., Xu, L., Zhang, Y., & Yang, Y. (2025). Enhancing plant drought tolerance through exogenous nitric oxide: a comprehensive meta-analysis. *BMC Plant Biology, 25*(1). https://doi.org/10.1186/s12870-025-06491-0.

Li, Y., Wu, Y., Liao, W., Hu, L., Dawuda, M. M., Jin, X., Tang, Z., Yang, J., & Yu, J. (2020). Nitric oxide is involved in the brassinolide-induced adventitious root development in cucumber. *BMC Plant Biology, 20*(1). https://doi.org/10.1186/s12870-020-2320-y.

Lian, H., Qin, C., Shen, J., & Ahanger, M. A. (2023). Alleviation of adverse effects of drought stress on growth and nitrogen metabolism in mungbean (*Vigna radiata)* by sulphur and nitric oxide involves up-regulation of antioxidant and osmolyte metabolism and gene expression. *Plants, 12*(17), 3082. https://doi.org/10.3390/plants12173082.

Liao, W. B., Huang, G. B., Yu, J. H., & Zhang, M. L. (2012). Nitric oxide and hydrogen peroxide alleviate drought stress in marigold explants and promote its adventitious root development. *Plant Physiology and Biochemistry, 58*, 6–15. https://doi.org/10.1016/j.plaphy.2012.06.012.

Liu, X., Shi, W., Zhang, S., Lou, C., et al. (2005). Nitric oxide is involved in the signal transduction of jasmonic acid-induced stomatal closure of *Vicia faba* L. *Chinese Science Bulletin, 50*, 520–525. https://doi.org/10.1007/BF02897475.

Liu, J., Hou, L. X., Liu, G. H., Liu, X., & Wang, X. C. (2011). Hydrogen sulfide induced by nitric oxide mediates ethylene-induced stomatal closure of *Arabidopsis thaliana. Chinese Science Bulletin, 56*(33), 3547–3553. https://doi.org/10.1007/s11434-011-4819-y.

Lubyanova, A. R., Bezrukova, M. V., & Shakirova, F. M. (2022). Involvement of nitric oxide in methyl jasmonate-mediated regulation of water metabolism in wheat plants under drought stress. *Stresses, 2*(4), 477–492.

Ma, Y., Wang, L., & Zhang, W. (2022). The role of hydrogen sulfide and its relationship with hydrogen peroxide and nitric oxide in brassinosteroid-induced stomatal closure of *Vicia faba* L. *South African Journal of Botany, 146*, 426–436. https://doi.org/10.1016/j.sajb.2021.11.012.

Magalhaes, J. R., Monte, D. C., & Durzan, D. (2000). Nitric oxide and ethylene emission in *Arabidopsis thaliana. Physiology and Molecular Biology of Plants, 6*, 117–127.

Mioto, P. T., & Mercier, H. (2013). Abscisic acid and nitric oxide signaling in two different portions of detached leaves of *Guzmania monostachia* with CAM up-regulated by drought. *Journal of Plant Physiology, 170*, 996–1002. https://doi.org/10.1016/j.jplph.2013.02.004.

Moeder, W., Barry, C. S., Tauriainen, A. A., et al. (2002). Ethylene synthesis regulated by biphasic induction of ACC synthase and ACC oxidase genes is required for H_2O_2 accumulation and cell death in ozone-exposed tomato. *Plant Physiology, 130*, 1918–1926. https://doi.org/10.1104/pp.009712.

Montilla-Bascón, G., Rubiales, D., Hebelstrup, K. H., Mandon, J., Harren, F. J. M., Cristescu, S. M., Mur, L. A. J., Prats, E., et al. (2017). Reduced nitric oxide levels during drought stress promote drought tolerance in barley and is associated with elevated polyamine biosynthesis. *Scientific Reports, 7*(1). https://doi.org/10.1038/s41598-017-13458-1.

Mukherjee, S., & Corpas, F. J. (2023). H_2O_2, NO, and H_2S networks during root development and signalling under physiological and challenging environments: Beneficial or toxic? *Plant Cell and Environment, 46*(3), 688–717. https://doi.org/10.1111/pce.14531.

Neill, S., Barros, R., Bright, J., Desikan, R., Hancock, J., Harrison, J., Morris, P., Ribeiro, D., & Wilson, I. (2008). Nitric oxide, stomatal closure, and abiotic stress. *Journal of Experimental Botany, 59*(2), 165–176. https://doi.org/10.1093/jxb/erm293.

Palmieri, M. C., Sell, S., Huang, X., Scherf, M., Werner, T., Durner, J., & Lindermayr, C. (2008). Nitric oxide-responsive genes and promoters in *Arabidopsis thaliana*: A bioinformatics approach. *Journal of Experimental Botany, 59*(2), 177–186. https://doi.org/10.1093/jxb/erm345.

Peng, P., Yan, Z., Zhu, Y., & Li, J. (2008). Regulation of the *Arabidopsis* GSK3-like kinase BRASSINOSTEROID-INSENSITIVE 2 through proteasome-mediated protein degradation. *Molecular Plant, 1*(2), 338–346. https://doi.org/10.1093/mp/ssn001.

Pimratch, S., Jogloy, S., Vorasoot, N., Toomsan, B., Patanothai, A., & Holbrook, C. C. (2008). Relationship between biomass production and nitrogen fixation under drought-stress conditions in peanut genotypes with different levels of drought resistance. *Journal of Agronomy and Crop Science, 194*(1), 15–25. https://doi.org/10.1111/j.1439-037x.2007.00286.x.

Rasheed, F., Mir, I. R., Sehar, Z., Fatma, M., Gautam, H., Khan, S., Anjum, N. A., Masood, A., Sofo, A., & Khan, N. A. (2022). Nitric oxide and salicylic acid regulate glutathione and ethylene production to enhance heat stress acclimation in wheat involving sulfur assimilation. *Plants, 11*(22), 3131. https://doi.org/10.3390/plants11223131.

Riyazuddin, R., Singh, K., Iqbal, N., Labhane, N., Ramteke, P., Singh, V. P., Gupta, R., et al. (2023). Unveiling the biosynthesis, mechanisms, and impacts of miRNAs in drought stress resilience in plants. *Plant Physiology and Biochemistry, 202*. https://doi.org/10.1016/j.plaphy.2023.107978.

Ru, C., Wang, K., Hu, X., Chen, D., Wang, W., Yang, H., et al. (2023). Nitrogen modulates the effects of heat, drought, and combined stresses on photosynthesis, antioxidant capacity, cell osmoregulation, and grain yield in winter wheat. *Journal of Plant Growth Regulation, 42*(3), 1681–1703. https://doi.org/10.1007/s00344-022-10650-0.

Sahay, S., Robledo-Arratia, L., Glowacka, K., & Gupta, M. (2021). Root NRT, NiR, AMT, GS, GOGAT and GDH expression levels reveal NO and ABA mediated drought tolerance in *Brassica juncea* L. *Scientific Reports, 11*(1). https://doi.org/10.1038/s41598-021-86401-0.

Salvi, P., Manna, M., Kaur, H., Thakur, T., Gandass, N., Bhatt, D., & Muthamilarasan, M. (2021). Phytohormone signaling and crosstalk in regulating drought stress response in plants. *Plant Cell Reports, 40*(8), 1305–1329. https://doi.org/10.1007/s00299-021-02683-8.

Shabbir, R. N., Waraich, E. A., Ali, H., Nawaz, F., Ashraf, M. Y., Ahmad, R., Awan, M. I., Ahmad, S., Irfan, M., Hussain, S., Ahmad, Z., et al. (2016). Supplemental exogenous NPK application alters biochemical processes to improve yield and drought tolerance in wheat (*Triticum aestivum* L.). *Environmental Science and Pollution Research, 23*(3), 2651–2662. https://doi.org/10.1007/s11356-015-5452-0.

Shan, C., & Yang, T. (2017). Nitric oxide acts downstream of hydrogen peroxide in the regulation of ascorbate and glutathione metabolism by jasmonic acid in *Agropyron cristatum* leaves. *Biologia Plantarum, 61*(4), 779–784. https://doi.org/10.1007/s10535-017-0708-9.

Shan, C., Zhou, Y., Liu, M., et al. (2015). Nitric oxide participates in the regulation of the ascorbate-glutathione cycle by exogenous jasmonic acid in the leaves of wheat seedlings under drought stress. *Protoplasma, 252*, 1397–1405. https://doi.org/10.1007/s00709-015-0756-y.

Shi, H., Ye, T., Zhu, J. K., & Chan, Z. (2014). Constitutive production of nitric oxide leads to enhanced drought stress resistance and extensive transcriptional reprogramming in *Arabidopsis*. *Journal of Experimental Botany, 65*(15), 4119–4131. https://doi.org/10.1093/jxb/eru184.

Silveira, N. M., Frungillo, L., Marcos, F. C. C., Pelegrino, M. T., Miranda, M. T., Seabra, A. B., Salgado, I., Machado, E. C., Ribeiro, R. V., et al. (2016). Exogenous nitric oxide improves sugarcane growth and photosynthesis under water deficit. *Planta, 244*(1), 181–190. https://doi.org/10.1007/s00425-016-2501-y.

Song, Y., Li, J., Liu, M., Meng, Z., Liu, K., Sui, N., et al. (2019). Nitrogen increases drought tolerance in maize seedlings. *Functional Plant Biology, 46*(4), 350–359. https://doi.org/10.1071/FP18186.

Sun, L., Li, Y., Miao, W., Piao, T., Hao, Y., Hao, F. S., et al. (2017). NADK2 positively modulates abscisic acid-induced stomatal closure by affecting accumulation of H_2O_2, Ca^{2+} and nitric oxide in *Arabidopsis* guard cells. *Plant Science, 262*, 81–90. https://doi.org/10.1016/j.plantsci.2017.06.003.

Sun, Y., Fan, X. Y., Cao, D. M., Tang, W., He, K., Zhu, J. Y., He, J. X., Bai, M. Y., Zhu, S., Oh, E., Patil, S., Kim, T. W., Ji, H., Wong, W. H., Rhee, S. Y., & Wang, Z. Y. (2010). Integration of brassinosteroid signal transduction with the transcription network for plant growth regulation in *Arabidopsis*. *Developmental Cell, 19*(5), 765–777. https://doi.org/10.1016/j.devcel.2010.10.010.

Tanaka, Y., Sano, T., Tamaoki, M., Nakajima, N., Kondo, N., & Hasezawa, S. (2005). Ethylene inhibits abscisic acid-induced stomatal closure in *Arabidopsis*. *Plant Physiology, 138*(4), 2337–2343. https://doi.org/10.1104/pp.105.063503.

Tossi, V. E., Acebedo, S. L., Cassia, R. O., Lamattina, L., Galagovsky, L. R., Ramírez, J. A., et al. (2015). A bioassay for brassinosteroid activity based on the in vitro fluorimetric detection of nitric oxide production. *Steroids, 102*, 46–52. https://doi.org/10.1016/j.steroids.2015.07.003.

Wang, P., Du, Y., Hou, Y. J., Zhao, Y., Hsu, C. C., Yuan, F., Zhu, X., Tao, W. A., Song, C. P., Zhu, J. K., et al. (2015). Nitric oxide negatively regulates abscisic acid signaling in guard cells by S-nitrosylation of OST1. *Proceedings of the National Academy of Sciences of the United States of America, 112*(2), 613–618. https://doi.org/10.1073/pnas.1423481112.

Wang, X., Ruan, M., Wan, Q., He, W., Yang, L., Liu, X., He, L., Yan, L., & Bi, Y. (2020). Nitric oxide and hydrogen peroxide increase glucose-6-phosphate dehydrogenase activities and expression upon drought stress in soybean roots. *Plant Cell Reports, 39*(1), 63–73. https://doi.org/10.1007/s00299-019-02473-3.

Wang, J. W., & Wu, J. Y. (2005). Nitric oxide is involved in methyl jasmonate-induced defense responses and secondary metabolism activities of *Taxus* cells. *Plant and Cell Physiology, 46*(6), 923–930. https://doi.org/10.1093/pcp/pci098.

Wilkinson, S., Kudoyarova, G. R., Veselov, D. S., Arkhipova, T. N., & Davies, W. J. (2012). Plant hormone interactions: Innovative targets for crop breeding and management. *Journal of Experimental Botany, 63*(9), 3499–3509. https://doi.org/10.1093/jxb/ers148.

Xia, H., Liu, X., Wang, Y., Lin, Z., Deng, H., Wang, J., Lin, L., Deng, Q., Lv, X., Xu, K., & Liang, D. (2022). 24-Epibrassinolide and nitric oxide combined to improve the drought tolerance in kiwifruit seedlings by proline pathway and nitrogen metabolism. *Scientia Horticulturae, 297*, 110929. https://doi.org/10.1016/j.scienta.2022.110929.

Xia, X. J., Zhou, Y. H., Shi, K., Zhou, J., Foyer, C. H., Yu, J. Q., et al. (2015). Interplay between reactive oxygen species and hormones in the control of plant development and stress tolerance. *Journal of Experimental Botany, 66*(10), 2839–2856. https://doi.org/10.1093/jxb/erv089.

Yastreb, T. O., Kolupaev, Y. E., Kokorev, A. I., Horielova, E. I., Dmitriev, A. P., et al. (2018). Methyl Jasmonate and nitric oxide in regulation of the stomatal apparatus of *Arabidopsis thaliana*. *Cytology and Genetics, 52*(6), 400–405. https://doi.org/10.3103/S0095452718060129.

Zhang, A., Zhang, J., Zhang, J., Ye, N., Zhang, H., Tan, M., & Jiang, M. (2011). Nitric oxide mediates brassinosteroid-induced aba biosynthesis involved in oxidative stress tolerance in maize leaves. *Plant and Cell Physiology, 52*(1), 181–192. https://doi.org/10.1093/pcp/pcq187.

Zhang, L.-X., Li, S.-X., Zhang, H., & Liang, Z.-S. (2007). Nitrogen rates and water stress effects on production, lipid peroxidation and antioxidative enzyme activities in two maize (*Zea mays* L.) Genotypes. *Journal of Agronomy and Crop Science, 193*(6), 387–397. https://doi.org/10.1111/j.1439-037x.2007.00276.x.

Zhou, Q., Tian, Y., Li, X., Wu, Z., Wang, X., & Dong, S. (2023). SNP application improves drought tolerance in soybean. *Scientific Reports*. https://doi.org/10.1038/s41598-023-38088-8.

Chapter 10

Role of nitric oxide in the mitigation of heavy metal stress

Hafiz Muhammad Ansab Jamil[1], Nadeem Iqbal[2], Muhammad Nauman[3], Gulnaz Ashraf[1], Moniba Zahid Mahmood[4], and Malik Tahir Hayat[5]

[1]*Department of Biotechnology, COMSATS University Islamabad, Abbottabad Campus, Abbottabad, Khyber Pakhtunkhwa, Pakistan,* [2]*School of Environmental Science and Engineering, Hunan University, Changsha, Hunan, P.R. China,* [3]*School of Metallurgical Engineering, Anhui University of Technology, Ma'anshan, Anhui, P.R. China,* [4]*Department of Ecology, Friedrich Schiller University, Jena, Germany,* [5]*Department of Environmental Sciences, COMSATS University Islamabad, Abbottabad Campus, Abbottabad, Khyber Pakhtunkhwa, Pakistan*

10.1 Introduction

Plants undergo several stress conditions during their lifespan. These are biotic (viruses, bacteria, fungi, weeds, and insects) or abiotic (drought, salinity, temperatures, heavy metals (HMs), mineral deficiency, or air/soil pollution) stresses. Due to industrialization and increasing urbanization, anthropogenic effects have increased air and soil pollution, ultimately causing ecological imbalance (Chen et al., 2022). Among these anthropogenic sources, HM contamination seriously threatens plant development, growth, and yield (Shahid et al., 2015). In many regions worldwide, HM toxicity results from agricultural pollution caused by improper irrigation, prolonged usage of phosphatic fertilizers, industrial waste, and wastewater treatment practices (Romero-Puertas et al., 2019; Iqbal et al., 2020). Among various stresses, HM stress is the most devastating abiotic stress, resulting in crop reduction, yielding losses, and leading to huge annual economic losses. Therefore, it is essential to tackle HM pollution and minimize its impact on plants and other living organisms, including humans, to prevent environmental and health hazards.

HMs like cadmium (Cd), arsenic (As), lead (Pb), copper (Cu), chromium (Cr), mercury (Hg), zinc (Zn), and nickel (Ni) are ubiquitous in all environments and pose severe health risks to humans upon consumption of HMs-contaminated food. However, HMs can harm plants even at low concentrations (Li et al., 2019). Pb, Hg, and Cd are toxic to plants even at inadequate concentrations and lead to changes in the physiology and morphology of plants under development, causing growth retardation (Chen et al., 2022). Plants normally produce reactive oxygen

Nitric Oxide in Plant Resilience to Abiotic Stress. DOI: https://doi.org/10.1016/B978-0-443-40497-9.00007-6

species (ROS) under abiotic stress; nevertheless, ROS have the potential to be toxic, and depending on their quantity, they may be either detrimental or helpful to plant cells (Iqbal et al., 2024). ROS production disrupts the redox balance within cells, leading to oxidative stress in plants. Furthermore, excessive ROS generation can harm cells and cause necrosis in various plant species (Petrov et al., 2015; Iqbal et al., 2023). NO corresponds to several signaling molecules during the generation of ROS to keep ROS levels constant. Although HMs produce ROS, NO is crucial for activating plants' antioxidant signaling response, which mitigates ROS's harmful effects (Sharma et al., 2020; Iqbal et al., 2023). As a result, NO treatment can help reduce ROS generation and alleviating plant HM stress.

NO is a reactive nitrogen species (RNS) that occurs in normal and stressful situations. It is a gaseous free radical that mitigates the deleterious effects caused by increased ROS production in plants during biotic and abiotic stress. NO also makes the plant tolerant to salt, HM, and drought stress (Lipton et al., 1993). NO is a complex bioactive molecule that plays a central part in regulating plant function, disease susceptibility, and stress tolerance and regulates multiple genes and the activity of proteins in stressed plants (Fancy et al., 2017). NO also contributes to the production of root hair, which allows interaction between plants and microorganisms during nitrogen fixation (Pande et al., 2021), regulates the ratio of auxin (IAA) to ROS, and maintains iron homeostasis (Parveen et al., 2022). According to a study, NO lessens ROS generation by increasing the amounts of ROS-scavenging enzymes. By forming less stable peroxynitrite from the superoxide radical ($O_2^{\cdot-}$), NO controls the excessive production of ROS during HM stress. Different studies showed NO production in plants, but the specific enzymatic mechanism that produces NO in plants remains challenging. However, the mitigating effect of NO in plants under HM toxicity has been thoroughly studied. It is an important molecule that acts as a signaling molecule and enhances the immune response (Nabi et al., 2019).

The chapter discusses the complex role of NO in reducing HM stress in plants, which negatively impacts the development, physiology, and yield. It also outlines how NO regulates ROS and RNS in restoring cellular redox reactions, gene expression, and protein activity, as well as its role in stress tolerance via metal chelation and antioxidant pathways. Enzymatic and nonenzymatic pathways of NO production were highlighted, and the roles of defense-, nutrition-, and hormone metabolism-related genes were addressed. Summarizing current literature, we emphasize the potential role of NO in reducing HM stress and its important biotechnological applications and sustainable agricultural practices.

10.2 Heavy metals and their impact on plants

Excessive consumption of HMs can be highly harmful to both plants and humans. Numerous human activities discharge significant amounts of these

HMs into the natural environment. Wastewater sludge from industrial wastewater outputs and groundwater runoff are the primary sources of these HMs and trace elements. Cd, Cu, Pb, As, and Zn are major elements considered harmful HMs (Sandeep et al., 2019).

10.2.1 Cadmium stress

Of all the metals, Cd is harmful to humans and plants and shows biological activities in both land and aquatic organisms. Cd in higher concentrations can be found in crop soils due to advanced practices in industries and agriculture (Chellaiah, 2018; Bojórquez et al., 2016). Excessive Cd accumulation and decreased uptake of key elements in plants can result from Cd, which can induce metabolic changes, mineral deficiencies, osmotic shifts, and slower development. The most easily recognized Cd poisoning symptoms include stunted growth and chlorosis that results in the inhibition of plant development and induces necrosis (Haider et al., 2021; Irfan et al., 2014). The main causes of Cd contamination are several human activities, such as mining and smelting, industries involved in metal manufacturing, battery production, pigment creation, agricultural practices, and atmospheric deposition. Cd is commonly found in insecticides sprayed on fields to protect plants from illnesses. The maximum limit for Cd in agricultural soil is 100 mg kg^{-1} (Niño-Savala et al., 2019). Inhibition of CO_2-fixation-related enzymes and decreased formation of chlorophyll affect photosynthesis (Ghori et al., 2019). Elevated levels of Cd also damage the nucleolus, resulting in chromosomal abnormalities and fragmentation. The destruction of mitochondria dramatically reduces plant respiration. Plants exposed to Cd toxicity exhibit reduced chlorophyll content, photosynthetic activity, and the ability to fix carbon (Gallego et al., 2012; Zulfiqar et al., 2022).

10.2.2 Chromium stress

Several industrial processes use hexavalent Cr (VI), including electroplating, textile dyeing, leather processing, steel production, and tanning (GracePavithra et al., 2019; Prasad et al., 2021). These procedures discharge effluents containing Cr, thereby considerably increasing ambient Cr levels (Joutey et al., 2015). Cr toxicity causes chlorosis, early development, top drooping, and root damage. A study on the effects of Cr in plants found that it inhibits the action of amylases, which impacts the flow of carbohydrates to the embryo's axis (Öztürk et al., 2015). Cr affects the synthesis of essential plant pigments, such as anthocyanins, as well as reduces photosynthesis. It also negatively impacts ROS homeostasis by altering the production of metabolites, including ascorbic acid and glutathione (Al-Huqail et al., 2020). Additionally, it stimulates the synthesis of alternative metabolites, such as histidine and phytochelatins, which support Cr stress tolerance (Ghori et al., 2019).

10.2.3 Lead stress

Naturally occurring weather processes, mining, and smelting all contribute to soil contamination with HMs, particularly Pb. After As, Pb is the second most hazardous HM not necessary for biological systems. Plants exposed to Pb poisoning suffer from a variety of harms, ranging from development to yield production; the toxicity of Pb depends on both time and concentration. Only 0.002% of the Earth's crust is made up of Pb, which is extremely toxic and nondisintegrative (Ashraf et al., 2016). Pb significantly impacts the growth and development of plants. Pb threshold levels for plants are approximately 2 mg kg^{-1}, whereas agricultural soils have levels between 50 and 300 mg kg^{-1} (Shakil et al., 2023). Pb pollution harms the anatomy and morphology of seeds and hinders the germination process and early development of plants. Its toxicity delays the development of radicals by increasing the amounts of carbohydrates and proteins (Seneviratne et al., 2019). This is achieved by interfering with the activities of peroxidase and polyphenol oxidases, inhibiting the roots' capacity to oxidize (Zulfiqar et al., 2019).

10.2.4 Arsenic stress

At a predicted quantity in the Earth's crust that varies from 1.5 to 3 mg kg^{-1}, As is the 20th most common element in the environment (Drewniak & Sklodowska, 2013). Its biological significance in plants or animals is unknown, but people have employed it for a very long time as poison. Considering how much As toxicity is available for plants to absorb is crucial to knowing the danger associated with it (Srivastava and Upadhyay, 2016). Both physical and chemical characteristics of the soil, the surrounding environment, the extent to which As accumulates in the body, and alterations to the soil's rhizosphere all influence how bioavailable As is to plants (Farooq et al., 2016). While it is a supplementary element, it is carried through channels of essential elements to pass through the plasmalemma of root cells. As is harmful because it replaces phosphate (P) ions in ATP generation, limiting cells' energy, and reaches root cells via P channels due to their chemical similarity (Chandrakar et al., 2018). Specifically, we have examined the impact on this significant commercial crop's growth, antioxidant response, and root ultrastructure. The suppression of root and shoot development, as well as cell death in the tips of roots has been linked to physiological changes (Armendariz et al., 2016).

10.2.5 Zinc stress

Plants need Zn as a necessary element for healthy development and proliferation. Protein, IAA, and carbohydrate metabolism are all impacted by Zn. It functions as a cofactor in enzyme processes, contributes to the synthesis of pollen, and confers resistance against numerous pathogens (Aziz et al., 2017). Zn, however, is only needed in small amounts, and an excess of it can have dangerous consequences due to Zn toxicity. Zn is involved in several cellular processes that are mediated by enzymes, and its toxicity modifies these processes. Chlorosis signs are evident in plants affected by Zn toxicity, particularly in younger leaves. If the toxicity persists, these symptoms may spread to older leaves. Other deposits of HMs in roots and shoots, such as Cu and manganese (Mn), are also affected by Zn toxicity (Nagajyoti et al., 2010).

10.3 Dynamic role of nitric oxide in signaling

Joseph Priestley discovered NO in 1772 and named it "nitrous air" (Priestley, 1776). A study of NO in animals was conducted initially; however, the attention was shifted toward plants after Klepper reported NO production in plants (Klepper, 1979). After NO compound's biological roles were discovered and clarified, it was a surprise in the 1980s. As an outcome, the journal "Science" named it "Molecule of the Year" in 1992. In 1997 it was shown that NO stimulates the transcription of genes associated with plant defense mechanisms and maintains chlorophyll levels in potato leaves infested by the disease *Phytophthora infestans* (Laxalt et al., 1997). It was postulated that NO could facilitate chlorophyll protection by scavenging ROS (Del Castello et al., 2019). NO is considered an old chemical in Earth's biological history, and its role in mitigating an upsurge in atmospheric O_2 and elevated ozone (O_3) levels has been postulated (Feelisch & Martin, 1995).

It is well recognized that HMs can change the regular metabolic events that occur inside plants in minute amounts. The root network is the first targeted by metalloids since it grows in soil and comes into close contact with HMs. HMs induce morphological, physiological, metabolic, and biochemical changes in plants, negatively impacting growth, photosynthesis, antioxidant activities, and cell differentiation (Kolbert, 2016). By combating invasive pathogens and tumor-affected cells, NO is regarded as a cytotoxic agent or treatment for susceptible plants (Trapet et al., 2015). It also plays a regulatory role in the plant immune system under pathogen attack. In response to HMs, NO endogenously accumulates in plants depending on the type and specificity of the metal (Kolbert et al., 2017). In Rose Mellow, aluminum causes a reduction of endogenous NO in roots under stress conditions. Similarly, a decrease in NO was also observed in barrel medic under Cd stress after 48 hours (Xu et al., 2010). However, in *Arabidopsis thaliana*, overproduction of NO was observed, causing enhancement in primary and lateral roots (Kolbert et al., 2017). NO is implicated in the regulation of plant growth and development as well as the defense responses against biotic and abiotic stress. In *Arabidopsis*, 6436 genes were upregulated and downregulated collectively when exposed to 1 mM S-nitroso cysteine and differentially expressed (Niyoifasha et al., 2023). Key physiological functions, including hormone regulation, other growth processes, and plant resistance against various biotic and abiotic stimuli, have been linked to these DEGs. Both endogenous and exogenous NO applications alleviate HM toxicity in several adaptive mechanisms, including plant growth, cell wall expansion, and relaxation (Oliveira et al., 2016). Both oxidative and reductive mechanisms are suggested for the generation of NO. The oxidative process for the NO generation uses polyamines (PAs), arginine, and hydroxylamine as substrates.

One important source of NO production in plants is the cytosolic enzyme nitrate reductase (NR). Some theories suggest that NR is necessary for the synthesis of NO during bacterial defense, when other pathogens are present, during severe drought, cold adaptation, stomatal regulation, the decrease of iron deficiency, and activities related to root development (Khan et al., 2023). The earlier findings demonstrate that NR tends to convert nitrate to nitrite under typical circumstances and that it promotes the generation of NO in specific situations, including anaerobic environments or at greater nitrate levels. The production of NO takes place inside mitochondria, chloroplasts, cytoplasm, cell walls, cell membranes, and peroxisomes (Hancock & Neill, 2019). The two most studied pathways of NO production will be discussed here.

10.4 Mechanisms of nitric oxide-mediated heavy metal stress mitigation

Over 40 years of plant research, a complex signaling network regulated by NO has been elucidated in numerous contexts; nonetheless, the origin of its generation remains mostly controversial. Both enzymatic and nonenzymatic mechanisms have been identified; nevertheless, there is no clarity about the primary source of NO in plants nor its regulation. Here, we will discuss briefly the synthesis of NO in plants from the knowledge we have gained over the years.

10.4.1 Nitric oxide synthesis pathways in plants

NO has an essential role in plant metabolism. There are 11 different mechanisms in plants for NO production (Astier et al., 2018). Two fundamental methods, oxidative and reductive, can produce NO in plants based on their chemical features. The oxidative pathway depends on oxidizing modified molecules, whereas the reductive pathway depends on reducing nitrites to NO (Kolbert et al., 2019) (Fig. 10.1).

The reduction of nitrate to nitrite is regulated by NADPH-dependent NR, a cytosolic nitrogen absorption enzyme and the best enzymatic source of NO generation in plants (Chamizo-Ampudia et al., 2017; León & Costa-Broseta, 2020). NR further reduces nitrite to NO, utilizing arginine as a substrate. Aside from the reductive mechanism using nitrite, many lines of evidence point to the presence of an oxidative pathway in plants that produces NO, similar to the

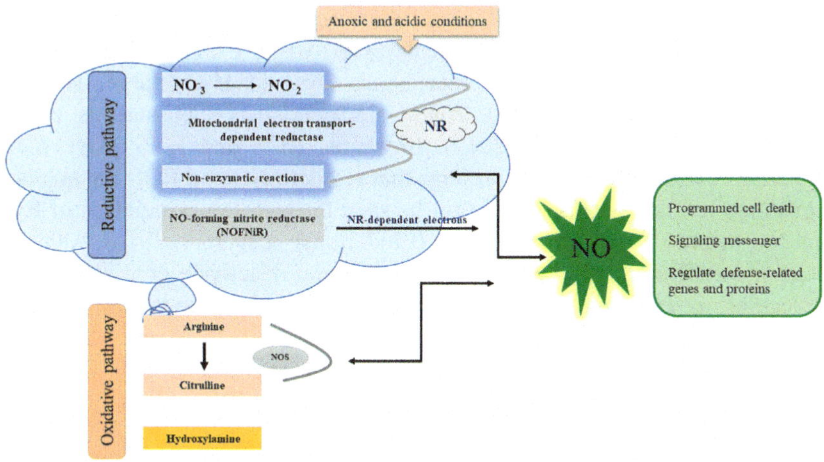

FIGURE 10.1 Schematic representation of oxidative and reductive pathways in NO production. NO production via reductive pathways takes place in anoxic and acidic conditions (*NO*, Nitric oxide).

major system outlined in animals (Gupta et al., 2025). The oxidative pathway plays a role in the formation of NO. Multiple lines of evidence demonstrate that plants can make NO by the oxidation of nitrogen molecules, a process that is comparable to the primary mechanism in animals (Lundberg & Weitzberg, 2022). In addition to mammals, it has been suggested that higher plants undergo the oxidation of arginine to citrulline and NO, which is mediated by three separate NO synthases (NOSs) (Zemojtel et al., 2006). Finding and describing NOS homologs in the kingdom of plants has been the focus of numerous investigations. A contender formerly designated AtNOS1 was found in *Arabidopsis* due to similarities with a protein that the snail *Helix pomatia* uses to produce NO. But later, scientists identified it as a functioning small GTPase, leading them to call it NO-associated 1 (*AtNOA1*) (Moreau et al., 2008). Several studies show increased NO production under HM stress after short-term treatments (Terrón-Camero et al., 2019). Plants can produce NO through oxidative means when PAs are broken down by enzymes called copper amine oxidases. Low-molecular compounds containing nitrogen known as PAs have a role in both stress adaptation and the development and growth of plants (Agurla et al., 2018).

10.4.2 Role of reactive oxygen species and reactive nitrogen species against heavy metal stress

ROS is a broad phrase that refers to chemical entities generated when molecular oxygen is partially reduced. The most prominent ROS are the hydroxyl radical ($^{\cdot}OH$), hydrogen peroxide (H_2O_2), and superoxide anion

($O^{\cdot-}_2$). ROS have unique biological characteristics, a short lifespan, and significant reactivity; hydroxyl radicals ($^{\cdot}OH$) exhibit nonspecific reactivity toward biomolecules. On the other hand, $O_2^{\cdot-}$ and H_2O_2 target selective biomolecules (Del Río, 2015). RNS jointly denotes NO and its related compounds (Rahman et al., 2012). In plants, ROS and RNS play crucial signaling roles. Both can alter proteins that regulate NO and ROS metabolism, signaling, and homeostasis. This can be seen through the association of ROS and NO to generate a variety of RNS (Lindermayr & Durner, 2015). Although ROS and RNS buildup can cause cell injury and deactivate crucial signaling molecules, maintaining an appropriate ROS/RNS equilibrium is vital to abiotic stress resistance and adaptability (Fancy et al., 2017).

10.4.3 Synthesis of reactive oxygen species in cells

NO and ROS interact in a variety of physiological activities, including the response to abiotic stimuli (Joudoi et al., 2013). Thus it is obvious that the changes caused by these signaling molecules differ substantially depending on the circumstances. The signaling pathways that ROS and NO trigger, how these molecules control one another and other aspects of the signaling route, and how antioxidants balance the creation and removal of reactive species, are poorly understood because of the intricate structure of this process (Farnese et al., 2016). These challenges have been the focus of an increasing number of studies, and the discipline has made significant advances. During stress-induced cell signaling, higher production of ROS and NO leads to the deactivation of antioxidant levels, rapidly altering the chemical composition of plant cells (Del Río, 2015). This results in a significant rise in the concentration of certain reactive species, which are essential for eliciting specific biological reactions. These reactions are protective mechanisms against abiotic stressors, leading to increased antioxidant system efficiency and concentration (Shi et al., 2014).

ROS and NO-producing enzymes are found in various areas of cells, and signaling is dependent on their timing and geographical placement. ROS and NO play various roles depending on where they originate, most likely because of interactions with neighboring molecules (Das & Roychoudhury, 2014). For example, it has been observed that the gene expression responses induced by H_2O_2 created in the chloroplasts differ from the transcriptional modifications caused by apoplastic H_2O_2 (Jiménez et al., 2019; Sierla et al., 2013). Likewise, NO produced by mitochondria and chloroplasts has a function in response to HMs, whereas NO produced by the plasma membrane (PM) is vital in hypoxic situations (Kumar & Trivedi, 2016; Rubio et al., 2019).

ROS can be produced directly from ROS-active metals via the Haber-Weiss/Fenton processes, followed by the activation of NADPH oxidases (NOXs), or by restricting enzymes by replacing necessary cations. ROS is primarily produced by the chloroplast, mitochondria, apoplast, and

TABLE 10.1 ROS and RNS species in plants (Farnese et al., 2016).

	Free radicals	Nonradicals
ROS	Superoxide ($O_2^{\cdot-}$)	Hydrogen peroxide (H_2O_2)
	Alkoxyl (RO)	Ozone (O_3)
	Hydroxyl ($^{\cdot}OH$)	Peroxynitrite ($ONOO^-$)
	Hydroperoxyl (HO_2)	Singlet oxygen ($^1O_2$)
RNS	Nitric oxide (NO)	Nitrous acid (HNO_2)
	Nitric dioxide (NO_2)	Peroxynitrite ($ONOO^-$)
	Nitrate radical (NO_3)	Nitrosyl anion (NO^-)
		Nitrosonium cation (NO^+)

ROS, Reactive oxygen species; *RNS*, reactive nitrogen species.

peroxisomes. The major ROS in plants include hydroxyl radicals ($^{\cdot}OH$), singlet oxygen ($^1O_2$), hydrogen peroxide (H_2O_2), and superoxide anion (O_2^-) (Table 10.1) (Zhao et al., 2020). Enzymatic antioxidants CAT, SOD, POD, APX, and GR are produced to scavenge ROS quickly. The generation of ROS primarily occurs in the PM due to the activity of proteins (*RBOH*) in the NOX family. Numerous investigations have demonstrated that stressful situations increase NOX expression and its activity, which results in an oxidative burst (Jajic et al., 2015; Wang, Zhang, et al., 2016). Although their roles in the apoplast's reaction to stressors are unclear, cell wall-associated oxidases and peroxidases also contribute to ROS production. Furthermore, it is evident from the studies that NO might eliminate ROS by controlling these antioxidant enzymes against HM stress.

10.5 Synergistic roles of melatonin and nitric oxide against heavy metal stress

Although NO and melatonin (MT) have independent functions that have been well investigated, little is known about how their combination affects plant development and stress tolerance. Numerous plant species express MT, including *Oryza sativa*, *Cucumis sativus*, *Arabidopsis*, *Malus domestica*, *Solanum lycopersicum*, and *Medicago sativa* (Gu et al., 2017). By dramatically reducing endogenous abscisic acid and jasmonic acid levels by suppressing *NCED3* and activating catabolic genes (*CYP707A1* and *CYP707A2*), both MT and NO reduced metal toxicity. Higher endo-MT and fluctuating endo-S-nitrosothiol levels brought on by higher expression of *gmNR* and *gmGSNOR* mRNAs were linked to MT/NO-induced alleviation of Cd and Pb stress. Interestingly, higher absorption of Ca^{2+} and K^{+} boosts the escape of organic acids into the root system; as a result, a combination of MT/NO decreased the uptake of Cd and Pb in the soil (Imran et al., 2022).

During drought, NO and MT treatments demonstrated lower lipid peroxidation and enhanced defense responses through substantially raising antioxidant enzyme activity compared to the control. MT treatment causes the downregulation of several miRNAs in watermelon, including *miR858, miR8029-3p, miR159-5p*, and novel-*m0048-3p*. In the presence of MT, the desired genes associated with these miRNAs, including *bHLA, WRKY, CDPK, DREB*, and others, continuously displayed elevated expressions (Imran et al., 2021). MT raised endogenous NO levels by enhancing the NR, which uses NADPH to convert nitrite into NO. Nitrosation of thiols, amino acids, and glutathione results in the formation of different mixtures. In plant cells, NO–MT is a necessary chemical (Sheikhalipour et al., 2024). In summary, by controlling the induction of molecular transcription factors and antioxidants, MT/NO modifies the stress-resistance system to reduce the harmful effects of HM stress.

10.6 Role of nitric oxide on antioxidant enzymes under heavy metal stress

Metal pollutants such as Zn, Cu, Ni, Cd, and Pb are a significant hazard to the ecosystem and human health. HM stress primarily affects plant development and growth through oxidative stress, gene, and protein modulation, which may result from the surplus generation of hazardous ROS (Soares et al., 2019). ROS are generated naturally by various metabolic routes, including chloroplastic and mitochondrial electron transport chains. Exposure to adverse conditions, such as metal toxicity, can lead to elevated ROS levels, disrupting cell homeostasis and causing oxidative injuries in plants (Kohli and Handa, 2019). However, plants adopted several mechanisms to regulate the oxidative damage caused by abiotic stress. NO production is one of the mechanisms that regulates cell signaling and protects plants from oxidative damage (Sharma et al., 2020). Different studies on the effect of NO against several HM stresses are presented in Table 10.2.

Numerous studies have explored the role of NO in regulating antioxidant enzymes in response to oxidative stress produced by excessive HMs (Roberts et al., 2018). The latest study showed pronounced activities of APX, SOD, and CAT in *S. lycopersicum* against Cd stress when exposed to 100 μM SNP, and NO can counteract Cd's harmful effects. Pb stress has been demonstrated to significantly increase SOD, PPO, and APX activities in rice. Adding NO reduces Pb toxicity by enhancing these actions (Rahim et al., 2022). Similarly, exogenous NO lowered As levels in plants exposed to As and NO donors, reducing ROS production and membrane damage while maintaining

TABLE 10.2 Effects of different concentrations of NO-donors on antioxidant levels in various plant species exposed to different heavy metals.

Plant species	Heavy metal	NO concentration	ROS	Antioxidant enzyme	Effect on antioxidant enzymes	References
Solanum lycopersicum (Tomato)	Cd	100 μM SNP	H_2O_2↓	SOD, CAT, APX	Increased	Ahmad et al., (2018)
Typha angustifolia (Narrow leaf cattail)	Cd	100 μM SNP	-	CAT, SOD	Decreased	Zhao et al. (2015)
Oryza sativa (Rice)	Pb	100 μM SNP	H_2O_2↓	APX, Polyphenol oxidase (PPO), SOD	Increased	Rahim et al. (2022)
Spirodela intermedia (Duckweed)	As	50 μM SNP	H_2O_2↓ $O_2^{\cdot-}$↓	CAT, Peroxdase (POX), SOD	Decreased	da-Silva et al. (2018)
Oryza sativa (Rice)	As	100 μM SNP	H_2O_2↓	Dehydroascorbate reductase (DHAR), APX	Decreased	Kushwaha et al. (2019)
Solanum melongena (Eggplant)	Ni	100 μM SNP	H_2O_2↓	SOD, Glutathione (GSH)	Increased	Soliman et al. (2019)
Solanum lycopersicum (Tomato)	Cu	100 μM SNP	H_2O_2↓	POD, SOD, APX	Increased	Sharma et al. (2020)
Helianthus annuus (Sunflower)	Zn	20 μM SNP	-	SOD, APX, GR,	Increased	Akladious and Mohamed (2017)

ROS, Reactive oxygen species; *NO*, nitric oxide; *SNP*, sodium nitroprusside.

antioxidant enzyme levels lower than in As-treated plants (da-Silva et al., 2018). The interplay between NO and ROS is implicated in the formation of new adventitious roots and the accumulation of primary root biomass in rice seedlings under As stress. As a NO donor, adding sodium nitroprusside (SNP) promoted new adventitious root (NAR) synthesis, increased primary root biomass accumulation (PRBA), and stabilized ascorbate's redox state and cell cycle activities (Kushwaha et al., 2019). Ni stress slowed growth, biomass, water content, and chlorophyll pigment synthesis, reducing photosynthetic efficiency. Exogenously given SNPs were efficient in overcoming Ni-induced growth constraints. NO-treated seedlings showed increased photosynthesis, stomatal conductance, and chlorophyll content, even at lower concentrations (100 μM SNP). SNP upregulated superoxide dismutase, GST, catalase, and ascorbate peroxidase activity (Soliman et al., 2019). In Cu-stressed *S. lycopersicum,* NO reduces oxidative damage by alleviating phytotoxicity and enhances SOD, APX, and POD activities. The interaction between Cu and NO is expected to cause a change in the plant's cell oxidative balance and to improve the antioxidant defense mechanism (Sharma et al., 2020). Similarly, NO-mediated defense against Zn intoxication correlates with increased antioxidant activity. Exogenous administration of SNP (20 μM) in *H. annuus* subjected to Zn resulted in increased tolerance and overexpression of key antioxidant enzymes, including SOD, APX, and GR (Akladious & Mohamed, 2017). Overall, it is evident from the above studies that NO mediates ROS-mediated responses and enhances antioxidant activities when exogenously applied to plants.

10.7 Modification of nitric oxide in gene expression during heavy metal stress

As previously stated, NO was shown to perform plant defensive action against HMs and was identified as a main contributor to the resistance response under HM stress (Nejamkin et al., 2025). As a result, numerous studies found that NO could control the expression levels of certain genes that are unavoidably involved in the response to excess HM (Wei et al., 2020). These genes include defense, nutrition, programmed cell death (PCD), and hormone metabolism (Fig. 10.2).

10.7.1 Defense genes

Recent studies have examined how HM stress in plants affects gene expression of enzymatic antioxidants. Rizwan et al. (2018) demonstrated the impact of NO on enzymatic antioxidant-related genes during Ni stress through qRT-PCR analysis. They discovered that excessive Ni treatment might raise the gene expression levels of the genes *CAT*, *POD*, *APX*, and *GR*, whereas *SOD* was shown to be downregulated in rice. Furthermore, in line with the shift in their

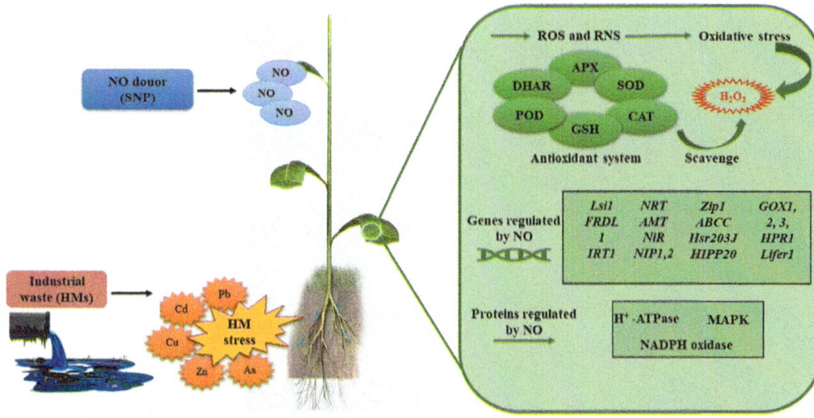

FIGURE 10.2 Schematic representation of the effect of NO on genes, antioxidants, and protein expression under HM stress in the plant cells.
The regulatory role of NO in the modulation of gene expression of antioxidants and proteins to combat HM stress (*NO*, Nitric oxide; *HM*, heavy metal).

enzyme activity, the insertion of SNP may further increase the gene expression of *APX, CAT, SOD, GR, and POD* (Rizwan et al., 2018). Zn stress might increase the expression of *pAPX, CAT1–2, pAPX*, and *cAPX* (Georgiadou et al., 2018). After 2 and 4 days of treatment, this upregulation was further strengthened by supplementing with 2-4-carboxyphenyl-4,4,5,5-tetramethylimidazoline-1-oxyl-3-oxide or NG-nitro-L-arginine methyl ester *(L-NAME)*, but it significantly decreased after 10 days of treatment in roots (Nejamkin et al., 2025). Additionally, Pb stress in the *A. thaliana* genome may cause *CAT1* expression to be upregulated, but Pb-stressed seedlings showed somewhat decreased expression of *CAT2, 3, GOX1, 2, 3*, and *HPR1* genes (Table 10.3) (Corpas & Barroso, 2017). Consequently, NO may contribute to regulating genes linked to enzymatic antioxidants, controlling the activity of antioxidant enzymes under HM stress.

10.7.2 Genes linked to hormone metabolism

It was also confirmed that NO controls the expression of genes involved in hormone metabolism, like gibberellin (GA), IAA, and cytokinin (CTK). IAA mediates several developmental stages in which the polar auxin transporter (PIN) participates. According to Praveen and Gupta (2018), treatment with As may suppress the gene expression of the majority of *PIN* genes in rice seedlings (*OsPIN1a, 1b, 1c*, and *1d, OsPIN2, OsPIN5c*, and *OsPIN10b*), which would prevent root growth. However, with SNP supplementation, these genes may become upregulated, encouraging rice root growth. The outcome was consistent with NO's functions in reducing toxicity (Praveen & Gupta, 2018).

TABLE 10.3 Genes that are regulated by NO in HM-exposed plants.

Heavy metals	Plant species	Genes in NO regulation	NO donor (Concentration)	Functions progress	References
Arsenic (As)	*Oryza sativa* roots	*OsLsi1, OsLsi2, OsYSL2, OsFRDL1, OsIRT1, OsIRO2*	SNP (100 μM)	Decrease As accumulation	Singh et al. (2016)
Arsenic (As)	*Oryza sativa* seedlings	*NRT, AMT, NiR, PHT, KTP*	SNP (100 μM)	Help in promoting nutrient transport	Praveen and Gupta (2018)
Arsenic (As)	*Oryza sativa* seedlings	*OsNIP3;2, OsABCG48, OsPDR20, OsNIP2;1, OsNIP3;5, OsABCB5*	SNP (30 μM)	Enhance levels of auxins	Praveen and Gupta (2018)
Cadmium (Cd)	*Nicotiana tabacum*	*Hsr203J*	SNP (0.5 μM)	Programmed cell death induction	Ma et al. (2010)
Cadmium (Cd)	*Arabidopsis thaliana*	*FIT, FRO2, IRT1,* SAG12, and *At5g03570,*	SNP (50 μM)	Controls Iron Uptake	Ma et al. (2010)
Cadmium (Cd)	*Hibiscus cannabinus*	*ZIP1, ZIP5, ABCG8, ABCC14, CAX18, VIT4, and HIPP20*	SNP (150 μM)	Enhance antioxidant enzyme activity	Cao et al. (2024)
Lead (Pb)	*Arabidopsis thaliana*	*CAT1, 2, 3, GOX1, 2, 3, HPR1*	Diethylamine NONOate	Responds to oxidative stress	Corpas and Barroso (2017)
Nickle (Ni)	*Oryza sativa*	*SOD, POD, CAT, APX, GR*	SNP (100-200 μM)	Responds to oxidative stress	Rizwan et al. (2018)
Iron (Fe)	*Lotus Japonicus*	*Ljfer1*	SNP (0.1, 0.5, 1 mM) GSNO (0.1 and 1 mM)	Controls the generation of ferritin.	Chungopast et al. (2017)

NO, Nitric oxide; *HM*, heavy metal; *SNP*, sodium nitroprusside; *KTP*, potassium transporter; *AMT*, ammonium transporter; *PHT*, phosphorus transporter; *NiR*, nitrite reductase; *NRT*, nitrate transporter.

Similar studies (Liu, Li, et al., 2015) indicated that salt stress reduced *Arabidopsis* expression of the *PIN* gene, which decreased IAA levels and inhibited the establishment of root meristems (Liu, Li, et al., 2015). Furthermore, (Singh et al., 2017) illumine sequencing was used to find OsSAUR39 in rice, which plays a role in IAA generation and transport, and OsPIN9, which is associated with IAA metabolism, were elevated in a combination of As and SNP as opposed to As alone treatment (Singh et al., 2017).

Researchers also discovered that NO may boost cell viability and root development during As stress. This shows that under As stress, NO could play an essential role in IAA signaling, cell dispersion, and proliferation. It is possible to conclude that NO governs plant development and evolution during HM stress by affecting the expression of genes associated with IAA (Bhat et al., 2021). Interestingly, Singh et al. (2017) discovered that concurrent treatment with As and SNP boosted the expression of the CTK biosynthetic gene (*isopentenyl transferase: IPT7*) compared to As treatment. This was linked to a shift in gene modulation implicated in GA metabolism (Singh et al., 2017). These substantial changes in gene expression involved in hormonal metabolism governed by NO indicate that it has an essential role during HM stress. Nevertheless, little research has been done on how NO regulates the expression of other hormonal genes during HM stress; more research is required to shed light on how NO functions to reduce HM stress.

10.7.3 Genes involved in nutrition

Most of the nutrient-associated genes, including potassium transporter, ammonium transporter, phosphorus transporter, and nitrite reductase, displayed decreased expression in the As-treated rice shoots and roots (Praveen & Gupta, 2018), whereas the gene for the high-affinity nitrate transporter was elevated. In contrast to As supplementation alone, NO supply combined with As stress may further stimulate these genes. One possible explanation is that plants under stress have poor nutrient availability, which can be modified by NO to improve nutrient uptake efficiency (Praveen & Gupta, 2018). Furthermore, Singh et al. (2016) verified how NO impacts the genes responsible for As absorption (*OsLsi1*) and transport (OsLsi2). They discovered that As treatment increased the expression levels of OsLsi1 and OsLsi2, whereas NO reduced the expression in As-exposed rice. Exogenous NO treatment in plants may lead to less As accumulation, which could support these findings (Singh et al., 2016).

Furthermore, Chungopast et al. (2017) explored how Fe-mediated NO production boosted ferritin (*Ljfer1*) gene expression during maturity in *Lotus japonicus* foci. As a result, NO promotes the effective uptake and accessibility of mineral-based nutrients in plants by regulating key genes responsible for nutrient absorption during HM stress. To better understand how NO impacts plants under HM stress, more testing with different developmental and growth phases is required.

10.7.4 Genes related to programmed cell death

Less investigation has been carried out to explore the regulatory role of NO in PCD and the expression of associated genes in plants exposed to HM stress. Ma et al. (2010) used TUNEL-positive nuclei to investigate how Cd exposure caused PCD in BY-2 cells and increased the PCD-associated gene expression of *Hsr203J*. Furthermore, NO was discovered to play a function in Cd-mediated PCD in *Arabidopsis* cultures by affecting the expression of senescence-associated gene 12 (SAG12). The scientists discovered that adding L-NMMA significantly reduced SAG12 expression. These findings underscore the critical roles that NO plays in modulating gene expression and Cd-induced programmed cell death (Ma et al., 2010). These findings reveal that NO helps to enhance HM tolerance by modulating the expression of numerous genes, providing a new perspective on improving HM tolerance and detoxification.

10.8 Modification of nitric oxide in protein expression under heavy metal stress

In addition to influencing gene expression under HM exposure, NO regulates the function of numerous proteins in plants.

10.8.1 H^+-transporting adenosine triphosphatase

H^+-adenosine triphosphatase, an essential enzyme in the tonoplast and PM, develops and maintains ion homeostasis in the presence of abiotic stress, therefore managing the system of secondary cellular transport (Zhang et al., 2020). According to a study, SNP mitigated the suppression of PM H^+-ATPase activity in the roots and shoots of white clover plants caused by Cd treatment (Liu, Yang, et al., 2015). Additionally, during Cd stress, SNP increased the activity of tonoplast H^+-ATPase in roots, which reduced cytoplasmic H^+ concentrations and altered the pH in the cytoplasm. The results demonstrated that Cd reduced PM H^+-ATPase activity in *T*rifolium *repens* roots and shoots but that SNP treatment increased this activity in contrast to Cd alone. Similar results were seen in the root tips of maize under severe drought, where NO triggered PM H^+-ATPase activity to increase resistance against drought stress (An et al., 2019). Surprisingly, Yang et al., (2016) demonstrated that adding NO to PM increased phosphatidic acid deposition driven by PLD activity in rice and Cd-induced PLD activity. According to these findings, NO aids proton pumps and creates an optical lipid environment in PM under HM stress by raising H^+-ATPase activity, which may be one way that NO increases plant resistance to HM stress (Yang et al., 2016).

10.8.2 Mitogen-activated protein kinase

In cell signal transduction, mitogen-activated protein kinase (MAPK) is a key player. It is also engaged in physiological and developmental signaling mechanisms in addition to biotic and abiotic stress signaling (Majeed et al., 2022). As treatment might raise the generation of NO and ROS in *O. sativa* leaves and roots, As-treated rice also showed a dose-related rise in MAPK activity (Wei et al., 2020). Remarkably, Jin et al. (2013) discovered that in contrast to wild-type *Arabidopsis*, the mpk6 mutant improved Cd tolerance and inhibited Cd-induced NO generation. As a result, we might hypothesize that Cd-induced MPK6 activity may have a detrimental effect on Cd tolerance. It is important to recall, however, that the specific process by which NO modulates the signaling cascade involving MAPK in HM-stressed plants is still not understood, and more scientific research is required (Jin et al., 2013). Therefore it can be hypothesized that Cd-mediated MPK6 activity may impair Cd tolerance. It is important to recall, however, that the specific procedure by which NO modulates the signaling pathway involving MAPK in HM-exposed plants is still not understood, and more investigation is required (Ye et al., 2013).

10.8.3 NADPH oxidase

$O_2^{\cdot-}$ is produced by NOX in plants. A study found that Ni treatment increased $O_2^{\cdot-}$ and H_2O_2 levels, likely due to increased NOX in wheat seedling roots (Hao et al., 2006). NOX activity was increased in *Panax ginseng* adventitious roots treated with Cu. In contrast, SNP reduced it, suggesting that NO may stop the production of O_2 by reducing NOX activity (Tewari et al., 2008). Furthermore, Valentovičová et al. (2010) showed that NADPH-diaphorase was elevated in root tips of barley with Cd supply and suppressed by L-NAME, demonstrating that NO played a role in controlling NADPH activity under Cd exposure. As a result, NO is necessary for controlling NOX in plants under HM stress (Qian et al., 2012).

Following Al-NO treatment, 92 proteins were found using comparative proteomic studies (Yang et al., 2013). By increasing the levels and protein expressions of plant antioxidants and S-nitroso glutathione reductase (GSNOR) in rice, researchers found that NO treatment lowered Al^{3+}-induced RNS and ROS nitrosative and oxidative damage, respectively (Han et al., 2014; Yang et al., 2013).

10.9 Application methods of nitric oxide donors

Tissue culture techniques have made extensive use of SNP, an NO^+ donor, as NO was recently identified as a crucial regulatory molecule (Khurana et al., 2014). While exposing cultures directly to NO is technically challenging, NO donors are typically added exogenously to cultures instead of NO itself in the gaseous state. SNP releases NO, pH-dependent, promoting plant development and growth and inhibiting senescence (Sundararajan et al., 2022). Several researchers have observed that exogenous NO reduced Cd levels in plants, protecting several plant species from Cd stress (Sterckeman & Thomine, 2020). According to one study, SNP pretreatment reduced Cd uptake and improved mineral nutrient absorption, delaying Cd-induced damage to various tomato morphophysiological characteristics. According to da-Silva et al. (2018), external NO decreased As uptake, decreased ROS generation and caused rupture of membranes in *Spirodela intermedia* during As-induced stress. Furthermore, according to Silveira et al. (2015), As treatment resulted in a notable rise in H_2O_2 as well as malondialdehyde levels in lettuce leaves due to As accumulation. However, SNP supplementation alleviated the oxidative stress and significantly disrupted the As translocation mechanism during prolonged As exposure, reducing oxidative stress (da-Silva et al., 2018; Silveira et al., 2015). Given the facts, it is deduced that exogenous NO could lessen As-induced cellular damage by lowering As, considering the previously mentioned data.

Furthermore, a study demonstrated that while SNP use may lessen HM damage by reducing Cu absorption under Cu + Cd treatment, it has no discernible impact on Cd translocation in tomato seedlings. This outcome may stem from exogenous NO's ability to give electrons, which causes it to interact with transition metals like Cu and Zn and to produce metal–nitrosyl complexes that lower Cu uptake in plants (Saxena & Shekhawat, 2013; Wang, Dong, et al., 2016). Yu et al. (2012) demonstrated that Pb content in root cells of *Pogonatherum crinitum* elevated when increased amounts of NO donors, that is, SNP (5 mM) and GSNO (3 mM), were applied but remained unchanged when low dosages of NO donors (0.1 mM SNP and 0.1 mM GSNO) were applied. This indicates that NO signaling is required for Pb absorption and that varying exogenous NO contents have varying effects on controlling Pb levels in plants (Yu et al., 2012). Interestingly, different research found that the 400 μM SNP increased Pb accumulation in roots while not lowering Pb content in shoots. In contrast, SNP (100 μM) demonstrated a reduced Pb deposition in ryegrass stems and a larger Pb concentration in plant roots (Bai et al., 2015; Rahim et al., 2022). When combined, the aforementioned findings highlighted the critical functions of external NO in reducing HM-induced stress by controlling HM absorption and deposition in plants. Therefore it is necessary to investigate more intricate mechanisms regarding the signaling routes of HM export and uptake.

10.10 Future prospects and conclusions

One of the main mediators controlling the uptake and buildup of HMs in plants under HM stress is NO. NO generally has the potential to reduce HM buildup, which can reduce HM toxicity and increase plant tolerance. It has been shown that NO controls the antioxidant system, which includes both the enzymatic and nonenzymatic antioxidant systems, to reduce or get rid of ROS and deal with HM stress. However, there are variations in NO-modulated antioxidant levels under various HM stressors, plant species, and tissues. Therefore it will be very fascinating to investigate if NO responds to HM stress in a constantly changing manner in various plant species. Although the percentage of environmental stresses comprising NO is expected to increase, our knowledge of how this gas affects HM stress remains in further progress. NO has the potential to control the gene expression levels, including those linked to defense, hormone metabolism, nutrition, and PCD. Thus they will give us a new instrument to investigate the possible mechanism by which NO reduces HM stress. Finding the NO-regulated mechanism that governs the transition between these linked genes during HM stress will also be crucial.

Because of its many roles in controlling gene expression, enzyme activity, and antioxidant defenses, NO is essential for reducing the adverse impacts of HMs on plants. NO lessens oxidative damage produced by reactive nitrogen and oxygen species under HM stress by preserving redox equilibrium and improving the effectiveness of both enzymatic and nonenzymatic antioxidant mechanisms. To further increase plant resilience, it also regulates hormone metabolism, food absorption, and programmed cell death. The complex processes via which NO interacts with certain HMs, plant cells, and species are still not entirely understood, despite tremendous advancements.

References

Agurla, S., Gayatri, G., & Raghavendra, A. S. (2018). Polyamines increase nitric oxide and reactive oxygen species in guard cells of *Arabidopsis thaliana* during stomatal closure. *Protoplasma, 255*(1), 153–162. https://doi.org/10.1007/s00709-017-1139-3, www.springer.co.at/springer.py?Page=40&Key=456&cat=9&id_journal=4.

Ahmad, P., Abass Ahanger, M., Nasser Alyemeni, M., Wijaya, L., Alam, P., & Ashraf, M. (2018). Mitigation of sodium chloride toxicity in*Solanum lycopersicum*L. by supplementation of jasmonic acid and nitric oxide. *Journal of Plant Interactions, 13*(1), 64–72. https://doi.org/10.1080/17429145.2017.1420830.

Akladious, S. A., & Mohamed, H. I. (2017). Physiological role of exogenous nitric oxide in improving performance, yield, and some biochemical aspects of sunflower plant under zinc stress. *Acta Biologica Hungarica, 68*(1), 101–114. https://doi.org/10.1556/018.68.2017.1.9, http://www.akademiai.com/doi/pdf/10.1556/018.68.2017.1.9.

Al-Huqail, A. A., Ali, H. M., Kushwaha, B. K., AL-Huqail, A. A., Singh, V. P., & Siddiqui, M. H. (2020). Ascorbic acid is essential for inducing chromium (VI) toxicity tolerance in tomato roots. *Journal of Biotechnology, 322*, 66–73. https://doi.org/10.1016/j.jbiotec.2020.07.011, www.elsevier.com/locate/jbiotec.

An, Z. F., Liu, F. J., Feng, D. Y., Xu, Q. L., & Zhou, C. J. (2019). Production of nitric oxide and phosphatidic acid is involved in activation of plasma membrane H+-ATPase in maize root tips in simulated drought stress. *Russian Journal of Plant Physiology, 66*(1), 50–58. https://doi.org/10.1134/s1021443719010035.

Armendariz, A. L., Talano, M. A., Travaglia, C., Reinoso, H., Wevar Oller, A. L., & Agostini, E. (2016). Arsenic toxicity in soybean seedlings and their attenuation mechanisms. *Plant Physiology and Biochemistry, 98*, 119–127. https://doi.org/10.1016/j.plaphy.2015.11.021, http://www.journals.elsevier.com/plant-physiology-and-biochemistry/.

Ashraf, M. Y., Roohi, M., Iqbal, Z., Ashraf, M., Öztürk, M., & Gücel, S. (2016). Cadmium (Cd) and lead (Pb) induced changes in growth, some biochemical attributes, and mineral accumulation in two cultivars of mung bean [*Vigna radiata* (L.) Wilczek]. *Communications in Soil Science and Plant Analysis, 47*(4), 405–413. https://doi.org/10.1080/00103624.2015.1118117, http://www.tandfonline.com/loi/lcss20.

Astier, J., Gross, I., & Durner, J. (2018). Nitric oxide production in plants: An update. *Journal of Experimental Botany, 69*(14), 3401–3411. https://doi.org/10.1093/jxb/erx420.

Aziz, M. A., Ahmad, H. R., Corwin, D. L., Sabir, M., Hakeem, K. R., & Öztürk, M. (2017). Influence of farmyard manure on retention and availability of nickel, zinc and lead in metal-contaminated calcareous loam soils. *Journal of Environmental Engineering and Landscape Management, 25*(3), 289–296. https://doi.org/10.3846/16486897.2016.1254639, http://www.tandfonline.com/loi/teel20.

Bai, X. Y., Dong, Y. J., Wang, Q. H., Xu, L. L., Kong, J., & Liu, S. (2015). Effects of lead and nitric oxide on photosynthesis, antioxidative ability, and mineral element content of perennial ryegrass. *Biologia Plantarum, 59*(1), 163–170. https://doi.org/10.1007/s10535-014-0476-8.

Bhat, J. A., Ahmad, P., & Corpas, F. J. (2021). Main nitric oxide (NO) hallmarks to relieve arsenic stress in higher plants. *Journal of Hazardous Materials, 406*. https://doi.org/10.1016/j.jhazmat.2020.124289, www.elsevier.com/locate/jhazmat.

Bojórquez, C., Frías Espericueta, M. G., & Voltolina, D. (2016). Removal of cadmium and lead by adapted strains of *Pseudomonas aeruginos*a AND Enterobacter cloacae. *Revista Internacional de Contaminacion Ambiental, 32*(4), 407–412. https://doi.org/10.20937/RICA.2016.32.04.04, http://www.revistascca.unam.mx/rica/index.php/rica/article/download/RICA.2016.32.04.04/46597.

Cao, S., Pan, J., Rehman, M., Luo, D., Wang, Q., Jin, G., Li, R., Chen, T., & Chen, P. (2024). Exogenous nitric oxide alleviates cadmium toxicity in kenaf (*Hibiscus cannabinus* L.) through modulating Cd deposition and regulating key genes and involved pathways. *Industrial Crops and Products, 221*, 119359. https://doi.org/10.1016/j.indcrop.2024.119359.

Chamizo-Ampudia, A., Sanz-Luque, E., Llamas, A., Galvan, A., & Fernandez, E. (2017). Nitrate reductase regulates plant nitric oxide homeostasis. *Trends in Plant Science, 22*(2), 163–174. https://doi.org/10.1016/j.tplants.2016.12.001.

Chandrakar, V., Pandey, N., & Keshavkant, S. (2018). Plant responses to arsenic toxicity: Morphology and physiology. In *Mechanisms of Arsenic Toxicity and Tolerance in Plants*. Springer Singapore, 27–48. http://dx.doi.org/10.1007/978–981-13–1292-2_2, https://www.springer.com/in/book/9789811312915.

Chellaiah, E. R. (2018). Cadmium (heavy metals) bioremediation by *Pseudomonas aeruginosa*: A minireview. *Applied Water Science, 8*(6). https://doi.org/10.1007/s13201-018-0796-5, https://www.springer.com/journal/13201.

Chen, L., Beiyuan, J., Hu, W., Zhang, Z., Duan, C., Cui, Q., Zhu, X., He, H., Huang, X., & Fang, L. (2022). Phytoremediation of potentially toxic elements (PTEs) contaminated soils using alfalfa (*Medicago sativa* L): A comprehensive review. *Chemosphere, 293*, 133577. https://doi.org/10.1016/j.chemosphere.2022.133577.

Chungopast, S., Duangkhet, M., Tajima, S., Ma, J. F., & Nomura, M. (2017). Iron-induced nitric oxide leads to an increase in the expression of ferritin during the senescence of Lotus japonicus nodules. *Journal of Plant Physiology, 208*, 40–46. https://doi.org/10.1016/j.jplph.2016.11.004, www.urbanfischer.de/journals/jpp/p_physio.htm.

Corpas, F. J., & Barroso, J. B. (2017). Lead-induced stress, which triggers the production of nitric oxide (NO) and superoxide anion (O2·-) in *Arabidopsis* peroxisomes, affects catalase activity. *Nitric Oxide - Biology and Chemistry, 68*, 103–110. https://doi.org/10.1016/j.niox.2016.12.010, http://www.elsevier.com/inca/publications/store/6/2/2/9/2/6/index.htt.

Das, K., & Roychoudhury, A. (2014). Reactive oxygen species (ROS) and response of antioxidants as ROS-scavengers during environmental stress in plants. *Frontiers in Environmental Science, 2*. https://doi.org/10.3389/fenvs.2014.00053.

da-Silva, C. J., Canatto, R. A., Cardoso, A. A., Ribeiro, C., & de Oliveira, J. A. (2018). Oxidative stress triggered by arsenic in a tropical macrophyte is alleviated by endogenous and exogenous nitric oxide. *Revista Brasileira de Botanica, 41*(1), 21–28. https://doi.org/10.1007/s40415-017-0431-y, http://rd.springer.com/journal/40415.

Del Castello, F., Nejamkin, A., Cassia, R., Correa-Aragunde, N., Fernández, B., Foresi, N., Lombardo, C., Ramirez, L., & Lamattina, L. (2019). The era of nitric oxide in plant biology: Twenty years tying up loose ends. *Nitric Oxide Biology and Chemistry / Official Journal of the Nitric Oxide Society, 85*, 17–27. https://doi.org/10.1016/j.niox.2019.01.013.

Del Río, L. A. (2015). ROS and RNS in plant physiology: An overview. *Journal of Experimental Botany, 66*(10), 2827–2837. https://doi.org/10.1093/jxb/erv099, http://jxb.oxfordjournals.org/.

Drewniak, L., & Sklodowska, A. (2013). Arsenic-transforming microbes and their role in biomining processes. *Environmental Science and Pollution Research, 20*(11), 7728–7739. https://doi.org/10.1007/s11356-012-1449-0.

Fancy, N. N., Bahlmann, A. K., & Loake, G. J. (2017). Nitric oxide function in plant abiotic stress. *Plant Cell and Environment, 40*(4), 462–472. https://doi.org/10.1111/pce.12707, http://www.blackwellpublishing.com/journals/PCE.

Farnese, F. S., Menezes-Silva, P. E., Gusman, G. S., & Oliveira, J. A. (2016). When bad guys become good ones: The key role of reactive oxygen species and Nitric Oxide in the plant responses to abiotic stress. *Frontiers in Plant Science, 7*, 2016. https://doi.org/10.3389/fpls.2016.00471, http://journal.frontiersin.org/article/10.3389/fpls.2016.00471/pdf.

Farooq, M. A., Islam, F., Ali, B., Najeeb, U., Mao, B., Gill, R. A., Yan, G., Siddique, K. H. M., & Zhou, W. (2016). Arsenic toxicity in plants: Cellular and molecular mechanisms of its transport and metabolism. *Environmental and Experimental Botany, 132*, 42–52. https://doi.org/10.1016/j.envexpbot.2016.08.004, www.elsevier.com/inca/publications/store/2/6/7.

Feelisch, M., & Martin, J. F. (1995). The early role of nitric oxide in evolution. *Trends in Ecology & Evolution, 10*(12), 496–499. https://doi.org/10.1016/S0169–5347(00)89206-X.

Gallego, S. M., Pena, L. B., Barcia, R. A., Azpilicueta, C. E., Iannone, M. F., Rosales, E. P., Zawoznik, M. S., Groppa, M. D., & Benavides, M. P. (2012). Unravelling cadmium toxicity and tolerance in plants: Insight into regulatory mechanisms. *Environmental and Experimental Botany, 83*, 33–46. https://doi.org/10.1016/j.envexpbot.2012.04.006.

Georgiadou, E. C., Kowalska, E., Patla, K., Kulbat, K., Smolińska, B., Leszczyńska, J., & Fotopoulos, V. (2018). Influence of heavy metals (Ni, Cu, and Zn) on nitro-oxidative stress responses, proteome regulation and allergen production in basil (*Ocimum basilicum* L.) plants. *Frontiers in Plant Science, 9*. https://doi.org/10.3389/fpls.2018.00862, http://journal.frontiersin.org/article/10.3389/fpls.2018.00862/pdf.

Ghori, N.-H., Ghori, T., Hayat, M. Q., Imadi, S. R., Gul, A., Altay, V., & Ozturk, M. (2019). Heavy metal stress and responses in plants. *International Journal of Environmental Science and Technology, 16*(3), 1807–1828. https://doi.org/10.1007/s13762-019-02215-8.

GracePavithra, K., Jaikumar, V., Kumar, P. S., & SundarRajan, P. S. (2019). A review on cleaner strategies for chromium industrial wastewater: Present research and future perspective.

Journal of Cleaner Production, 228, 580–593. https://doi.org/10.1016/j.jclepro.2019.04.117, https://www.journals.elsevier.com/journal-of-cleaner-production.

Gu, Q., Chen, Z., Yu, X., Cui, W., Pan, J., Zhao, G., Xu, S., Wang, R., & Shen, W. (2017). Melatonin confers plant tolerance against cadmium stress via the decrease of cadmium accumulation and reestablishment of microRNA-mediated redox homeostasis. *Plant Science, 261*, 28–37. https://doi.org/10.1016/j.plantsci.2017.05.001.

Gupta, P., Saxena, G., & Gupta, R. (2025). Nitrate reductase-mediated nitric oxide synthesis in shaping stress resilience in plants. *Journal of Experimental Botany, 76*(22), 6634–6656. https://doi.org/10.1093/jxb/eraf353.

Haider, F. U., Liqun, C., Coulter, J. A., Cheema, S. A., Wu, J., Zhang, R., Wenjun, M., & Farooq, M. (2021). Cadmium toxicity in plants: Impacts and remediation strategies. *Ecotoxicology and Environmental Safety, 211*. https://doi.org/10.1016/j.ecoenv.2020.111887, http://www.elsevier.com/inca/publications/store/6/2/2/8/1/9/index.htt.

Han, B., Yang, Z., Xie, Y., Nie, L., Cui, J., & Shen, W. (2014). Arabidopsis HY1 confers cadmium tolerance by decreasing nitric oxide production and improving iron homeostasis. *Molecular Plant, 7*(2), 388–403. https://doi.org/10.1093/mp/sst122.

Hancock, J. T., & Neill, S. J. (2019). Nitric oxide: Its generation and interactions with other reactive signaling compounds. *Plants, 8*(2). https://doi.org/10.3390/plants8020041, https://www.mdpi.com/2223–7747/8/2/41/pdf.

Hao, F., Wang, X., & Chen, J. (2006). Involvement of plasma-membrane NADPH oxidase in nickel-induced oxidative stress in roots of wheat seedlings. *Plant Science, 170*(1), 151–158. https://doi.org/10.1016/j.plantsci.2005.08.014.

Iqbal, N., Ördög, A., Koprivanacz, P., Kukri, A., Czékus, Z., & Poór, P. (2024). Salicylic acid- and ethylene-dependent effects of the ER stress-inducer tunicamycin on the photosynthetic light reactions in tomato plants. *Journal of Plant Physiology, 295*, 154222. https://doi.org/10.1016/j.jplph.2024.154222.

Iqbal, N., Czékus, Z., Poór, P., & Ördög, A. (2023). Ethylene-dependent regulation of oxidative stress in the leaves of fusaric acid-treated tomato plants. *Plant Physiology and Biochemistry, 196*, 841–849. https://doi.org/10.1016/j.plaphy.2023.02.047.

Iqbal, N., Nazir, N., Nauman, M., Hayat, M. T., & Waqar-un-Nisa (2020). Agronomic Crop Responses and Tolerance to Metals/Metalloids Toxicity. In *Agronomic Crops*. Springer Singapore, 191–208. https://doi.org/10.1007/978-981-15-0025-1_12.

Imran, M., Shazad, R., Bilal, S., Imran, Q. M., Khan, M., Kang, S. M., Khan, A. L., Yun, B. W., & Lee, I. J. (2021). Exogenous melatonin mediates the regulation of endogenous nitric oxide in Glycine max L. to reduce effects of drought stress. *Environmental and Experimental Botany, 188*. https://doi.org/10.1016/j.envexpbot.2021.104511, www.elsevier.com/inca/publications/store/2/6/7.

Imran, M., Khan, A. L., Mun, B. G., Bilal, S., Shaffique, S., Kwon, E. H., Kang, S. M., Yun, B. W., & Lee, I. J. (2022). Melatonin and nitric oxide: Dual players inhibiting hazardous metal toxicity in soybean plants via molecular and antioxidant signaling cascades. *Chemosphere, 308*. https://doi.org/10.1016/j.chemosphere.2022.136575, www.elsevier.com/locate/chemosphere.

Irfan, M., Ahmad, A., & Hayat, S. (2014). Effect of cadmium on the growth and antioxidant enzymes in two varieties of *Brassica juncea*. *Saudi Journal of Biological Sciences, 21*(2), 125–131. https://doi.org/10.1016/j.sjbs.2013.08.001.

Jajic, I., Sarna, T., & Strzalka, K. (2015). Senescence, stress, and reactive oxygen species. *Plants, 4*(3), 393–411. https://doi.org/10.3390/plants4030393.

Jiménez, A., Martí, M.ª C., Camejo, D., & Sevilla, F. (2019). *Hydrogen peroxide and nitric oxide metabolism in chloroplasts*. Springer Science and Business Media LLC, 107–144. http://dx.doi.org/10.1007/978-3-030–11129-8_6.

Jin, C. W., Mao, Q. Q., Luo, B. F., Lin, X. Y., & Du, S. T. (2013). Mutation of mpk6 enhances cadmium tolerance in *Arabidopsis* plants by alleviating oxidative stress. *Plant and Soil, 371*(1–2), 387–396. https://doi.org/10.1007/s11104-013-1699-8.

Joudoi, T., Shichiri, Y., Kamizono, N., Akaike, T., Sawa, T., Yoshitake, J., Yamada, N., & Iwai, S. (2013). Nitrated Cyclic GMP modulates guard cell signaling in *Arabidopsis*. *The Plant Cell, 25*(2), 558–571. https://doi.org/10.1105/tpc.112.105049.

Joutey, N. T., Sayel, H., Bahafid, W., & Ghachtouli, N. E. (2015). Mechanisms of hexavalent chromium resistance and removal by microorganisms. *Reviews of Environmental Contamination and Toxicology, 233*, 45–69. https://doi.org/10.1007/978-3-319–10479-9_2, http://www.springer.com/series/398.

Khan, M., Ali, S., Al Azzawi, T. N. I., Saqib, S., Ullah, F., Ayaz, A., & Zaman, W. (2023). The key roles of ROS and RNS as a signaling molecule in plant–microbe interactions. *Antioxidants, 12*(2). https://doi.org/10.3390/antiox12020268, www.mdpi.com/journal/antioxidants.

Khurana, A., Kumar, R., & Babbar, S. B. (2014). Nitric oxide is involved in salicylic acid-induced flowering of *Lemna aequinoctialis* Welw. *Acta Physiologiae Plantarum, 36*(10), 2827–2833. https://doi.org/10.1007/s11738-014-1600-2, https://rd.springer.com/journal/11738.

Klepper, L. (1979). Nitric oxide (NO) and nitrogen dioxide (NO_2) emissions from herbicide-treated soybean plants. *Atmospheric Environment (1967), 13*(4), 537–542. https://doi.org/10.1016/0004–6981(79)90148-3.

Kohli, S. K., Handa, N., Bali, S., Khanna, K., Arora, S., Sharma, A., & Bhardwaj, R. (2019). *Current scenario of Pb toxicity in plants: unraveling plethora of physiological responses* (249). Springer Science and Business Media LLC, 153–197. http://dx.doi.org/10.1007/398_2019_25.

Kolbert, Z., Barroso, J. B., Brouquisse, R., Corpas, F. J., Gupta, K. J., Lindermayr, C., Loake, G. J., Palma, J. M., Petřivalský, M., Wendehenne, D., & Hancock, J. T. (2019). A forty year journey: The generation and roles of NO in plants. *Nitric Oxide Biology and Chemistry / Official Journal of the Nitric Oxide Society, 93*, 53–70. https://doi.org/10.1016/j.niox.2019.09.006.

Kolbert, Z. (2016). Implication of nitric oxide (NO) in excess element-induced morphogenic responses of the root system. *Plant Physiology and Biochemistry, 101*, 149–161. https://doi.org/10.1016/j.plaphy.2016.02.003.

Kolbert, Z., Feigl, G., Bordé, Á., Molnár, Á., & Erdei, L. (2017). Protein tyrosine nitration in plants: Present knowledge, computational prediction and future perspectives. *Plant Physiology and Biochemistry, 113*, 56–63. https://doi.org/10.1016/j.plaphy.2017.01.028.

Kumar, S., & Trivedi, P. K. (2016). *Heavy metal stress signaling in plants*. Elsevier BV, 585–603. http://dx.doi.org/10.1016/b978-0-12–803158-2.00025–4.

Kushwaha, B. K., Singh, S., Tripathi, D. K., Sharma, S., Prasad, S. M., Chauhan, D. K., Kumar, V., & Singh, V. P. (2019). New adventitious root formation and primary root biomass accumulation are regulated by nitric oxide and reactive oxygen species in rice seedlings under arsenate stress. *Journal of Hazardous Materials, 361*, 134–140. https://doi.org/10.1016/j.jhazmat.2018.08.035, www.elsevier.com/locate/jhazmat.

Laxalt, A. M., Beligni, M. V., & Lamattina, L. (1997). Nitric oxide preserves the level of chlorophyll in potato leaves infected by *Phytophthora infestans*. *European Journal of Plant Pathology, 103*(7), 643–651. https://doi.org/10.1023/A:1008604410875.

León, J., & Costa-Broseta, Á. (2020). Present knowledge and controversies, deficiencies, and misconceptions on nitric oxide synthesis, sensing, and signaling in plants. *Plant, Cell & Environment, 43*(1), 1–15. https://doi.org/10.1111/pce.13617.

Li, C., Zhou, K., Qin, W., Tian, C., Qi, M., Yan, X., & Han, W. (2019). A review on heavy metals contamination in soil: Effects, sources, and remediation techniques. *Soil and Sediment*

Contamination: An International Journal, 28(4), 380–394. https://doi.org/10.1080/15320383.2019.1592108.

Lindermayr, C., & Durner, J. (2015). Interplay of reactive oxygen species and nitric oxide: Nitric oxide coordinates reactive oxygen species homeostasis. *Plant Physiology, 167*(4), 1209–1210. https://doi.org/10.1104/pp.15.00293.

Lipton, S. A., Choi, Y. B., Pan, Z. H., Lei, S. Z., Chen, H. S. V., Sucher, N. J., Loscalzo, J., Singel, D. J., & Stamler, J. S. (1993). A redox-based mechanism for the neuroprotective and neurodestructive effects of nitric oxide and related nitroso-compounds. *Nature, 364*(6438), 626–632. https://doi.org/10.1038/364626a0.

Liu, S., Yang, R., Pan, Y., Ma, M., Pan, J., Zhao, Y., Cheng, Q., Wu, M., Wang, M., & Zhang, L. (2015). Nitric oxide contributes to minerals absorption, proton pumps and hormone equilibrium under cadmium excess in *Trifolium repens* L. plants. *Ecotoxicology and Environmental Safety, 119*, 35–46. https://doi.org/10.1016/j.ecoenv.2015.04.053.

Liu, W., Li, R. J., Han, T. T., Cai, W., Fu, Z. W., & Lu, Y. T. (2015). Salt stress reduces root meristem size by nitric oxidemediated modulation of auxin accumulation and signaling in *Arabidopsis*. *Plant Physiology, 168*(1), 343–356. https://doi.org/10.1104/pp.15.00030, http://www.plantphysiol.org/content/168/1/321.full.pdf.

Lundberg, J. O., & Weitzberg, E. (2022). Nitric oxide signaling in health and disease. *Cell, 185*(16), 2853–2878. https://doi.org/10.1016/j.cell.2022.06.010, https://www.sciencedirect.com/journal/cell.

Ma, W., Xu, W., Xu, H., Chen, Y., He, Z., & Ma, M. (2010). Nitric oxide modulates cadmium influx during cadmium-induced programmed cell death in tobacco BY-2 cells. *Planta, 232*(2), 325–335. https://doi.org/10.1007/s00425-010-1177-y.

Majeed, Y., Zhu, X., Zhang, N., Rasheed, A., Tahir, M. M., & Si, H. (2022). Functional analysis of mitogen-activated protein kinases (MAPKs) in potato under biotic and abiotic stress. *Molecular Breeding, 42*(6). https://doi.org/10.1007/s11032-022-01302-y, https://www.springer.com/journal/11032.

Moreau, M., Gyu, I. L., Wang, Y., Crane, B. R., & Klessig, D. F. (2008). AtNOS/AtNOA1 is a functional *Arabidopsis thaliana* cGTPase and not a nitric-oxide synthase. *Journal of Biological Chemistry, 283*(47), 32957–32967. https://doi.org/10.1074/jbc.M804838200, http://www.jbc.org/cgi/reprint/283/47/32957.

Nabi, R. B. S., Tayade, R., Hussain, A., Kulkarni, K. P., Imran, Q. M., Mun, B. G., & Yun, B. W. (2019). Nitric oxide regulates plant responses to drought, salinity, and heavy metal stress. *Environmental and Experimental Botany, 161*, 120–133. https://doi.org/10.1016/j.envexpbot.2019.02.003, www.elsevier.com/inca/publications/store/2/6/7.

Nagajyoti, P. C., Lee, K. D., & Sreekanth, T. V. M. (2010). Heavy metals, occurrence and toxicity for plants: A review. *Environmental Chemistry Letters, 8*(3), 199–216. https://doi.org/10.1007/s10311-010-0297-8.

Nejamkin, A., Del Castello, F., Lamattina, L., Correa-Aragunde, N., & Foresi, N. (2025). Nitric oxide is required for primary nitrate response in *Arabidopsis:* Evidence for S-nitrosation of NLP7. *Antioxidants and Redox Signaling, 42*(4–6), 280–291. https://doi.org/10.1089/ars.2022.0210, www.liebertonline.com/ars.

Niño-Savala, A. G., Zhuang, Z., Ma, X., Fangmeier, A., Li, H., Tang, A., & Liu, X. (2019). Cadmium pollution from phosphate fertilizers in arable soils and crops: An overview. *Frontiers of Agricultural Science and Engineering, 6*(4), 419. https://doi.org/10.15302/j-fase-2019273.

Niyoifasha, C. J., Borena, B. M., Ukob, I. T., Minh, P. N., Al Azzawi, T. N. I., Imran, M., Ali, S., Inthavong, A., Mun, B. G., Lee, I. J., Khan, M., & Yun, B. W. (2023). Alleviation of Hg-, Cr-,

Cu-, and Zn-induced heavy metals stress by exogenous sodium nitroprusside in rice plants. *Plants, 12*(6). https://doi.org/10.3390/plants12061299, www.mdpi.com/journal/plants.

Oliveira, H. C., Gomes, B. C. R., Pelegrino, M. T., & Seabra, A. B. (2016). Nitric oxide-releasing chitosan nanoparticles alleviate the effects of salt stress in maize plants. *Nitric Oxide - Biology and Chemistry, 61*, 10–19. https://doi.org/10.1016/j.niox.2016.09.010, http://www.elsevier.com/inca/publications/store/6/2/2/9/2/6/index.htt.

Öztürk, M., Ashraf, M., Aksoy, A., Ahmad, M. S. A., & Rehman Hakeem, K. (2015). *Plants, pollutants and remediation*. Springer Netherlands. http://dx.doi.org/10.1007/978-94-017-7194-8.

Pande, A., Mun, B. G., Lee, D. S., Khan, M., Lee, G. M., Hussain, A., & Yun, B. W. (2021). NO network for plant–microbe communication underground: A review. *Frontiers in Plant Science, 12*. https://doi.org/10.3389/fpls.2021.658679, https://www.frontiersin.org/journals/plant-science.

Parveen, N., Kandhol, N., Sharma, S., Singh, V. P., Chauhan, D. K., Ludwig-Müller, J., Corpas, F. J., & Tripathi, D. K. (2022). Auxin crosstalk with reactive oxygen and nitrogen species in plant development and abiotic stress. *Plant and Cell Physiology, 63*(12), 1814–1825. https://doi.org/10.1093/pcp/pcac138, http://pcp.oxfordjournals.org/.

Petrov, V., Hille, J., Mueller-Roeber, B., & Gechev, T. S. (2015). ROS-mediated abiotic stress-induced programmed cell death in plants. *Frontiers in Plant Science, 6*. https://doi.org/10.3389/fpls.2015.00069, http://journal.frontiersin.org/Article/10.3389/fpls.2015.00069/pdf.

Prasad, S., Yadav, K. K., Kumar, S., Gupta, N., Cabral-Pinto, M. M. S., Rezania, S., Radwan, N., & Alam, J. (2021). Chromium contamination and effect on environmental health and its remediation: A sustainable approaches. *Journal of Environmental Management, 285*. https://doi.org/10.1016/j.jenvman.2021.112174, https://www.sciencedirect.com/journal/journal-of-environmental-management.

Praveen, A., & Gupta, M. (2018). Nitric oxide confronts arsenic stimulated oxidative stress and root architecture through distinct gene expression of auxin transporters, nutrient related genes and modulates biochemical responses in *Oryza sativa* L. *Environmental Pollution, 240*, 950–962. https://doi.org/10.1016/j.envpol.2018.04.096.

Priestley, J. (1776). Experiments and observations on different kinds of air.

Qian, J., Chen, F., Kovalenkov, Y., Pandey, D., Moseley, M. A., Foster, M. W., Black, S. M., Venema, R. C., Stepp, D. W., & Fulton, D. J. R. (2012). Nitric oxide reduces NADPH oxidase 5 (Nox5) activity by reversible S-nitrosylation. *Free Radical Biology and Medicine, 52*(9), 1806–1819. https://doi.org/10.1016/j.freeradbiomed.2012.02.029.

Rahim, W., Khan, M., Al Azzawi, T. N. I., Pande, A., Methela, N. J., Ali, S., Imran, M., Lee, D. S., Lee, G. M., Mun, B. G., Moon, Y. S., Lee, I. J., & Yun, B. W. (2022). Exogenously applied sodium nitroprusside mitigates lead toxicity in rice by regulating antioxidants and metal stress-related transcripts. *International Journal of Molecular Sciences, 23*(17). https://doi.org/10.3390/ijms23179729, http://www.mdpi.com/journal/ijms.

Rahman, T., Hosen, I., Islam, M. M. T., & Shekhar, H. U. (2012). Oxidative stress and human health. *Advances in Bioscience and Biotechnology, 03*(07), 997–1019. https://doi.org/10.4236/abb.2012.327123.

Rizwan, M., Mostofa, M. G., Ahmad, M. Z., Imtiaz, M., Mehmood, S., Adeel, M., Dai, Z., Li, Z., Aziz, O., Zhang, Y., & Tu, S. (2018). Nitric oxide induces rice tolerance to excessive nickel by regulating nickel uptake, reactive oxygen species detoxification and defense-related gene expression. *Chemosphere, 191*, 23–35. https://doi.org/10.1016/j.chemosphere.2017.09.068, www.elsevier.com/locate/chemosphere.

Roberts, J. A., Planchet, E., & Rümer, S. (2018). Nitrate reductase and nitric oxide. *Wiley, Annual Plant Reviews online, 42*, 127–145. https://doi.org/10.1002/9781119312994.apr0453.

Romero-Puertas, M. C., Terrón-Camero, L. C., Peláez-Vico, M.Á., Olmedilla, A., & Sandalio, L. M. (2019). Reactive oxygen and nitrogen species as key indicators of plant responses to Cd stress. *Environmental and Experimental Botany, 161*, 107–119. https://doi.org/10.1016/j.envexpbot.2018.10.012, www.elsevier.com/inca/publications/store/2/6/7.

Rubio, M. C., Calvo-Begueria, L., Díaz-Mendoza, M., Elhiti, M., Moore, M., Matamoros, M. A., James, E. K., Díaz, I., Pérez-Rontomé, C., Villar, I., Sein-Echaluce, V. C., Hebelstrup, K. H., Dietz, K. J., & Becana, M. (2019). Phytoglobins in the nuclei, cytoplasm and chloroplasts modulate nitric oxide signaling and interact with abscisic acid. *Plant Journal, 100*(1), 38–54. https://doi.org/10.1111/tpj.14422, http://onlinelibrary.wiley.com/journal/10.1111/(ISSN)1365-313X.

Sandeep, G., Vijayalatha, K. R., & Anitha, T. (2019). Heavy metals and its impact in vegetable crops. *International Journal of Chemical Studies, 7*(1), 1612–1621.

Saxena, I., & Shekhawat, G. S. (2013). Nitric oxide (NO) in alleviation of heavy metal induced phytotoxicity and its role in protein nitration. *Nitric Oxide Biology and Chemistry / Official Journal of the Nitric Oxide Society, 32*, 13–20. https://doi.org/10.1016/j.niox.2013.03.004.

Seneviratne, M., Rajakaruna, N., Rizwan, M., Madawala, H. M. S. P., Ok, Y. S., & Vithanage, M. (2019). Heavy metal-induced oxidative stress on seed germination and seedling development: A critical review. *Environmental Geochemistry and Health, 41*(4), 1813–1831. https://doi.org/10.1007/s10653-017-0005–8, www.wkap.nl/journalhome.htm/0269–4042.

Shahid, M., Khalid, S., Abbas, G., Shahid, N., Nadeem, M., Sabir, M., Aslam, M., & Dumat, C. (2015). Heavy metal stress and crop productivity. In *Crop production and global environmental issues*. Springer International Publishing, 1–25. http://dx.doi.org/10.1007/978-3-319–23162-4, http://dx.doi.org/10.1007/978-3-319–23162-4_1.

Shakil, S., Nawaz, K., & Sadef, Y. (2023). Evaluation and environmental risk assessment of heavy metals in the soil released from e-waste management activities in Lahore, Pakistan. *Environmental Monitoring and Assessment, 195*(1). https://doi.org/10.1007/s10661-022-10701-9.

Sharma, A., Soares, C., Sousa, B., Martins, M., Kumar, V., Shahzad, B., Sidhu, G. P. S., Bali, A. S., Asgher, M., Bhardwaj, R., Thukral, A. K., Fidalgo, F., & Zheng, B. (2020). Nitric oxide-mediated regulation of oxidative stress in plants under metal stress: A review on molecular and biochemical aspects. *Physiologia Plantarum, 168*(2), 318–344. https://doi.org/10.1111/ppl.13004, http://onlinelibrary.wiley.com/journal/10.1111/(ISSN)1399–3054.

Sheikhalipour, M., Kulak, M., Mohammadi, S. A., Esmaielpour, B., Nouraein, M., Kocak, M. Z., Farajzadeh, S. M., Gohari, G., Fotopoulos, V., & Vita, F. (2024). Foliar application of either melatonin or sodium nitpoprusside regulates the antioxidant status, and the morpho-physiological attributes and essential oil production in sage (*Salvia officinalis* L.) under salinity stress. *Scientia Horticulturae, 323*. https://doi.org/10.1016/j.scienta.2023.112526, www.elsevier.com/inca/publications/store/5/0/3/3/1/6.

Shi, H., Ye, T., Zhu, J. K., & Chan, Z. (2014). Constitutive production of nitric oxide leads to enhanced drought stress resistance and extensive transcriptional reprogramming in *Arabidopsis*. *Journal of Experimental Botany, 65*(15), 4119–4131. https://doi.org/10.1093/jxb/eru184, http://jxb.oxfordjournals.org/.

Sierla, M., Rahikainen, M., Salojärvi, J., Kangasjärvi, J., & Kangasjärvi, S. (2013). Apoplastic and chloroplastic redox signaling networks in plant stress responses. *Antioxidants & Redox Signaling, 18*(16), 2220–2239. https://doi.org/10.1089/ars.2012.5016.

Silveira, N. M., De Oliveira, J. A., Ribeiro, C., Canatto, R. A., Siman, L., Cambraia, J., & Farnese, F. (2015). Nitric oxide attenuates oxidative stress induced by arsenic in lettuce (*Lactuca sativa*) leaves. *Water, Air, and Soil Pollution, 226*(11), 226–379. https://doi.org/10.1007/s11270-015-2630-0, http://www.kluweronline.com/issn/0049–6979/.

Singh, A. P., Dixit, G., Kumar, A., Mishra, S., Singh, P. K., Dwivedi, S., Trivedi, P. K., Chakrabarty, D., Mallick, S., Pandey, V., Dhankher, O. P., & Tripathi, R. D. (2016). Nitric oxide alleviated arsenic toxicity by modulation of antioxidants and thiol metabolism in rice (*Oryza sativa* L). *Frontiers in Plant Science, 6*. https://doi.org/10.3389/fpls.2015.01272, https://www.frontiersin.org/journals/plant-science.

Singh, P. K., Indoliya, Y., Chauhan, A. S., Singh, S. P., Singh, A. P., Dwivedi, S., Tripathi, R. D., & Chakrabarty, D. (2017). Nitric oxide mediated transcriptional modulation enhances plant adaptive responses to arsenic stress. *Scientific Reports, 7*(1). https://doi.org/10.1038/s41598-017-03923-2, www.nature.com/srep/index.html.

Soares, C., Carvalho, M. E. A., Azevedo, R. A., & Fidalgo, F. (2019). Plants facing oxidative challenges—A little help from the antioxidant networks. *Environmental and Experimental Botany, 161*, 4–25. https://doi.org/10.1016/j.envexpbot.2018.12.009, www.elsevier.com/inca/publications/store/2/6/7.

Soliman, M., Alhaithloul, H. A., Hakeem, K. R., Alharbi, B. M., El-Esawi, M., & Elkelish, A. (2019). Exogenous nitric oxide mitigates nickel-induced oxidative damage in eggplant by upregulating antioxidants, osmolyte metabolism, and glyoxalase systems. *Plants, 8*(12). https://doi.org/10.3390/plants8120562, https://www.mdpi.com/2223–7747/8/12/562/pdf.

Srivastava, S., Upadhyay, M. K., Tripathi, R. D., & Dhankher, O. P. (2016). Arsenic transport, metabolism and toxicity in plants. *International Journal of Plant and Environment, 2*(1 and 2), 17–28. https://doi.org/10.18811/ijpen.v2i1-2.6614.

Sterckeman, T., & Thomine, S. (2020). Mechanisms of cadmium accumulation in plants. *Critical Reviews in Plant Sciences, 39*(4), 322–359. https://doi.org/10.1080/07352689.2020.1792179.

Sundararajan, S., Rajendran, V., Sivakumar, H. P., Kumariah, M., & Ramalingam, S. (2022). Growth modulation by nitric oxide donor sodium nitroprusside in in vitro plant tissue cultures – A review. *Biologia (Lahore, Pakistan), 77*(7), 1699–1711. https://doi.org/10.1007/s11756-022-01027-5, http://www.springer.com/journal/11756.

Terrón-Camero, L. C., Peláez-Vico, M.Á., Del-Val, C., Sandalio, L. M., & Romero-Puertas, M. C. (2019). Role of nitric oxide in plant responses to heavy metal stress: Exogenous application versus endogenous production. *Journal of Experimental Botany, 70*(17), 4477–4488. https://doi.org/10.1093/jxb/erz184, http://jxb.oxfordjournals.org/.

Tewari, R. K., Hahn, E. J., & Paek, K. Y. (2008). Modulation of copper toxicity-induced oxidative damage by nitric oxide supply in the adventitious roots of *Panax ginseng*. *Plant Cell Reports, 27*(1), 171–181. https://doi.org/10.1007/s00299-007-0423-7.

Trapet, P., Kulik, A., Lamotte, O., Jeandroz, S., Bourque, S., Nicolas-Francès, V., Rosnoblet, C., Besson-Bard, A., & Wendehenne, D. (2015). NO signaling in plant immunity: A tale of messengers. *Phytochemistry, 112*(1), 72–79. https://doi.org/10.1016/j.phytochem.2014.03.015.

Valentovičová, K., Halušková, L.'ubica, Huttová, J., Mistrík, I., & Tamás, L. (2010). Effect of cadmium on diaphorase activity and nitric oxide production in barley root tips. *Journal of Plant Physiology, 167*(1), 10–14. https://doi.org/10.1016/j.jplph.2009.06.018.

Wang, X., Zhang, M.-M., Wang, Y.-J., Gao, Y.-T., Li, R., Wang, G.-F., Li, W.-Q., Liu, W.-T., & Chen, K.-M. (2016). The plasma membrane NADPH oxidase OsRbohA plays a crucial role in developmental regulation and drought-stress response in rice. *Physiologia Plantarum, 156*(4), 421–443. https://doi.org/10.1111/ppl.12389.

Wang, Yj, Dong, Y. X., Wang, J., & Cui, Xm (2016). Alleviating effects of exogenous NO on tomato seedlings under combined Cu and Cd stress. *Environmental Science and Pollution Research, 23*(5), 4826–4836. https://doi.org/10.1007/s11356-015-5525-0, http://www.springerlink.com/content/0944-1344.

Wei, L., Zhang, M., Wei, S., Zhang, J., Wang, C., & Liao, W. (2020). Roles of nitric oxide in heavy metal stress in plants: Cross-talk with phytohormones and protein S-nitrosylation. *Environmental Pollution, 259*, 113943. https://doi.org/10.1016/j.envpol.2020.113943.

Xu, J., Wang, W., Yin, H., Liu, X., Sun, H., & Mi, Q. (2010). Exogenous nitric oxide improves antioxidative capacity and reduces auxin degradation in roots of *Medicago truncatula* seedlings under cadmium stress. *Plant and Soil, 326*(1-2), 321–330. https://doi.org/10.1007/s11104-009-0011-4.

Yang, L., Tian, D., Todd, C. D., Luo, Y., & Hu, X. (2013). Comparative proteome analyses reveal that nitric oxide is an important signal molecule in the response of rice to aluminum toxicity. *Journal of Proteome Research, 12*(3), 1316–1330. https://doi.org/10.1021/pr300971n, http://pubs.acs.org/journal/jprobs.

Yang, L., Ji, J., Harris-Shultz, K. R., Wang, H., Wang, H., Abd-Allah, E. F., Luo, Y., & Hu, X. (2016). The dynamic changes of the plasma membrane proteins and the protective roles of nitric oxide in rice subjected to heavy metal cadmium stress. *Frontiers in Plant Science, 7*(2016). https://doi.org/10.3389/fpls.2016.00190, http://journal.frontiersin.org/article/10.3389/fpls.2016.00190/pdf.

Ye, Y. U. N., Li, Z. H. E., & Xing, D. A. (2013). Nitric oxide promotes MPK6-mediated caspase-3-like activation in cadmium-induced *Arabidopsis thaliana* programmed cell death. *Plant, Cell & Environment, 36*(1), 1–15. https://doi.org/10.1111/j.1365-3040.2012.02543.x.

Yu, Q., Sun, L., Jin, H., Chen, Q., Chen, Z., & Xu, M. (2012). Lead-induced nitric oxide generation plays a critical role in lead uptake by pogonatherum crinitum root cells. *Plant and Cell Physiology, 53*(10), 1728–1736. https://doi.org/10.1093/pcp/pcs116.

Zemojtel, T., Frohlich, A., Palmieri, M., Kolanczyk, M., Mikula, I., Wyrwicz, L., Wanker, E., Mundlos, S., Vingron, M., & Martasek, P. (2006). Plant nitric oxide synthase: A never-ending story? *Trends in Plant Science, 11*(11), 524–525. https://doi.org/10.1016/j.tplants.2006.09.008.

Zhang, H., Deng, C., Wu, X., Yao, J., Zhang, Y., Zhang, Y., Deng, S., Zhao, N., Zhao, R., Zhou, X., Lu, C., Lin, S., & Chen, S. (2020). Populus euphratica remorin 6.5 activates plasma membrane H+-ATPases to mediate salt tolerance. *Tree Physiology, 40*(6), 731–745. https://doi.org/10.1093/treephys/tpaa022.

Zhao, C., Zhang, H., Song, C., Zhu, J.-K., & Shabala, S. (2020). Mechanisms of plant responses and adaptation to soil salinity. *The Innovation, 1*(1), 100017. https://doi.org/10.1016/j.xinn.2020.100017.

Zhao, H., Jin, Q., Wang, Y., Chu, L., Li, X., & Xu, Y. (2015). Effects of nitric oxide on alleviating cadmium stress in Typha angustifolia. *Plant Growth Regulation, 78*(2), 243–251. https://doi.org/10.1007/s10725-015-0089-z.

Zulfiqar, U., Ayub, A., Hussain, S., Waraich, E. A., El-Esawi, M. A., Ishfaq, M., Ahmad, M., Ali, N., & Maqsood, M. F. (2022). Cadmium toxicity in plants: Recent progress on morpho-physiological effects and remediation strategies. *Journal of Soil Science and Plant Nutrition, 22*(1), 212–269. https://doi.org/10.1007/s42729-021-00645-3, https://www.springer.com/life+sciences/plant+sciences/journal/42729.

Zulfiqar, U., Farooq, M., Hussain, S., Maqsood, M., Hussain, M., Ishfaq, M., Ahmad, M., & Anjum, M. Z. (2019). Lead toxicity in plants: Impacts and remediation. *Journal of Environmental Management, 250*, 109557. https://doi.org/10.1016/j.jenvman.2019.109557.

Chapter 11

Role of nitric oxide in the mitigation of salinity stress

Muhammad Mughees[1], Nadeem Iqbal[2], and Malik Tahir Hayat[3]
[1]*Department of Biotechnology, COMSATS University Islamabad, Abbottabad Campus, Abbottabad, Khyber Pakhtunkhwa, Pakistan,* [2]*School of Environmental Science and Engineering, Hunan University, Changsha, Hunan, P.R. China,* [3]*Department of Environmental Sciences, COMSATS University Islamabad, Abbottabad Campus, Abbottabad, Khyber Pakhtunkhwa, Pakistan*

11.1 Introduction

Plants continually experience various environmental stressors that can adversely impact their growth, development, and productivity (Liu et al., 2016). These stresses are broadly categorized into abiotic and biotic factors. Abiotic stresses are due to extreme environmental conditions such as salinity, drought, toxicity of heavy metals and organic pollutants, and extreme temperatures, while biotic stresses are caused by pathogens, insects, and weeds (Suzuki et al., 2014). Salinity is a crucial limiting factor that reduces land usability and crop yields due to unsustainable irrigation practices, excessive fertilizer use, and seawater intrusion caused by rising sea levels. Over 8 billion hectares of land worldwide are impacted by salinity levels, accounting for over 6% of the total available cultivated land area, according to the Food and Agriculture Organization. This proportion is rising daily as a result of multiple human activities and climate change (Sherin et al., 2022). The rising incidence of soil salinization can be attributed to unsustainable irrigation practices, overuse of chemical fertilizers, and the intrusion of seawater due to rising sea levels, all of which exacerbate challenges to sustainable agriculture (Zörb et al., 2019; EL Sabagh et al., 2021). Addressing these intertwined stress factors is crucial for enhancing crop resilience and ensuring sustainable agricultural practices.

Understanding the processes for reducing salt stress is essential to preserving agricultural yields and guaranteeing food security, as salinity affects 20% of the land under cultivation (Sehar et al., 2019). In addition to this ecological problem, there is an imperative need to double crop production

Nitric Oxide in Plant Resilience to Abiotic Stress. DOI: https://doi.org/10.1016/B978-0-443-40497-9.00012-X

to satisfy the demands of the anticipated 50%–70% increase in the world's population by 2050 (Tester & Langridge, 2010; Shabala, 2013). This impending circumstance emphasizes developing innovative methods to increase agricultural productivity in saline regions. Mitigating the negative effects on agricultural output requires effective salinity control. Innovative strategies are essential for addressing this urgent problem, such as genetically enhancing crop tolerance, using sustainable irrigation techniques, and using plant growth regulators such as nitric oxide (NO) (EL Sabagh et al., 2021). Advances in plant genetics, sustainable irrigation techniques, and the use of plant growth regulators provide intriguing approaches for improving resilience in areas impacted by salt and meeting the growing need for food brought on by population expansion.

NO, among other strategies, has become a crucial signaling molecule in reducing salt stress. NO enhances plant tolerance to salt stress by controlling important physiological reactions such as photosynthesis, ion balance, and antioxidant defense. NO reduces oxidative damage caused by the overproduction of reactive oxygen species (ROS) under salinity exposure by boosting the activity of antioxidant enzymes (Iqbal et al., 2023; Iqbal et al., 2023; Iqbal et al., 2023). Similarly, Hasanuzzaman et al. (2018) showed that by adjusting Na^+/K^+ ratios in salt-sensitive crops, NO treatment enhanced ion homeostasis. NO has been demonstrated to improve water use capacity and photosynthetic efficiency in wheat, reducing growth inhibition due to high salinity (Ahmad et al., 2016). Zhang et al. (2023) emphasized how NO and sulfur work together to reduce oxidative stress brought on by salt. The detrimental effects of salinity on plant metabolism, adaptive strategies for managing salt stress, and the contribution of NO to improving crop resilience are all covered in this chapter. It also looks at how NO works in concert with sulfur and nitrogen, highlighting how it might be used in sustainable agriculture to fulfill future food demands. A possible way to increase crop output in salty conditions is to combine NO signaling with nutrient management techniques.

11.2 Salinity-mediated effects on plants

11.2.1 Osmotic effects on water uptake and cell expansion

The initial impact of salinity on plants is osmotic stress, which limits water availability. High soil salt concentrations reduce soil water potential, making it more difficult for plants to absorb water. This condition triggers a drought-like response, even when water is physically present. As a result, cell expansion is restricted, a critical factor for early plant growth stages. Seed germination and root elongation are particularly sensitive to osmotic stress (Feng et al., 2016). Salinity-induced osmotic effects reduce water uptake, leading to slower growth rates and impaired development. Stunted seedlings and reduced leaf area are common symptoms of osmotic stress in salt-affected plants (Syed et al., 2021).

In rice (*Oryza sativa*), osmotic stress caused by salinity results in delayed germination, reduced shoot growth, and decreased leaf water potential that significantly reduces shoot biomass and increases leaf rolling, a common symptom of water deficit. Similarly, in wheat (*Triticum aestivum*), salinity affects early seedling vigor, reducing water uptake and limiting root and shoot elongation (Farooq et al., 2015). To cope with osmotic stress, plants activate various adaptive mechanisms, including the accumulation of osmolytes such as proline and glycine betaine. These compounds help maintain cell turgor by balancing the osmotic pressure within cells. Additionally, root architecture adjustments and changes in aquaporin expression regulate water transport under saline conditions. However, these adaptations are often insufficient to fully mitigate the negative impacts of prolonged salinity exposure (Wijayasinghe et al., 2017).

11.2.2 Ionic toxicity and nutrient imbalance

Prolonged exposure to salinity leads to ionic toxicity, primarily due to the excessive uptake and accumulation of sodium (Na^+) and chloride (Cl^-) ions in plant tissues. Sodium ions disrupt the uptake of essential nutrients, such as potassium (K^+) and calcium (Ca^{2+}), by competing for binding sites on transport proteins. This results in nutrient imbalances that compromise cellular metabolism and membrane stability. The toxic effects of Na^+ and Cl^- accumulation lead to protein denaturation, enzyme inhibition, and oxidative damage (Gupta et al., 2019). Excess sodium also interferes with photosynthesis by affecting chloroplast structure and function, reducing energy production and carbohydrate synthesis.

Salinity stress induces sodium accumulation in maize leaves, disrupting the K^+/Na^+ ratio that is critical for enzyme activation and stomatal regulation. This imbalance affects photosynthetic efficiency, leading to reduced biomass production. Similarly, in tomato (*Solanum lycopersicum*), high salinity causes leaf necrosis and premature senescence due to ionic toxicity, which directly impacts fruit yield and quality (Anjum et al., 2023). One key strategy plants use to mitigate ionic stress involves selective ion transport and compartmentalization. Specialized transporters, such as SOS1 (salt overly sensitive 1) and HKT (high-affinity potassium transporter), help regulate ion homeostasis by sequestering sodium into vacuoles or limiting its transport to shoots (Ismail & Horie, 2017). Additionally, the synthesis of antioxidants and activation of detoxifying enzymes counteract oxidative stress caused by salinity.

11.2.3 Cumulative impact on agricultural productivity

The combined osmotic and ionic effects of salinity significantly reduce agricultural productivity. Yield losses in salt-sensitive crops, such as rice and wheat, are directly linked to reduced water use efficiency, impaired photosynthesis, and diminished reproductive success. In rice, salinity stress

TABLE 11.1 Effects of abiotic and biotic stresses on plants.

Type of stress	Example	Effect on plants
Abiotic stressesors		
Drought	Water deficiency	Wilting, reduced growth, lower photosynthesis, and early senescence (Farooq et al., 2012)
Salinity	High salt concentration in soil	Osmotic stress, ion toxicity, and reduced nutrient uptake (Munns & Tester, 2008)
Temperature extremes	Heat or cold	Disruption of cellular metabolism, protein denaturation, and membrane instability (Hasanuzzaman et al., 2013)
Heavy metals	Lead, cadmium, and arsenic	Chlorosis, inhibition of root and shoot growth, and oxidative damage (Shahid et al., 2017)
Biotic stressesors		
Pathogens	Bacteria, fungi, and viruses	Leaf spots, wilting, root rot, and necrosis (Glazebrook, 2005; Iqbal, Czékus, Ördög, et al., 2023; Iqbal, Czékus, Angeli, et al., 2023; Iqbal, Czékus, Poór, et al., 2023; Iqbal et al., 2024)
Insects and pests	Herbivorous insects	Leaf damage, reduced photosynthetic area, and sap depletion (Scholthof et al., 2011)
Weeds	Competitive plants	Nutrient, water, and light competition reduce crop yield (Oerke et al., 2016)

reduces tiller number, panicle size, and grain filling, resulting in lower yields. Wheat experiences reduced spikelet fertility and grain weight under high salinity conditions (Table 11.1). In horticultural crops such as cucumber (*Cucumis sativus*), salinity not only decreases fruit yield but also alters fruit taste and appearance by affecting sugar and organic acid content (Chen & Wang, 2024). Salinity stress affects flowering, pollen viability, and seed set, further compounding yield reductions. In saline soils, root-zone management, irrigation with low-salinity water, and the application of organic amendments are common practices used to alleviate salt stress.

11.2.4 Oxidative damage

One of the critical biochemical impacts of salinity stress is the overproduction of ROS, including superoxide radicals (O_2^-) and hydrogen peroxide (H_2O_2). These compounds damage main cell components such as DNA, proteins, and lipids, resulting in oxidative stress (Iqbal et al., 2024). ROS-induced damage is mitigated by plants via the activation of enzymatic antioxidants such as superoxide dismutase (SOD), catalase (CAT), and peroxidase, along with nonenzymatic antioxidants like ascorbate and glutathione (Iqbal, Czékus, Ördög, et al., 2023; Iqbal, Czékus, Angeli, et al., 2023; Iqbal, Czékus, Poór, et al., 2023). The efficiency of these defense mechanisms determines the capability of salt stress tolerance in numerous plant species and their different parts (Fancy et al., 2017). The effects of various stresses on plants are shown in Table 11.1. Salinity triggers the generation of ROS in plant cells, which inflict oxidative injuries on cellular macromolecules, including lipids and proteins, thereby exacerbating growth and productivity limitations (Mandal et al., 2022).

11.2.5 Effects on plant growth and development

Salt stress can considerably affect photosynthesis by closing stomata, limiting the supply of CO_2 for metabolic activities. This decline reduces chlorophyll concentration and net efficiency (Parida & Das, 2005). Enhancing photosynthetic parameters is essential for maintaining plant growth under saline conditions. High ionic concentration in the soil around roots results in decreased water-holding capacity, which affects the roots' ability to absorb water. Salinity causes reduced cell turgor pressure, which impacts cell elongation and, eventually, growth with a reduced or no transpiration rate by stomatal closure. Plant growth and development are hampered by salinity in many ways because it affects essential metabolic functions. A high level of salt in the rhizosphere can have deleterious impacts on plant growth and production by disrupting cytosolic ion homeostasis, which in turn affects metabolic dysfunction and other important physiological and biochemical processes (Khanna et al., 2021).

11.3 Adaptive strategies of plants under salinity stress

Stress factors in agriculture, including biotic and abiotic, significantly impact crop productivity and sustainability, which directly or indirectly harm crops or weaken their resilience to environmental challenges, necessitating integrated management strategies. Understanding the effects of salinity stress and developing strategies to mitigate its adverse impacts are critical for sustainable agriculture. Salinity presents a dual challenge by imposing osmotic stress, which limits water uptake, and ionic stress, which leads to nutrient imbalances and toxicity. These combined stresses reduce crop growth and productivity, posing significant threats to food security (Munns & Gilliham, 2015). Understanding the physiological responses to salinity is vital to developing reliable and sustainable approaches to increase crop tolerance and mitigate yield losses.

Using a variety of physiological and molecular defensive mechanisms, including mineral homeostasis, ion compartmentalization, solute accumulation, antioxidant system upregulation, and phytohormonal control, plants reduce the damage caused by salt. Though the accumulation of most osmolytes in the cytosol is nontoxic, it is energetically more expensive (Ahmed et al., 2022; Hasanuzzaman, 2022). Research indicates that NO application can improve chlorophyll content and photosystem II activity, thereby enhancing photosynthetic performance in salt-stressed plants (Garcia-Mata & Lamattina, 2007).

11.3.1 Ion homeostasis and compartmentalization

Salt stress is a major abiotic factor limiting plant growth and productivity worldwide, primarily due to ionic imbalances and osmotic stress caused by excessive Na^+ and Cl^- ions in the soil. High salt concentrations disrupt water uptake, induce oxidative stress, and lead to ion toxicity, ultimately inhibiting photosynthesis and reducing crop yields. To survive and maintain cellular function under saline conditions, plants employ several mechanisms, including ion homeostasis and ion compartmentalization. Ion balance (homeostasis) is sustained by regulating the uptake and distribution of ions using specialized transporters. For instance, SOS1 is an antiporter of Na^+/H^+ present in the plasma membrane that pumps sodium ions out of the cytoplasm, while HKT1 (High-Affinity Potassium Transporter 1) limits Na^+ entry into cells by controlling selective ion uptake (Zhang et al., 2016). These transport systems help mitigate cytosolic Na^+ toxicity and maintain K^+/Na^+ balance, both of which are crucial for cellular metabolism and enzyme activity.

In addition to ion homeostasis, plants manage salt stress by compartmentalizing toxic ions into vacuoles to reduce cytoplasmic sodium concentration. Tonoplast-localized Na^+/H^+ antiporters, such as NHX1, transport excess sodium ions into vacuoles, where they are sequestered away from sensitive cellular processes (Bassil et al., 2018). This vacuolar compartmentalization not only alleviates ion toxicity but also contributes to osmotic adjustment by maintaining cellular turgor. Recent advancements in molecular biology have highlighted the potential of genetic engineering to enhance these mechanisms. For instance, Jiang et al. (2024) reported that overexpression of key ion transporters significantly improves salt tolerance in plant species, while Zhao et al. (2025) used transcriptomic analyses to identify novel salt-responsive genes that regulate ion transport and signaling pathways. These findings underscore the importance of ion homeostasis and compartmentalization as critical strategies for plant adaptation to salinity and provide new horizons to develop salt-resistant varieties through biotechnology.

11.3.2 Compatible solute accumulation

Salt stress imposes significant challenges on plant growth by creating osmotic and ionic imbalances that hinder water absorption and disrupt cellular functions. To counter these effects, plants accumulate compatible solutes—small, nontoxic molecules that stabilize cellular structures, maintain osmotic balance, and protect metabolic processes during stress conditions (Ashraf & Foolad, 2007). Unlike inorganic ions, these solutes do not interfere with normal biochemical activities and help in osmotic adjustment. Key compatible solutes include proline, glycine betaine, sugars, and polyols, each contributing uniquely to plant salt tolerance mechanisms. Proline acts as a multifunctional osmoprotectant, stabilizing proteins and membranes, scavenging ROS, and buffering cellular osmotic pressure. Its accumulation is linked to increased expression of pyrroline-5-carboxylate synthase (P5CS), a key enzyme in its biosynthesis (Meng et al., 2023). Genetic studies highlight the effectiveness of enhancing proline pathways to improve salt tolerance, with notable success in transgenic cereals. Similarly, glycine betaine helps protect membrane integrity and enzyme function under saline conditions. Its synthesis involves a pathway regulated by choline monooxygenase and betaine aldehyde dehydrogenase (BADH), making it a common target for genetic modification in salt-sensitive crops (Ashraf & Foolad, 2007).

Sugars, such as trehalose, and polyols, such as mannitol and sorbitol, further contribute to osmotic adjustment. Trehalose not only stabilizes proteins and lipids during desiccation but also acts as a molecular signal influencing stress-related genes (Zhu et al., 2023). Mannitol and sorbitol enhance salt tolerance by reducing oxidative stress and maintaining osmotic stability. Advances in plant biotechnology have successfully increased the production of these solutes by manipulating biosynthetic genes, as seen in experiments on rice, maize, and tomato (Ashraf & Foolad, 2007; Zhu et al., 2023). These strategies enhance cellular resilience, improve water retention, and reduce ionic toxicity, presenting a promising approach for sustainable crop production in saline environments. In summary, compatible solute accumulation represents a vital adaptive mechanism that integrates osmotic regulation and oxidative stress management in plants under salt stress. By understanding the biosynthetic pathways and regulatory mechanisms of these solutes, future innovations in genetic engineering and breeding programs can unlock new potential for developing resilient crops capable of thriving in high-salinity soils.

11.3.3 Protein regulation in response to salinity stress

Salinity affects protein regulation through changes in synthesis, degradation, and posttranslational modifications. Proteins involved in ionic transport, antioxidant defense, osmotic balance, and stress signaling undergo significant changes to help plants adapt to harsh environments, such as salinity. SOS1 is implicated in exporting excess Na^+ out of the cell to reduce ion toxicity. Its activity is regulated by SOS2 (a kinase) and SOS3 (a calcium sensor). This coordination helps prevent the accumulation of toxic sodium levels in plant cells (Gámez-Arjona et al., 2024). In addition, HKT1 (High-Affinity Potassium

Transporter 1) assists in retrieving Na^+ from the xylem sap, maintaining ionic balance, and preventing sodium accumulation in leaves. This protein plays a crucial role in sustaining plant health under saline conditions by ensuring the proper distribution of sodium and potassium ions (Roy & Soni, 2022).

11.3.4 Antioxidant enzymes

Proteins such as SOD, CAT, and ascorbate peroxidase (APX) mitigate ROS accumulation caused by oxidative stress under salinity (Athar et al., 2022). These enzymes protect cell components from oxidative injuries, thereby enhancing stress tolerance. Recent studies have further elucidated the mechanisms and significance of these antioxidant enzymes in stress responses. For instance, SOD increased salt tolerance in plants by reducing superoxide radicals and preventing lipid peroxidation (Zhang et al., 2021). Similarly, CAT activity has been linked to improved stress resilience in crops under saline conditions, as it degrades hydrogen peroxide into water and oxygen, reducing oxidative damage (Hasanuzzaman et al., 2021). APX, which is part of the ascorbate–glutathione cycle, has been demonstrated to scavenge hydrogen peroxide effectively, maintaining redox homeostasis under salinity exposure (Gill et al., 2021). Additionally, the overexpression of antioxidant enzymes has exhibited enhanced salt tolerance in transgenic plants (Li et al., 2022). Recent research also highlights the synergistic role of APX, CAT, and SOD in mitigating oxidative stress in halophytes, which are naturally adapted to high-salinity environments (Vikram et al., 2022). Furthermore, molecular studies have identified key regulatory pathways that modulate the expression of these enzymes under salt stress, providing insights into potential targets for crop improvement (Wang et al., 2023).

11.3.5 Osmoprotectant biosynthetic enzymes

Enzymes responsible for synthesizing compatible solutes, such as proline and glycine betaine, stabilize proteins and membranes under osmotic stress, preserving cell integrity (Alzahrani, 2021). For example, P5CS catalyzes proline biosynthesis, which helps maintain osmotic balance (Turchetto-Zolet et al., 2009). In a study on wheat, overexpression of the BADH gene was shown to enhance salt tolerance by increasing glycine betaine accumulation, which improved water retention and reduced oxidative damage (Zhang et al., 2022). Another example is the enzyme MTK (mannitol-1-phosphate dehydrogenase), which synthesizes mannitol, a sugar alcohol that acts as an osmoprotectant. Research on transgenic tobacco plants revealed that increased MTK activity led to higher mannitol levels, which improved salt tolerance by protecting cellular structures and maintaining osmotic balance (Liu et al., 2021). Additionally, ornithine aminotransferase, another enzyme in the proline biosynthesis pathway, has been shown to enhance proline accumulation in rice under salt stress, contributing to improved tolerance under salinity (Patel et al., 2020). These examples highlight the importance of enzymes involved in compatible solute synthesis as potential candidates in plant adaptation to osmotic stress.

11.3.6 Role of transcription factors in plants under salinity stress

Transcription factors (TFs) such as DREB (Dehydration-Responsive Element-Binding Protein), NAC (NAM, ATAF, and CUC), and MYB (MYeloBlastosis) play pivotal roles in regulating environmental stresses, such as salinity. These TFs act as molecular switches, activating or repressing the expression of stress-responsive genes that help plants adapt to adverse conditions. DREB transcription factors are particularly well-known for their role in osmoprotection. They bind to dehydration-responsive elements (DRE) in the promoters of stress-related genes, activating pathways that lead to the accumulation of osmoprotectants, including sugars and proline, contents responsible for cellular osmotic balance. For example, Patel et al., 2020 demonstrated that over-expression of DREB2A in rice significantly enhanced drought and salt tolerance through the upregulation of genes implicated in osmoprotection and ROS scavenging. Similarly, a study on wheat by Zhang et al. (2021) showed that DREB1 overexpression improved cold tolerance by activating genes responsible for membrane stabilization and stress signaling. Another study on *Arabidopsis* revealed that DREB1C plays a dual role in both drought and cold stress responses, highlighting its importance in multiple stress pathways (Wang et al., 2023).

NAC transcription factors are another critical group of regulators involved in stress adaptation. They regulate several processes, such as cell wall reinforcement, senescence, and secondary metabolism. For instance, NAC072 showed the potential to enhance drought stress tolerance in rice by promoting the biosynthesis of lignin, a key component of cell walls that provides structural integrity under stress (Forlani 2025). In rice, SNAC1 (Stress-Responsive NAC1) was found to improve drought and salt tolerance by regulating genes involved in stomatal closure and water retention (Hu et al., 2023). Additionally, NAC016 has been linked to delayed leaf senescence under drought conditions, allowing plants to maintain photosynthesis and productivity for longer periods (Kim et al., 2013). MYB transcription factors can also play an integral part in stress responses, particularly in regulating secondary metabolism and stress signaling pathways. For example, MYB44 exhibited pronounced salt tolerance in *Arabidopsis* by upregulating genes implicated in ion homeostasis and ROS detoxification (Du et al., 2024). In maize, MYB48 was found to improve drought tolerance by regulating the biosynthesis of flavonoids, which act as antioxidants and protect cells from oxidative damage (Li et al., 2022). Another study on ryegrass revealed that MYB1R1 plays a major part in cold stress adaptation by activating genes involved in membrane stabilization and energy metabolism (Feng et al., 2022).

The interplay between these TFs and their target genes creates a complex regulatory network that enables plants to respond dynamically to stress. For example, DREB and NAC often work together to activate overlapping sets of stress-responsive genes, while MYB transcription factors fine-tune secondary metabolic pathways to enhance stress resilience. Recent studies have also explored the potential of engineering these TFs to improve crop tolerance to abiotic stresses. For instance, transgenic rice overexpressing DREB1A and SNAC1 showed remarkable improvements in drought and salt tolerance, demonstrating the practical applications of these findings (Li et al., 2022; Chen et al., 2023). Similarly, coexpression of MYB44 and NAC072 in *Arabidopsis* resulted in synergistic improvements in stress tolerance, highlighting the potential of combining multiple TFs for crop improvement (Chen et al., 2023). In conclusion, DREB, NAC, and MYB transcription factors are considered central key players in inducing stress responses in plants, regulating a wide array of genes involved in osmoprotection, cell wall reinforcement, secondary metabolism, and stress signaling. Understanding the mechanisms by which these TFs operate provides valuable insights into plant stress biology and offers promising avenues for developing stress-resilient crops through genetic engineering and breeding.

11.3.7 Phytohormonal regulation

Phytohormones are key signaling molecules that regulate plant growth, development, and stress responses. Under salt stress, the dynamic interplay of hormones such as abscisic acid (ABA), jasmonic acid (JA), salicylic acid (SA), ethylene (ET), and auxins (IAA) plays a regulatory role in modulating physiological and molecular mechanisms to enhance salt tolerance. These hormones act synergistically or antagonistically to fine-tune stress responses, enabling plants to survive under harsh environments such as salinity.

ABA is often referred to as the "stress hormone" due to its central role in mediating plant defense against abiotic stresses, including salinity. Under salinity stress, ABA levels increase rapidly, triggering the closure of stomata to decrease excess water loss and activating stress-responsive genes. For example, ABA-responsive element-binding factors are TFs that regulate the expression of genes involved in osmoprotection and ROS scavenging. A study by Zhang et al. (2021) demonstrated that ABA-deficient mutants of *Arabidopsis* exhibited reduced salt tolerance, while exogenous application of ABA improved stress resilience by increasing antioxidant enzyme activity and osmolyte accumulation. Similarly, in rice, overexpression of OsABI5, an ABA-responsive TF, significantly improved salt tolerance by upregulating genes responsible for ion homeostasis and stress signaling (Li et al., 2022). JA is another important hormone implicated in salt stress responses. It regulates the biosynthesis of secondary metabolites, such as flavonoids and anthocyanins, which act as antioxidants to mitigate oxidative damage. JA also interacts with other phytohormones, including ABA and ET, to initiate and regulate stress responses in plants. For instance, a study on tomato plants revealed that JA application enhanced salt tolerance by increasing the activity of antioxidant enzymes, such as SOD and CAT, while also improving root architecture for better water and nutrient uptake (Wang et al., 2022). Additionally, JA signaling has been shown to enhance the expression of JAZ proteins (a negative regulator of JA signaling), which are degraded under stress to activate defense mechanisms (Hu et al., 2021).

SA plays a crucial role in the regulation and activation of plant defense systems under pathogenic attacks, but it also plays a significant role in abiotic stress tolerance. SA enhances salt tolerance by modulating ion uptake, improving antioxidant defense systems, and maintaining membrane integrity. For example, a study on wheat showed that the exogenous application of SA improved salt tolerance by reducing sodium (Na^+) accumulation and enhancing potassium (K^+) uptake, thereby maintaining ion homeostasis (Khan et al., 2022). SA also induces the expression of pathogenesis-related proteins, which contribute to stress tolerance by stabilizing cellular structures and scavenging ROS (Chen et al., 2021). ET is a gaseous hormone that regulates various aspects of plant growth and stress responses. Under salt stress, ET production increases, influencing root growth, leaf senescence, and stress signaling pathways. ET modulates stress responses by interacting with ABA and JA. For instance, ET-responsive transcription factors (ERFs) play a vital part in activating stress-related genes. A study on *Arabidopsis* revealed that *ERF1* overexpression enhanced salt stress tolerance by upregulating genes implicated in osmoprotection and ROS detoxification (Zhou et al., 2022). In rice, ET signaling was found to improve salt stress tolerance via promoting root elongation and decreasing Na^+ accumulation in shoots (Liu et al., 2022). Auxins, particularly indole-3-acetic acid (IAA), are essential for root development and architecture, which are critical for salt tolerance. Under salt stress, auxin distribution is altered, leading to changes in root growth patterns that help plants cope with osmotic stress. For instance, a study on *Arabidopsis* revealed that PIN2, an auxin efflux carrier, plays a key part in root gravitropism under salinity, enabling plants to explore deeper soil layers for water and nutrients (Sun et al., 2026). In maize, auxin signaling was found to enhance salt tolerance by improving lateral root development and improving water uptake efficiency (Zhang et al., 2022).

The interplay between different phytohormones is essential for fine-tuning salt stress responses. For instance, ABA and ET often work antagonistically, with ABA promoting stomatal closure and ET promoting stomatal opening. However, under salt stress, their interaction can lead to a balanced response that optimizes water use efficiency and stress tolerance. Similarly, JA and SA often act synergistically to enhance antioxidant defense systems and mitigate oxidative damage. A study on *Arabidopsis* revealed that the MYC2 transcription factor, a key regulator of JA signaling, regulates stress responses by interacting with ABA signaling components (Khan et al., 2022). Understanding phytohormonal regulation in salt tolerance has significant implications for crop improvement. For example, genetic engineering approaches have been used to manipulate hormone biosynthesis and signaling pathways to enhance stress tolerance. Transgenic rice overexpressing OsABA8ox, a gene involved in ABA catabolism, showed improved salt tolerance by maintaining higher ABA levels under stress (Cai et al., 2015). Similarly, exogenous application of hormones such as SA and JA has been shown to improve salt tolerance in crops such as wheat, tomato, and maize, offering a cost-effective strategy for stress management.

11.4 Impact of nitric oxide on plant acclimation to salt stress

The ability of NO to influence salt-responsive genes through multiple mechanisms makes it a central player in enhancing plant resilience under salt stress. NO contributes to salt tolerance by maintaining ionic balance, enhancing antioxidant capacity, sustaining photosynthetic efficiency, and activating osmolyte biosynthesis. These processes work in concert to alleviate the detrimental effects of salt stress and improve plant survival and productivity. NO enhances salt tolerance by maintaining ionic balance, particularly by regulating sodium (Na^+) and potassium (K^+) homeostasis. Under salt stress, excessive Na^+ accumulation in the cytoplasm can disrupt cellular functions, while K^+ deficiency impairs enzyme activity and osmotic adjustment. NO addresses this issue by enhancing the activity of Na^+/H^+ antiporters, such as SOS1, which export excess Na^+ from the cytoplasm into the apoplast or vacuole. For example, a study by Zhang et al. (2021) reported that NO treatment in *Arabidopsis* enhanced the expression of *SOS1* and *NHX1* (Na^+/H^+ Exchanger 1), leading to improved Na^+ exclusion and K^+ retention under salt stress. Similarly, in rice, NO was found to upregulate OsHKT1;5, a gene responsible for Na^+ exclusion from shoots, thereby improving salt tolerance (Li et al., 2022). Another study on wheat showed that NO application enhanced the activity of plasma membrane H^+-ATPase, which drives Na^+/H^+ antiporters, resulting in better ion homeostasis and reduced Na^+ accumulation (Wang et al., 2023). These findings highlight the critical role of NO in maintaining ionic balance under salt stress.

In addition to regulating ion homeostasis, NO plays a crucial role in enhancing the plant's antioxidant capacity. Salt stress often leads to the overproduction of ROS, which can damage cellular components such as DNA, proteins, and lipids. NO mitigates this oxidative damage by boosting the activities of ROS-detoxifying enzymes such as SOD, CAT, and APX. For example, a study on tomato plants revealed that NO treatment increased the activity of SOD and CAT, reducing ROS levels and improving salt tolerance

(Hu et al., 2021). In maize, NO was found to enhance the expression of APX1 and GR (glutathione reductase), which are key components of the ascorbate–glutathione cycle, thereby mitigating oxidative stress under salinity (Chen et al., 2023). Additionally, NO has been shown to induce the biosynthesis of nonenzymatic antioxidants, such as glutathione and ascorbate, further enhancing the plant's capability to thrive under oxidative stress (Khan et al., 2022). These studies underscore the importance of NO in strengthening the antioxidant defense system under salt stress.

Another critical function of NO is its ability to sustain photosynthetic efficiency under salt stress. Salt stress often leads to the inhibition of photosynthesis due to oxidative damage to chloroplasts and the degradation of photosynthetic pigments. NO helps protect chloroplasts from oxidative damage and maintains the integrity of photosynthetic machinery. For instance, a study on wheat showed that NO application prevented the degradation of chlorophyll and carotenoids under salt stress, thereby maintaining photosynthetic activity (Liu et al., 2021). In *Arabidopsis*, NO was found to upregulate the expression of PsbA and PsbD, genes encoding key components of photosystem II, which are critical for maintaining photosynthetic efficiency under stress (Zhou et al., 2022). Another study on rice demonstrated that NO treatment improved the stability of thylakoid membranes, enhancing light-harvesting efficiency and reducing photoinhibition under salt stress (Zhang et al., 2023). These findings highlight the role of NO in preserving photosynthetic efficiency, which is essential for plant growth and productivity under stress conditions.

NO also promotes the biosynthesis of compatible solutes, that is, proline, glycine betaine, and sugars, which are responsible for maintaining osmotic balance and protecting cellular structures under salt stress. Compatible solutes are involved in osmoprotection, stabilization of proteins and membranes, and ROS scavenging. For example, NO has been shown to upregulate the expression of P5CS, a key enzyme in proline biosynthesis, leading to elevated levels of proline in salt-stressed plants (Li et al., 2021). In barley, NO treatment enhanced the activity of BADH, an enzyme involved in glycine betaine synthesis, resulting in improved osmotic adjustment and stress tolerance (Wang et al., 2022). Additionally, NO has been found to promote the accumulation of soluble sugars, such as sucrose and trehalose, which act as osmoprotectants and stabilize membranes under stress conditions (Chen et al., 2021). These studies demonstrate the role of NO in activating osmolyte biosynthesis, which is crucial for osmotic adjustment and stress tolerance.

Recent studies have uncovered additional mechanisms by which NO enhances salt tolerance. For instance, NO has been shown to modulate the expression of aquaporins, which are water channel proteins that regulate water uptake and transport under stress conditions. A study on soybeans revealed that NO treatment increased the expression of PIP2;1, a plasma membrane aquaporin, improving water use efficiency under salt stress (Pan et al., 2023). Furthermore, NO has been found to interact with other signaling molecules, such as calcium (Ca^{2+}) and hydrogen sulfide (H_2S), to amplify stress responses. For example, in *Arabidopsis*, NO and Ca^{2+} signaling were found to synergistically enhance the expression of genes associated with stress responses, leading to improved salt tolerance (Zhou et al., 2016). These results reveal the complexity of NO-mediated stress responses and interactions with other signaling pathways. In conclusion, NO plays a multifaceted role in enhancing salt tolerance in plants. By maintaining ionic balance, enhancing antioxidant capacity, sustaining photosynthetic efficiency, and activating osmolyte biosynthesis, NO helps plants cope with the detrimental effects of salt stress. Recent studies have provided valuable insights into the mechanisms underlying NO-mediated stress responses, offering potential strategies for improving crop resilience under salt exposure. Understanding the role of NO in salt tolerance not only advances our knowledge of plant stress biology but also opens new avenues for developing stress-resistant crops through genetic engineering and the exogenous application of NO donors.

11.5 Role of nitric oxide in salinity management

Micronutrients derived from minerals are crucial for agricultural productivity, as they influence plant development, growth, and adaptability to various stressors, such as salt stress (Santos et al., 2020). These nutrients regulate the production and release of vital plant hormones that enhance stress tolerance. Among the key macronutrients, nitrogen (N) and sulfur (S) are pivotal in determining plant physiological responses and stress adaptations (Zheng et al., 2016). NO serves as a significant signaling compound in plants, playing a role in nutrient assimilation and developmental processes. It also strengthens plant resilience to environmental challenges, including saline conditions (Patel et al., 2020). Similarly, ET, a gaseous hormone, influences plant mechanisms for managing salt-induced stress by modulating relevant pathways (Wang et al., 2021). Extensive research has explored the combined effects of NO and ET with nitrogen and sulfur under abiotic stress. For example, studies show that NO interacts synergistically with N and S to enhance plant tolerance to stress, while ETH's relationship with these macronutrients affects adaptive responses (Santos et al., 2020).

Nitrogen availability has been observed to regulate ET synthesis, improving salt tolerance in mustard plants. Additionally, NO improves photosynthetic nitrogen and sulfur use efficiencies, optimizing plant metabolic performance during stress conditions (Parray et al., 2023). The dynamic interplay between nitrogen and sulfur is fundamental for synthesizing proteins, amino acids, nucleic acids, and other essential biomolecules that support plant growth and productivity (Ren et al., 2024). This interconnected relationship underscores the need for integrated nutrient management strategies to reduce abiotic stress impacts and boost agricultural yield.

11.6 Nitric oxide as a signaling molecule

NO is a signaling molecule that has been studied over the past 20 years for its involvement in plant stress responses. Aerial NO treatment to different crops has been demonstrated to promote plant growth exposed to saline conditions (Wu et al., 2011). One of the explanations for NO's beneficial effect during salt stress is the upregulation of antioxidant enzyme activity (Khan et al., 2023). Plants employ multifaceted strategies to mitigate the detrimental effects of abiotic stress, such as salinity, by orchestrating molecular and physiological adaptations. A critical mechanism is ion homeostasis, involving the selective uptake and compartmentalization of ions such as Na^+ and K^+ to prevent cytotoxic accumulation. For instance, plants enhance vacuolar compartmentalization by upregulating transporters such as NHX1 (Na^+/H^+ exchangers), which sequester excess Na^+ into vacuoles, thereby reducing cytoplasmic toxicity and maintaining cellular function under salinity stress (Manai et al., 2014). Concurrently, plants accumulate compatible solutes, such as proline, glycine betaine, and sugars, which act as osmolytes to stabilize cellular turgor, protect enzyme activity, and prevent protein denaturation (Sohag et al., 2020). These solutes also serve as ROS scavengers, complementing the regulation of antioxidant systems that neutralize oxidative stress. Enhanced activity of enzymes like SOD, CAT, and APX ensures the detoxification of ROS, safeguarding cellular integrity (Alnusairi et al., 2021). Together, these interconnected mechanisms—ion homeostasis, vacuolar sequestration, osmolyte synthesis, and antioxidant activation—form a coordinated defense network that enhances plant resilience to salinity and other abiotic stresses. Plant adaptation to salt stress is mostly influenced by nitrogen (N) uptake and ion homeostasis, but sulfur (S), which is present in many compounds, is essential for boosting stress endurance. These minerals interact with phytohormones that affect the body's physiological and biochemical reactions to salt, such as ETH and NO (Yu et al., 2014). As ETH increases salt stress reactivity and modifies N and S interactions, NO, a reactive nitrogen molecule, promotes photosynthetic efficiency, nutritional absorption, and overall salt tolerance. Osmotic stress, oxidative damage, and nutritional imbalance are all caused by salt stress's disruption of ion homeostasis, which lowers crop production and plant development. Gaining knowledge of how NO and ETH control the uptake of N and S can help reduce the negative impacts of salinity and improve agricultural sustainability (Sami et al., 2021).

Produced via enzymatic and nonenzymatic pathways, NO regulates numerous biological processes, including stress adaptation, root formation, and seed germination, by interacting with ROS and phytohormones to modulate gene expression and protein function (Georgieva & Vassileva, 2023). Its effects depend on the concentration and cellular context, enabling either growth promotion or stress defense. Understanding NO's role in mitigating salt stress, which affects 20% of cultivated land, is vital for enhancing plant salt tolerance and ensuring food security. Advances in molecular genetics highlight potential strategies such as NO-releasing compounds and genetic modifications to boost crop productivity, contributing to sustainable agriculture and salinity management (Shokat & Großkinsky, 2019). It is challenging to offer a comprehensive recommendation for farmers based on the numerous research studies conducted to ascertain the impact of exogenous NO on mitigating salt stress. It is challenging to assess the overall beneficial effects because of the varied doses, frequency, timing, and route of application.

11.7 Role of nitric oxide in protein regulation under salinity

NO is a dynamic gaseous signaling molecule that plays a central role in regulating plant growth, development, and stress adaptation. Its versatility stems from its capability to interact with proteins, DNA, and other signaling molecules, enabling plants to fine-tune responses to environmental challenges such as salinity, drought, pathogens, and heavy metals.

11.7.1 S-nitrosation of proteins

NO enhances the activity of antioxidant enzymes, such as *APX and CAT*, by S-nitrosation, boosting their capacity to scavenge ROS under stress (Khator et al., 2024). NO modifies cysteine residues in target proteins through S-nitrosation, affecting their stability, activity, and interactions. S-nitrosation of antioxidant enzymes, such as CAT, enhances their capacity to detoxify ROS, providing better oxidative stress management (Wei et al., 2022). In saline conditions, NO S-nitrosates the *SOS1*, Na^+/H^+ antiporter, improving its activity to exclude cytotoxic sodium ions (Na^+) from the cytoplasm (Mangal et al., 2023). NO also modifies TFs, such as *WRKY* and *NAC*, through PTMs, enabling them to activate genes involved in osmoprotection and stress signaling (Wang et al., 2022).

11.7.2 Regulation of gene expression

NO influences gene expression by interacting with TFs and modifying their activity or DNA-binding capacity. NO modulates TFs such as WRKY, DREB, NAC, and MYB, enhancing the transcription of genes encoding ion transporters, osmoprotectant biosynthetic enzymes, and stress-protective proteins, boosting salt tolerance (Sarkar & Sadhukhan, 2022). NO enhances the activity of TFs that bind to promoter regions of stress-responsive genes. For example, NO upregulates *WRKY70* in rice, which activates genes encoding osmoprotectants (e.g., proline) and antioxidant enzymes such as SOD under salinity (Verma et al., 2022). In *Arabidopsis*, NO stabilizes the HSFA2 (Heat Shock Factor A2) TF under heat stress, promoting the expression of heat shock proteins that protect cellular proteins (Zhang et al., 2022).

11.7.3 Signal transduction

NO acts as a secondary messenger, integrating various stress signals to activate downstream responses. It significantly contributes to plant resilience under salinity stress by enhancing protein regulation. It interacts with ROS signaling and hormonal pathways such as ABA to fine-tune plant adaptive mechanisms (Song et al., 2023).

11.8 Nitric oxide-regulated functional genes implicated in salt stress tolerance

The SOS pathway regulates Na^+ exclusion. SOS1, a plasma membrane Na^+/H^+ antiporter, reduces cytosolic sodium levels. NHX1 sequesters Na^+ into vacuoles, contributing to ion compartmentalization (Fig. 11.1).

Stress-responsive TFs (DREB, NAC) bind to cis-regulatory elements, orchestrating transcriptional responses. Antioxidant enzymes (APX, SOD) scavenge ROS, mitigating oxidative stress. NO modulates physiological responses, enhancing ion transport regulation, antioxidant defenses, and osmotic adjustment.

Plants employ a sophisticated network of genes to combat salt stress, which can be classified into ion transporters, osmoprotectant synthesis genes, TFs, and antioxidant enzymes. Each category plays a distinct yet interconnected and complex role in improving salt tolerance. For instance, ion transporters are critical for maintaining ion balance under saline conditions. The SOS1 exports surplus Na^+ from the cytoplasm, reducing Na^+ toxicity (Guo et al., 2022)

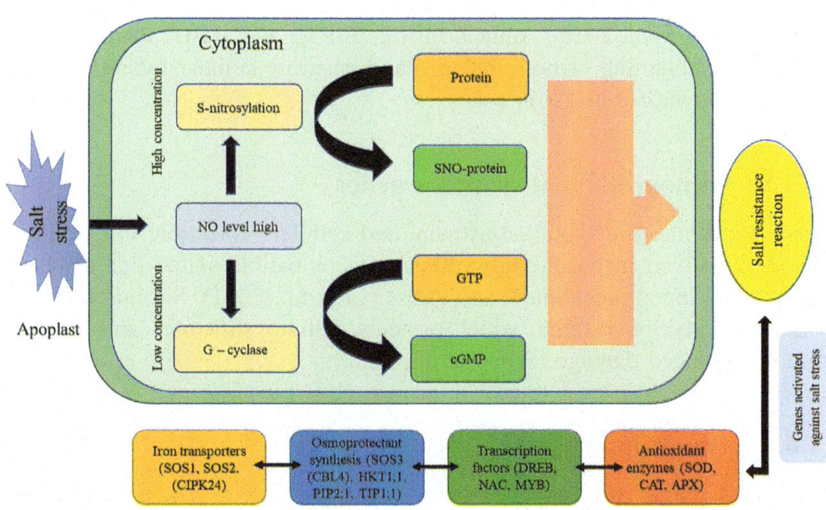

FIGURE 11.1 Schematic representation of plants' response to salt stress at the cellular level. Activation of iron transporters, antioxidant enzymes, and synthesis of osmoprotectants is facilitated by NO under salinity stress (*NO*, Nitric oxide).

TABLE 11.2 Proteins and genes regulated by NO under salt stress.

Gene/Protein	Regulation by NO	Function	References
SOS1	Upregulated	Facilitates Na^+ efflux	(Hasanuzzaman, 2022)
NHX1	Enhanced	Sequesters Na^+ into vacuoles	(Zhu et al., 2019)
P5CS	Induced	Promotes proline synthesis	(Wang et al., 2021)
APX	Increased	Detoxifies H_2O_2	(Hasanuzzaman et al., 2013)
NR	ActivatedNO	Participates in NO production	(Munns & Tester, 2008)

NO, Nitric oxide.

(Table 11.2). This process is regulated by the SOS2 (CIPK24) gene, a serine/threonine protein kinase that activates SOS1 under salt stress (Yin et al., 2019). Additionally, SOS3 (CBL4), a calcium-binding protein, detects cytosolic Ca^{2+} levels and activates SOS2, forming a signaling cascade that enhances Na^+ exclusion (Gámez-Arjona et al., 2024). Another key transporter, HKT1;1, retrieves Na^+ from the xylem, preventing excessive sodium accumulation in shoots (Roy & Soni, 2022). Aquaporins like PIP2;1 and TIP1;1 facilitate water transport, maintaining osmotic balance and ensuring cellular hydration under stress (Pawłowicz et al., 2017).

11.8.1 Osmoprotectant synthesis genes

These contribute to osmotic adjustment and cellular protection. For example, P5CS is involved in proline biosynthesis, a compatible solute that stabilizes proteins and membranes under salt stress (Li et al., 2021). Similarly, BADH synthesizes glycine betaine, which protects cellular structures and enzymes from salt-induced damage (Wang et al., 2022).

11.8.2 Transcription factors

TFs act as master regulators of stress-responsive genes. The DREB family activates genes involved in osmoprotection and stress signaling, enhancing overall tolerance (Hassan et al., 2021). NAC transcription factors regulate processes such as cell death and senescence, ensuring that plants can adapt to prolonged stress (Li et al., 2021). Meanwhile, MYB transcription factors control the production of secondary metabolites, which play a role in stress adaptation and ROS scavenging (Wu et al., 2019).

Antioxidant enzymes are essential for mitigating oxidative stress caused by salt-induced ROS accumulation. Genes encoding SOD, CAT, and APX detoxify ROS, protecting cellular components from oxidative damage (Athar et al., 2022). For instance, SOD converts superoxide radicals into hydrogen peroxide, which is then broken down by CAT and APX into water and oxygen, maintaining redox homeostasis. It is concluded that the coordinated action of ion transporters, osmoprotectant synthesis genes, TFs, and antioxidant enzymes enables plants to withstand salt stress by maintaining ion balance, stabilizing cellular structures, and mitigating oxidative damage. Understanding these genetic mechanisms provides valuable insights for developing salt-tolerant crops through genetic engineering and breeding.

11.9 Applications of exogenous nitric oxide in plants under salinity stress

Exogenous application of NO donors improves plant tolerance to salinity by modulating protein expression and activity (Table 11.3). Research has demonstrated the effectiveness of applying NO exogenously to mitigate salinity stress under diverse conditions. Recent studies have explored various application methods to enhance plant growth and stress tolerance, each tailored to specific growth conditions and frequencies. Seed priming is a widely used technique where seeds are pretreated with bioactive compounds or solutions to improve germination and early seedling vigor. For instance, a study demonstrated that seed priming under greenhouse conditions, applied as a one-time treatment, significantly enhanced plant growth and stress resilience (AL-Quraan et al., 2024). Another effective method is foliar pretreatment, which involves spraying leaves with growth-promoting or stress-alleviating solutions. This method is particularly useful in controlled growth chambers, where treatments can be applied at regular intervals to ensure sustained benefits. (Brito et al., 2020) reported that foliar pretreatment improved plant performance under stress by enhancing antioxidant activity and nutrient uptake.

For field conditions, root medium application is often preferred due to its practicality and continuous delivery of nutrients or stress-mitigating agents. Egamberdieva et al. (2019) highlighted the effectiveness of this method in field settings, where continuous application to the root medium improved water and nutrient availability, leading to better crop yields under stress. These methods, that is, seed priming, foliar pretreatment, and root medium application, offer versatile strategies for enhancing plant growth and stress tolerance, each suited to specific environmental and experimental conditions. Although these methods show promise, the variability in concentration, timing, and mode of application

TABLE 11.3 Effects of application of NO donors on different plant species.

Plant species	NO donor	Concentration	Effect on plants	References
Rice (Oryza sativa)	Sodium nitroprusside (SNP)	100 μM	Enhanced SOD and CAT activities, increased proline accumulation, reduced Na^+ uptake	(Fang et al., 2026
Tomato (Solanum lycopersicum)	SNP	200 μM	Enhanced activities of antioxidant enzymes and decreased lipid peroxidation	(Huang et al., 2024)
Wheat (Triticum aestivum)	S-nitrosoglutathione (GSNO)	50 μM	Increased expression of HKT1 and SOS1, improved K^+/Na^+ ratio	(Khan et al., 2023)
Maize (Zea mays)	SNP	150 μM	Upregulated aquaporin expression, improved water use efficiency	(Majeed et al., 2018)
Arabidopsis thaliana	GSNO	100 μM	Enhanced expression of DREB and NAC transcription factors, increased osmoprotectant synthesis	()(Mittler et al., 2022)
Barley (Hordeum vulgare)	SNP	100 μM	Improved root growth, reduced oxidative damage via APX and GR activation	(Thabet et al., 2022)
Soybean (Glycine max)	GSNO	75 μM	Enhanced antioxidant enzymes (SOD, CAT), reduced H_2O_2 accumulation	(Msarie et al., 2025)
Cucumber (Cucumis sativus)	SNP	250 μM	Increased chlorophyll content, improved photosynthetic efficiency under salinity	(Zhang et al., 2020)
Mustard (Brassica juncea)	GSNO	150 μM	Enhanced proline and glycine betaine accumulation, reduced membrane leakage	(Song et al., 2022)
Poplar (Populus spp.)	SNP	200 μM	Reduced ROS levels via SOD activation, improved lignin biosynthesis for cell wall strength	(Han et al., 2024)

Sunflower (Helianthus annuus)	S-nitroso-N-acetylpenicillamine (SNAP)	100 μM	Improved ion homeostasis (Na^+/K^+ balance), upregulated NHX1 transporters	(Shan et al., 2023)
Potato (Solanum tuberosum)	SNP	300 μM	Enhanced osmolyte synthesis (trehalose), reduced electrolyte leakage	(Chourasia et al., 2021)
Spinach (Spinacia oleracea)	GSNO	125 μM	Improved stomatal regulation, increased aquaporin (PIP2) activity for water balance	(Casajús et al., 2023)
Pea (Pisum sativum)	SNP	180 μM	Enhanced flavonoid biosynthesis, reduced lipid peroxidation	(Hasanuzzaman, 2022)

NO, Nitric oxide; *SOD*, superoxide dismutase; *CAT*, catalase.

remains a challenge in establishing standardized guidelines for agricultural practices.

11.10 Challenges in the application of nitric oxide in agriculture

Exogenous NO enhances antioxidant capacity, stabilizes ion transport proteins, and increases osmoprotectant accumulation. The choice of NO donor, concentration, and timing is critical for achieving optimal effects. Optimal NO donor concentrations vary by species. Excessive NO can be toxic, necessitating further research to refine its application for sustainable agriculture. Despite extensive research, several challenges persist in harnessing NO for agricultural applications.

Research on NO application has shown significant variability in effective concentrations. Studies report a range from nanomolar to micromolar levels, complicating dosage standardization. Excessive NO can induce oxidative stress and phytotoxicity, while insufficient amounts fail to confer beneficial effects on plant growth (Maurya & Rani, 2017). Salinity stress varies significantly among crops. For instance, NO application improved salinity tolerance in rice at salinity levels up to 100 mM NaCl, but was ineffective beyond that threshold for wheat (Nayak et al., 2022). Establishing crop-specific thresholds remains a challenge. Different plant species and growth stages exhibit variable responses to NO. Environmental factors, such as temperature, humidity, and soil type, further complicate uniform application strategies (Hasanuzzaman, 2022). These factors illustrate the complexity of NO research and underscore the need for a unified analytical framework to inform practical guidelines for its agricultural use.

11.11 Meta-analysis tools for investigating nitric oxide functions in crop resilience

Metaanalysis provides a robust approach to synthesizing diverse research findings, yielding comprehensive insights that address the complexities inherent in NO research. These metaanalytical tools refine our understanding of NO's multifaceted role in plant resilience and productivity. A metaanalysis identifies optimal NO delivery techniques by pooling data and comparing foliar sprays, soil amendments, and hydroponic systems, and reveals significant differences in efficacy (Sun et al., 2021). A large dataset calculates optimal dosages and treatment intervals, which vary by plant species and stress conditions. Metaanalytical studies highlight best practices for sustainable use (Patel et al., 2020). Systematic reviews and metaanalyses identify under-researched areas, such as NO interactions with other plant hormones, such as auxins or ABA, under drought conditions (Gurevitch et al., 2018).

Metaanalytical approaches offer significant advantages in synthesizing research on NO applications in plant stress physiology. One key benefit is the identification of effective NO delivery methods, such as comparing the efficacy of foliar sprays versus soil amendments. For example, Zhang et al. (2019) used a metaanalysis to demonstrate that foliar application of NO donors often outperforms soil amendments in improving nutrient uptake and stress resilience under drought conditions. Another advantage lies in determining optimized dosages and timing for NO treatments. Another study aggregated data from 45 studies to identify optimal NO concentrations (e.g., 100–200 μM) that maximize salinity mitigation without causing phytotoxicity (Tufail et al., 2011). The metaanalysis also helps pinpoint research gaps, such as the limited understanding of hormonal interactions (e.g., NO-ABA or NO-ethylene cross-talk) under combined abiotic stresses. Eagle et al. (2017) highlighted this gap, emphasizing the need for studies exploring how NO synergizes with phytohormones to regulate stress responses. Collectively, these metaanalytical insights guide future research and practical applications of NO in agriculture.

Consolidated findings enable the development of clear, data-driven guidelines for crop management. For example, metaanalyses of salinity tolerance mechanisms identify genetic markers critical for breeding salt-tolerant rice and wheat varieties (Gurevitch et al., 2018). Integrating data from geographically and temporally diverse studies helps discern broader patterns, offering insights into genotype–environment interactions affecting yield and stress tolerance (Pan et al., 2020). Metaanalyses also facilitate evaluations of plant–microbe associations that enhance salt tolerance, integrating soil health into sustainable agricultural solutions (Rai, 2023). The application of metaanalysis in NO research provides a powerful tool for advancing evidence-based improvements in crop resilience and sustainable farming practices. By consolidating data, identifying patterns, and integrating soil–microbe interactions, metaanalyses contribute to the development of innovative solutions for mitigating abiotic stresses such as salinity. These efforts are crucial to ensure global food security and health safety in the face of climate change, offering scalable strategies to enhance agricultural productivity and environmental sustainability.

11.12 Conclusions and future perspectives

Salinity stress significantly impairs plant growth, metabolism, and productivity, posing a critical challenge for sustainable agriculture. Addressing this issue requires a comprehensive understanding of plant adaptive mechanisms and the development of innovative mitigation strategies. NO plays a pivotal role in enhancing salinity tolerance by regulating antioxidant activity, maintaining ion homeostasis, and improving photosynthetic efficiency. Its interaction with essential macronutrients, such as nitrogen and sulfur, further optimizes plant stress responses. Harnessing NO-based solutions and integrated nutrient management strategies can strengthen crop resilience, ensuring agricultural sustainability and food security in the face of rising global salinization.

The future of NO research in agriculture holds immense potential for addressing salinity stress and other abiotic challenges. A key priority is the translation of laboratory findings into practical, field-level applications. Developing cost-effective and scalable NO delivery methods, such as nanoencapsulated NO donors or foliar sprays, will be critical for widespread adoption in agriculture. Additionally, exploring synergistic approaches that combine NO with other stress mitigation techniques, such as biochar application, microbial inoculants, and precision irrigation, can significantly enhance crop performance under saline conditions. Advances in genetic engineering, particularly CRISPR–Cas9 technology, offer exciting opportunities to develop NO-overproducing crop varieties. By modulating genes involved in NO biosynthesis (e.g., NO synthase-like) or signaling pathways (e.g., S-nitrosation targets), researchers can create salt-tolerant crops with improved yields. Furthermore, investigating the role of NO in climate-resilient crops is essential, especially in regions prone to combined stresses like salinity, drought, and heat. Future studies should also focus on the integration of soil health, exploring the interplay between NO, soil microbiota, and nutrient cycling to develop holistic solutions that enhance both crop resilience and soil fertility. Finally, fostering global collaboration through international research partnerships and data-sharing platforms will enable the development of region-specific strategies for salinity mitigation, leveraging NO-based technologies to address diverse agricultural challenges. By addressing these future directions, NO-based strategies can play a transformative role in sustainable agriculture, ensuring food security and environmental resilience in a rapidly changing world.

References

Ahmad, P., Abdel Latef, A. A., Hashem, A., Abd_Allah, E. F., Gucel, S., & Tran, L.-S. P. (2016). Nitric oxide mitigates salt stress by regulating levels of osmolytes and antioxidant enzymes in chickpea. *Frontiers in Plant Science, 7*(2016). https://doi.org/10.3389/fpls.2016.00347.

Ahmed, H. G. M. D., Zeng, Y., Raza, H., Muhammad, D., Iqbal, M., Uzair, M., Khan, M. A., Iqbal, R., & EL Sabagh, A. (2022). Characterization of wheat (*Triticum aestivum* L.) accessions using morpho-physiological traits under varying levels of salinity stress at seedling stage. *Frontiers in Plant Science, 13*. https://doi.org/10.3389/fpls.2022.953670, https://www.frontiersin.org/journals/plant-science.

Alnusairi, G. S. H., Mazrou, Y. S. A., Qari, S. H., Elkelish, A. A., Soliman, M. H., Eweis, M., Abdelaal, K., El-Samad, G. Abd, Ibrahim, M. F. M., & ElNahhas, N. (2021). Exogenous Nitric Oxide Reinforces Photosynthetic Efficiency, Osmolyte, Mineral Uptake, Antioxidant, Expression of Stress-Responsive Genes and Ameliorates the Effects of Salinity Stress in Wheat. *Plants, 10*(8), 1693. https://doi.org/10.3390/plants10081693.

AL-Quraan, N. A., Samarah, N. H., & AL-Fawaz, A. M. (2024). The physiological effect of GABA priming on pepper (*Capsicum annuum* L.) during seed germination under salt stress. *Jordan Plant Growth Regulation, 104*(2), 953–974. https://doi.org/10.1007/s10725-024-01207-0, https://www.springer.com/journal/10725.

Anjum, N., Masood, A., Thangavel, P., & Khan, N. A. (2023). Making plant life easier and productive under salinity-updates and prospects: updates and prospects. *BoD–Books on Demand.*

Alzahrani, O. (2021). Metabolic engineering of osmoprotectants to elucidate the mechanism (s) of salt stress tolerance in crop plants. *Planta, 253*.

Ashraf, M., & Foolad, M. R. (2007). Roles of glycine betaine and proline in improving plant abiotic stress resistance. *Environmental and Experimental Botany, 59*(2), 206–216. https://doi.org/10.1016/j.envexpbot.2005.12.006.

Athar, H.-ur-R., Zulfiqar, F., Moosa, A., Ashraf, M., Zafar, Z. U., Zhang, L., Ahmed, N., Kalaji, H. M., Nafees, M., Hossain, M. A., Islam, M. S., El Sabagh, A., & Siddique, K. H. M. (2022). Salt stress proteins in plants: An overview. *Frontiers in Plant Science, 13*. https://doi.org/10.3389/fpls.2022.999058.

Bassil, E., Zhang, S., Gong, H., Tajima, H., & Blumwald, E. (2018). Cation Specificity of Vacuolar NHX-Type Cation/H + Antiporters. *Plant Physiology, 179*(2), 616–629. https://doi.org/10.1104/pp.18.01103.

Brito, C., Dinis, L. T., Ferreira, H., Moutinho-Pereira, J., & Correia, C. M. (2020). Foliar pre-treatment with abscisic acid enhances olive tree drought adaptability. *Plants, 9*(3). https://doi.org/10.3390/plants9030341, https://www.mdpi.com/2223-7747/9/3/341/pdf.

Cai, S., Jiang, G., Ye, N., Chu, Z., Xu, X., Zhang, J., & Zhu, G. (2015). A Key ABA Catabolic Gene, OsABA8ox3, Is Involved in Drought Stress Resistance in Rice. *PLOS ONE, 10*(2), e0116646. https://doi.org/10.1371/journal.pone.0116646.

Casajús, V., Howe, K., Fish, T., Civello, P., Thannhauser, T., Li, L., Gómez Lobato, M., & Martínez, G. (2023). Evidence of glucosinolates translocation from inflorescences to stems during postharvest storage of broccoli. *Plant Physiology and Biochemistry, 195*, 322–329. https://doi.org/10.1016/j.plaphy.2023.01.012.

Chen, J., & Wang, Y. (2024). Understanding the salinity resilience and productivity of halophytes in saline environments. *Plant Science, 346*, 112171. https://doi.org/10.1016/j.plantsci.2024.112171.

Chen, H., Qin, Y., Pu, J., Hu, J., & Wen, Y. (2021). Phytotoxicity of the chiral herbicide dichlorprop: Cross-talk between nitric oxide, reactive oxygen species and phytohormones. *Science of The Total Environment, 788*, 147866. https://doi.org/10.1016/j.scitotenv.2021.147866.

Chen, S., Liu, H., Yangzong, Z., Gardea-Torresdey, J. L., White, J. C., & Zhao, L. (2023). Seed Priming with Reactive Oxygen Species-Generating Nanoparticles Enhanced Maize Tolerance to Multiple Abiotic Stresses. *Environmental Science & Technology, 57*(48), 19932–19941. https://doi.org/10.1021/acs.est.3c07339.

Chen, Y., Li, X., Xie, X., Liu, L., Fu, J., & Wang, Q. (2023). Maize transcription factor ZmNAC2 enhances osmotic stress tolerance in transgenic Arabidopsis. *Journal of Plant Physiology, 282*, 153948. https://doi.org/10.1016/j.jplph.2023.153948.

Chourasia, K. N., Lal, M. K., Tiwari, R. K., Dev, D., Kardile, H. B., Patil, V. U., Kumar, A., Vanishree, G., Kumar, D., Bhardwaj, V., Meena, J. K., Mangal, V., Shelake, R. M., Kim, J. Y., & Pramanik, D. (2021). Salinity stress in potato: Understanding physiological, biochemical and molecular responses. *Life (Chicago, Ill: 1978), 11*(6). https://doi.org/10.3390/life11060545, https://www.mdpi.com/2075-1729/11/6/545/pdf.

Du, S. X., Wang, L. L., Yu, W.-P., Xu, S.-X., Chen, L., & Huang, W. (2024). Appropriate induction of TOC1 ensures optimal MYB44 expression in ABA signaling and stress response in *Arabidopsis*. *Plant, Cell & Environment, 47*(8), 3046–3062. https://doi.org/10.1111/pce.14922.

Eagle, A. J., Christianson, L. E., Cook, R. L., Harmel, R. D., Miguez, F. E., Qian, S. S., & Ruiz Diaz, D. A. (2017). Meta-analysis constrained by data: Recommendations to improve relevance of nutrient management research. *Agronomy Journal, 109*(6), 2441–2449. https://doi.org/10.2134/agronj2017.04.0215, https://dl.sciencesocieties.org/publications/aj/pdfs/109/6/2441.

Egamberdieva, D., Wirth, S., Bellingrath-Kimura, S. D., Mishra, J., & Arora, N. K. (2019). Salt-tolerant plant growth promoting rhizobacteria for enhancing crop productivity of saline soils.

Frontiers in Microbiology, 10. https://doi.org/10.3389/fmicb.2019.02791, https://www.frontiersin.org/journals/microbiology#.

Fancy, N. N., Bahlmann, A. K., & Loake, G. J. (2017). Nitric oxide function in plant abiotic stress. *Plant Cell and Environment, 40*(4), 462–472. https://doi.org/10.1111/pce.12707, http://www.blackwellpublishing.com/journals/PCE.

Fang, W., Raza, A., Zhu, Q., Wang, Q., Ren, Q., Liu, M., Wang, S., & Hassan, M. A. (2026). Regulatory networks and molecular mechanisms underlying salt stress tolerance in rice. *Frontiers in Plant Science, 17*. https://doi.org/10.3389/fpls.2026.1757448.

Farooq, M., Hussain, M., Wahid, A., & Siddique, K. H. M. (2012). Drought stress in plants: An overview. *Plant Responses to Drought Stress: From Morphological to Molecular Features, 26530*, 1–33. https://doi.org/10.1007/978-3-642-32653-0_1, http://doi.org/10.1007/978-3-642-32653-0.

Farooq, M., Hussain, M., Wakeel, A., & Siddique, K. H. M. (2015). Salt stress in maize: Effects, resistance mechanisms, and management. A review. *Agronomy for Sustainable Development, 35*(2), 461–481. https://doi.org/10.1007/s13593-015-0287-0, http://www.springerlink.com/content/1773-0155.

Feng, W., Lindner, H., Robbins, N. E., & Dinneny, J. R. (2016). Growing out of stress: The role of cell- and organ-scale growth control in plant water-stress responses. *The Plant Cell, 28*(8), 1769–1782. https://doi.org/10.1105/tpc.16.00182, http://www.plantcell.org/content/28/8/1769.full.pdf.

Feng, G., Xiao, P., Wang, X., Huang, L., Nie, G., Li, Z., Peng, Y., Li, D., & Zhang, X. (2022). Comprehensive Transcriptome Analysis Uncovers Distinct Expression Patterns Associated with Early Salinity Stress in Annual Ryegrass (*Lolium Multiflorum* L.). *International Journal of Molecular Sciences, 23*(6), 3279. https://doi.org/10.3390/ijms23063279.

Forlani (2025). Enhancement of Antioxidant Enzymatic Defenses in Salt-Adapted Rice Seedlings. *Agriculture, 15*(12), 1234. https://doi.org/10.3390/agriculture15121234.

Garcia-Mata, C., & Lamattina, L. (2007). Abscisic acid (ABA) inhibits light-induced stomatal opening through calcium- and nitric oxide-mediated signaling pathways. *Nitric Oxide: Biology and Chemistry / Official Journal of the Nitric Oxide Society, 17*(3-4), 143–151. https://doi.org/10.1016/j.niox.2007.08.001.

Georgieva, M., & Vassileva, V. (2023). Stress management in plants: examining provisional and unique dose-dependent responses. *International Journal of Molecular Sciences, 24*(6). https://doi.org/10.3390/ijms24065105.

Gill, R., Ahmar, S., Ali, B., Saleem, M. H., Khan, M. U., Zhou, W., & Liu, S. (2021). The Role of Membrane Transporters in Plant Growth and Development, and Abiotic Stress Tolerance. *International Journal of Molecular Sciences, 22*(23), 12792. https://doi.org/10.3390/ijms222312792.

Glazebrook, J. (2005). Contrasting mechanisms of defense against biotrophic and necrotrophic pathogens. *Annual Review of Phytopathology, 43*(1), 205–227. https://doi.org/10.1146/annurev.phyto.43.040204.135923.

Guo, S., Zhang, Y., Li, M., Zeng, P., Zhang, Q., Li, X., Xu, Q., Li, T., Wang, X., Kang, Z., & Zhang, X. (2022). TaBln1, a member of the Blufensin family, negatively regulates wheat resistance to stripe rust by reducing Ca2+ influx. *Plant Physiology, 189*(3), 1380–1396. https://doi.org/10.1093/plphys/kiac112.

Gupta, A., Singh, S. K., Singh, M. K., Singh, V. K., Modi, A., Singh, P. K., & Kumar, A. (2019). *Plant growth-promoting rhizobacteria and their functional role in salinity stress management Abatement of Environmental Pollutants: Trends and Strategies*. Elsevier, 151–160. http://doi.org/10.1016/B978-0-12-818095-2.00007-2, http://www.sciencedirect.com/science/book/9780128180952.

Gurevitch, J., Koricheva, J., Nakagawa, S., & Stewart, G. (2018). Meta-analysis and the science of research synthesis. *Nature, 555*(7695), 175–182. https://doi.org/10.1038/nature25753.

Gámez-Arjona, F., Park, H. J., García, E., Aman, R., Villalta, I., Raddatz, N., Carranco, R., Ali, A., Ali, Z., Zareen, S., De Luca, A., Leidi, E. O., Daniel-Mozo, M., Xu, Z. Y., Albert, A., Kim, W. Y., Pardo, J. M., Sánchez-Rodriguez, C., Yun, D. J., & Quintero, F. J. (2024). Inverse regulation of SOS1 and HKT1 protein localization and stability by SOS3/CBL4 in Arabidopsis thaliana. *Proceedings of the National Academy of Sciences of the United States of America, 121*(9). https://doi.org/10.1073/pnas.2320657121, https://www.pnas.org/doi/epdf/10.1073/pnas.2320657121.

Han, S., Cheng, Y., Wu, G., He, X., & Zhao, G. (2024). Enhancing salt tolerance in poplar seedlings through arbuscular mycorrhizal fungi symbiosis. *Plants, 13*(2), 233. https://doi.org/10.3390/plants13020233.

Hasanuzzaman, M., Inafuku, M., Nahar, K., Fujita, M., & Oku, H. (2021). Nitric Oxide Regulates Plant Growth, Physiology, Antioxidant Defense, and Ion Homeostasis to Confer Salt Tolerance in the Mangrove Species, Kandelia obovata. *Antioxidants, 10*(4), 611. https://doi.org/10.3390/antiox10040611.

Hasanuzzaman, M., Nahar, K., Alam, M. M., Roychowdhury, R., & Fujita, M. (2013). Physiological, biochemical, and molecular mechanisms of heat stress tolerance in plants. *International Journal of Molecular Sciences, 14*(5), 9643–9684. https://doi.org/10.3390/ijms14059643Bangladesh, http://www.mdpi.com/1422-0067/14/5/9643/pdf.

Hasanuzzaman, M., Oku, H., Nahar, K., Bhuyan, M. H. M. B., Mahmud, J. A., Baluska, F., & Fujita, M. (2018). Nitric oxide-induced salt stress tolerance in plants: ROS metabolism, signaling, and molecular interactions. *Plant Biotechnology Reports, 12*(2), 77–92. https://doi.org/10.1007/s11816-018-0480-0.

Hasanuzzaman, M. (2022). *Salt stress tolerance in rice and wheat: physiological and molecular mechanism*. IntechOpen. http://doi.org/10.5772/intechopen.101529.

Hassan, S., Berk, K., & Aronsson, H. (2021). Evolution and identification of DREB transcription factors in the wheat genome: modeling, docking and simulation of DREB proteins associated with salt stress. *Journal of Biomolecular Structure and Dynamics, 40*(16), 7191–7204. https://doi.org/10.1080/07391102.2021.1894980.

Hu, S., Yang, H., Gao, H., Yan, J., & Xie, D. (2021). Control of seed size by jasmonate. *Science China Life Sciences, 64*(8), 1215–1226. https://doi.org/10.1007/s11427-020-1899-8.

Hu, Y., Li, C., Zhou, R., Song, Y., Lv, Z., Wang, Q., Dong, X., Liu, S., Feng, C., Zhou, Y., Zeng, X., Zhang, L., Wang, Z., & Di, H. (2023). The Transcription Factor ZmNAC89 Gene Is Involved in Salt Tolerance in Maize (*Zea mays* L.). *International Journal of Molecular Sciences, 24*(20), 15099. https://doi.org/10.3390/ijms242015099.

Huang, D., Chen, X., Yun, F., Fang, H., Wang, C., & Liao, W. (2024). Nitric oxide alleviates programmed cell death induced by cadmium in *Solanum lycopersicum* seedlings through protein S-nitrosylation. *Science of The Total Environment, 931*, 172812. https://doi.org/10.1016/j.scitotenv.2024.172812.

Iqbal, N., Czékus, Z., Ördög, A., & Poór, P. (2023). *The main fungal pathogens and defense-related hormonal signaling in crops plant hormones in crop improvement*. Elsevier, 307–331. http://doi.org/10.1016/B978-0-323-91886-2.00014-8, https://www.sciencedirect.com/book/9780323918862.

Iqbal, N., Czékus, Z., Angeli, C., Bartók, T., Poór, P., & Ördög, A. (2023). Fumonisin B1-induced oxidative burst perturbed photosynthetic activity and affected antioxidant enzymatic response in tomato plants in ethylene-dependent manner. *Journal of Plant Growth Regulation, 42*(3), 1865–1878. https://doi.org/10.1007/s00344-022-10665-7.

Iqbal, N., Czékus, Z., Poór, P., & Ördög, A. (2023). Ethylene-dependent regulation of oxidative stress in the leaves of fusaric acid-treated tomato plants. *Plant Physiology and Biochemistry, 196*, 841–849. https://doi.org/10.1016/j.plaphy.2023.02.047.

Iqbal, N., Czékus, Z., Ördög, A., & Poór, P. (2024). Fusaric acid-evoked oxidative stress affects plant defence system by inducing biochemical changes at subcellular level. *Plant Cell Reports, 43*(1). https://doi.org/10.1007/s00299-023-03084-9.

Ismail, A. M., & Horie, T. (2017). Genomics, physiology, and molecular breeding approaches for improving salt tolerance. *Annual Review of Plant Biology, 68*, 405–434. https://doi.org/10.1146/annurev-arplant-042916-040936, http://www.annualreviews.org/journal/arplant.

Jiang, W., Wang, Z., Li, Y., Liu, X, Ren, Y, Li, C, Luo, S, Singh, RM., Li, Y, Kim, C, & Zhao, C (2024). FERONIA regulates salt tolerance in Arabidopsis by controlling photorespiratory flux. *The Plant Cell, 36*(11), 4732–4751. https://doi.org/10.1093/plcell/koae246.

Khan, M. I., Ali, N., Jan, G., Hamayun, M., Jan, F. G., Iqbal, A., Hussain, A., & Lee, I. J. (2022). Salt Stress Alleviation in Triticum aestivum Through Primary and Secondary Metabolites Modulation by Aspergillus terreus BTK-1. *Frontiers in Plant Science, 13*. https://doi.org/10.3389/fpls.2022.779623.

Khan, M., Ali, S., Al Azzawi, T. N. I., & Yun, B. W. (2023). Nitric oxide acts as a key signaling molecule in plant development under stressful conditions. *International Journal of Molecular Sciences, 24*(5). https://doi.org/10.3390/ijms24054782, http://www.mdpi.com/journal/ijms.

Khan, M., Al Azzawi, T. N., Ali, S., Yun, B.-W., & Mun, B.-G. (2023). Nitric Oxide, a Key Modulator in the Alleviation of Environmental Stress-Mediated Damage in Crop Plants: A Meta-Analysis. *Plants, 12*(11), 2121. https://doi.org/10.3390/plants12112121.

Khanna, R. R., Jahan, B., Iqbal, N., Khan, N. A., AlAjmi, M. F., Tabish Rehman, M., & Khan, M. I. R. (2021). GABA reverses salt-inhibited photosynthetic and growth responses through its influence on NO-mediated nitrogen-sulfur assimilation and antioxidant system in wheat. *Journal of Biotechnology, 325*, 73–82. https://doi.org/10.1016/j.jbiotec.2020.11.015, www.elsevier.com/locate/jbiotec.

Khator, K., Parihar, S., Jasik, J., & Shekhawat, G. S. (2024). Nitric oxide in plants: An insight on redox activity and responses toward abiotic stress signaling. *Plant Signaling and Behavior, 19*(1). https://doi.org/10.1080/15592324.2023.2298053, http://www.tandfonline.com/loi/kpsb20.

Kim, Y.-S., Sakuraba, Y., Han, S.-H., Yoo, S.-C., & Paek, N.-C. (2013). Mutation of the *Arabidopsis* NAC016 Transcription Factor Delays Leaf Senescence. *Plant and Cell Physiology, 54*(10), 1660–1672. https://doi.org/10.1093/pcp/pct113.

Li, H., Tang, X., Yang, X., & Zhang, H. (2021). Comprehensive transcriptome and metabolome profiling reveal metabolic mechanisms of *Nitraria sibirica* Pall. to salt stress. *Scientific Reports, 11*(1). https://doi.org/10.1038/s41598-021-92317-6.

Li, J., Yu, B., Ma, C., Li, H., Jiang, D., Nan, J., Xu, M., Liu, H., Chen, S., Duanmu, H., & Li, H. (2022). Functional Characterization of Sugar Beet M14 Antioxidant Enzymes in Plant Salt Stress Tolerance. *Antioxidants, 12*(1), 57. https://doi.org/10.3390/antiox12010057.

Liu, Z., Ding, Y., Wang, F., Ye, Y., & Zhu, C. (2016). Role of salicylic acid in resistance to cadmium stress in plants. *Plant Cell Reports, 35*(4), 719–731. https://doi.org/10.1007/s00299-015-1925-3.

Liu, X., Gao, Y., Guo, Z., Wang, N., Wegner, A., Wang, J., Zou, X., Hu, J., Liu, M., Zhang, H., Zheng, X., Wang, P., Schaffrath, U., & Zhang, Z. (2022). MoIug4 is a novel secreted effector promoting rice blast by counteracting host OsAHL1-regulated ethylene gene transcription. *New Phytologist, 235*(3), 1163–1178. https://doi.org/10.1111/nph.18169.

Liu, X., Yang, X., & Zhang, B. (2021). Transcriptome analysis and functional identification of GmMYB46 in soybean seedlings under salt stress. *PeerJ, 9*, e12492. https://doi.org/10.7717/peerj.12492.

Majeed, S., Nawaz, F., Naeem, M., & Ashraf, M. Y. (2018). Effect of exogenous nitric oxide on sulfur and nitrate assimilation pathway enzymes in maize (*Zea mays* L.) under drought stress. *Acta Physiologiae Plantarum, 40*(12). https://doi.org/10.1007/s11738-018-2780-y.

Manai, J., Kalai, T., Gouia, H., & Corpas, F. J. (2014). Exogenous nitric oxide (NO) ameliorates salinity-induced oxidative stress in tomato (Solanum lycopersicum) plants. *Journal of soil science and plant nutrition*(ahead). https://doi.org/10.4067/s0718-95162014005000034.

Mandal, M., Sarkar, M., Khan, A., Biswas, M., Masi, A., Rakwal, R., Agrawal, G. K., Srivastava, A., & Sarkar, A. (2022). Reactive oxygen species (ROS) and reactive nitrogen species (RNS) in plants– maintenance of structural individuality and functional blend. *Advances in Redox Research, 5*, 100039. https://doi.org/10.1016/j.arres.2022.100039.

Mangal, V., Lal, M. K., Tiwari, R. K., Altaf, M. A., Sood, S., Kumar, D., Bharadwaj, V., Singh, B., Singh, R. K., & Aftab, T. (2023). Molecular insights into the role of reactive oxygen, nitrogen and sulphur species in conferring salinity stress tolerance in plants. *Journal of Plant Growth Regulation, 42*(2), 554–574. https://doi.org/10.1007/s00344-022-10591-8, https://www.springer.com/journal/344.

Maurya, A. K., & Rani, A. (2017). *Nitric oxide (NO) and physio-biochemical adaptation in plants against stress plant adaptation strategies in changing environment.* Springer Singapore, 347–386. http://doi.org/10.1007/978-981-10-6744-0_15, http://www.springer.com/in/book/9789811067433.

Meng, Q., Yan, M., Zhang, J., Zhang, Q., Zhang, X., Yang, Z., Luo, Y., & Wu, W. (2023). Humic acids enhance salt stress tolerance associated with pyrroline 5-carboxylate synthetase gene expression and hormonal alteration in perennial ryegrass (*Lolium perenne* L.). *Frontiers in Plant Science, 14*. https://doi.org/10.3389/fpls.2023.1272987.

Mittler, R., Zandalinas, S. I., Fichman, Y., & Van Breusegem, F. (2022). Reactive oxygen species signalling in plant stress responses. *Nature Reviews Molecular Cell Biology, 23*(10), 663–679. https://doi.org/10.1038/s41580-022-00499-2.

Msarie, M. W., Methela, N. J., Islam, M. S., An, T. H., Das, A. K., Lee, D. S., Mun, B. G., & Yun, B. W. (2025). Enhancing soybean salt tolerance with GSNO and silicon: a comprehensive physiological, biochemical, and genetic study. *International Journal of Molecular Sciences, 26*(2). https://doi.org/10.3390/ijms26020609, http://www.mdpi.com/journal/ijms.

Munns, R., & Tester, M. (2008). Mechanisms of salinity tolerance. *Annual Review of Plant Biology, 59*(1), 651–681. https://doi.org/10.1146/annurev.arplant.59.032607.092911.

Munns, R., & Gilliham, M. (2015). Salinity tolerance of crops – what is the cost? *New Phytologist, 208*(3), 668–673. https://doi.org/10.1111/nph.13519.

Nayak, H. S., Silva, J. V., Parihar, C. M., Kakraliya, S. K., Krupnik, T. J., Bijarniya, D., Jat, M. L., Sharma, P. C., Jat, H. S., Sidhu, H. S., & Sapkota, T. B. (2022). Rice yield gaps and nitrogen-use efficiency in the Northwestern Indo-gangetic plains of India: Evidence based insights from heterogeneous farmers' practices. *Field Crops Research, 275*. https://doi.org/10.1016/j.fcr.2021.108328, www.elsevier.com/inca/publications/store/5/0/3/3/0/8.

Oerke, E. C., Herzog, K., & Toepfer, R. (2016). Hyperspectral phenotyping of the reaction of grapevine genotypes to *Plasmopara viticola*. *Journal of Experimental Botany, 67*(18), 5529–5543. https://doi.org/10.1093/jxb/erw318.

Pan, J., Peng, F., Tedeschi, A., Xue, X., Wang, T., Liao, J., Zhang, W., & Huang, C. (2020). Do halophytes and glycophytes differ in their interactions with arbuscular mycorrhizal fungi under salt stress? A meta-analysis. *Botanical Studies, 61*(1). https://doi.org/10.1186/s40529-020-00290-6.

Pan, Y., Xu, X., Li, L., Sun, Q., Wang, Q., Huang, H., Tong, Z., & Zhang, J. (2023). Melatonin-mediated development and abiotic stress tolerance in plants. *Frontiers in Plant Science, 14*. 10.3389/fpls.2023.1100827.

Parida, A. K., & Das, A. B. (2005). Salt tolerance and salinity effects on plants: A review. *Ecotoxicology and Environmental Safety, 60*(3), 324–349. https://doi.org/10.1016/j.ecoenv.2004.06.010, http://www.elsevier.com/inca/publications/store/6/2/2/8/1/9/index.htt.

Parray, J. A., Shameem, N., Egamberdieva, D., & Sayyed, R. Z. (2023). *Rhizobiome: Ecology, management and application rhizobiome: Ecology, management and application*. Elsevier, 1–480. http://doi.org/10.1016/C2022-0-02288-1, https://www.sciencedirect.com/book/9780443160301.

Patel, M. K., Kumar, M., Li, W., Luo, Y., Burritt, D. J., Alkan, N., & Tran, L.-S. P. (2020). Enhancing salt tolerance of plants: From metabolic reprogramming to exogenous chemical treatments and molecular approaches. *Cells, 9*(11), 2492. https://doi.org/10.3390/cells9112492.

Pawłowicz, I., Rapacz, M., Perlikowski, D., Gondek, K., & Kosmala, A. (2017). Abiotic stresses influence the transcript abundance of PIP and TIP aquaporins in *Festuca* species. *Journal of Applied Genetics, 58*(4), 421–435. https://doi.org/10.1007/s13353-017-0403-8.

Rai, K. (2023). *The role of salicylic acid and nitric oxide in plant heat response*. Cambridge Scholars Publishing.

Ren, L., Pan, X., Deng, Y., Ge, Z., Li, S., Su, D., Zhao, G., Tang, H., & Wang, X. (2024). Combined Transcriptomic and metabolomic analyses reveal the mechanisms by which the interaction between sulfur and nitrogen affects garlic yield and quality. *Horticulturae, 10*(11), 1203. https://doi.org/10.3390/horticulturae10111203.

Roy, S., & Soni, P. (2022). Unraveling the epigenetic landscape for salt tolerance in plants. *International Journal of Plant Biology, 13*(4), 443–462. https://doi.org/10.3390/ijpb13040036.

EL Sabagh, A., Islam, M. S., Skalicky, M., Ali Raza, M., Singh, K., Anwar Hossain, M., Hossain, A., Mahboob, W., Iqbal, M. A., Ratnasekera, D., Singhal, R. K., Ahmed, S., Kumari, A., Wasaya, A., Sytar, O., Brestic, M., ÇIG, F., Erman, M., Habib Ur Rahman, M., ... Arshad, A. (2021). Salinity stress in wheat (*Triticum aestivum* L.) in the changing climate: Adaptation and management strategies. *Frontiers in Agronomy, 3*. https://doi.org/10.3389/fagro.2021.661932, www.frontiersin.org/journals/agronomy#.

Sami, F., Siddiqui, H., Alam, P., & Hayat, S. (2021). Nitric oxide mitigates the salt-induced oxidative damage in mustard by upregulating the activity of various enzymes. *Journal of Plant Growth Regulation, 40*(6), 2409–2432. https://doi.org/10.1007/s00344-021-10331-4.

Santos, M. P., Zandonadi, D. B., de Sá, A. F. L., Costa, E. P., de Oliveira, C. J. L., Perez, L. E. P., Façanha, A. R., & Bressan-Smith, R. (2020). Abscisic acid-nitric oxide and auxin interaction modulates salt stress response in tomato roots. *Theoretical and Experimental Plant Physiology, 32*(4), 301–313. https://doi.org/10.1007/s40626-020-00187-6, http://link.springer.com/journal/40626.

Sarkar, A. K., & Sadhukhan, S. (2022). Imperative role of trehalose metabolism and trehalose-6-phosphate signaling on salt stress responses in plants. *Physiologia Plantarum, 174*(1). https://doi.org/10.1111/ppl.13647, http://onlinelibrary.wiley.com/journal/10.1111/(ISSN)1399-3054.

Scholthof, K. B. G., Adkins, S., Czosnek, H., Palukaitis, P., Jacquot, E., Hohn, T., Hohn, B., Saunders, K., Candresse, T., Ahlquist, P., Hemenway, C., & Foster, G. D. (2011). Top 10 plant viruses in molecular plant pathology. *Molecular Plant Pathology, 12*(9), 938–954. https://doi.org/10.1111/j.1364-3703.2011.00752.x.

Sehar, Z., Masood, A., & Khan, N. A. (2019). Nitric oxide reverses glucose-mediated photosynthetic repression in wheat (*Triticum aestivum* L.) under salt stress. *Environmental and Experimental Botany, 161*, 277–289. https://doi.org/10.1016/j.envexpbot.2019.01.010.

Shabala, S. (2013). Learning from halophytes: Physiological basis and strategies to improve abiotic stress tolerance in crops. *Annals of Botany, 112*(7), 1209–1221. https://doi.org/10.1093/aob/mct205.

Shahid, M., Dumat, C., Khalid, S., Niazi, N. K., & Antunes, P. M. C. (2017). Cadmium bioavailability, uptake, toxicity and detoxification in soil-plant system. *Reviews of Environmental Contamination and Toxicology, 241*, 73–137. https://doi.org/10.1007/398_2016_8, http://www.springer.com/series/398.

Shan, F., Wu, Y., Du, R., Yang, Q., Liu, C., Wang, Y., Zhang, C., & Chen, Y. (2023). Evolutionary analysis of the OSCA gene family in sunflower (*Helianthus annuus* L) and expression analysis under NaCl stress. *PeerJ, 11*, e15089. https://doi.org/10.7717/peerj.15089.

Sherin, G., Aswathi, K. P. R., & Puthur, J. T. (2022). Photosynthetic functions in plants subjected to stresses are positively influenced by priming. *Plant Stress, 4*. https://doi.org/10.1016/j.stress.2022.100079, www.journals.elsevier.com/plant-stress.

Shokat, S., & Großkinsky, D. K. (2019). Tackling salinity in sustainable agriculture—What developing countries may learn from approaches of the developed world. *Sustainability, 11*(17), 4558. https://doi.org/10.3390/su11174558.

Sohag, A. A. M., Tahjib-Ul-Arif, M., Afrin, S., Khan, M. K., Hannan, M. A., Skalicky, M., Mortuza, M. G., Brestic, M., Hossain, M. A., & Murata, Y. (2020). Insights into nitric oxide-mediated water balance, antioxidant defence and mineral homeostasis in rice (*Oryza sativa* L.) under chilling stress. *Nitric Oxide, 100-101*, 7–16. https://doi.org/10.1016/j.niox.2020.04.001.

Song, W., Loik, M. E., Cui, H., Fan, M., & Sun, W. (2022). Effect of nitrogen addition on leaf photosynthesis and water use efficiency of the dominant species *Leymus chinensis* (Trin.) Tzvelev in a semi-arid meadow steppe. *Plant Growth Regulation, 98*(1), 91–102. https://doi.org/10.1007/s10725-022-00835-8, http://www.springeronline.com/sgw/cda/frontpage/0,11855,1-40109-70-35547532-0,00.html.

Song, W., Shao, H., Zheng, A., Zhao, L., & Xu, Y. (2023). Advances in roles of salicylic acid in plant tolerance responses to biotic and abiotic stresses. *Plants, 12*(19), 3475. https://doi.org/10.3390/plants12193475.

Sun, C., Ali, K., Yan, K., Fiaz, S., Dormatey, R., Bi, Z., & Bai, J. (2021). Exploration of epigenetics for improvement of drought and other stress resistance in crops: A review. *Plants, 10*(6), 1226. https://doi.org/10.3390/plants10061226.

Sun, J., Agneessens, J., Topping, J. F., & Lindsey, K. (2026). The vesicle trafficking R- SNARE VAMP714 connects auxin responses, ROS and ion homeostasis in *Arabidopsis* under salt stress. *New Phytologist, 251*(1), 207–227. https://doi.org/10.1111/nph.71202.

Suzuki, N., Rivero, R. M., Shulaev, V., Blumwald, E., & Mittler, R. (2014). Abiotic and biotic stress combinations. *New Phytologist, 203*(1), 32–43. https://doi.org/10.1111/nph.12797, http://www.wiley.com/bw/editors.asp?ref=0028-646X&site=1.

Syed, A., Sarwar, G., Shah, S. H., & Muhammad, S. (2021). Soil salinity research in 21st century in Pakistan: Its impact on availability of plant nutrients, growth and yield of crops. *Communications in Soil Science and Plant Analysis, 52*(3), 183–200. https://doi.org/10.1080/00103624.2020.1854294, http://www.tandfonline.com/loi/lcss20.

Tester, M., & Langridge, P. (2010). Breeding technologies to increase crop production in a changing world. *Science (New York, N.Y.), 327*(5967), 818–822. https://doi.org/10.1126/science.1183700.

Thabet, S. G., Alomari, D. Z., Börner, A., Brinch-Pedersen, H., & Alqudah, A. M. (2022). Elucidating the genetic architecture controlling antioxidant status and ionic balance in barley under salt stress. *Plant Molecular Biology, 110*(3), 287–300. https://doi.org/10.1007/s11103-022-01302-8, https://link.springer.com/journal/11103.

T.A. Tufail, Bejarano, A. Shakoor, Can bacterial endophytes be used as a promising bio-inoculant for the mitigation of salinity stress in crop plants?—A global meta-analysis of the last decade. 9 (2011).

Turchetto-Zolet, A. C., Margis-Pinheiro, M., & Margis, R. (2009). The evolution of pyrroline-5-carboxylate synthase in plants: A key enzyme in proline synthesis. *Molecular Genetics and Genomics, 281*(1), 87–97. https://doi.org/10.1007/s00438-008-0396-4.

Verma, S., Verma, P. K., & Chakrabarty, D. (2022). *Plant transcription factors: important factors controlling oxidative stress in plants plant transcription factors: Contribution in development, metabolism, and environmental stress*. Elsevier, 383–417. http://doi.org/10.1016/B978-0-323-90613-5.00006-6, https://www.sciencedirect.com/book/9780323906135.

Vikram, Pooja, Sharma, J., & Sharma, A. (2022). Role of Brassinosteroids in plants responses to salinity stress: A review. *Journal of Applied and Natural Science, 14*(2), 582–599. https://doi.org/10.31018/jans.v14i2.3466.

Wang, M., Ren, L. T., Wei, X. Y., Ling, Y. M., Gu, H. T., Wang, S. S., Ma, X. F., & Kong, G. C. (2022). NAC Transcription factor TwNAC01 positively regulates drought stress responses in *Arabidopsis* and *Triticale*. *Frontiers in Plant Science, 13*. https://doi.org/10.3389/fpls.2022.877016, https://www.frontiersin.org/journals/plant-science.

Wang, G., Zhang, L., Zhang, S., Li, B., Li, J., Wang, X., Zhang, J., Guan, C., & Ji, J. (2023). The combined use of a plant growth promoting *Bacillus* sp. strain and GABA promotes the growth of rice under salt stress by regulating antioxidant enzyme system, enhancing photosynthesis and improving soil enzyme activities. *Microbiological Research, 266*, 127225. https://doi.org/10.1016/j.micres.2022.127225.

Wang, T., Xun, H., Wang, W., Ding, X., Tian, H., Hussain, S., Dong, Q., Li, Y., Cheng, Y., Wang, C., Lin, R., Li, G., Qian, X., Pang, J., Feng, X., Dong, Y., Liu, B., & Wang, S. (2021). Mutation of GmAITR genes by CRISPR/Cas9 genome editing results in enhanced salinity stress tolerance in soybean. *Frontiers in Plant Science, 12*. https://doi.org/10.3389/fpls.2021.779598.

Wei, L., Zhang, J., Wei, S., Wang, C., Deng, Y., Hu, D., Liu, H., Gong, W., Pan, Y., & Liao, W. (2022). Nitric oxide alleviates salt stress through protein S-nitrosylation and transcriptional regulation in tomato seedlings. *Planta, 256*(6). https://doi.org/10.1007/s00425-022-04015-w.

Wijayasinghe, Y. S., Tyagi, A., & Poddar, N. K. (2017). *Regulation of cell volume by osmolytes Cellular Osmolytes: From chaperoning protein folding to clinical perspectives*. Springer Singapore, 195–228. http://doi.org/10.1007/978-981-10-3707-8_9, http://www.springer.com/in/book/9789811037061.

Wu, H., D'Alessio, A. C., Ito, S., Wang, Z., Cui, K., Zhao, K., Sun, Y. E., & Zhang, Y. (2011). Genome-wide analysis of 5-hydroxymethylcytosine distribution reveals its dual function in transcriptional regulation in mouse embryonic stem cells. *Genes and Development, 25*(7), 679–684. https://doi.org/10.1101/gad.2036011, http://genesdev.cshlp.org/content/25/7/679.full.pdf+html.

Wu, J., Jiang, Y., Liang, Y., Chen, L., Chen, W., & Cheng, B. (2019). Expression of the maize MYB transcription factor ZmMYB3R enhances drought and salt stress tolerance in transgenic plants. *Plant Physiology and Biochemistry, 137*, 179–188. https://doi.org/10.1016/j.plaphy.2019.02.010.

Yin, X., Xia, Y., Xie, Q., Cao, Y., Wang, Z., Hao, G., Song, J., Zhou, Y., & Jiang, X. (2019). The protein kinase complex CBL10–CIPK8–SOS1 functions in Arabidopsis to regulate salt tolerance. *Journal of Experimental Botany, 71*(6), 1801–1814. https://doi.org/10.1093/jxb/erz549.

Yu, M., Lamattina, L., Spoel, S. H., & Loake, G. J. (2014). Nitric oxide function in plant biology: A redox cue in deconvolution. *New Phytologist, 202*(4), 1142–1156. https://doi.org/10.1111/nph.12739, http://www.wiley.com/bw/editors.asp?ref=0028-646X&site=1.

Zhang, F., Zhu, G., Du, L., Shang, X., Cheng, C., Yang, B., Hu, Y., Cai, C., & Guo, W. (2016). Genetic regulation of salt stress tolerance revealed by RNA-Seq in cotton diploid wild species, Gossypium davidsonii. *Scientific Reports, 6*(1). https://doi.org/10.1038/srep20582.

Zhang, H., Zhao, Z., Zhang, X., Zhang, W., Huang, L., Zhang, Z., Yuan, L., & Liu, X. (2019). Effects of foliar application of selenate and selenite at different growth stages on Selenium accumulation and speciation in potato (*Solanum tuberosum* L). *Food Chemistry, 286*, 550–556. https://doi.org/10.1016/j.foodchem.2019.01.185.

Zhang, L., Liu, Y., Zhang, Z., & Fang, S. (2023). Physiological response and molecular regulatory mechanism reveal a positive role of nitric oxide and hydrogen sulfide applications in salt tolerance of *Cyclocarya paliurus*. *Frontiers in Plant Science, 14*. https://doi.org/10.3389/fpls.2023.1211162.

Zhao, H., Dong, X., Yang, D., Ge, Q., Lu, P., & Liu, C. (2025). New insights into the salt-responsive regulation in eelgrass at transcriptional and post-transcriptional levels. *Frontiers in Plant Science, 16*. https://doi.org/10.3389/fpls.2025.1497064.

Zhang, M., Li, C., Zhang, Y., Pan, J., Huang, S., He, L., & Jin, G. (2022). Impact of salt content and hydrogen peroxide-induced oxidative stress on protein oxidation, conformational/morphological changes, and micro-rheological properties of porcine myofibrillar proteins. *Food Chemistry, 370*, 131074. https://doi.org/10.1016/j.foodchem.2021.131074.

Zhang, Z., Wu, P., Zhang, W., Yang, Z., Liu, H., Ahammed, G. J., & Cui, J. (2020). Calcium is involved in exogenous NO-induced enhancement of photosynthesis in cucumber (*Cucumis sativus* L.) seedlings under low temperature. *Scientia Horticulturae, 261*, 108953. https://doi.org/10.1016/j.scienta.2019.108953.

Zhang, X., Zhang, L., Chen, Y., Wang, S., Fang, Y., Zhang, X., Wu, Y., & Xue, D. (2021). Genome-wide identification of the SOD gene family and expression analysis under drought and salt stress in barley. *Plant Growth Regulation, 94*(1), 49–60. https://doi.org/10.1007/s10725-021-00695-8.

Zheng, P., Bai, X., Long, J., Li, K., & Xu, H. (2016). Nitric oxide enhances the nitrate stress tolerance of spinach by scavenging ROS and RNS. *Scientia Horticulturae, 213*, 24–33. https://doi.org/10.1016/j.scienta.2016.10.008.

Zhou, J., Mu, Q., Wang, X., Zhang, J., Yu, H., Huang, T., He, Y., Dai, S., & Meng, X. (2022). Multilayered synergistic regulation of phytoalexin biosynthesis by ethylene, jasmonate, and MAPK signaling pathways in *Arabidopsis*. *The Plant Cell, 34*(8), 3066–3087. https://doi.org/10.1093/plcell/koac139.

Zhu, Y., Gao, H., Lu, M., Hao, C., Pu, Z., Guo, M., Hou, D., Chen, L.-Y., & Huang, X. (2019). Melatonin-Nitric Oxide Crosstalk and Their Roles in the Redox Network in Plants. *International Journal of Molecular Sciences, 20*(24), 6200. https://doi.org/10.3390/ijms20246200.

Zhou, S., Jia, L., Chu, H., Wu, D., Peng, X., Liu, X., Zhang, J., Zhao, J., Chen, K., & Zhao, L. (2016). Arabidopsis CaM1 and CaM4 Promote Nitric Oxide Production and Salt Resistance by Inhibiting S-Nitrosoglutathione Reductase via Direct Binding. *PLOS Genetics, 12*(9), e1006255. https://doi.org/10.1371/journal.pgen.1006255.

Zhu, Z., Dai, Y., Yu, G., Zhang, X., Chen, Q., Kou, X., Mehareb, E. M., Raza, G., Zhang, B., Wang, B., Wang, K., & Han, J. (2023). Dynamic physiological and transcriptomic changes reveal memory effects of salt stress in maize. *BMC Genomics, 24*(1). https://doi.org/10.1186/s12864-023-09845-w.

Zörb, C., Geilfus, C.-M., Dietz, K.-J., & Weber, A. (2019). Salinity and crop yield. *Plant Biology, 21*(S1), 31–38. https://doi.org/10.1111/plb.12884.

Chapter 12

Role of nitric oxide in the mitigation of mineral stress

Selahattin Kondak, Dóra Kondak, Oluwatosin Peace Adedokun, and Zsuzsanna Kolbert
Department of Plant Biology, University of Szeged, Szeged, Hungary

12.1 Introduction

Mineral nutrient stress constitutes a major limitation to plant growth, development, and productivity in both natural and agricultural systems (Mir et al., 2023). Plants depend on coordinated supplies of essential macronutrients, including nitrogen (N), phosphorus (P), potassium (K), calcium (Ca), magnesium (Mg), and sulfur (S), together with micronutrients such as iron (Fe), zinc (Zn), manganese (Mn), copper (Cu), boron (B), molybdenum (Mo), nickel (Ni), and chlorine (Cl), to sustain metabolic activity and cellular structure (Hänsch and Mendel, 2009; Maathuis, 2009; Tewari et al., 2021). Deficiency, excess, or imbalance of these elements perturbs ion homeostasis, enzyme function, and osmotic regulation, ultimately constraining biomass accumulation and yield formation. Nitric oxide (NO) has emerged as an important signaling component in plant responses to mineral nutrient stress (Buet et al., 2019; Simontacchi et al., 2015). Produced through both enzymatic and nonenzymatic routes in response to nutritional cues, NO modulates transcriptional programs, posttranslational protein modifications, and cellular redox balance, thereby influencing nutrient acquisition and internal redistribution. Unlike classical phytohormones such as abscisic acid or ethylene, which are synthesized through defined biosynthetic pathways, transported to target tissues, and perceived by specific receptor systems, NO lacks dedicated receptors and long-distance transport mechanisms (Beligni & Lamattina, 2001). Its signaling function is instead characterized by rapid diffusion, short half-life, and strong dependence on local redox conditions.

This mode of action places NO alongside a group of emerging regulatory molecules, including hydrogen sulfide and carbon monoxide, that shape stress signaling without conforming to the structural and functional criteria of traditional plant hormones (Freschi, 2013). Through redox-based interactions,

Nitric Oxide in Plant Resilience to Abiotic Stress. DOI: https://doi.org/10.1016/B978-0-443-40497-9.00010-6

particularly *S*-nitrosation and coordination with reactive oxygen species (ROS), NO integrates mineral nutrient signals with hormonal and metabolic networks to support adaptive growth responses under nutrient-limited or imbalanced conditions (Singhal et al., 2021;Khan et al., 2023a,b). The present chapter focuses on NO-mediated mechanisms involved in the mitigation of stress associated with essential mineral nutrients. While certain minerals may exert toxic effects at elevated concentrations, nonessential heavy metals such as cadmium (Cd) and lead (Pb) are excluded from this discussion, as their modes of toxicity and signaling responses are mechanistically distinct and have been addressed elsewhere (Terrón-Camero et al., 2019). This framework provides a basis for examining how NO contributes to plant adjustment under macronutrient and micronutrient constraints and how mineral-derived signals are coordinated with broader stress-response pathways.

12.2 Mineral nutrient stress in plants: an overview

12.2.1 Essential mineral nutrients and their physiological roles

Essential mineral nutrients sustain plant growth through their incorporation into biomolecules and participation in core metabolic processes. N is a structural component of amino acids, nucleic acids, chlorophyll, and numerous coenzymes, thereby supporting protein synthesis, genome replication, and photosynthetic carbon assimilation. P is integral to ATP, NADPH, phospholipids, and phosphorylated intermediates, enabling energy transfer, membrane organization, and carbohydrate metabolism. K functions predominantly as an enzyme activator and osmotic regulator, maintaining membrane potentials, stomatal function, and turgor-driven cell expansion. Ca contributes to cell wall stability through pectin cross-linking and supports membrane integrity and compartmentalization. Mg occupies the central position in chlorophyll and activates ATP-dependent enzymes, facilitating light harvesting and phosphate transfer reactions. S is incorporated into cysteine- and methionine-containing proteins and redox-active molecules such as glutathione and ferredoxin, supporting protein structure and electron transport processes (Maathuis, 2009).

Micronutrients are required at substantially lower concentrations but play indispensable catalytic and structural roles in plant metabolism. Fe participates in electron transfer reactions as a component of heme-containing cytochromes and Fe–S proteins involved in photosynthesis and respiration. Zn stabilizes protein structures and catalytic sites in enzymes, including carbonic anhydrase and alcohol dehydrogenase. Cu mediates electron flow through plastocyanin and contributes to the activity of oxidative enzymes involved in lignification. Mn forms the catalytic core of the oxygen-evolving complex in photosystem II and activates enzymes of amino acid metabolism. B ensures cell wall integrity by cross-linking rhamnogalacturonan II domains. Mo is required for nitrate reductase and nitrogenase activity through its association with the molybdopterin cofactor, while Ni activates urease, enabling urea hydrolysis and N recycling (Hänsch & Mendel, 2009).

12.2.2 Types of mineral stress

Mineral stress arises when the availability of essential nutrients deviates from the concentrations required to sustain optimal metabolic and structural functions, imposing constraints on plant growth and resource allocation (Mir et al., 2023). Nutrient deficiencies commonly develop under conditions of low soil nutrient pools, restricted ion mobility, or impaired root acquisition, leading to depletion of cellular cofactors and metabolic substrates (Maathuis, 2009; Hänsch & Mendel, 2009). Insufficient N limits amide synthesis and transamination reactions, reducing protein production and chloroplast development, while P deficiency constrains the formation of phosphorylated intermediates required for energy transfer and membrane assembly. Inadequate K disrupts enzyme activation and stomatal regulation, whereas shortages of Ca or Mg weaken cell wall stability and chlorophyll integrity, respectively. Mineral excess represents the opposite end of the nutritional spectrum, where essential elements accumulate beyond physiological requirements due to elevated external availability or insufficient regulatory control. Excessive concentrations of macronutrients or micronutrients disturb ionic equilibrium, alter membrane properties, and interfere with coordinated metabolic activity. In the case of micronutrients, the narrow margin between nutritional requirement and toxicity is particularly evident, as elements such as Cu and Zn readily transition from indispensable cofactors to disruptors of protein function when present in excess. This duality underscores the borderline nature of micronutrient stress, in which beneficial and detrimental effects are closely linked and highly concentration-dependent (Chrysargyris et al., 2022).

Mineral stress is frequently exacerbated by nutrient imbalance arising from antagonistic interactions among elements during uptake, transport, or utilization. Competitive relationships at shared transport systems or binding sites in the rhizosphere can favor one ion at the expense of another, resulting in secondary deficiencies even when total nutrient supply is adequate. K acquisition is reduced under high sodium availability due to overlapping transport selectivity, while elevated Ca can suppress Mg absorption at the root surface. Similarly, interactions between macro- and micronutrients, such as Fe–Zn antagonism, influence long-distance translocation and tissue-specific nutrient distribution, thereby constraining nutrient-use efficiency and overall plant performance (Maathuis, 2009;Kumar et al., 2021).

12.2.3 Plant responses to mineral stress

Plants exposed to mineral nutrient deficiencies or excesses exhibit pronounced plasticity in root system architecture, enabling optimization of resource acquisition while minimizing energetic costs. Under P limitation, primary root elongation is often reduced, whereas lateral root formation and density increase, thereby enhancing exploration of the surface soil layers, where phosphate availability is typically higher. Root hair proliferation represents another adaptive response, particularly under Fe or Zn limitation, expanding the absorptive surface area available for micronutrient uptake (Kudoyarova et al., 2015;Pandey et al., 2021). These root-level adjustments are commonly accompanied by restrained shoot growth, resulting in altered biomass allocation that prioritizes belowground foraging over aerial expansion. Such developmental reprioritization supports survival under nutrient limitation by balancing carbon investment between vegetative maintenance and reproductive potential (Mir et al., 2023).

In parallel with morphological changes, plants undergo metabolic and physiological adjustments to stabilize internal nutrient status and sustain essential processes. Enhanced expression and activity of nutrient transport systems facilitate increased ion uptake during deficiency, while efflux mechanisms and intracellular sequestration limit cytosolic accumulation under excess conditions (Kumar et al., 2021). Perturbations in mineral availability frequently affect photosynthetic performance through constraints on chlorophyll synthesis, enzyme activity, or electron transport efficiency, leading to broader adjustments in respiratory and metabolic fluxes. Redistribution of nutrients from mature tissues to actively growing sinks contributes to internal homeostasis, while vacuolar compartmentalization serves as a buffering mechanism to isolate excess ions and protect sensitive cellular processes (Maathuis, 2009; Hänsch & Mendel, 2009).

These structural and metabolic adjustments are often accompanied by disturbances in cellular redox balance. Disruption of electron transport reactions under mineral stress can increase the generation of ROS, challenging the capacity of antioxidant defense systems (Pandey et al., 2021). In response, enzymatic and nonenzymatic antioxidants are mobilized to limit oxidative damage to lipids, proteins, and nucleic acids (Mir et al., 2023). Beyond their cytotoxic effects, elevated ROS act as secondary signals that contribute to stress perception and the coordination of downstream transcriptional and physiological responses across cellular compartments.

12.3 Nitric oxide biosynthesis and signaling under mineral stress

12.3.1 Sources of nitric oxide in plants

Plants generate NO through multiple enzymatic pathways whose activity depends on substrate availability and cellular conditions. Nitrate reductase represents a major enzymatic source of NO, catalyzing the reduction of nitrite to NO in the cytosol. This reaction is favored under low oxygen availability, acidic conditions, and elevated nitrite-to-nitrate ratios, which commonly arise during environmental or nutritional constraints (Chamizo-Ampudia et al., 2017; Rockel et al., 2002). In addition to nitrate reductase, NOS-like activities have been reported in plants, despite the absence of genes encoding canonical animal-type NO synthases. Such activities have been detected primarily in peroxisomes, where L-arginine oxidation yields NO and citrulline in reactions that appear to depend on Ca and NADPH availability (Corpas & Barroso, 2017). Mitochondria also contribute to NO formation, as components of the electron transport chain, particularly complexes III and IV, can reduce nitrite to NO under hypoxic or anoxic conditions (Gupta et al., 2011). The efficiency of these enzymatic routes is closely linked to mineral nutrient status, as nitrate and nitrite supply determine substrate availability, while Mo is required for the activity of nitrate reductase and related molybdoenzymes (Kolbert et al., 2019a,b).

Nonenzymatic pathways complement enzymatic NO production by enabling chemical nitrite reduction independently of protein catalysts. In the apoplast, nitrite can be converted to NO under acidic pH, a process enhanced by the presence of reducing compounds such as ascorbate or phenolic metabolites in the extracellular space (Bethke et al., 2004; Rümer et al., 2009). Similar redox- and pH-dependent reactions may occur within vacuoles, where nitrite protonation and subsequent chemical transformations yield NO alongside nitrate. These nonenzymatic routes become particularly relevant when enzymatic NO production is constrained or when nitrite accumulates as a result of imbalanced N nutrition, thereby contributing to localized increases in NO levels (Astier et al., 2018).

12.3.2 Nitric oxide as a signaling molecule in nutrient stress conditions

NO accumulation varies with the type, intensity, and duration of mineral nutrient stress, shaping acclimation responses through concentration-dependent signaling that reflects both nutrient identity and developmental context. Under nutrient limitation, NO levels frequently increase in root tissues, where localized signaling contributes to adjustments in nutrient acquisition and internal allocation. N deficiency has been associated with elevated root NO production, coinciding with shifts in N metabolism and uptake capacity, whereas P limitation induces transient NO elevations that accompany altered carbon partitioning and root system expansion (Buet et al., 2019; Khan et al., 2023a,b). In contrast, exposure to excessive concentrations of essential micronutrients such as Zn or Cu can stimulate NO accumulation beyond adaptive ranges, leading to a transition from regulatory signaling to interference with ion transport and cellular coordination (Simontacchi et al., 2015). These observations underscore the importance of NO concentration thresholds, as low-level NO signals support acclimation, whereas sustained or excessive accumulation may compromise physiological balance (Kolbert et al., 2019a,b).

Spatial regulation further defines NO signaling under mineral stress, with roots representing primary sites of NO generation during nutrient limitation. Rapid NO accumulation has been detected in root apices and elongation zones under Fe or Mn deficiency, coinciding with developmental reprogramming that enhances exploratory growth and nutrient uptake capacity (Buet et al., 2019). In contrast, NO signals within vascular tissues tend to persist over longer periods, reflecting roles in coordinating long-distance nutrient transport and redistribution. Shoot tissues generally display lower NO accumulation during mineral stress, consistent with a signaling strategy that prioritizes root-based adjustment while preserving photosynthetic competence and metabolic stability in aerial organs (Khan et al., 2023a,b). This spatial partitioning of NO signaling is modulated by developmental stage, with young seedlings exhibiting pronounced root-localized NO responses compared with more distributed signaling patterns in mature plants.

Temporal dynamics add an additional layer of specificity to NO-mediated signaling during mineral stress. Early stress perception is often associated with rapid and transient NO increases that precede morphological or metabolic adjustments, allowing prompt activation of nutrient-responsive gene expression programs (Simontacchi et al., 2015;Kolbert et al., 2019a,b). Under prolonged nutrient limitation or excess, NO signaling frequently adopts sustained or oscillatory patterns, integrating environmental persistence with internal metabolic status over extended time scales. Such temporal modulation enables plants to distinguish transient fluctuations in nutrient availability from chronic stress conditions, coordinating immediate responses with longer-term acclimatory strategies while maintaining flexibility within broader regulatory networks.

12.3.3 Interactions of nitric oxide with reactive oxygen species and redox homeostasis

Mineral nutrient imbalances frequently disturb cellular redox homeostasis by altering electron flow through chloroplasts and mitochondria, leading to increased production of ROS. Under these conditions, NO is often generated in parallel with ROS, establishing a coupled redox environment in which both molecules influence stress perception and acclimation. Rather than acting independently, NO and ROS operate within overlapping concentration ranges, where their relative abundance and persistence determine whether redox signaling supports adaptive adjustment or contributes to oxidative injury. Low and transient increases in NO generally coincide with controlled ROS accumulation and are associated with stabilization of cellular redox balance, whereas prolonged or excessive accumulation of either molecule may overwhelm buffering capacity and compromise membrane integrity and metabolic coordination (Farnese et al., 2016; Khator et al., 2024; Wani et al., 2021).

The regulatory role of NO within redox networks is closely linked to its capacity to modulate antioxidant potential without eliminating ROS altogether. At moderate levels, NO contributes to the maintenance of redox buffering systems, supporting antioxidant metabolite pools and preserving redox flexibility under mineral stress. This interaction allows ROS to retain its signaling function while limiting uncontrolled propagation. When NO accumulates beyond adaptive thresholds, however, its interaction with ROS can intensify oxidative pressure, shifting redox balance toward cellular dysfunction. Such dual behavior underscores the importance of spatial and temporal control of NO production in determining redox outcomes during mineral stress (Farnese et al., 2016; Khator et al., 2024).

A central mechanism through which NO integrates into redox regulation is *S*-nitrosation, a reversible posttranslational modification involving the covalent attachment of NO to cysteine thiol groups. *S*-nitrosation modulates protein activity, stability, and interaction potential in a redox-dependent manner, allowing NO to fine-tune metabolic enzymes, membrane transport systems, and signaling components without broadly suppressing ROS generation. The specificity of this modification is governed by local redox conditions, thiol accessibility, and the balance between NO and ROS, enabling selective regulation rather than indiscriminate inhibition. Through this mechanism, NO contributes to dynamic redox homeostasis that adapts to the identity, intensity, and duration of mineral stress (Lin et al., 2025; Wani et al., 2021)

Collectively, NO–ROS interactions do not function to eliminate oxidative signals but to adjust redox thresholds that determine cellular responsiveness. By integrating *S*-nitrosation with ROS-dependent signaling, NO supports a flexible redox framework that preserves metabolic function while allowing stress signals to remain informative. This coordination ensures that mineral stress responses remain reversible and proportionate, enabling plants to balance protection with plasticity across fluctuating nutritional environments.

12.4 Role of nitric oxide in mitigating macronutrient stress

12.4.1 Nitrogen stress

N availability exerts a strong influence on NO production, reflecting the close biochemical relationship between N metabolism and NO generation. Under N limitation, NO formation is closely associated with nitrite reduction mediated by nitrate reductase, creating a regulatory link between N uptake, assimilation, and signaling. Reduced nitrate availability often coincides with transient increases in root NO levels, which are associated with adjustments in nitrate uptake capacity and root system architecture that enhance soil exploration (Buet et al., 2019; Sun et al., 2015). Through this coupling, NO contributes to the coordination of N acquisition with downstream metabolic demand, supporting amino acid synthesis under limiting conditions. In contrast, excessive ammonium supply is generally accompanied by restrained NO accumulation, reflecting the need to prevent nitrite build-up and maintain intracellular pH stability and N assimilation capacity.

Beyond its role in uptake regulation, nitrate reductase occupies a central position in integrating N assimilation with NO signaling. In addition to catalyzing the reduction of nitrate to nitrite, nitrate reductase can generate NO under conditions favoring elevated nitrite availability or reduced oxygen supply. This dual functionality links N assimilation efficiency with NO-dependent regulatory processes that adjust carbon–N balance and metabolic allocation (Chamizo-Ampudia et al., 2017; Frungillo et al., 2014). The capacity of nitrate and nitrite to serve both as nutrients and NO precursors allows plants to modulate signaling outputs in direct proportion to N status, providing a responsive yet metabolically integrated control mechanism. Under prolonged N stress, NO contributes to N-use efficiency by influencing internal N redistribution rather than uptake alone. NO-associated signaling has been linked to the remobilization of Nous compounds from older tissues to actively growing sinks, supporting chloroplast function and delaying senescence during vegetative growth under deficiency (Costa-Broseta et al., 2020). Conversely, conditions of excess N are associated with altered NO dynamics that favor shoot growth over extensive root proliferation, contributing to balanced biomass allocation and prevention of excessive dilution of N-containing compounds. While transient NO accumulation supports adaptive adjustments in N assimilation and allocation, sustained or excessive NO signaling under chronic N stress can interfere with nitrate reductase activity and assimilation efficiency, ultimately constraining growth and productivity (Buet et al., 2019; Kolbert et al., 2019a, b).

12.4.2 Phosphorus stress

P deficiency is commonly associated with localized increases in NO accumulation in root tissues, reflecting the low mobility of phosphate in soils and the consequent reliance on morphological strategies for nutrient acquisition. Under low phosphate availability, elevated NO levels are detected in root apices and pericycle regions, where they are associated with reduced primary root elongation and enhanced formation of lateral roots and root hairs. These architectural adjustments expand the absorptive surface area and improve access to poorly available P pools in the rhizosphere (Galatro et al., 2020; Wang et al., 2010). Through these responses, NO contributes to the reallocation of carbon resources toward root development, supporting phosphate acquisition efficiency while maintaining overall growth balance. NO acts in close coordination with auxin signaling to regulate root system plasticity under P limitation. Altered NO accumulation influences auxin distribution and sensitivity within developing root tissues, thereby modulating lateral root initiation and emergence. This interaction appears to involve NO-dependent regulation of auxin-responsive components, allowing fine control of root developmental programs in response to phosphate scarcity (Nawaz et al., 2017;Das et al., 2025). In parallel, ethylene levels often increase under P deficiency and interact with NO to reinforce root hair formation and cell wall modifications that enhance P mobilization. Rather than acting independently, NO, auxin, and ethylene form an integrated signaling network that adjusts root architecture according to stress intensity and duration, enabling coordinated developmental adaptation under P-limited conditions.

12.4.3 Potassium, calcium, and magnesium stress

K, Ca, and Mg imbalances disrupt membrane potentials and ionic gradients, imposing constraints on cellular stability and metabolic coordination. Under such conditions, NO production is frequently induced as part of adaptive responses that restore ion homeostasis and preserve membrane function. In K-limited plants, altered NO accumulation is associated with improved intracellular K^+ retention and stabilization of osmotic balance, supporting enzyme activity and turgor maintenance under fluctuating external availability (Buet et al., 2019). Evidence from ion-stress studies further indicates that NO, particularly in coordination with N-derived metabolites such as arginine, contributes to the regulation of cellular ion balance and membrane-associated transport processes under stress conditions, reinforcing ionic stability and physiological resilience (Riaz et al., 2025). Ca deficiency similarly elicits NO-dependent adjustments in intracellular Ca^{2+} distribution, contributing to the maintenance of Ca-dependent signaling processes essential for membrane integrity, vesicle trafficking, and cell wall organization. In the case of Mg deficiency, NO signaling has been linked to regulatory responses that prevent disturbances in ribosome stability and ATP-dependent reactions that rely on adequate Mg availability. NO-mediated modulation of root surface properties and ion distribution supports Mg^{2+} acquisition and alleviates membrane strain associated with deficiency conditions (Liu et al., 2017; Shah et al., 2023). Beyond ion balance, NO contributes to membrane stability under K, Ca, and Mg stress by limiting the secondary effects of ionic disequilibrium. Stabilization of membrane-associated proton gradients and improved compartmentalization of ions reduce the risk of depolarization-induced dysfunction and support recovery of growth once nutrient availability is restored (Buet et al., 2019; Riaz et al., 2025).

12.5 Role of nitric oxide in mitigating micronutrient stress

12.5.1 Iron deficiency and homeostasis

Fe is a redox-active micronutrient required for electron transfer reactions in photosynthesis and respiration, yet its bioavailability in aerobic soils is inherently limited by the predominance of sparingly soluble Fe^{3+} forms.

Under Fe deficiency, nongraminaceous plants activate Strategy I acquisition responses in roots, characterized by rhizosphere acidification, enzymatic reduction of Fe^{3+} to Fe^{2+}, and subsequent uptake through ferrous Fe transport systems. NO functions as an important regulatory signal within this framework, linking cellular redox status with root-based Fe acquisition. Rapid NO accumulation in root epidermal tissues has been consistently observed under Fe limitation, coinciding with activation of physiological responses that enhance Fe mobilization and uptake capacity (Graziano & Lamattina, 2007;Ramírez et al., 2008). The contribution of NO to Fe deficiency responses includes modulation of processes that facilitate Fe solubilization and reduction at the root surface. Enhanced rhizosphere acidification under Fe limitation has been associated with NO-dependent stimulation of plasma membrane H^+-ATPase activity, promoting conditions favorable for Fe^{3+} mobilization (Graziano & Lamattina, 2005). In parallel, NO accumulation is closely linked with increased ferric chelate reductase activity, which constitutes a critical step in Strategy I Fe acquisition. Elevated NO levels under Fe deficiency correlate with enhanced FCR-mediated reduction of Fe^{3+}, thereby expanding the pool of bioavailable Fe^{2+} for root uptake (Chen et al., 2010; Rahman et al., 2021). Through this coordinated regulation, NO supports reductive capacity at the root–soil interface under redox imbalance induced by Fe scarcity.

Beyond its influence on biochemical acquisition mechanisms, NO contributes to the integration of Fe uptake with developmental and morphological adjustments. Fe-deficient plants exhibiting increased NO accumulation frequently display enhanced root hair formation, expanding the effective absorptive surface area and improving contact with Fe-containing soil fractions (Ramírez et al., 2008). By synchronizing redox signaling with both enzymatic and structural components of Strategy I responses, NO reinforces Fe acquisition efficiency and mitigates the metabolic consequences of Fe limitation, thereby supporting Fe homeostasis under conditions of restricted availability.

12.5.2 Zinc, copper, and manganese stress

Zn, Cu, and Mn function as essential cofactors in numerous enzymes involved in redox regulation, including superoxide dismutases and components of electron transfer pathways. Deficiencies in these micronutrients compromise enzymatic efficiency and disrupt redox-dependent metabolic processes, whereas excess accumulation perturbs metal homeostasis by displacing other ions from binding sites and enhancing oxidative pressure through uncontrolled redox reactions. Zn deficiency has been shown to induce elevated NO levels in moderately to highly sensitive species such as *Arabidopsis thaliana*, *Brassica napus*, and *Zea mays* (Kondak et al., 2022; Molnár et al., 2022; Oláh et al., 2023). NO contributes to the regulation of cellular responses to Zn, Cu, and Mn imbalance by modulating micronutrient distribution and compartmentalization, thereby limiting cytosolic toxicity while maintaining availability for essential biochemical functions (Rather et al., 2020; Xu et al., 2010). Through this regulatory role, NO supports balanced micronutrient handling under both deficient and excessive supply conditions.

Under micronutrient limitation, elevated NO levels are associated with enhanced root acquisition and internal redistribution of Zn, Cu, and Mn, facilitating the maintenance of redox-active processes required for growth and stress adaptation. Conversely, when micronutrient availability exceeds physiological demand, NO signaling contributes to detoxification strategies that restrict free metal ion accumulation, including enhanced sequestration into intracellular compartments and association with metal-binding pools that reduce reactivity (Mahjoubi et al., 2022; Rather et al., 2020). These adjustments mitigate the adverse effects of excess micronutrients without compromising their essential roles in cellular metabolism. NO further supports tolerance to Zn-, Cu-, and Mn-induced stress by stabilizing cellular redox balance under conditions of micronutrient imbalance. Elevated metal availability frequently intensifies oxidative pressure due to disrupted electron transfer and redox cycling, and NO-associated responses contribute to maintaining antioxidant capacity and preserving redox buffering systems (Srivastava & Dubey, 2012). By integrating micronutrient status with redox regulation, NO enables plants to accommodate fluctuations in Zn, Cu, and Mn availability while preventing oxidative damage and sustaining metabolic homeostasis.

12.5.3 Boron and other essential micronutrients

B plays a critical structural role in plants through its involvement in cross-linking rhamnogalacturonan II within the cell wall matrix and stabilizing membrane-associated glycoproteins, processes that underpin root elongation, vascular integrity, and reproductive development. B deficiency disrupts these structural functions, resulting in impaired meristem activity, reduced pollen viability, and compromised tissue elasticity, whereas excess B induces membrane perturbations, localized necrosis, and oxidative stress. Emerging evidence suggests that NO participates in plant responses to B imbalance, primarily through modulation of redox-related processes rather than direct regulation of B transport. Studies reporting exogenous NO application indicate attenuation of B-induced oxidative damage and partial normalization of physiological performance under both deficient and toxic conditions, supporting a role for NO in buffering redox disturbances associated with altered B availability (Farag et al., 2017; Kolbert & Ördög, 2021). Nonetheless, mechanistic understanding remains limited, and current data do not support a direct role for NO in primary B uptake or allocation.

For other essential micronutrients such as Mo and Ni, evidence for NO involvement in mineral stress responses is sparse and often indirect. Mo deficiency constrains nitrate assimilation through reduced activity of molybdoenzymes, and NO has been implicated in redox-related adjustments associated with Mo-dependent metabolic processes, particularly under combined nutrient and environmental stress conditions (Wu et al., 2017). Similarly, Ni deficiency impairs urease function and N metabolism, and NO-related responses observed under Ni-associated stress appear to reflect general mitigation of oxidative pressure rather than nutrient-specific signaling mechanisms (Siddiqui et al., 2013). In the case of Cl, which contributes to osmotic regulation and photosynthetic function, clear evidence for NO-mediated regulation under deficiency or excess remains lacking. Evidence from NiO nanoparticle studies indicates that Ni-associated NO responses may, under specific physicochemical contexts, involve regulated nitrosative signaling rather than solely reflecting generalized oxidative stress, with protein nitration and GSNOR-related modulation correlating with growth-related physiological responses (Kondak, Janovszky, et al., 2024; Kondak, Kondak, et al., 2024). Collectively, these observations underscore substantial knowledge gaps in our understanding of NO involvement in less-characterized micronutrient stresses and position B, Mo, Ni, and Cl as promising targets for future research aimed at elucidating whether NO functions as a broader integrator of structural, metabolic, and redox-derived signals under diverse mineral imbalance scenarios.

12.6 Nitric oxide crosstalk with hormonal and metabolic pathways during mineral stress

NO functions as a central signaling integrator during mineral stress, coordinating hormonal cues with redox and metabolic adjustments to support adaptive plant responses. Unlike classical phytohormones, NO lacks dedicated receptors and operates through local, transient signaling events that are tightly coupled to cellular redox status, frequently mediated by reversible posttranslational modifications. Through these properties, NO modulates hormone responsiveness rather than acting as an independent growth regulator. Interactions between NO and auxin are particularly relevant under nutrient limitation, where changes in NO accumulation are associated with altered auxin sensitivity and redistribution, contributing to lateral root formation and root hair development that enhance soil exploration (Freschi, 2013). Crosstalk with ethylene further shapes stress-adaptive growth patterns, as ethylene-induced NO production has been linked to root hair responses and localized growth restraint under adverse nutritional conditions (Khan et al., 2023a,b).

Interactions between NO and other hormonal pathways refine resource allocation during mineral stress. NO–abscisic acid interplay contributes to balancing stress signaling with continued growth, allowing plants to restrict excessive inhibition while maintaining physiological activity necessary for nutrient acquisition. Cytokinins provide an additional layer of modulation, often acting antagonistically to NO-associated signals to preserve meristem function and coordinate growth–stress trade-offs. These interactions are highly context-dependent and vary with nutrient identity, stress duration, and developmental stage, underscoring the role of NO as a modulator rather than a dominant regulator of hormonal signaling (Nawaz et al., 2017).

Hormonal crosstalk mediated by NO converges on the regulation of root system architecture, enabling developmental plasticity in response to heterogeneous mineral availability. Accumulation of NO in root apical and differentiation zones under nutrient limitation integrates auxin and ethylene signaling to promote branching and root hair proliferation, optimizing absorptive capacity while restricting unnecessary biomass investment (Simontacchi et al., 2015). These structural adjustments enhance nutrient-use efficiency by aligning root growth with spatial nutrient gradients and internal metabolic capacity. Beyond morphological responses, NO participates in broader metabolic reprogramming during mineral stress by coordinating hormonal inputs with cellular energy status. Adjustments in carbon and N utilization associated with NO signaling support growth restraint under acute stress and facilitate recovery following nutrient resupply. By integrating mineral availability with hormonal and metabolic cues, NO contributes to maintaining physiological flexibility and optimizing resource use under fluctuating nutrient conditions (Sun et al., 2021). An integrative overview of NO-mediated coordination between mineral stress, hormonal signaling, and nutrient-use efficiency is presented in Fig. 12.1.

12.7 Distinction between mineral nutrient stress and heavy metal stress

Essential mineral nutrients, including micronutrients such as Fe, Zn, Cu, and Mn, sustain plant metabolism within defined concentration ranges but elicit stress responses when accumulated in excess. At elevated levels, these essential elements share physicochemical characteristics with heavy metals, including redox activity and competition for binding sites in proteins and transport systems, resulting in dose-dependent toxicity. Such overlap gives rise to common physiological outcomes, including ionic imbalance, oxidative pressure, and disruption of photosynthetic and respiratory processes. However, the functional context differs fundamentally: mineral nutrient stress reflects a breakdown of homeostatic regulation, whereas heavy metal stress arises from exposure to elements that lack physiological utility and exert toxicity irrespective of metabolic demand (Jarin et al., 2025).

Despite partial convergence in early stress perception and signaling, key mechanistic distinctions separate mineral nutrient stress from heavy metal stress. Responses to excess essential minerals primarily involve adaptive recalibration of uptake, redistribution, and compartmentalization to restore optimal intracellular concentrations. In contrast, heavy metal stress is

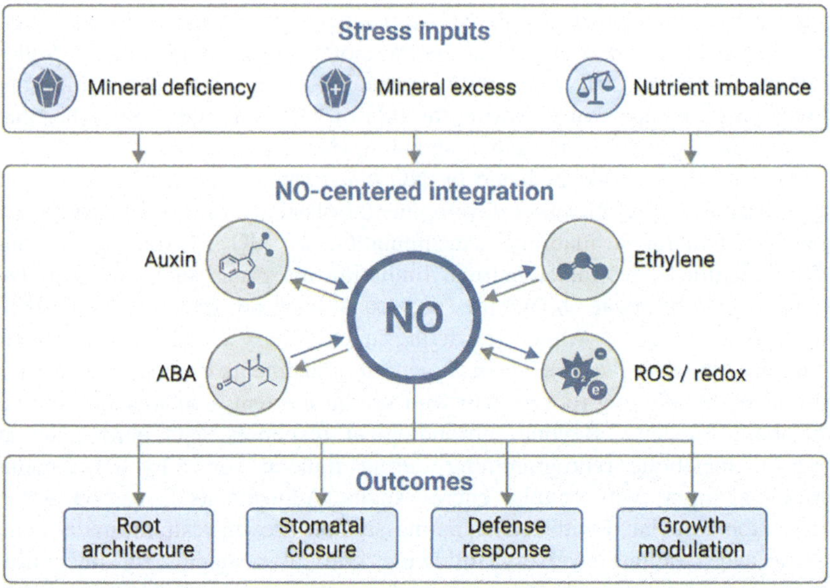

FIGURE. 12.1 Integrative role of NO in coordinating mineral nutrient stress responses from signaling to outcomes. Mineral nutrient–related stress inputs (mineral deficiency, mineral excess, and nutrient imbalance) are shown as upstream drivers feeding into an NO-centered integration module. Within this integration layer, NO is depicted as a central signaling node interconnected with key regulatory components, including ABA, auxin, ethylene, and ROS/redox processes. Bidirectional arrows indicate reciprocal interactions and potential feedback among NO and these pathways rather than a strictly linear cascade. Downstream arrows connect the NO-centered module to representative physiological and developmental outcomes, including root architecture, stomatal closure, defense response, and growth modulation, summarizing how integrated signaling may translate into stress-response phenotypes (*NO*, Nitric oxide; *ROS*, reactive oxygen species).

dominated by exclusion and detoxification strategies, including restricted root uptake, enhanced sequestration, and ligand-mediated immobilization to minimize cellular interference. Nonessential heavy metals such as Cd, Pb, and mercury (Hg) fall outside the scope of this chapter and are not discussed further here. NO signaling exhibits shared features across both stress types, including redox modulation and posttranslational regulation of stress-responsive proteins; however, in the context of mineral nutrient stress, NO functions predominantly to support adaptive homeostasis rather than irreversible detoxification pathways (Kolbert et al., 2019a,b).

12.8 Agricultural and biotechnological implications

Exogenous application of NO donors, including commonly used compounds such as sodium nitroprusside and S-nitrosothiols, has been widely employed to manipulate NO availability in plants and examine its influence on nutrient acquisition under mineral stress. Across controlled studies, NO supplementation is frequently associated with improved nutrient uptake efficiency and enhanced physiological performance under limited nutrient supply, although outcomes are strongly dependent on dose, timing, and growth conditions. Low concentrations typically support adaptive responses linked to mineral stress tolerance, whereas excessive NO levels may exert inhibitory effects due to disruption of cellular redox balance. Experimental evidence further indicates that pretreatment with NO donors often produces more consistent benefits than application during advanced stress, underscoring the importance of temporal control when targeting nutrient-use efficiency (Sun et al., 2021).

From an agronomic perspective, NO-modulating strategies hold potential as complementary tools to improve nutrient-use efficiency and support sustainable fertilization practices. By facilitating internal nutrient redistribution and enhancing plant capacity to cope with suboptimal mineral availability, NO signaling may contribute to maintaining productivity under reduced fertilizer inputs. Such effects are particularly relevant in efforts to limit nutrient losses associated with leaching or volatilization, thereby mitigating environmental impacts while sustaining crop performance. However, these benefits should be viewed as context-dependent and integrated with existing nutrient management approaches rather than as stand-alone interventions (Simontacchi et al., 2015).

Differences between controlled and field environments further constrain the direct translation of NO-based strategies into agricultural practice. Controlled systems, including hydroponic setups and vertical farming, allow precise regulation of NO delivery and nutrient availability, often resulting in reproducible improvements in nutrient efficiency. In contrast, field conditions introduce substantial variability arising from soil heterogeneity, microbial activity, and fluctuating climatic factors, which can alter NO persistence and effectiveness. These disparities highlight the need for environment-specific optimization and cautious scaling when considering NO-modulating approaches for field applications (Khan et al., 2023a,b).

12.9 Conclusions and future perspectives

NO functions as a central regulatory signal in plant responses to mineral nutrient stress, coordinating nutrient acquisition with cellular homeostasis and redox balance. Through modulation of transport activity, enzymatic function, and local redox conditions, NO supports the maintenance of essential mineral availability while limiting imbalances that impair metabolic processes. Its prominent activity in roots links nutrient sensing to developmental plasticity, including adjustments in root architecture that enhance resource acquisition under heterogeneous nutrient distributions. This regulatory role underscores a key distinction between mineral nutrient stress, which involves adaptive recalibration of physiological demand, and heavy metal stress, which is dominated by exclusion and detoxification of nonessential elements without homeostatic benefit.

Despite substantial progress, several aspects of NO-mediated mineral stress signaling remain unresolved. NO responses display strong context dependence, with concentration, timing, and tissue specificity shaping outcomes in ways that are not yet fully defined. The absence of dedicated NO receptors complicates the understanding of signal perception and specificity, as NO relies on transient and reversible modifications within highly interconnected signaling networks. Moreover, knowledge remains uneven across mineral elements, with comparatively limited mechanistic insight into NO involvement in stresses associated with B, Mo, and Ni. Translational challenges persist as well, as environmental variability, soil complexity, and microbial interactions modify NO dynamics and constrain direct extrapolation from controlled experimental systems to field conditions.

Future progress will depend on integrative approaches that combine physiological, metabolic, and environmental perspectives to resolve the spatial and temporal dimensions of NO signaling under mineral stress. Greater emphasis on in situ analyses and cross-scale validation will be essential to define the boundaries of NO-mediated regulation in agricultural contexts. While NO-modulating strategies show promise as supportive tools for improving nutrient-use efficiency, their deployment should remain complementary to established nutrient management practices, guided by a nuanced understanding of NO's regulatory capacity rather than expectations of universal applicability.

AI disclosure

During the preparation of this work, the author(s) used ChatGPT in order to translate to help study and figure organization. After using this tool/service, the author(s) reviewed and edited the content as needed and take full responsibility for the content of the publication.

References

Astier, J., Gross, I., & Durner, J. (2018). Nitric oxide production in plants: An update. *Journal of Experimental Botany, 69*(14), 3401–3411. https://doi.org/10.1093/jxb/erx420.

Beligni, M. V., & Lamattina, L. (2001). Nitric oxide: A non-traditional regulator of plant growth. *Trends in Plant Science, 6*(11), 508–509. https://doi.org/10.1016/s1360-1385(01)02156-2.

Bethke, P. C., Badger, M. R., & Jones, R. L. (2004). Apoplastic synthesis of nitric oxide by plant tissues. *The Plant Cell, 16*(2), 332–341. https://doi.org/10.1105/tpc.017822, http://www.plantcell.org/.

Buet, A., Galatro, A., Ramos-Artuso, F., Simontacchi, M., & Brouquisse, R. (2019). Nitric oxide and plant mineral nutrition: Current knowledge. *Journal of Experimental Botany, 70*(17), 4461–4476. https://doi.org/10.1093/jxb/erz129.

Chamizo-Ampudia, A., Sanz-Luque, E., Llamas, A., Galvan, A., & Fernandez, E. (2017). Nitrate reductase regulates plant nitric oxide homeostasis. *Trends in Plant Science, 22*(2), 163–174. https://doi.org/10.1016/j.tplants.2016.12.001, www.elsevier.com/inca/publications/store/3/0/9/6/0/index.htt.

Chen, W. W., Yang, J. L., Qin, C., Jin, C. W., Mo, J. H., Ye, T., & Zheng, S. J. (2010). Nitric oxide acts downstream of auxin to trigger root ferric-chelate reductase activity in response to iron deficiency in *Arabidopsis*. *Plant Physiology, 154*(2), 810–819. https://doi.org/10.1104/pp.110.161109, http://www.plantphysiol.org/cgi/reprint/154/2/810.pdf.

Chrysargyris, A., Höfte, M., Tzortzakis, N., Petropoulos, S. A., & Di Gioia, F. (2022). Editorial: Micronutrients: The borderline between their beneficial role and toxicity in plants. *Frontiers in Plant Science, 13*. https://doi.org/10.3389/fpls.2022.840624, https://www.frontiersin.org/journals/plant-science.

Corpas, F. J., & Barroso, J. B. (2017). Nitric oxide synthase-like activity in higher plants. *Nitric Oxide - Biology and Chemistry, 68*, 5–6. https://doi.org/10.1016/j.niox.2016.10.009, http://www.elsevier.com/inca/publications/store/6/2/2/9/2/6/index.htt.

Costa-Broseta, Á., Castillo, M. C., & León, J. (2020). Nitrite reductase 1 is a target of nitric oxide-mediated post-translational modifications and controls nitrogen flux and growth in *Arabidopsis*. *International Journal of Molecular Sciences, 21*(19), 7270. https://doi.org/10.3390/ijms21197270.

Das, A., Pal, S., Sarkar, A. K., & Adak, M. K. (2025). Mechanistic insights into nitric oxide signaling in shaping root architecture under challenging environments. *Plant Science, 334*, 112903. https://doi.org/10.1016/j.plantsci.2025.112903.

Farag, M., Najeeb, U., Yang, J., Hu, Z., & Fang, Z. M. (2017). Nitric oxide protects carbon assimilation process of watermelon from boron-induced oxidative injury. *Plant Physiology and Biochemistry, 111*, 166–173. https://doi.org/10.1016/j.plaphy.2016.11.024, http://www.journals.elsevier.com/plant-physiology-and-biochemistry/.

Farnese, F. S., Menezes-Silva, P. E., Gusman, G. S., & Oliveira, J. A. (2016). When bad guys become good ones: The key role of reactive oxygen species and nitric oxide in the plant responses to abiotic stress. *Frontiers Media S.A., Brazil Frontiers in Plant Science, 7*, 2016. https://doi.org/10.3389/fpls.2016.00471 Available from: http://journal.frontiersin.org/article/10.3389/fpls.2016.00471/pdf.

Freschi, L. (2013). Nitric oxide and phytohormone interactions: Current status and perspectives. *Frontiers in Plant Science, 4*. https://doi.org/10.3389/fpls.2013.00398.

Frungillo, L., Skelly, M. J., Loake, G. J., Spoel, S. H., & Salgado, I. (2014). S-nitrosothiols regulate nitric oxide production and storage in plants through the nitrogen assimilation pathway. *Nature Communications, 5*. https://doi.org/10.1038/ncomms6401, http://www.nature.com/ncomms/index.html.

Galatro, A., Ramos-Artuso, F., Luquet, M., Buet, A., & Simontacchi, M. (2020). An update on nitric oxide production and role under phosphorus scarcity in plants. *Frontiers in Plant Science, 11*. https://doi.org/10.3389/fpls.2020.00413.

Graziano, M., & Lamattina, L. (2007). Nitric oxide accumulation is required for molecular and physiological responses to iron deficiency in tomato roots. *The Plant Journal, 52*(5), 949–960. https://doi.org/10.1111/j.1365-313x.2007.03283.x.

Graziano, M., & Lamattina, L. (2005). Nitric oxide and iron in plants: An emerging and converging story. *Trends in Plant Science, 10*(1), 4–8. https://doi.org/10.1016/j.tplants.2004.12.004.

Gupta, K. J., Fernie, A. R., Kaiser, W. M., & van Dongen, J. T. (2011). On the origins of nitric oxide. *Trends in Plant Science, 16*(3), 160–168. https://doi.org/10.1016/j.tplants.2010.11.007.

Hänsch, R., & Mendel, R. R. (2009). Physiological functions of mineral micronutrients (Cu, Zn, Mn, Fe, Ni, Mo, B, Cl). *Current Opinion in Plant Biology, 12*(3), 259–266. https://doi.org/10.1016/j.pbi.2009.05.006.

Jarin, A. S., Khan, M. A. R., Apon, T. A., Islam, M. A., Rahat, A., Akter, M., Anik, T. R., Nguyen, H. M., Nguyen, T. T., Ha, C. V., & Tran, L. S. P. (2025). Plant responses to heavy metal stresses: Mechanisms, defense strategies, and nanoparticle-assisted remediation. *Plants, 14*(24). https://doi.org/10.3390/plants14243834, www.mdpi.com/journal/plants.

Khan, M., Al Azzawi, T. N. I., Ali, S., Yun, B.-W., & Mun, B.-G. (2023a). Nitric oxide, a key modulator in the alleviation of environmental stress-mediated damage in crop plants: A meta-analysis. *Plants, 12*(11), 2121. https://doi.org/10.3390/plants12112121.

Khan, M., Ali, S., Al Azzawi, T. N. I., & Yun, B.-W. (2023b). Nitric oxide acts as a key signaling molecule in plant development under stressful conditions. *International Journal of Molecular Sciences, 24*(5), 4782. https://doi.org/10.3390/ijms24054782.

Khator, K., Parihar, S., Jasik, J., & Shekhawat, G. S. (2024). Nitric oxide in plants: An insight on redox activity and responses toward abiotic stress signaling. *Plant Signaling & Behavior, 19*(1). https://doi.org/10.1080/15592324.2023.2298053.

Kolbert, Z., Barroso, J. B., Brouquisse, R., Corpas, F. J., Gupta, K. J., Lindermayr, C., Loake, G. J., Palma, J. M., Petřivalský, M., Wendehenne, D., & Hancock, J. T. (2019a). A forty year journey: The generation and roles of NO in plants. *Nitric Oxide: Biology and Chemistry / Official Journal of the Nitric Oxide Society, 93*, 53–70. https://doi.org/10.1016/j.niox.2019.09.006.

Kolbert, Z., Feigl, G., Freschi, L., & Poór, P. (2019b). A forty year journey: The generation and roles of NO in plants. *Nitric Oxide: Biology and Chemistry / Official Journal of the Nitric Oxide Society, 93*, 53–70. https://doi.org/10.1016/j.niox.2019.09.006.

Kolbert, Z., & Ördög, A. (2021). Involvement of nitric oxide (NO) in plant responses to metalloids. *Journal of Hazardous Materials, 420*, 126606. https://doi.org/10.1016/j.jhazmat.2021.126606.

Kondak, S., Kondak, D., Kabadayi, O., Erdei, L., Rónavári, A., Kónya, Z., & Kolbert, Z. (2024a). Current insights into the green synthesis, in planta characterization and phytoeffects of nickel nanoparticles and their agricultural implications. *Environmental Research, 260*, 119665. https://doi.org/10.1016/j.envres.2024.119665.

Kondak, S., Janovszky, P., Szőllősi, R., Molnár, Á., Oláh, D., Adedokun, O. P., Dimitrakopoulos, P. G., Rónavári, A., Kónya, Z., Erdei, L., Galbács, G., & Kolbert, Z. (2024b). Nickel oxide nanoparticles induce cell wall modifications, root anatomical changes, and nitrosative signaling in ecotypes of Ni hyperaccumulator Odontarrhena lesbiaca. *Hungary Environmental Pollution, 341*. https://doi.org/10.1016/j.envpol.2023.122874, https://www.sciencedirect.com/science/journal/02697491.

KONDAK, S., MOLNÁR, Á., OLÁH, D., & KOLBERT, Z. (2022). The role of nitric oxide (NO) in plant responses to disturbed zinc homeostasis. *Plant Stress, 4*, 100068. https://doi.org/10.1016/j.stress.2022.100068.

Kudoyarova, G. R., Dodd, I. C., Veselov, D. S., Rothwell, S. A., & Yu. Veselov, S. (2015). Common and specific responses to availability of mineral nutrients and water. *Federation Journal of Experimental Botany, 66*(8), 2133–2144. https://doi.org/10.1093/jxb/erv017, http://jxb.oxfordjournals.org/.

Kumar, S., Kumar, S., & Mohapatra, T. (2021). Interaction between macro- and micro-nutrients in plants. *Frontiers in Plant Science, 12*. https://doi.org/10.3389/fpls.2021.665583.

Lin, W., Shang, J. X., Li, X. Y., Zhou, X. F., & Zhao, L. Q. (2025). Nitric Oxide regulates multiple signal pathways in plants via protein S-nitrosylation. *Current Issues in Molecular Biology, 47*(6). https://doi.org/10.3390/cimb47060407, https://www.mdpi.com/journal/cimb.

Liu, M., Liu, X. X., He, X. L., Liu, L. J., Wu, H., Tang, C. X., Zhang, Y. S., & Jin, C. W. (2017). Ethylene and nitric oxide interact to regulate the magnesium deficiency-induced root hair development in *Arabidopsis*. *New Phytologist, 213*(3), 1242–1256. https://doi.org/10.1111/nph.14259, http://onlinelibrary.wiley.com/journal/10.1111/(ISSN)1469-8137.

Maathuis, F. J. (2009). Physiological functions of mineral macronutrients. *Current Opinion in Plant Biology, 12*(3), 250–258. https://doi.org/10.1016/j.pbi.2009.04.003.

Mahjoubi, Y., Rzigui, T., Kharbech, O., Mohamed, S. N., Abaza, L., Chaoui, A., Nouairi, I., & Djebali, W. (2022). Exogenous nitric oxide alleviates manganese toxicity in bean plants by modulating photosynthesis in relation to leaf lipid composition. *Protoplasma, 259*(4), 949–964. https://doi.org/10.1007/s00709-021-01713-2, www.springer.co.at/springer.py?Page=40&Key=456&cat=9&id_journal=4.

Mir, W. A., Mir, S. A., Townsend, T., Sofi, P. A., Dar, M. H., Bhat, M. A., Bhat, R. A., Dar, S. A., & Dar, S. A. (2023). Mineral nutrients in plants under changing environments: A road to sustainable crop production. *The Plant Genome, 16*(4), e20362. https://doi.org/10.1002/tpg2.2036.

Molnár, A., Kondak, S., Benkő, P., Janovszky, P., Kovács, K., Szőllősi, R., Gondor, O. K., Oláh, D., Gémes, K., Galbács, G., Janda, T., & Kolbert, Z. (2022). Limited Zn supply affects nutrient distribution, carbon metabolism and causes nitro-oxidative stress in sensitive *Brassica napus*. *Environmental and Experimental Botany, 202*, 105032. https://doi.org/10.1016/j.envexpbot.2022.105032.

Nawaz, F., Shabbir, R. N., Shahbaz, M., Majeed, S., Raheel, M., Hassan, W., & Sohail, M. A. (2017). Crosstalk between nitric oxide and phytohormones regulate plant development during abiotic stresses. In In. M. A. El-Esawi (Ed.). *Phytohormones – Signaling mechanisms and crosstalk in plant development and stress responses*IntechOpenhttps://doi.org/10.5772/intechopen.69812.

Oláh, D., Kondak, S., Molnár, Á., Adedokun, O. P., Czékus, Z., Gémes, K., Galbács, G., & Kolbert, Z. (2023). Suboptimal zinc supply affects the S-nitrosoglutathione reductase enzyme and nitric oxide signaling in Arabidopsis. *Plant Stress, 10*, 100250. https://doi.org/10.1016/j.stress.2023.100250.

Pandey, R., Vengavasi, K., & Hawkesford, M. J. (2021). Plant adaptation to nutrient stress. *Springer, India Plant Physiology Reports, 26*(4), 583–586. https://doi.org/10.1007/s40502-021-00636-7, https://link.springer.com/journal/volumesAndIssues/40502.

Rahman, M. A., Kabir, A. H., Song, Y., Lee, S. H., Hasanuzzaman, M., & Lee, K. W. (2021). Nitric oxide prevents fe deficiency-induced photosynthetic disturbance, and oxidative stress in alfalfa by regulating fe acquisition and antioxidant defense. *Antioxidants, 10*(10). https://doi.org/10.3390/antiox10101556, https://www.mdpi.com/2076-3921/10/10/1556/pdf.

Ramírez, L., Graziano, M., & Lamattina, L. (2008). Decoding plant responses to iron deficiency: Is nitric oxide a central player? *Plant Signaling & Behavior, 3*(10), 795–797. https://doi.org/10.4161/psb.3.10.5874.

Rather, B. A., Masood, A., Sehar, Z., Majid, A., Anjum, N. A., & Khan, N. A. (2020). Mechanisms and role of nitric oxide in phytotoxicity-mitigation of copper. *Frontiers in Plant Science, 11*. https://doi.org/10.3389/fpls.2020.00675, https://www.frontiersin.org/journals/plant-science.

Riaz, M., Noreen, Z., Usman, S., Shah, A. A., Zafar, F., Ashraf, S., Shaffique, S., Mahmoud, E. A., Elhindi, K. M., & Elansary, H. O. (2025). Nitric oxide and arginine mitigate salt stress through physio-biochemical modulations and ions regulation in *Brassica napus* L. *Scientific Reports, 15*(1). https://doi.org/10.1038/s41598-025-17887-1, https://www.nature.com/srep/.

Rockel, P., Strube, F., Rockel, A., Wildt, J., & Kaiser, W. M. (2002). Regulation of nitric oxide (NO) production by plant nitrate reductase in vivo and in vitro. *Journal of Experimental Botany, 53*(366), 103–110. https://doi.org/10.1093/jexbot/53.366.103.

Rümer, S., Gupta, K. J., & Kaiser, W. M. (2009). Plant cells oxidize hydroxylamines to NO. *Journal of Experimental Botany, 60*(7), 2065–2072. https://doi.org/10.1093/jxb/erp077, http://jxb.oxfordjournals.org/.

Shah, S., Chen, C., Sun, Y., Wang, D., Nawaz, T., El-Kahtany, K., & Fahad, S. (2023). Mechanisms of nitric oxide involvement in plant-microbe interaction and its enhancement of stress resistance. *Plant Stress, 10*. https://doi.org/10.1016/j.stress.2023.100191, www.journals.elsevier.com/plant-stress.

Siddiqui, M. H., Al-Whaibi, M. H., Ali, H. M., Sakran, A. M., Basalah, M. O., & AlKhaishany, M. Y. Y. (2013). Mitigation of nickel stress by the exogenous application of salicylic acid and nitric oxide in wheat. *Australian Journal of Crop Science, 7*(11), 1780–1788. http://www.cropj.com/siddique_7_11_2013_1780_1788.pdf.

Simontacchi, M., Galatro, A., Ramos-Artuso, F., & Santa-María, G. E. (2015). Plant survival in a changing environment: The role of nitric oxide in plant responses to abiotic stress. *Frontiers Research Foundation, Argentina Frontiers in Plant Science, 6*(NOVEMBER). https://doi.org/10.3389/fpls.2015.00977, http://journal.frontiersin.org/article/10.3389/fpls.2015.00977/pdf.

Singhal, R. K., Jatav, H. S., Aftab, T., Pandey, S., Mishra, U. N., Chauhan, J., Chand, S., Saha, I. D., Dadarwal, B. K., Chandra, K., Khan, M. A., Rajput, V. D., Minkina, T., Narayana, E. S., Sharma, M. K., & Ahmed, S. (2021). Roles of nitric oxide in conferring multiple abiotic stress tolerance in plants and crosstalk with other plant growth regulators. *Journal of Plant Growth Regulation, 40*(6), 2303–2328. https://doi.org/10.1007/s00344-021-10446-8, https://www.springer.com/journal/344.

Srivastava, S., & Dubey, R. S. (2012). Nitric oxide alleviates manganese toxicity by preventing oxidative stress in excised rice leaves. *Acta Physiologiae Plantarum, 34*(2), 819–825. https://doi.org/10.1007/s11738-011-0863-0.

Sun, H., Li, J., Song, W., Tao, J., Huang, S., Chen, S., Hou, M., Xu, G., & Zhang, Y. (2015). Nitric oxide generated by nitrate reductase increases nitrogen uptake capacity by inducing lateral root formation and inorganic nitrogen uptake under partial nitrate nutrition in rice. *Journal of Experimental Botany, 66*(9), 2449–2459. https://doi.org/10.1093/jxb/erv030.

Sun, C., Zhang, Y., Liu, L., Liu, X., Li, B., Jin, C., & Lin, X. (2021). Molecular functions of nitric oxide and its potential applications in horticultural crops. *Horticulture Research, 8*(1). https://doi.org/10.1038/s41438-021-00500-7.

Terrón-Camero, L. C., Peláez-Vico, M.Á., Del-Val, C., Sandalio, L. M., & Romero-Puertas, M. C. (2019). Role of nitric oxide in plant responses to heavy metal stress: Exogenous application versus endogenous production. *Journal of Experimental Botany, 70*(17), 4477–4488. https://doi.org/10.1093/jxb/erz184, http://jxb.oxfordjournals.org/.

Tewari, R. K., Horemans, N., & Watanabe, M. (2021). Evidence for a role of nitric oxide in iron homeostasis in plants. *Journal of Experimental Botany, 72*(4), 990–1006. https://doi.org/10.1093/jxb/eraa484.

Wang, B. L., Tang, X. Y., Cheng, L. Y., Zhang, A. Z., Zhang, W. H., Zhang, F. S., Liu, J. Q., Cao, Y., Allan, D. L., Vance, C. P., & Shen, J. B. (2010). Nitric oxide is involved in phosphorus deficiency-induced cluster-root development and citrate exudation in white lupin. *New Phytologist, 187*(4), 1112–1123. https://doi.org/10.1111/j.1469-8137.2010.03323.x.

Wani, K. I., Naeem, M., Castroverde, C. D. M., Kalaji, H. M., Albaqami, M., & Aftab, T. (2021). Molecular mechanisms of nitric oxide (No) signaling and reactive oxygen species (ros) homeostasis during abiotic stresses in plants. *International Journal of Molecular Sciences, 22*(17). https://doi.org/10.3390/ijms22179656, https://www.mdpi.com/1422-0067/22/17/9656/pdf.

Wu, S., Hu, C., Tan, Q., Xu, S., & Sun, X. (2017). Nitric oxide mediates molybdenum-induced antioxidant defense in wheat under drought stress. *Frontiers in Plant Science, 8*. https://doi.org/10.3389/fpls.2017.01085.

Xu, J., Yin, H., Li, Y., & Liu, X. (2010). Nitric oxide is associated with long-term zinc tolerance in *Solanum nigrum*. *Plant Physiology, 154*(3), 1319–1334. https://doi.org/10.1104/pp.110.162982.

Chapter 13

Nitric oxide in the mitigation of ultraviolet stress in plants

Zalán Czékus, Attila Ördög, and Péter Poór
Department of Plant Biology, University of Szeged, Szeged, Hungary

13.1 Introduction

Solar radiation incident on terrestrial ecosystems includes ultraviolet (UV) wavelengths that exert a measurable influence on plant growth and development (Frohnmeyer & Staiger, 2003). Of the UV spectrum, UV-A (315–400 nm) and UV-B (280–315 nm) reach the Earth's surface under ambient conditions, whereas UV-C (<280 nm) is effectively excluded by stratospheric ozone (Kakani et al., 2003; Hideg et al., 2013). Although UV-A accounts for the majority of ground-level UV irradiance and often exceeds UV-B in intensity, UV-B is generally considered more biologically active due to its stronger absorption by nucleic acids, proteins, and photosynthetic components (Frohnmeyer & Staiger, 2003). The relative contribution of these wavebands varies with latitude, altitude, season, and atmospheric composition, generating pronounced spatial and temporal heterogeneity in UV exposure across natural and managed ecosystems (Kakani et al., 2003). Periodic anthropogenic depletion of stratospheric ozone has further amplified the ecological relevance of UV-B, with documented consequences for plant community structure and crop productivity, particularly in UV-sensitive species such as legumes and cereals, where reductions in biomass accumulation and reproductive performance have been reported (Kakani et al., 2003). Plants perceive and respond to these variations in light environment through a diverse suite of photoreceptors that decode both light quantity and spectral composition. These include red/far-red light-sensing phytochromes (phyA-phyE), blue light-responsive cryptochromes (cry1–3) and phototropins (phot1 and phot2), members of the Zeitlupe family (ZTL, FKF1, and LKP2), as well as the UV-B-specific photoreceptor UVR8, which plays a central role in initiating UV-B-dependent signaling cascades. Activation of these photoreceptors enables plants to translate changes in UV irradiance into coordinated transcriptional, metabolic, and developmental responses, thereby linking

Nitric Oxide in Plant Resilience to Abiotic Stress. DOI: https://doi.org/10.1016/B978-0-443-40497-9.00009-X

environmental light cues to physiological acclimation processes (Czékus et al., 2021; Paik & Huq, 2019; Demarsy et al., 2018).

Sustained or elevated UV exposure elicits a suite of physiological and developmental adjustments in plants, including restrained stem elongation, increased leaf thickness, and altered developmental timing, responses commonly interpreted as protective acclimation rather than generalized growth inhibition (Frohnmeyer & Staiger, 2003; Hideg et al., 2013). At the cellular level, UV radiation can impair photosynthetic efficiency and primary metabolism through damage to photosystem components and associated metabolic enzymes, contributing to reduced carbon assimilation and growth (Hideg et al., 2013). Concomitantly, metabolic resources are frequently reallocated toward the synthesis of UV-absorbing and antioxidant compounds, notably phenolics and related secondary metabolites, often at the expense of growth-associated processes (Frohnmeyer & Staiger, 2003). These responses reflect a broader disturbance of cellular redox homeostasis, as UV-driven excitation events promote the formation of reactive oxygen species (ROS) via photosensitization and perturbation of electron transport processes (Hideg et al., 2013). When ROS production exceeds buffering capacity, oxidative pressure can propagate changes in gene expression, enzyme activity, and metabolic fluxes, linking localized UV perception to systemic physiological responses (Frohnmeyer & Staiger, 2003; Hideg et al., 2013).

Within this redox-centered framework of UV stress, NO has emerged as a transient and context-dependent signaling component operating alongside ROS and other redox-active mediators (Shi et al., 2005; Simontacchi et al., 2015). Rather than functioning as a dedicated UV signal, NO appears to participate in the modulation of oxidative balance and stress acclimation in a manner dependent on concentration, timing, tissue specificity, and interaction with concurrent signaling pathways (Simontacchi et al., 2015). In some systems, application of NO donors has been associated with attenuation of UV-induced oxidative damage, including reduced lipid peroxidation and modulation of antioxidant capacity, yet the contribution of endogenous NO remains variable across species and experimental contexts (Shi et al., 2005). Interpretation of these effects is further complicated due to methodological limitations in detecting NO dynamics in vivo and by its rapid reactivity within complex redox networks (Simontacchi et al., 2015). These considerations position NO as a relevant but not universally protective component of UV stress responses and underscore the need for cautious evaluation of its role within integrated signaling frameworks.

13.2 Ultraviolet-induced redox perturbations and nitric oxide dynamics

UV-A and UV-B radiation penetrate plant tissues to differing extents and perturb cellular redox homeostasis primarily through excitation of endogenous chromophores, promoting electron-transfer reactions that favor ROS formation (Hideg et al., 2013). Depending on wavelength, intensity, and tissue optical properties, ROS generation may arise from the apoplast via the activity of NADPH oxidases or from disruptions in electron transport linked to photosynthetic and respiratory metabolism (Czarnocka & Karpiński, 2018). Moderate UV exposure commonly results in transient ROS elevations that can be accommodated by antioxidant systems and contribute to acclimatory adjustments, whereas sustained or high-intensity UV fluxes can exceed buffering capacity and lead to oxidative damage to lipids, proteins, and metabolic functions (Hideg et al., 2013). The spatial distribution of ROS production across intracellular and apoplastic compartments further shapes downstream signaling and stress outcomes, influencing the balance between adaptive reprogramming and irreversible injury (Czarnocka & Karpiński, 2018).

Against this redox-perturbed background, changes in NO levels have been reported following UV exposure, although the direction, magnitude, and temporal characteristics of these responses vary considerably across species and experimental systems (Kataria et al., 2020). Several studies describe transient increases in NO levels following UV-B treatment, often localized to specific tissues such as leaves or epidermal structures, whereas prolonged or high-dose exposures may attenuate NO accumulation or uncouple it from adaptive outcomes. Such variability reflects differences in UV regime, developmental stage, and background light conditions, as well as the plasticity of NO-generating pathways under stress (Kataria et al., 2020). Rather than indicating a uniform UV-induced NO response, these observations support the view that NO dynamics are contingent upon local redox cues and physiological context, without invoking dedicated receptors or hormone-like long-distance signaling (Simontacchi et al., 2015).

Interactions between NO and ROS under UV stress are dependent on concentration thresholds, reaction kinetics, and spatial constraints (Maurya et al., 2025). At relatively low and transient levels, NO may intersect with ROS-driven signaling processes associated with acclimation, whereas excessive or prolonged NO accumulation, particularly under sustained UV exposure, can contribute to nitrosative pressure and exacerbate cellular dysfunction (Simontacchi et al., 2015; Maurya et al., 2025). This duality underscores the importance of timing and localization, as early-phase NO responses may coincide with redox buffering, while later or excessive NO production can shift the balance toward damage. Importantly, these effects are integrated within broader redox networks that include antioxidant systems and other signaling mediators, limiting simplistic interpretations of NO as either protective or detrimental under UV stress (Czarnocka & Karpiński, 2018).

Interpretation of NO dynamics during UV exposure is further constrained by methodological limitations. Commonly used fluorescent probes are susceptible to interference from ROS, pH fluctuations, and photochemical artifacts, complicating efforts to resolve spatially and temporally discrete NO signals (Simontacchi et al., 2015). Experimental design factors, including UV dose relative to natural fluence, background spectral composition, and tissue heterogeneity, also exert a strong influence on reported outcomes (Pandey et al., 2023). As a result, conclusions drawn from NO donor treatments or acute UV exposures must be interpreted cautiously, particularly when extrapolating to endogenous signaling processes operating under environmentally relevant conditions (Pandey et al., 2023).

13.3 Integration of nitric oxide with antioxidant systems under ultraviolet stress

Plant acclimation to UV radiation relies on coordinated adjustments of enzymatic and nonenzymatic antioxidant systems that buffer redox perturbations rather than eliminate ROS outright (Czarnocka & Karpiński, 2018; Hideg et al., 2013). Enzymatic components such as superoxide dismutase, catalase, ascorbate peroxidase, and glutathione reductase contribute to maintaining redox balance across cellular compartments by modulating superoxide and hydrogen peroxide levels generated during UV-induced excitation events (Hideg et al., 2013). Nonenzymatic antioxidants, including ascorbate, glutathione, and phenolic compounds, complement these processes by stabilizing redox couples and limiting oxidative propagation, particularly under fluctuating UV regimes (Czarnocka & Karpiński, 2018). The accumulation of these metabolites frequently correlates with UV tolerance across genotypes, reflecting an integrated defense strategy that constrains oxidative pressure while sustaining metabolic activity (Frohnmeyer & Staiger, 2003). Evidence from rice exposed to supplemental UV-B supports this view, linking enhanced glutathione-associated redox buffering with improved physiological stability under elevated UV conditions (Shahzad et al., 2024).

Within this antioxidant framework, NO has been reported to modulate redox responses under UV exposure, although the extent and consistency of these effects vary substantially across species and experimental contexts (Pandey et al., 2023; Simontacchi et al., 2015). Changes in NO availability, whether arising endogenously or following donor application, have been associated with altered antioxidant enzyme activities and metabolite pools in several systems, often in a tissue- and regime-dependent manner (Pandey et al., 2023). Under moderate UV-B regimes, NO-associated shifts in enzymatic antioxidant capacity have been reported in crops such as soybean, coinciding in some cases with reduced indices of lipid peroxidation (Kataria et al., 2020). Donor-based studies in photosynthetic microorganisms and higher plants further suggest that NO can intersect with both enzymatic and nonenzymatic antioxidant components, including phenolic and flavonoid accumulation, without inducing uniform activation across antioxidant networks (Singh et al., 2024). Collectively, these observations support a modulatory rather than directive role for NO within UV-responsive antioxidant systems.

Importantly, NO-antioxidant interactions under UV stress are constrained by dose, timing, and exposure regime and do not uniformly favor protection (Maurya et al., 2025). Under conditions of intense or prolonged UV exposure, elevated NO levels may contribute to redox imbalance by increasing nitrosative pressure, particularly where antioxidant regeneration capacity is limited (Simontacchi et al., 2015). In such cases, interactions between NO and ROS can shift from signaling-associated modulation toward chemical interference with redox homeostasis, compromising enzymatic function and metabolic stability (Maurya et al., 2025). These threshold-dependent outcomes underscore the integration of NO within broader redox networks, where its influence on antioxidant systems remains contingent on cellular context rather than inherently beneficial. Interpretation of these dynamics is further complicated by variability in donor stability, species-specific antioxidant capacity, and limitations in resolving endogenous NO fluctuations under UV stress (Pandey et al., 2023; Maurya et al., 2025).

13.4 Nitric oxide crosstalk with ultraviolet-responsive signaling pathways

UV radiation induces rapid changes in cellular redox status, with ROS acting as primary early signals that shape downstream metabolic and transcriptional adjustments (Czarnocka & Karpiński, 2018; Hideg et al., 2013). Within this redox-responsive framework, NO operates as a context-dependent modulator that interacts bidirectionally with ROS rather than functioning as a dominant upstream signal (Simontacchi et al., 2015). ROS accumulation under UV exposure has been associated with concurrent changes in NO levels, while NO availability can influence the magnitude and persistence of oxidative signals through effects on redox buffering and reaction kinetics (Lopes-Oliveira et al., 2021). These interactions support a dynamic signaling environment in which NO contributes to the coordination of early UV responses without imposing linear control over ROS-driven pathways. Studies in *Arabidopsis* have linked UV-B exposure to localized NO accumulation that coincides temporally with ROS dynamics during photomorphogenic adjustment, reinforcing the view that NO participates in redox-based signal integration rather than acting as a UV-specific regulator (Latorre et al., 2024).

UV exposure also activates calcium (Ca^{2+})-dependent signaling events and phosphorylation-based cascades that contribute to signal propagation and response specificity (Pandey et al., 2023). Evidence suggests that NO can intersect with these processes indirectly, most plausibly through modulation of cellular redox conditions that influence Ca^{2+} fluxes or kinase activity, rather than through direct engagement with signaling components (Simontacchi et al., 2015). While NO-related modifications of signaling proteins, including kinases, have been reported in broader abiotic stress contexts, consistent patterns of activation or inhibition under UV stress remain limited (Sarma et al., 2025). As such, NO is best viewed as a conditional participant whose influence on Ca^{2+}- and phosphorylation-dependent signaling depends on exposure regime, tissue type, and developmental stage, rather than as a fixed component of UV signal transduction hierarchies (Pandey et al., 2023).

Crosstalk between NO and hormonal pathways implicated in UV responses further illustrates the indirect and variable nature of NO signaling. Interactions involving abscisic acid, auxin, and ethylene have been described in relation to UV-induced growth modulation and acclimation, yet these relationships are neither uniform nor unidirectional (Shoaib et al., 2024). NO may influence hormone-associated responses through redox-mediated adjustments that affect signaling sensitivity or downstream responsiveness, rather than through direct regulation of hormone biosynthesis or perception (Simontacchi et al., 2015). For example, ethylene accumulation under UV-B stress has been associated with growth restraint, with NO contributing to the modulation of this response in a manner that varies among tissues and developmental stages (Pandey et al., 2023). Such observations emphasize coordination within signaling networks rather than dominance of discrete pathways (Shoaib et al., 2024). UV-A and UV-B exposure is depicted as a dose- and duration-dependent driver of redox perturbation via ROS, while NO is presented as a local, transient modulator with conditional, bidirectional interactions across antioxidant, Ca^{2+}/phosphorylation, hormonal, and developmental modules (Fig. 13.1).

Spatial and temporal features further constrain NO involvement in UV-responsive signaling. NO responses are typically transient and localized, occurring in directly exposed tissues during early phases of UV stress and, in some cases, contributing to broader systemic adjustments at later stages (Lopes-Oliveira et al., 2021). This behavior is consistent with the diffusible and short-lived nature of NO, which enables localized signal integration without sustained systemic control (Simontacchi et al., 2015). Interpretation of NO crosstalk under UV stress remains complicated by methodological limitations, including reliance on donor treatments, challenges in resolving endogenous NO dynamics, and difficulties in separating signaling functions from generalized redox chemistry (Sarma et al., 2025). Consequently, current evidence supports a role for NO as a flexible integrator within UV-responsive signaling networks, while underscoring the need for cautious interpretation of mechanistic relationships.

13.5 Growth and developmental adjustments induced by nitric oxide and ultraviolet radiation

UV radiation induces characteristic growth and morphological adjustments in plants, including reduced hypocotyl and stem elongation, altered leaf expansion and orientation, and changes in tissue thickness, responses that have been widely reported across species exposed to elevated UV-B radiation (Frohnmeyer & Staiger, 2003; Hideg et al., 2013). These morphological shifts

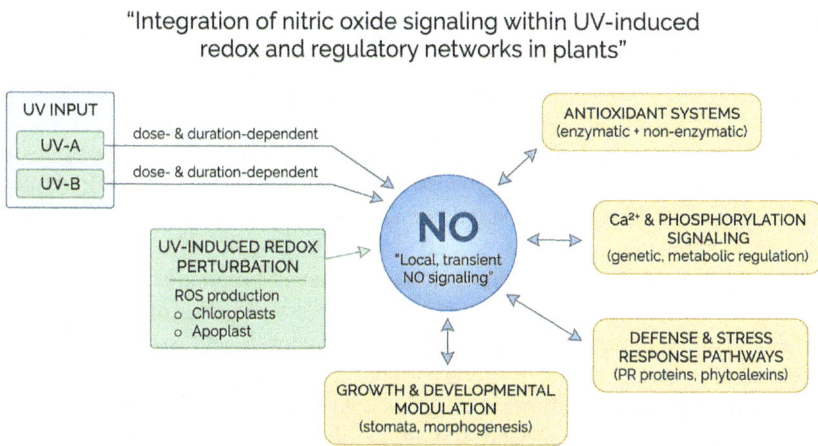

FIGURE 13.1 Integration of NO signaling within UV-induced redox and regulatory networks in plants. UV-A and UV-B represent dose- and duration-dependent inputs associated with redox perturbation, summarized here as ROS production in chloroplasts and the apoplast. NO is shown as a local, transient signaling node within this redox environment. Surrounding modules include antioxidant systems (enzymatic and nonenzymatic), Ca^{2+}- and phosphorylation-dependent signaling, hormonal signaling (ABA, auxin, and ethylene), and growth/developmental plasticity (hypocotyl elongation, leaf morphology). Solid thin arrows indicate crosstalk connections among modules; dashed bidirectional arrows indicate conditional and reciprocal modulation involving NO. The threshold bar depicts outcome dependence on NO dynamics: low/transient NO is associated with adaptive signaling, whereas high/sustained NO is associated with nitrosative interference. Outcome modifiers include UV dose, tissue type, developmental stage, and environmental context (*NO*, Nitric oxide; *UV*, ultraviolet).

are generally interpreted as adaptive adjustments that balance light perception, photoprotection, and resource allocation rather than simple manifestations of damage (Hideg et al., 2013). At the whole-plant level, UV exposure is frequently associated with reduced shoot biomass accumulation and modified architectural traits, while effects on root systems remain variable and species-dependent (Frohnmeyer & Staiger, 2003). Importantly, the magnitude and direction of these growth responses depend on UV wavelength, dose, duration, and genetic background, reinforcing the plastic nature of UV-mediated developmental regulation (Frohnmeyer & Staiger, 2003; Hideg et al., 2013).

Evidence linking NO to UV-induced growth modulation supports an associative and context-dependent role rather than a deterministic one (Kumar & Ohri, 2023). In *Arabidopsis*, UV-B exposure has been associated with localized increases in NO levels in hypocotyl tissues that temporally coincide with restrained elongation and altered photomorphogenic outcomes (Latorre et al., 2024). Perturbation of NO availability, through genetic or pharmacological approaches, can modify these UV-related growth responses, although the direction and magnitude of such effects differ substantially between endogenous NO dynamics and donor-based manipulations (Khan et al., 2023). Studies relying on NO donors have reported both amplification and attenuation of UV-associated growth phenotypes depending on concentration and timing, underscoring the difficulty of extrapolating donor effects to endogenous signaling processes (Khan et al., 2023; Kumar & Ohri, 2023). Collectively, available evidence suggests that NO operates within broader redox- and light-responsive networks that influence developmental outcomes under UV exposure, rather than acting as an independent regulator of growth (Kumar & Ohri, 2023).

The contribution of NO to UV-induced growth and developmental responses also exhibits pronounced tissue and developmental-stage specificity. NO-associated effects are most consistently observed in aerial tissues during early developmental phases, particularly in hypocotyls and expanding leaves, whereas responses in root systems are more variable and often less pronounced (Kumar & Ohri, 2023; Latorre et al., 2024). At later vegetative or reproductive stages, UV exposure has been linked to altered timing of developmental transitions, including flowering, in a manner that can intersect with NO signaling in some species but not others (Ageeva-Kieferle et al., 2021). Importantly, these outcomes depend strongly on UV regime, genetic background, and the temporal dynamics of NO accumulation, with both insufficient and excessive NO levels associated with neutral or maladaptive growth responses (Kumar & Ohri, 2023). Such patterns reinforce the view that NO contributes to fine-tuning developmental plasticity under UV stress in a highly conditional manner, rather than functioning as a general growth-promoting or growth-inhibiting signal.

13.6 Protective versus detrimental roles of nitric oxide under ultraviolet stress conditions

NO displays a dual and threshold-dependent character in plant responses to UV radiation, with outcomes shaped by the magnitude and persistence of NO accumulation within cellular environments (Kumar & Ohri, 2023; Mata-Pérez et al., 2023). Under conditions where NO production is transient and moderate, NO can integrate into redox-responsive networks that support acclimatory adjustments, partly by influencing ROS dynamics and redox-sensitive regulatory processes (Pandey et al., 2023; Simontacchi et al., 2015). In contrast, sustained or excessive NO accumulation has been associated with nitrosative modifications that interfere with protein function and compromise metabolic balance, thereby intensifying UV-induced redox perturbations rather than alleviating them (Kumar & Ohri, 2023; Mata-Pérez et al., 2023). This shift from signaling to direct chemical interference has also been seen under many other types of environmental stress. It shows that the effects of NO depend on how much of it is present, rather than acting only as a protective factor (Mata-Pérez et al., 2023).

Whether NO contributes to protection or cellular damage under UV exposure further depends on the radiation regime and biological context.

Under moderate UV-B intensities, NO accumulation has been associated with constrained oxidative injury and improved maintenance of cellular homeostasis, consistent with a role in fine-tuning redox signaling during acclimation (Simontacchi et al., 2015). By contrast, prolonged exposure or high-intensity UV treatments can overwhelm antioxidant buffering systems, creating conditions under which NO-derived reactions exacerbate cellular injury (Pandey et al., 2023). Developmental stage further modulates these outcomes, as seedlings often exhibit heightened sensitivity to NO-related imbalances as compared to mature plants, reflecting differences in antioxidant capacity and redox plasticity (Kumar & Ohri, 2023).

Interpretation of these contrasting roles is complicated by methodological factors that strongly influence experimental outcomes. NO donor compounds differ widely in release kinetics and stability, making it difficult to align exogenous NO supply with endogenous signaling dynamics or UV exposure timing (Pandey et al., 2023). In parallel, commonly used NO detection approaches, including fluorescent probes, are susceptible to interference from ROS and other reactive species, limiting confidence in absolute NO quantification under UV stress (Mata-Pérez et al., 2023). These constraints contribute to divergent conclusions across studies and underscore the need to evaluate NO effects within integrated experimental frameworks that account for UV regime, redox context, and methodological limitations rather than attributing outcomes to NO in isolation (Simontacchi et al., 2015).

13.7 Ecophysiological and agronomic perspectives

In natural environments, NO participates in plant responses to spatial and temporal variability in UV radiation, intersecting with diurnal fluctuations in light intensity and temperature that collectively shape redox homeostasis and metabolic regulation (Sarma et al., 2025; Simontacchi et al., 2015). Under ecologically relevant UV-B regimes, NO has been implicated in the modulation of stomatal behavior, photosynthetic adjustment, and redox signaling processes that support acclimation in species adapted to highlight or high-altitude habitats, although these effects remain strongly context-dependent (Mmbando & Hidema, 2025). Interactions with water availability further complicate these dynamics, as combined UV and drought exposure can alter NO-associated signaling in ways that influence resource allocation and conservation strategies, yet the direction and magnitude of these responses vary substantially among genotypes with differing UV sensitivity (Sarma et al., 2025). Such variability underscores that NO integration into UV responses reflects environment- and genotype-specific physiological adjustment rather than a conserved ecological strategy operating uniformly across species (Kumari et al., 2024).

In crop systems, evidence for the involvement of NO in UV stress responses is most consistent under controlled environmental conditions, where

UV exposure and background variables can be tightly regulated (Wang et al., 2021). Experimental studies conducted in growth chambers have reported associations between NO-related signaling and reduced photoinhibition or altered physiological performance under supplemental UV-B application in selected crops, although these observations often rely on indirect inference or donor-based manipulation (Pandey et al., 2023; Wang et al., 2021). By contrast, field environments introduce substantial complexity, including fluctuating UV irradiance, soil heterogeneity, and biotic interactions, which frequently obscure or diminish NO-associated effects observed under controlled conditions (Sarma et al., 2025). Comparative studies in cereals such as maize and sorghum further highlight pronounced genotypic variation in UV sensitivity and morphophysiological adjustment, reinforcing that NO-related responses cannot be readily extrapolated across crop species or production systems (Pitz et al., 2025).

Translating insights from laboratory-based NO-UV studies into agronomic contexts, therefore, remains highly constrained. Environmental heterogeneity, cooccurring stresses, and temporal mismatches between transient NO signaling events and prolonged UV exposure all limit the predictability of NO-associated outcomes in the field (Sarma et al., 2025). Consequently, conceptual exploration of NO precursors or modulatory approaches, when discussed, should be viewed strictly as complementary components within integrated stress physiology frameworks rather than as standalone mitigation strategies (Pandey et al., 2023). Progress in this area will depend on context-specific validation across environments and genotypes, rather than assumptions of broadly transferable NO-mediated UV tolerance (Kumari et al., 2024; Mmbando & Hidema, 2025).

13.8 Concluding remarks and future directions

NO emerges from this chapter as a redox-sensitive signaling component that participates in plant responses to UV stress through integration with broader redox and signaling networks rather than through isolated or directive action (Kumar & Ohri, 2023; Simontacchi et al., 2015). Across physiological contexts, NO aligns temporally and spatially with ROS dynamics, contributing to acclimation processes that span metabolic adjustment, signaling coordination, and developmental plasticity (Kumar & Ohri, 2023; Sarma et al., 2025). Evidence from UV-focused studies further supports the view that NO involvement in growth and morphological responses is tissue- and stage-dependent, reinforcing its role in fine-tuning adaptive outcomes rather than governing developmental programs (Khan et al., 2023; Latorre et al., 2024).

Despite increasing recognition of NO as an integrative signal under UV exposure, substantial uncertainties remain regarding the specificity and resolution of its action. The absence of a defined perception mechanism, combined with strong dependence on concentration, timing, and cellular context, complicates the interpretation of NO function in UV stress responses (Kumari et al., 2024; Allagulova et al., 2023). These challenges are exacerbated by reliance on donor-based approaches and limitations in current detection strategies, which frequently obscure endogenous NO dynamics and spatial heterogeneity (Simontacchi et al., 2015; Allagulova et al., 2023). Addressing these conceptual limitations requires greater emphasis on spatially and temporally resolved analyses that prioritize localized signaling events over bulk measurements.

Importantly, UV stress responses do not occur in isolation, and NO-associated signaling must be interpreted within multistress environments that include fluctuating light, temperature, and water availability (Sarma et al., 2025). Under such conditions, NO participation reflects adaptive integration across overlapping stress signals rather than UV-specific regulation, with outcomes shaped by environmental variability and genotype-dependent resilience (Kumari et al., 2024; Sarma et al., 2025). Placing NO in a broader ecophysiological context highlights the need to move beyond single-stressor approaches and better reflect the complex conditions found in natural and agricultural systems (Mmbando & Hidema, 2025).

AI disclosure

During the preparation of this work, the author(s) used ChatGPT in order to assist with translation tasks and to support efficient studying by helping organize, clarify, and structure complex information. After using this tool/service, the author(s) reviewed and edited the content as needed and take full responsibility for the content of the publication.

References

Ageeva-Kieferle, A., Georgii, E., Winkler, B., Ghirardo, A., Albert, A., Hüther, P., Mengel, A., Becker, C., Schnitzler, J. P., Durner, J., & Lindermayr, C. (2021). Nitric oxide coordinates growth, development, and stress response via histone modification and gene expression. *Plant Physiology, 187*(1), 336–360. https://doi.org/10.1093/plphys/kiab222, https://academic.oup.com/plphys/article/187/1/336/6277902?login=true.

Allagulova, C. R., Lubyanova, A. R., & Avalbaev, A. M. (2023). Multiple ways of nitric oxide production in plants and its functional activity under abiotic stress conditions. *International Journal of Molecular Sciences, 24*(14), 11637. https://doi.org/10.3390/ijms241411637.

Czarnocka, W., & Karpiński, S. (2018). Friend or foe? Reactive oxygen species production, scavenging and signaling in plant response to environmental stresses. *Free Radical Biology and Medicine, 122*, 4–20. https://doi.org/10.1016/j.freeradbiomed.2018.01.011.

Czékus, Z., Kukri, A., Hamow, K.Á., Szalai, G., Tari, I., Ördög, A., & Poór, P. (2021). Activation of local and systemic defence responses by Flg22 is dependent on daytime and ethylene in intact tomato plants. *International Journal of Molecular Sciences, 22*(15), 8354. https://doi.org/10.3390/ijms22158354.

Demarsy, E., Goldschmidt-Clermont, M., & Ulm, R. (2018). Coping with 'dark sides of the sun'through photoreceptor signaling. *Trends in Plant Science, 23*(3), 260–271. https://doi.org/10.1016/j.tplants.2017.11.007.

Frohnmeyer, H., & Staiger, D. (2003). Ultraviolet-B radiation-mediated responses in plants. balancing damage and protection. *Plant Physiology, 133*(4), 1420–1428. https://doi.org/10.1104/pp.103.030049.

Hideg, É., Jansen, M. A. K., & Strid, Å. (2013). UV-B exposure, ROS, and stress: Inseparable companions or loosely linked associates? *Trends in Plant Science, 18*(2), 107–115. https://doi.org/10.1016/j.tplants.2012.09.003.

Kakani, V. G., Reddy, K. R., Zhao, D., & Sailaja, K. (2003). Field crop responses to ultraviolet-B radiation: A review. *Agricultural and Forest Meteorology, 120*(1-4), 191–218. https://doi.org/10.1016/j.agrformet.2003.08.015.

Kataria, S., Rastogi, A., Bele, A., & Jain, M. (2020). Role of nitric oxide and reactive oxygen species in static magnetic field pre-treatment induced tolerance to ambient UV-B stress in soybean. *Physiology and Molecular Biology of Plants, 26*(5), 931–945. https://doi.org/10.1007/s12298-020-00802-5.

Khan, M., Ali, S., Al Azzawi, T. N. I., & Yun, B. W. (2023). Nitric oxide acts as a key signaling molecule in plant development under stressful conditions. *International Journal of Molecular Sciences, 24*(5). https://doi.org/10.3390/ijms24054782, http://www.mdpi.com/journal/ijms.

Kumar, D., & Ohri, P. (2023). Say "NO" to plant stresses: unravelling the role of nitric oxide under abiotic and biotic stress. *Nitric Oxide Biology and Chemistry / Official Journal of the Nitric Oxide Society, 130*, 36–57. https://doi.org/10.1016/j.niox.2022.11.004.

Kumari, R., Kapoor, P., Mir, B. A., Singh, M., Parrey, Z. A., Rakhra, G., Parihar, P., Khan, M. N., & Rakhra, G. (2024). Unlocking the versatility of nitric oxide in plants and insights into its molecular interplays under biotic and abiotic stress. *Nitric Oxide - Biology and Chemistry, 150*, 1–17. https://doi.org/10.1016/j.niox.2024.07.002, https://www.sciencedirect.com/science/journal/10898603.

Latorre, L., Fernández, M. B., & Cassia, R. (2024). Nitric oxide is a key part of the UV-B-induced photomorphogenesis in *Arabidopsis*. *Environmental and Experimental Botany, 217*, 105538. https://doi.org/10.1016/j.envexpbot.2023.105538.

Lopes-Oliveira, P. J., Oliveira, H. C., Kolbert, Z., & Freschi, L. (2021). The light and dark sides of nitric oxide: Multifaceted roles of nitric oxide in plant responses to light. *Journal of Experimental Botany, 72*(3), 885–903. https://doi.org/10.1093/jxb/eraa504, http://jxb.oxfordjournals.org/.

Mata-Pérez, C., Sánchez-Vicente, I., Arteaga, N., Gómez-Jiménez, S., Fuentes-Terrón, A., Oulebsir, C. S., Calvo-Polanco, M., Oliver, C., & Lorenzo, Ó. (2023). Functions of nitric oxide-mediated post-translational modifications under abiotic stress. *Frontiers in Plant Science, 14*. https://doi.org/10.3389/fpls.2023.1158184, https://www.frontiersin.org/journals/plant-science.

Maurya, A. K., Agarwal, R., & Gupta, R. (2025). Unraveling the crosstalk among ethylene, nitric oxide, and polyamines in tailoring the abiotic stress resilience in plants. *Stress Biology, 5*, 20. https://doi.org/10.1007/s44154-024-00198-2.

Mmbando, G. S., & Hidema, J. (2025). Hormonal regulation of crop adaptation to UV-B radiation stress: Implications for UV-B-Resistant crop varieties and sustainable agriculture. *Discover Agriculture, 3*(1). https://doi.org/10.1007/s44279-025-00267-8.

Paik, I., & Huq, E. (2019). Plant photoreceptors: Multi-functional sensory proteins and their signaling networks. *Seminars in Cell and Developmental Biology, 92*, 114–121. https://doi.

org/10.1016/j.semcdb.2019.03.007, http://www.elsevier.com/inca/publications/store/6/2/2/9/4/4/index.htt.

Pandey, A., Khan, M. K., Hamurcu, M., Athar, T., Yerlikaya, B. A., Yerlikaya, S., Kavas, M., Rustagi, A., Zargar, S. M., Sofi, P. A., Chaudhry, B., Topal, A., & Gezgin, S. (2023). Role of exogenous nitric oxide in protecting plants against abiotic stresses. *Agronomy, 13*(5). https://doi.org/10.3390/agronomy13051201, www.mdpi.com/journal/agronomy/.

Pitz, A., Jócsák, I., Varga, C., & Somfalvi-Tóth, K. (2025). Comparative UV-B stress responses in maize and sorghum based on biophoton emission measurements and morphophysiological traits. *Agronomy, 15*(9), 2224. https://doi.org/10.3390/agronomy15092224.

Sarma, H., Ramchiary, C., Sen, B., Daimary, M., & Prasad, R. (2025). Integration of nitric oxide signaling in plant stress responses: Unveiling its role in enhancing crop tolerance to abiotic stress. *Discover Plants, 2*(1). https://doi.org/10.1007/s44372-025-00102-w.

Shahzad, R., Harlina, P. W., Khan, S. U., Ihtisham, M., Khan, A. H., Husain, F. M., & Karuniawan, A. (2024). Physio-biochemical and transcriptomics analyses reveal molecular mechanisms of enhanced UV-B stress tolerance in rice induced by titanium dioxide nanoparticles. *Journal of Plant Interactions, 19*(1). https://doi.org/10.1080/17429145.2024.2328713, http://www.tandf.co.uk/journals/titles/17429145.asp.

Shi, S., Wang, G., Wang, Y., Zhang, L., & Zhang, L. (2005). Protective effect of nitric oxide against oxidative stress under ultraviolet-B radiation. *Nitric Oxide Biology and Chemistry / Official Journal of the Nitric Oxide Society, 13*(1), 1–9. https://doi.org/10.1016/j.niox.2005.04.006.

Shoaib, N., Pan, K., Mughal, N., Raza, A., Liu, L., Zhang, J., Wu, X., Sun, X., Zhang, L., & Pan, Z. (2024). Potential of UV-B radiation in drought stress resilience: A multidimensional approach to plant adaptation and future implications. *Plant, Cell & Environment, 47*(2), 387–407. https://doi.org/10.1111/pce.14774.

Simontacchi, M., Galatro, A., Ramos-Artuso, F., & Santa-María, G. E. (2015). Plant survival in a changing environment: The role of nitric oxide in plant responses to abiotic stress. *Frontiers in Plant Science, 6*. NOVEMBER https://doi.org/10.3389/fpls.2015.00977, http://journal.frontiersin.org/article/10.3389/fpls.2015.00977/pdf.

Singh, G., Patel, A., Tiwari, S., Pandey, A., Kumar, S., & Prasad, S. M. (2024). Nitric oxide and 5-aminolevulinic acid up-regulate growth, PS II photochemistry and antioxidant activity in UV-B stressed cyanobacterium *Nostoc muscorum*. *Plant Stress, 11*, 100332. https://doi.org/10.1016/j.stress.2023.100332.

Wang, J., Li, M., Feng, J., Yan, X., Chen, H., & Han, R. (2021). Effects of TiO_2-NPs pretreatment on UV-B stress tolerance in *Arabidopsis thaliana*. *Chemosphere, 281*, 130809. https://doi.org/10.1016/j.chemosphere.2021.130809.

Section III

NO crosstalk with other signaling molecules for sustainable agriculture

Chapter 14

Crosstalk between nitric oxide and emerging plant growth regulators in shaping abiotic stress tolerance

Ágnes Szepesi and Oluwatosin Peace Adedokun
Department of Plant Biology, University of Szeged, Szeged, Hungary

14.1 Introduction

Abiotic stresses, including drought, salinity, elevated temperature, and heavy metal toxicity, impose persistent constraints on plant growth and productivity across diverse agroecosystems. Such stresses perturb cellular homeostasis by altering ion balance, constraining photosynthetic performance, and promoting the accumulation of reactive oxygen species (ROS), thereby increasing the risk of oxidative damage and metabolic imbalance (Waqas et al., 2019). Plant responses to these challenges are mediated through complex signaling networks that integrate hormonal regulation with redox-based cues to coordinate stress perception and downstream adaptive adjustments (Simontacchi et al., 2015).

Nitric oxide (NO) is increasingly recognized as an integral component of these networks. Endogenous NO levels are frequently altered under abiotic stress conditions, and experimental evidence associates NO accumulation with modulation of antioxidant enzyme activities, osmotic regulation, and stomatal responses, effects that may collectively contribute to limiting stress-associated cellular damage in a context-dependent manner (Kumar and Ohri, 2023). In parallel, NO has been linked to changes in the expression of stress-responsive genes through redox-sensitive mechanisms, providing a potential interface between early stress perception and transcriptional reprogramming (Fancy et al., 2017; Simontacchi et al., 2015).

From a signaling perspective, NO is distinct from classical phytohormones in both chemical properties and mode of action. Canonical phytohormones, such as auxin (IAA), gibberellins (GAs), cytokinins (CKs), abscisic acid

Nitric Oxide in Plant Resilience to Abiotic Stress. DOI: https://doi.org/10.1016/B978-0-443-40497-9.00015-5

(ABA), and ethylene (ET), are synthesized through defined biosynthetic pathways, perceived by characterized receptor systems, and often distributed via regulated transport processes to support coordinated systemic responses. By contrast, NO is a short-lived, diffusible radical generated through multiple enzymatic and nonenzymatic routes, typically acting locally and transiently without reliance on dedicated receptors or long-distance transport (Freschi, 2013; Gupta et al., 2011). Its signaling capacity is largely mediated through reversible posttranslational modifications, including S-nitrosation and protein tyrosine nitration, which can modulate protein activity, stability, or interaction capacity in a stimulus- and tissue-dependent manner (Corpas et al., 2011).

Beyond classical phytohormones, a range of emerging plant growth regulators (PGRs) has broadened the regulatory landscape governing plant stress responses. Compounds such as brassinosteroids (BRs) and melatonin (MEL) have been associated with enhanced stress tolerance through effects on membrane properties, redox balance, and metabolic adjustment, frequently in coordination with established hormonal pathways rather than through isolated signaling routes (Singhal et al., 2021). Many of these regulators influence cellular redox status, thereby creating physiological contexts in which NO production and NO-dependent signaling outputs may be modulated.

Within this framework, NO can be viewed as operating at the intersection of hormonal, redox, and emerging PGR-mediated regulatory layers. Through reversible posttranslational modifications and interactions with antioxidant systems, NO appears to modulate the activity of enzymes and regulatory proteins involved in stress adjustment, contributing to flexible physiological responses under adverse conditions (Sarma et al., 2025; Singhal et al., 2021). Rather than functioning as a conventional hormone within a linear pathway, NO is more appropriately considered a context-dependent integrator that links environmental inputs with adaptive physiological processes.

14.2 Nitric oxide as a central regulator in plant stress responses

Early observations of NO in biological systems emphasized its presence as an atmospheric pollutant associated with ozone-related processes and acid rain, leading to its initial perception as an external stress factor rather than an endogenous signal. This view shifted with the detection of NO production in plants responding to pathogens and environmental cues. Subsequent studies associated NO with disease resistance, stomatal regulation, and growth modulation, supporting its recognition as an intrinsic signaling component in plant systems (Gupta et al., 2011; Simontacchi et al., 2015). These observations contributed to a conceptual transition in which NO was no longer regarded solely as a by-product of stress but as a regulated participant in plant signaling networks.

The classification of NO as a plant signal derives primarily from its physicochemical properties rather than from hormone-like behavior. As a small, gaseous free radical, NO diffuses readily across membranes and typically acts locally without reliance on dedicated transport systems or receptor-mediated perception. In contrast to classical phytohormones that operate through defined biosynthetic routes and signal transduction cascades, NO signaling is strongly influenced by concentration, redox state, and cellular context. These features enable NO to convey environmental information directly to cellular targets in a manner that is highly conditional and transient, consistent with its role as a signaling mediator rather than a conventional hormone (Gupta et al., 2011; Khan et al., 2023).

Plants generate NO through multiple enzymatic and nonenzymatic pathways that are differentially engaged under abiotic stress. Nitrate reductase has been identified as a major contributor to NO formation via nitrite reduction, particularly under hypoxia, salinity, or water deficit, with enzyme activity subject to regulation by phosphorylation and redox status (Gupta et al., 2011; Kumar and Ohri, 2023). Mitochondrial electron transport can further support NO production through nitrite reduction under low-oxygen conditions, linking NO generation to respiratory adjustment and energy metabolism. In parallel, polyamine turnover has been associated with NO formation through oxidative pathways involving amine oxidases, especially under stress conditions that enhance polyamine metabolism (Kumar and Ohri, 2023). Nonenzymatic NO production may also occur in acidic apoplastic environments with elevated nitrite availability, conditions frequently associated with stress-induced pH shifts (Gupta et al., 2011).

The presence of multiple NO-producing routes enables plants to sustain NO signaling across diverse and fluctuating stress scenarios. Abiotic stress factors such as drought, salinity, and heavy metal exposure influence substrate availability, redox balance, and compartment-specific NO production, allowing signaling capacity to be maintained even when individual pathways are constrained (Simontacchi et al., 2015; Waqas et al., 2019). This multiplicity underscores the adaptive flexibility of NO biosynthesis within stress physiology rather than implying a single dominant source.

NO functions predominantly as a redox-active signaling molecule through dynamic interactions with ROS and other reactive nitrogen species. At controlled concentrations, NO has been associated with modulation of redox homeostasis, including attenuation of lipid peroxidation and interaction with radical species, while low-molecular-weight NO derivatives such as S-nitrosoglutathione can extend signaling capacity within the cell (Corpas et al., 2021; Fancy et al., 2017). Central to these effects is the capacity of NO to mediate posttranslational protein modifications (PTMs), particularly S-nitrosation of cysteine residues. Through these PTMs, NO can influence enzyme activity, protein stability, and subcellular localization, enabling rapid and reversible adjustment of stress-related pathways (Corpas et al., 2021;

Fancy et al., 2017). Additional modifications, including protein tyrosine nitration, have been associated with altered function of metabolic enzymes and signaling components under stress, linking NO-dependent redox chemistry with broader regulatory outcomes (Corpas et al., 2011).

A defining feature of NO signaling is its pronounced spatial and temporal plasticity. The short half-life and diffusible nature of NO support rapid responses to changing stress intensity, while localized production within organelles such as mitochondria and chloroplasts permits close coupling between signaling, energy metabolism, and redox balance (Gupta et al., 2011; Simontacchi et al., 2015). Under hypoxic conditions, NO has been associated with respiratory adjustments that contribute to metabolic maintenance, whereas under drought or salinity stress, it has been linked to modulation of photosynthetic performance and cellular redox status (Kumar and Ohri, 2023).

NO signaling operates in a concentration- and context-dependent manner. Evidence indicates that low to moderate NO levels are commonly associated with activation of antioxidant systems and protective adjustments, whereas sustained or excessive NO accumulation can contribute to growth inhibition or cell death under severe or prolonged stress (Kumar and Ohri, 2023; Singhal et al., 2021). This duality allows plants to scale physiological responses in accordance with stress severity, supporting metabolic reprogramming, osmotic adjustment, and ion homeostasis in heterogeneous environments.

14.3 Mechanisms of nitric oxide signaling under abiotic stress

NO contributes to redox regulation during abiotic stress through tightly coupled interactions with ROS that are strongly dependent on concentration, subcellular localization, and stress intensity. NO can react with superoxide to form peroxynitrite, a reaction that has been associated with attenuation of lipid peroxidation when NO levels remain constrained but with nitrosative damage when this balance is disrupted (Corpas et al., 2011; Simontacchi et al., 2015). Conversely, ROS accumulation under stress has been shown to influence NO generation, in part by modulating nitrate reductase activity and nitrite availability, thereby establishing feedback relationships that shape redox signaling thresholds rather than linear cause–effect pathways (Gupta et al., 2011; Sarma et al., 2025). Under drought and salinity, maintenance of an appropriate NO–ROS balance has been associated with restriction of uncontrolled oxidative bursts, whereas loss of this equilibrium correlates with impaired cellular function and oxidative injury (Simontacchi et al., 2015; Waqas et al., 2019). These observations reinforce the concentration-dependent nature of NO action, with moderate NO levels generally associated with cytoprotective adjustments and sustained NO accumulation linked to growth inhibition or cell death under severe stress (Singhal et al., 2021). Beyond individual redox reactions, nitric oxide signaling under abiotic stress has been conceptualized as part of a broader gasotransmitter network that integrates multiple stress-responsive cues rather than acting in isolation. Reviews by Kolbert and colleagues emphasize that NO-mediated responses are strongly context-dependent, shaped by interactions with other gaseous signals such as ethylene and hydrogen sulfide, as well as by spatial and temporal constraints on NO production and turnover (Kolbert, Barroso, et al., 2019; Kolbert, Feigl, et al., 2019; Kolbert, Chapter 8).

PTMs constitute a principal mechanism through which NO exerts regulatory effects during stress. S-nitrosation, involving reversible modification of cysteine residues, has been shown to alter enzyme activity, protein–protein interactions, and subcellular localization in a redox-sensitive manner (Corpas et al., 2021). In mitochondria, S-nitrosation of metabolic enzymes such as aconitase has been associated with altered tricarboxylic acid cycle activity under hypoxic conditions, potentially contributing to metabolic adjustment during oxygen limitation (Simontacchi et al., 2015). Tyrosine nitration, frequently mediated by peroxynitrite, has been reported to modify enzymes involved in nitrogen metabolism and photosynthesis under salinity and other abiotic stresses, often coinciding with reduced catalytic efficiency rather than complete inactivation (Corpas et al., 2011). In addition, metal nitrosation allows NO to interact with heme-containing proteins, including cytochrome c oxidase, thereby influencing electron flow within respiratory chains and linking NO signaling to mitochondrial function (Graska et al., 2023).

NO-mediated posttranslational modifications also intersect with antioxidant regulation. S-nitrosation has been associated with the modulation of antioxidant enzymes such as superoxide dismutase and ascorbate peroxidase, contributing to controlled detoxification of ROS under conditions including heavy metal exposure and osmotic stress (Fancy et al., 2017; Sarma et al., 2025). In contrast, nitration-dependent modification of catalase has been linked to partial inhibition of activity, which may permit localized accumulation of hydrogen peroxide compatible with signaling functions rather than complete scavenging (Corpas et al., 2011; Corpas et al., 2021). Redox buffering capacity is further influenced by S-nitrosation of glutathione, which stabilizes cellular thiol pools and limits irreversible oxidation during prolonged stress exposure (Fancy et al., 2017). At the transcriptional level, NO-associated redox changes have been linked to altered expression of genes encoding antioxidant enzyme isoforms, supporting sustained defense capacity under chronic stress conditions (Sarma et al., 2025).

Beyond redox regulation, NO interfaces with cellular energy metabolism through effects on mitochondrial respiration under abiotic constraints. NO-dependent inhibition of cytochrome c oxidase has been associated with redirection of electron flow toward alternative oxidase, a shift that can reduce mitochondrial ROS formation during hypoxia associated with salinity or waterlogging (Simontacchi et al., 2015; Graska et al., 2023). In parallel, nitrite-dependent NO respiration has been proposed to function as an alternative electron sink under low-oxygen conditions, contributing to maintenance of redox balance and limited ATP production without full engagement of the classical electron transport chain (Gupta et al., 2011). Through these interconnected mechanisms, NO links redox modulation with energy homeostasis, supporting metabolic flexibility and ion transport processes in stressed tissues (Kumar and Ohri, 2023; Simontacchi et al., 2015).

14.4 Nitric oxide interactions with classical phytohormones under stress conditions

NO interacts closely with abscisic acid (ABA) signaling during drought and salinity stress, particularly in the regulation of stomatal behavior. ABA-associated signaling has been linked to enhanced NO accumulation in guard cells via redox-mediated processes involving hydrogen peroxide, and this NO accumulation has been associated with modulation of ion channel activity that supports stomatal closure and reduced transpirational water loss (Freschi, 2013; Xu et al., 2024). In addition, redox-sensitive posttranslational modification by NO, including S-nitrosation of ABA signaling components such as protein phosphatases, has been reported to influence signal sensitivity and downstream responsiveness (Freschi, 2013; Rai et al., 2021). Evidence also suggests that NO may participate in feedback regulation of ABA homeostasis by modifying ABA biosynthetic enzymes, thereby limiting excessive hormone accumulation during prolonged stress exposure (Rai et al., 2021). Together, these observations position NO as a context-dependent modulator acting downstream of ABA perception while enabling reciprocal fine-tuning of dehydration responses.

NO also exhibits bidirectional interactions with ET signaling that influence growth restraint and senescence under abiotic stress. Experimental studies have associated NO with suppression of ethylene biosynthesis through redox-dependent modification of enzymes involved in 1-aminocyclopropane-1-carboxylic acid metabolism, coinciding with reduced ethylene accumulation and delayed stress-induced senescence under salinity or drought (Freschi, 2013; Singhal et al., 2021). Conversely, ET signaling has been shown to affect NO availability, particularly under hypoxic conditions, where ET-mediated NO depletion has been linked to metabolic preadaptation and improved stress tolerance (Hartman et al., 2019). This reciprocal regulation appears especially relevant during oxygen limitation, where coordinated NO–ET interactions contribute to maintenance of cellular integrity while restraining growth-related carbon expenditure.

Interactions between NO and salicylic acid have also been associated with convergence of stress adaptation and defense-related pathways. NO has been linked to modulation of salicylic acid-associated signaling through redox effects on enzymes involved in phenylpropanoid metabolism, supporting antioxidant capacity and glutathione-dependent protection under both abiotic and biotic stress contexts (Rai et al., 2021; Sarkar et al., 2023). In parallel, NO influences auxin-regulated growth responses by affecting auxin transport and signaling dynamics. S-nitrosation of components associated with auxin perception and transport has been shown to alter auxin distribution, thereby contributing to adjustments in root growth and architecture under stress conditions (Fernández-Marcos et al., 2011; Terrile et al., 2012). These effects are strongly concentration- and tissue-dependent, with lower NO levels generally associated with stabilization of auxin signaling in meristematic regions and higher levels linked to growth restraint, enabling flexible developmental responses aligned with stress intensity.

Taken together, interactions between NO and classical phytohormones constitute core regulatory layers in plant stress adaptation but do not fully explain the complexity of responses observed under multifactorial stress scenarios. This limitation points to the involvement of additional regulatory components capable of integrating redox cues with broader metabolic and transcriptional control, providing a conceptual transition toward emerging plant growth regulators that operate in concert with NO rather than in isolation (Freschi, 2013; Singhal et al., 2021).

14.5 Brassinosteroids and nitric oxide-mediated abiotic stress tolerance

BRs are steroidal phytohormones with established roles in plant growth and development and have also been associated with the modulation of stress responses. The identification of the leucine-rich repeat receptor kinase BRI1 provided an early molecular framework for BR perception and signal initiation (Li and Chory, 1997). Subsequent studies have demonstrated that BR signaling influences diverse physiological processes, including cell expansion, vascular differentiation, and stress acclimation, supporting a role for BRs in shaping plant performance under adverse environmental conditions (Nolan et al., 2020).

Under abiotic stress, BR application has been associated with enhanced tolerance through modulation of redox balance and activation of stress-related signaling pathways. One of the early responses linked to BR treatment is the controlled accumulation of hydrogen peroxide, which functions as a signaling intermediate rather than as a marker of oxidative injury (Zhu, Deng, et al., 2016; Zhu, Tan, et al., 2016). In tomato, BR-induced salt tolerance has been shown to require coordinated involvement of ethylene signaling and hydrogen peroxide production, indicating that redox signaling is integral to BR-mediated stress responses (Zhu, Deng, et al., 2016; Zhu, Tan, et al., 2016). Genetic disruption of this process, through silencing of nicotinamide adenine dinucleotide phosphate oxidase and MAP kinase components, abolishes BR-associated hydrogen peroxide accumulation and stress tolerance, highlighting the importance of ROS-generating and signaling systems in BR action under salinity (Nie et al., 2019).

NO plays a complementary and functionally relevant role in BR-associated stress responses. BR treatment has been linked to enhanced NO production under salt stress, where NO has been associated with modulation of antioxidant defenses and respiratory metabolism (Zhu, Deng, et al., 2016; Zhu, Tan, et al., 2016). In *Nicotiana benthamiana*, BR-induced activation of the alternative respiratory pathway was shown to depend on NO availability, contributing to limitation of oxidative damage under saline conditions (Zhu, Deng, et al., 2016; Zhu, Tan, et al., 2016). These observations are consistent with broader evidence positioning NO as a stress-related signaling modulator acting downstream of multiple hormonal cues rather than as a primary regulator of hormone biosynthesis or perception (Nawaz et al., 2017; Khan et al., 2023).

BR signaling during stress operates within a broader hormonal network rather than in isolation. In Arabidopsis, BR-mediated stress tolerance has been shown to involve functional interactions with abscisic acid, ethylene, and salicylic acid signaling pathways, indicating that BRs reshape existing hormonal networks to support stress adjustment (Divi et al., 2010). Regulation of ethylene biosynthesis represents one point of interaction, as BRs influence the stability of 1-aminocyclopropane-1-carboxylate synthase proteins, thereby modulating ethylene levels under stress conditions (Hansen et al., 2009). Consistent with this integrative view, analyzes of endogenous hormone profiles under drought and salinity reveal coordinated changes among BRs and other phytohormones as part of an adaptive response (Llanes et al., 2016).

Protein tyrosine nitration and related NO-derived modifications have been associated with altered activity of stress-responsive enzymes and antioxidant systems, providing a mechanistic link between BR-induced NO production and downstream physiological adjustment (Corpas et al., 2021). The presence of multiple NO-generating routes further enables flexible modulation of NO levels in response to stress intensity and developmental context (Allagulova et al., 2023). Evidence from integrative analyzes indicates that functional interactions between NO and phytohormones, including BRs, contribute to stress resilience by coordinating redox, metabolic, and hormonal regulation rather than through linear signaling hierarchies (Kolupaev et al., 2024).

Experimental observations across species support the view that NO frequently operates downstream of BR signaling rather than engaging in extensive reciprocal regulation of BR biosynthesis or perception. NO-associated enhancement of antioxidant capacity, glutathione homeostasis, and protection against temperature or salinity stress has been reported in multiple systems, reinforcing its role as a convergence point for BR-induced stress signaling (Singhal et al., 2021; Song et al., 2023). In parallel, exogenous manipulation of NO availability has been shown to improve plant performance under salinity and other abiotic stresses, underscoring the functional relevance of NO-dependent pathways associated with BR responses (Hajihashemi & Jahantigh, 2023; Youssef & Adawy, 2023).

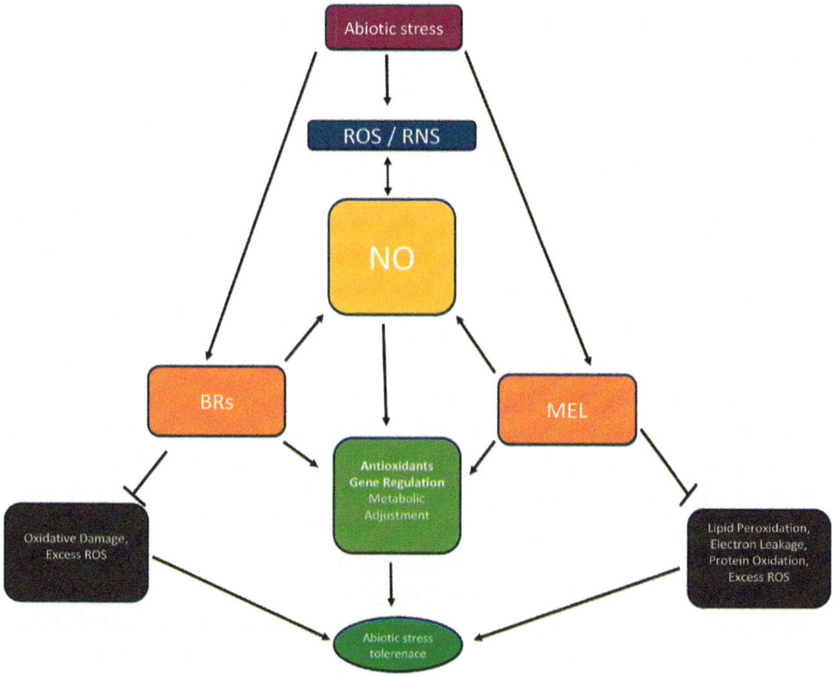

FIGURE 14.1 Model of brassinosteroid (*BR*), melatonin (*MEL*), and nitric oxide (*NO*) interactions in the regulation of abiotic stress tolerance in plants.

Overall, BRs contribute to abiotic stress tolerance through mechanisms involving controlled ROS and NO production, modulation of hormonal balance, and metabolic adjustment. Rather than extensive reciprocal crosstalk, BR-induced NO signaling appears to function primarily as a downstream integrative layer that reinforces stress tolerance pathways initiated by BR perception in a context-dependent manner (Fig. 14.1).

14.6 Melatonin as an emerging plant growth regulator in stress tolerance

MEL was initially characterized as a regulator of circadian rhythms and physiological homeostasis in animals, but its identification and functional characterization in plants revealed a substantially broader regulatory scope. In plant systems, MEL has been associated with modulation of seed germination, circadian and photoperiodic responses, flowering, leaf senescence, postharvest performance, and tolerance to biotic and abiotic stresses, with many of these effects linked to redox-associated signaling processes (Ikram et al., 2024; Shi et al., 2024). Rather than acting through a single linear pathway, MEL appears to operate through integration with reactive oxygen- and nitrogen-related signaling, positioning it as a multifunctional modulator of plant stress physiology (Ahmad et al., 2023).

Under abiotic stress conditions, MEL-associated stress tolerance has been linked to coordinated regulation of antioxidant defenses, hormonal balance, metabolic adjustment, and energy homeostasis. Experimental evidence across stress types, including temperature extremes and water deficit, indicates that MEL modulates antioxidant enzyme activity, sulfur metabolism, mitochondrial function, and expression of stress-responsive genes in a context-dependent manner (Gharanjik et al., 2025). In fruit crops and other agronomically relevant species, MEL-related stress resilience has been associated with improved physiological stability and maintenance of yield-related traits, supporting its applied relevance under stress-prone conditions (Shi et al., 2024).

MEL signaling exhibits close functional overlap with NO-dependent pathways. Both molecules are generated in subcellular compartments such as mitochondria and chloroplasts, enabling coordinated modulation of redox balance and stress signaling outputs (Ahmad et al., 2023; Sarma et al., 2025). Similar to NO, MEL has been associated with the regulation of germination, flowering, senescence, and stress responses, and their interaction frequently converges on redox homeostasis and posttranslational modification of target proteins rather than on receptor-mediated perception (Ahmad et al., 2023). This spatial and functional convergence creates multiple opportunities for interaction between MEL- and NO-mediated signaling during stress adaptation.

One outcome of MEL–NO interaction is the formation of N-nitrosomelatonin (NOMela), a nitrosated derivative of MEL that has been proposed as a relatively stable reservoir for NO. NOMela has been shown to release NO in a controlled manner while retaining antioxidant properties associated with MEL, potentially extending the spatial and temporal range of NO signaling (Hussain et al., 2024; Martínez-lorente et al., 2022). Comparative analyzes using ozone chemiluminescence-based detection suggest that NOMela exhibits NO-release kinetics distinct from commonly used NO donors such as S-nitrosoglutathione, indicating that its physiological behavior under stress conditions may differ from that of classical NO-releasing compounds (Hussain et al., 2024).

The functional consequences of MEL–NO crosstalk extend beyond redox buffering alone. Regulation of NO homeostasis through phytoglobin-mediated scavenging has been associated with altered ethylene levels and improved drought tolerance, linking NO availability with broader hormonal regulation during stress acclimation (Montilla-Bascon et al., 2025). In parallel, combined MEL and NO signaling has been associated with coordinated regulation of antioxidant gene expression, energy metabolism, and PTM of stress-related proteins, contributing to maintenance of cellular stability under prolonged or combined stress conditions (Beltrán-García et al., 2024; Sarma et al., 2025).

Emerging evidence further suggests that MEL–NO interactions intersect with additional signaling molecules, including β-aminobutyric acid, hydrogen sulfide, and classical phytohormones, forming multilayered regulatory networks rather than isolated signaling axes (Devi et al., 2023). Experimental manipulation of MEL levels, NO donors, or related precursors has been shown to improve plant performance under water deficit, salinity, and temperature stress, underscoring the translational potential of these pathways when appropriately managed (do Carmo et al., 2025; Zulfiqar et al., 2024).

Despite increasing evidence for cooperation between MEL and NO, the molecular determinants governing their interaction remain incompletely resolved. While individual signaling roles of MEL and NO are well documented, NOMela represents an emerging regulatory entity whose contribution to stress signaling integration is still being defined (Hussain et al., 2024; Martínez-lorente et al., 2022). Current evidence supports the view that MEL does not replace classical phytohormones but instead functions as an additional regulatory layer that refines stress responses through redox-based signaling and interaction with NO-dependent pathways (Ahmad et al., 2023).

In this context, MEL can be viewed as an emerging regulatory component in plant stress responses that operates in close association with NO-mediated signaling. Their interaction supports coordinated regulation of redox balance, hormonal integration, and metabolic adjustment during abiotic stress, contributing to flexible and context-dependent stress adaptation (Fig. 14.1).

Abiotic stress induces ROS/RNS accumulation and stimulates BR- and MEL-associated signaling. NO functions as a central integrator downstream of BR and MEL pathways, coordinating antioxidant defense, stress-responsive gene regulation, and metabolic adjustment. MEL additionally suppresses lipid peroxidation, electron leakage, protein oxidation, and excess ROS through direct antioxidant activity. BR-mediated stress protection operates largely through NO-dependent redox signaling. The coordinated actions of BRs, MEL, and NO contribute to the maintenance of redox homeostasis and enhancement of abiotic stress tolerance. Solid arrows indicate established regulatory relationships, and inhibition bars represent suppression of oxidative damage processes (Fig. 14.1).

14.7 Agricultural and biotechnological implications

Exogenous application of plant growth regulators has demonstrated potential for mitigating abiotic stress across diverse crop systems, although outcomes remain strongly context-dependent. Brassinosteroid treatments applied as foliar sprays or soil drenches at defined concentrations have been shown to alleviate flooding and salinity stress in *Garcinia humilis*, an effect associated with reduced ROS accumulation and partial maintenance of photosynthetic performance (Sanchez et al., 2025). Similarly, seed priming with brassinosteroids at micromolar concentrations has been reported to mitigate chromium toxicity in rice cultivars, coinciding with enhanced antioxidant enzyme activity and reduced lipid peroxidation, consistent with protection against metal-induced oxidative stress (Basit et al., 2021).

Melatonin has been extensively evaluated as an exogenous regulator under multiple abiotic stress conditions. Applications via seed priming, foliar spraying, or root irrigation have been associated with improved tolerance to drought, salinity, and low-temperature stress in a range of crop species, effects frequently linked to modulation of osmolyte accumulation and stabilization of redox balance (Colombage et al., 2023). Nitric oxide donors have likewise been investigated for stress mitigation, with foliar application shown to enhance drought tolerance in *Calendula officinalis*, coinciding with increased proline accumulation and moderated oxidative stress indicators (Mahabadi et al., 2026). In addition, soil-based application of NO donors has been associated with reduced heavy metal uptake and improved ion balance in crops exposed to salinity and drought, although these effects vary with environmental and management conditions (Pandey et al., 2023).

Despite these promising outcomes, the effectiveness of exogenous regulators is strongly influenced by dosage, timing, and developmental stage. NO donors exhibit relatively narrow effective concentration ranges, with both insufficient and excessive application failing to optimize osmoprotectant accumulation or stress tolerance, as observed in *Calendula officinalis* under drought stress (Mahabadi et al., 2026). Timing of application further constrains efficacy, as prestress brassinosteroid treatments confer measurable protection in *Garcinia humilis*, whereas poststress applications provide limited benefit (Sanchez et al., 2025). Developmental sensitivity also affects melatonin responsiveness, with vegetative-stage applications generally enhancing stress tolerance, while applications at later stages may disrupt reproductive processes in susceptible genotypes (Colombage et al., 2023). Excessive application of NO donors has been associated with redox imbalance, including elevated hydrogen peroxide levels and electrolyte leakage, while overapplication of brassinosteroids can alter growth patterns under metal stress, underscoring the need for careful dose optimization (Basit et al., 2021; Pandey et al., 2023).

Environmental context further determines the translational success of these approaches. Controlled environments often facilitate reproducibility, as uniform growth conditions support consistent stress mitigation outcomes, including improved salinity tolerance following plant growth regulator priming under stable chamber conditions (Quamruzzaman et al., 2021). In contrast, field environments introduce variability arising from soil heterogeneity, microbial interactions, and climatic fluctuations, factors that reduce the predictability of melatonin efficacy and dilute foliar NO donor responses under erratic precipitation regimes (Colombage et al., 2023; Pandey et al., 2023). Genotype-specific responses and environment-dependent interactions further restrict the practical application window for brassinosteroids across cropping systems, emphasizing the need for context-specific optimization strategies rather than generalized application protocols (Sanchez et al., 2025; Quamruzzaman et al., 2021).

14.8 Conclusions and future perspectives

The stress responses discussed in this chapter converge on NO-centered signaling as an integrative regulatory layer linking hormonal, redox, and metabolic adjustments under adverse environmental conditions. Across both classical phytohormones and emerging plant growth regulators, NO has been associated with modulation of redox balance, posttranslational regulation, and stress-responsive pathways in a manner that is strongly dependent on developmental stage, tissue context, and stress severity (Freschi, 2013; Singhal et al., 2021). Rather than acting as a universal driver of stress tolerance, NO appears to function as a conditional signaling mediator whose effects are shaped by local biochemical and physiological constraints (Simontacchi et al., 2015).

Interactions between NO and brassinosteroids primarily reflect hierarchical signal integration. BR perception activates redox- and kinase-associated pathways that converge on NO production, with downstream effects on antioxidant capacity, respiratory flexibility, and metabolic adjustment during stress exposure (Divi et al., 2010; Zhu, Deng, et al., 2016; Zhu, Tan, et al., 2016). Available evidence suggests that this relationship is largely downstream rather than strongly reciprocal, distinguishing BR–NO interactions from more classical forms of hormonal crosstalk (Singhal et al., 2021; Kolupaev et al., 2024). In contrast, interactions between NO and melatonin display greater functional overlap and apparent bidirectionality, consistent with shared subcellular localization, redox-related functions, and the formation of nitrosated derivatives such as N-nitrosomelatonin (Hussain et al., 2024; Martínez-lorente et al., 2022).

Despite substantial progress, important gaps remain in understanding how specificity is achieved within these signaling networks. The mechanisms governing selective targeting of NO-dependent posttranslational modifications during BR signaling, as well as the physiological relevance, stability, and mobility of nitrosated melatonin derivatives under complex stress combinations, remain incompletely resolved (Corpas et al., 2021; Hussain et al., 2024). In addition, pronounced species- and stress-specific variability complicates extrapolation from controlled experimental systems to field environments, particularly under multifactorial stress conditions (Sarma et al., 2025).

Future research will benefit from spatially and temporally resolved analyzes of NO-dependent regulation, particularly at the level of posttranslational modification and organelle-specific signaling dynamics (Allagulova et al., 2023; Corpas et al., 2021). Comparative studies across crop species and combined stress scenarios will be essential for distinguishing conserved regulatory features from context-specific adaptations (Sarma et al., 2025; Singhal et al., 2021). From an applied perspective, improved control over dosage, timing, and delivery of NO-related regulators will be critical for translating mechanistic insights into reproducible and reliable stress management strategies in agricultural systems (Pandey et al., 2023).

Overall, positioning stress tolerance within an NO-integrated signaling framework provides a coherent basis for linking hormonal regulation with redox control and metabolic adjustment, while also defining clear priorities for future fundamental and translational research.

AI disclosure

During the preparation of this work, the author(s) used ChatGPT in order to imrpove language. After using this tool/service, the author(s) reviewed and edited the content as needed and take(s) full responsibility for the content of the publication.

References

Ahmad, I., Song, X., Hussein Ibrahim, M. E., Jamal, Y., Younas, M. U., Zhu, G., Zhou, G., & Adam Ali, A. Y. (2023). The role of melatonin in plant growth and metabolism, and its interplay with nitric oxide and auxin in plants under different types of abiotic stress. *Frontiers in Plant Science, 14*. https://doi.org/10.3389/fpls.2023.1108507, https://www.frontiersin.org/journals/plant-science.

Allagulova, C. R., Lubyanova, A. R., & Avalbaev, A. M. (2023). Multiple ways of nitric oxide production in plants and its functional activity under abiotic stress conditions. *Federation International Journal of Molecular Sciences, 24*(14). https://doi.org/10.3390/ijms241411637, http://www.mdpi.com/journal/ijms.

Basit, F., Chen, M., Ahmed, T., Shahid, M., Noman, M., Liu, J., An, J., Hashem, A., Fahad Al-Arjani, A.-B., Alqarawi, A. A., Fahad, M., Alsayed, S., Abd_Allah, E. F., Hu, J., & Guan, Y. (2021). Seed priming with brassinosteroids alleviates chromium stress in rice cultivars via improving ROS metabolism and antioxidant defense response at biochemical and molecular levels. *Antioxidants, 10*(7), 1089. https://doi.org/10.3390/antiox10071089.

Beltrán-García, M. J., Martínez-Rodríguez, A., Beltran-García, C., Miranda-Rivera, J. V., Valdez-Salas, B., Di Mascio, P., & White, J. F. (2024). Multifactorial stress combination, plant microbiome recruitment, and reactive oxygen species/antioxidant feedbacks for plant stress alleviation. In *Sustainable agricultural practices*. Elsevier, 1–32. http://doi.org/10.1016/B978-0-443-19150-3.00007-2, https://www.sciencedirect.com/book/9780443191503.

Colombage, R., Singh, M. B., & Bhalla, P. L. (2023). Melatonin and abiotic stress tolerance in crop plants. *International Journal of Molecular Sciences, 24*(8). https://doi.org/10.3390/ijms24087447, http://www.mdpi.com/journal/ijms.

Corpas, F. J., Leterrier, M., Valderrama, R., Airaki, M., Chaki, M., Palma, J. M., & Barroso, J. B. (2011). Nitric oxide imbalance provokes a nitrosative response in plants under abiotic stress. *Plant Science, 181*(5), 604–611. https://doi.org/10.1016/j.plantsci.2011.04.005.

Corpas, F. J., González-Gordo, S., & Palma, J. M. (2021). Protein nitration: A connecting bridge between nitric oxide (NO) and plant stress. *Plant Stress, 2*. http://doi.org/10.1016/j.stress.2021.100026, www.journals.elsevier.com/plant-stress.

Devi, H. S., Chongtham, R., & Roy, S. S. (2023). Exogenous Melatonin in Crop Improvement: A Promising Strategy for Enhanced Productivity. *Agricluture and food e-letter, 5*, 187–189.

Divi, U. K., Rahman, T., & Krishna, P. (2010). Brassinosteroid-mediated stress tolerance in *Arabidopsis* shows interactions with abscisic acid, ethylene and salicylic acid pathways. *BMC Plant Biology, 10*(1). https://doi.org/10.1186/1471-2229-10-151.

do Carmo, G. C., de Mello, J. V. S., Preisler, A. C., de Lima, J. D. O., Pereira, R. H., Pieretti, J. C., Stolf-Moreira, R., Pimenta, J. A., Seabra, A. B., & Oliveira, H. C. (2025). Effects of nanoencapsulated nitric oxide donor on *Cecropia pachystachya* Trécul and *Cariniana estrellensis* (Raddi) Kuntze seedlings subjected to short and long-term water deficit. *Trees - Structure and Function, 39*(4). https://doi.org/10.1007/s00468-025-02656-4, https://www.springer.com/journal/468.

Fancy, N. N., Bahlmann, A. K., & Loake, G. J. (2017). Nitric oxide function in plant abiotic stress. *Plant Cell and Environment, 40*(4), 462–472. https://doi.org/10.1111/pce.12707, http://www.blackwellpublishing.com/journals/PCE.

Fernández-Marcos, M., Sanz, L., Lewis, D. R., Muday, G. K., & Lorenzo, O. (2011). Nitric oxide causes root apical meristem defects and growth inhibition while reducing PIN-FORMED 1 (PIN1)-dependent acropetal auxin transport. *Proceedings of the National Academy of Sciences, 108*(45), 18506–18511. https://doi.org/10.1073/pnas.1108644108.

Freschi, L. (2013). Nitric oxide and phytohormone interactions: Current status and perspectives. *Frontiers in Plant Science, 4*. https://doi.org/10.3389/fpls.2013.00398.

Gharanjik, S., Ebrahimi, A., Somee, L. R., Chaghakaboodi, Z., & Alipour, H. (2025). Systemic role of melatonin in enhancing temperature stress tolerance in fenugreek. *BMC Plant Biology, 25*(1), 1131. https://doi.org/10.1186/s12870-025-07224-z.

Graska, J., Fidler, J., Gietler, M., Prabucka, B., Nykiel, M., & Labudda, M. (2023). Nitric Oxide in Plant Functioning: Metabolism, Signaling, and Responses to Infestation with Ecdysozoa Parasites. *Biology, 12*(7), 927. https://doi.org/10.3390/biology12070927.

Gupta, K. J., Fernie, A. R., Kaiser, W. M., & van Dongen, J. T. (2011). On the origins of nitric oxide. *Trends in Plant Science, 16*(3), 160–168. https://doi.org/10.1016/j.tplants.2010.11.007.

Hajihashemi, S., & Jahantigh, O. (2023). Nitric Oxide Effect on Growth, Physiological and Biochemical Processes, Flowering, and Postharvest Performance of Narcissus tazzeta. *Journal of Plant Growth Regulation, 42*, 892–907. https://doi.org/10.1007/s00344-022-10596-3.

Hansen, M., Chae, H. S., & Kieber, J. J. (2009). Regulation of ACS protein stability by cytokinin and brassinosteroid. *Plant Journal, 57*(4), 606–614. https://doi.org/10.1111/j.1365-313X.2008.03711.x.

Hartman, S., Liu, Z., van Veen, H., Vicente, J., Reinen, E., Martopawiro, S., Zhang, H., van Dongen, N., Bosman, F., Bassel, G. W., Visser, E. J. W., Bailey-Serres, J., Theodoulou, F. L., Hebelstrup, K. H., Gibbs, D. J., Holdsworth, M. J., Sasidharan, R., & Voesenek, L. A. C. J. (2019). Ethylene-mediated nitric oxide depletion pre-adapts plants to hypoxia stress. *Nature Communications, 10*(1). https://doi.org/10.1038/s41467-019-12045-4.

Hussain, A., Faheem, B., Jang, H. S., Lee, D. S., Mun, B. G., Rolly, N. K., & Yun, B. W. (2024). Melatonin–Nitric oxide crosstalk in plants and the prospects of NOMela as a nitric oxide donor. *International Journal of Molecular Sciences, 25*(15). https://doi.org/10.3390/ijms25158535, http://www.mdpi.com/journal/ijms.

Ikram, M., Mehran, M. T., Rehman, H. U., Ullah, S., Bakhsh, M. Z. M., Tahira, M., ... Sabagh, A. E. (2024). Mechanistic review of melatonin metabolism and signaling pathways in plants. *Plant Stress, 14*, 100685. https://doi.org/10.1016/j.stress.2024.100685.

Khan, M., Ali, S., Al Azzawi, T. N. I., & Yun, B. W. (2023). Nitric Oxide Acts as a Key Signaling Molecule in Plant Development under Stressful Conditions. *International journal of molecular sciences, 24*(5), 4782. https://doi.org/10.3390/ijms24054782.

Kumar, D., & Ohri, P. (2023). Say "NO" to plant stresses: Unravelling the role of nitric oxide under abiotic and biotic stress. *Nitric Oxide: Biology and Chemistry / Official Journal of the Nitric Oxide Society, 130*, 36–57. https://doi.org/10.1016/j.niox.2022.11.004.

Kolbert, Z., Feigl, G., Freschi, L., & Poór, P. (2019). Gasotransmitters in action: Nitric oxide-ethylene crosstalk during plant growth and abiotic stress responses. *Antioxidants, 8*(6), 167. https://doi.org/10.3390/antiox8060167.

Kolbert, Z. S., Barroso, J. B., Brouquisse, R., Corpas, F. J., Gupta, K. J., Lindermayr, C., & Hancock, J. T. (2019). A forty year journey: The generation and roles of NO in plants. *Nitric Oxide: Biology and Chemistry / Official Journal of the Nitric Oxide Society, 93*, 53–70. https://doi.org/10.1016/j.niox.2019.09.006.

Kolupaev, Y. E., Shkliarevskyi, M. A., Pyshchalenko, M. A., & Dmitriev, A. P. (2024). Nitric oxide: functional interaction with phytohormones and applications in crop production. *Agriculture and Forestry, 70*(1), 379–411. https://doi.org/10.17707/AgricultForest.70.1.24.

Llanes, A., Andrade, A., Alemano, S., & Luna, V. (2016). Alterations of endogenous hormonal levels in plants under drought and salinity. *American Journal of Plant Sciences, 07*(09), 1357–1371. https://doi.org/10.4236/ajps.2016.79129.

Li, J., & Chory, J. (1997). A putative leucine-rich repeat receptor kinase involved in brassinosteroid signal transduction. *Cell, 90*(5), 929–938. https://doi.org/10.1016/s0092-8674(00)80357-8.

Mahabadi, A. H., Eghlima, G., Seyed Hajizadeh, H., Aghamir, F., Mohammadi, M., & Kaya, O. (2026). Exogenous nitric oxide donor mitigates drought stress in pot marigold (*Calendula officinalis*). *Journal of Plant Growth Regulation,* 1–19. https://doi.org/10.1007/s00344-025-12029-3.

Martínez-lorente, S. E., Pardo-hernández, M., Martí-guillén, J. M., López-delacalle, M., & Rivero, R. M. (2022). Interaction between Melatonin and NO: Action mechanisms, main targets and putative roles of the emerging molecule NOmela. *International Journal of Molecular Sciences, 23*(12). https://doi.org/10.3390/ijms23126646, https://www.mdpi.com/1422-0067/23/12/6646/pdf?version=1655210573.

Montilla-Bascon, G., Cristescu, S. M., Mur, L. A. J., & Prats, E. (2025). Phytoglobin scavenging of nitric oxide is associated with ethylene reduction and drought tolerance in oat (*Avena sativa*). *Physiologia Plantarum, 177*(6). https://doi.org/10.1111/ppl.70597, http://onlinelibrary.wiley.com/journal/10.1111/(ISSN)1399-3054.

Nawaz, F., Shabbir, R. N., Shahbaz, M., Majeed, S., Raheel, M., Hassan, W., & Sohail, M. A. (2017). Cross Talk between Nitric Oxide and Phytohormones Regulate Plant Development during Abiotic Stresses. In *Phytohormones - Signaling Mechanisms and Crosstalk in Plant Development and Stress Responses*. InTech. https://doi.org/10.5772/intechopen.69812.

Nie, S., Huang, S., Wang, S., Mao, Y., Liu, J., Ma, R., & Wang, X. (2019). Enhanced brassinosteroid signaling intensity via SlBRI1 overexpression negatively regulates drought resistance in a manner opposite of that via exogenous BR application in tomato. *Plant Physiology and Biochemistry, 138*, 36–47. https://doi.org/10.1016/j.plaphy.2019.02.014.

Nolan, T. M., Vukašinović, N., Liu, D., Russinova, E., & Yin, Y. (2020). Brassinosteroids: Multidimensional regulators of plant growth, development, and stress responses. *The Plant Cell, 32*(2), 295–318. https://doi.org/10.1105/tpc.19.00335.

Pandey, A., Khan, M. K., Hamurcu, M., Athar, T., Yerlikaya, B. A., Yerlikaya, S., Kavas, M., Rustagi, A., Zargar, S. M., Sofi, P. A., Chaudhry, B., Topal, A., & Gezgin, S. (2023). Role of exogenous nitric oxide in protecting plants against abiotic stresses. *Agronomy, 13*(5). https://doi.org/10.3390/agronomy13051201, www.mdpi.com/journal/agronomy/.

Quamruzzaman, M., Manik, S. M. N., Shabala, S., & Zhou, M. (2021). Improving Performance of Salt-Grown Crops by Exogenous Application of Plant Growth Regulators. *Biomolecules, 11*(6), 788. https://doi.org/10.3390/biom11060788 PMID: 34073871; PMCID: PMC8225067.

Rai, K. K., Pandey, N., Rai, N., Rai, S. K., & Pandey-Rai, S. (2021). Salicylic acid and nitric oxide: Insight into the transcriptional regulation of their metabolism and regulatory functions in plants. *Frontiers in Agronomy, 3.* https://doi.org/10.3389/fagro.2021.781027, www.frontiersin.org/journals/agronomy#.

Sanchez, F. W., Crane, J. H., Bayabil, H. K., Sarkhosh, A., Shahid, M. A., & Schaffer, B. (2025). Brassinosteroid priming mitigates negative physiological responses of *Garcina humilis* to flooding and salinity. *Plant Stress, 16.* https://doi.org/10.1016/j.stress.2025.100892, https://www.sciencedirect.com/science/journal/2667064X.

Sarma, H., Ramchiary, C., Sen, B., Daimary, M., & Prasad, R. (2025). Integration of nitric oxide signaling in plant stress responses: Unveiling its role in enhancing crop tolerance to abiotic stress. *Discover Plants, 2*(1). https://doi.org/10.1007/s44372-025-00102-w.

Sarkar, M. M., Subba, R., Roy, S., & Mathur, P. (2023). Understanding the Role of Nitric Oxide and Its Interactive Effects with Phytohormones in Mitigation of Salinity Stress. In *Gasotransmitters Signaling in Plants under Challenging Environment.* Cham: Springer International Publishing, 121–145.

Shi, L., Chen, Y., Dong, W., Li, S., Chen, W., Yang, Z., & Cao, S. (2024). Melatonin delayed senescence by modulating the contents of plant signalling molecules in postharvest okras. *Front. Plant Sci. 15*, 1304913. https://doi.org/10.3389/fpls.2024.1304913.

Simontacchi, M., Galatro, A., Ramos-Artuso, F., & Santa-María, G. E. (2015). Plant survival in a changing environment: The role of nitric oxide in plant responses to abiotic stress. *Frontiers in Plant Science, 6*(NOVEMBER). https://doi.org/10.3389/fpls.2015.00977, http://journal.frontiersin.org/article/10.3389/fpls.2015.00977/pdf.

Singhal, R. K., Jatav, H. S., Aftab, T., Pandey, S., Mishra, U. N., Chauhan, J., Chand, S., Indu, Saha, D., Dadarwal, B. K., Chandra, K., Khan, M. A., Rajput, V. D., Minkina, T., Narayana, E. S., Sharma, M. K., & Ahmed, S. (2021). Roles of nitric oxide in conferring multiple abiotic stress tolerance in plants and crosstalk with other plant growth regulators. *Journal of Plant Growth Regulation, 40*(6), 2303–2328. https://doi.org/10.1007/s00344-021-10446-8, https://www.springer.com/journal/344.

Song, X., Xu, Z., Zhang, J., Liang, L., Xiao, J., Liang, Z., Yu, G., Sun, B., Huang, Z., Tang, Y., Lai, Y., & Li, H. (2023). NO and GSH Alleviate the inhibition of low-temperature stress on cowpea seedlings. *Plants, 12*(6), 1317. https://doi.org/10.3390/plants12061317.

Terrile, M. C., París, R., Calderõn-Villalobos, L. I. A., Iglesias, M. J., Lamattina, L., Estelle, M., & Casalongué, C. A. (2012). Nitric oxide influences auxin signaling through S-nitrosylation of the *Arabidopsis* TRANSPORT INHIBITOR RESPONSE 1 auxin receptor. *Plant Journal, 70*(3), 492–500. https://doi.org/10.1111/j.1365-313X.2011.04885.x.

Waqas, M. A., Kaya, C., Riaz, A., Farooq, M., Nawaz, I., Wilkes, A., & Li, Y. (2019). Potential mechanisms of abiotic stress tolerance in crop plants induced by thiourea. *Frontiers in Plant Science, 10.* https://doi.org/10.3389/fpls.2019.01336, https://www.frontiersin.org/journals/plant-science.

Xu, J., Lu, X., Liu, Y., Lan, W., Wei, Z., Yu, W., & Li, C. (2024). Interaction between ABA and NO in plants under abiotic stresses and its regulatory mechanisms. *Frontiers in Plant Science, 15.* https://doi.org/10.3389/fpls.2024.1330948, https://www.frontiersin.org/journals/plant-science.

Youssef, S., & Adawy, A. (2023). Exogenous Application of Arginine Alleviates the Adverse Effects of Nacl-Salt Stress on Calendula Officinalis L. Plants. *Scientific Journal of Flowers and Ornamental Plants, 10*(3), 191–215. https://doi.org/10.21608/sjfop.2023.238192.1022.

Zhu, T., Deng, X., Zhou, X., Zhu, L., Zou, L., Li, P., & Lin, H. (2016). Ethylene and hydrogen peroxide are involved in brassinosteroid-induced salt tolerance in tomato. *Scientific Reports, 6*(1), 35392. https://doi.org/10.1038/srep35392.

Zhu, T., Deng, X. G., Tan, W. R., Zhou, X., Luo, S. S., Han, X. Y., Zhang, D. W., & Lin, H. H. (2016). Nitric oxide is involved in brassinosteroid-induced alternative respiratory pathway in Nicotiana benthamiana seedlings' response to salt stress. *Physiologia Plantarum, 156*(2), 150–163. https://doi.org/10.1111/ppl.12392.

Zulfiqar, F., Nafees, M., Moosa, A., Ferrante, A., & Darras, A. (2024). Melatonin induces proline, secondary metabolites, sugars and antioxidants activity to regulate oxidative stress and ROS scavenging in salt stressed sword lily. *Heliyon, 10*(11), e32569. https://doi.org/10.1016/j.heliyon.2024.e32569 PMID: 38961974; PMCID: PMC11219490.

Chapter 15

Potential role of nitric oxide in sustainable agriculture

Kamal Kant[1], Aniruddhabhai Khuman[2], Ambreen Tahir Shah[3], and Abu Imran Baba[4]

[1]*Institute of Plant Biology, HUN-REN Biological Research Centre, Szeged, Hungary,* [2]*Department of Bioinformatics, Marwadi University, Rajkot, Gujarat, India,* [3]*Department of Botany, University of Kashmir, Srinagar, Jammu and Kashmir, India,* [4]*Department of Forest Genetics and Plant Physiology, Umeå Plant Science Centre (UPSC), Swedish University of Agricultural Sciences, Umeå, Sweden*

15.1 Introduction

Nitric oxide (NO) is a diatomic gaseous molecule known to be important in various physiological processes, and in certain aspects, its synthesis, signaling, and impact are specific in plants (Freschi, 2013). It was initially characterized as an air contaminant, but is now a well-known multifunctional bioactive substance that combines environmental, hormonal, and redox signals to modulate plant growth and development (Naaz et al., 2025). It was first discovered as a gaseous signaling molecule in eukaryotes, however, its formation and signaling exhibit several characteristics that are unique to the plant kingdom (Kolbert et al., 2021). In plants it is primarily produced endogenously by nitrate reductase (NR) and NO synthase-like (NOS-like) activities under physiologically relevant conditions (Freschi, 2013). Several studies highlight the synthesis of NO in plants from nitrite, polyamines, and L-arginine, both from nonenzymatic or enzymatic mechanisms (Astier & Lindermayr, 2012). Moreover, plants can also produce NO from nitrite through the mitochondrial electron transport chain (Freschi, 2013), an anaerobic mechanism that depends on the availability of nitrite (Gupta & Igamberdiev, 2011; Planchet et al., 2005). Different organelles, such as peroxisomes, chloroplasts, and mitochondria, are all potential sites of NO production that function in plant metabolism (Gupta et al., 2022; Kolbert et al., 2019). Despite the lack of known plant NOS homologs, peroxisomes are thought to exhibit NOS-like activity. Transfer of NO bioactivity is known to be regulated by several posttranslational modifications (PTMs) that regulate different

Nitric Oxide in Plant Resilience to Abiotic Stress. DOI: https://doi.org/10.1016/B978-0-443-40497-9.00006-4

functions, such as plant protein activity, stress responses, and signaling (Kolbert et al., 2021). The most prominent of these is protein S-nitrosation, which involves the reversible covalent attachment of a NO group to cysteine thiols, generating S-nitrosothiols (Astier & Lindermayr, 2012 ; Lindermayr et al., 2005). This dynamic redox-based alteration controls enzyme activity, the function of transcription factors, protein–protein interactions, and shields thiols from irreversible oxidation (Mur et al., 2011). In addition to S-nitrosylation, NO causes tyrosine nitration, which is typically irreversible and mediated by peroxynitrite, and metal nitrosylation, which impacts metalloprotein activity and redox signaling (Corpas et al., 2013). As the production of NO is facilitated by the enzymatic activities of NR and NOS-like activities, its increased activity is observed during certain early physiological processes in plants, such as during seed imbibition (Corpas et al., 2009). Seed germination is one of the crucial and complex mechanisms that involve various biochemical and physiological cues such as light, water, and phytohormones (Sen, 2010). NO serves an essential role in seed germination and early development of seedlings, and it is a crucial part of signaling networks that regulate cellular growth and various responses in conjugation with calcium and cyclic guanosine monophosphate (cGMP) (Šírová et al., 2011). Exogenous NO improves seed germination and seedling vigor under hexavalent chromium (Cr(VI)) toxicity by improving α-amylase and protease activities, raising antioxidants (such as ascorbate and glutathione (GSH)), and regulating proline metabolism (Khan et al., 2021). In seed germination, NO is generated by both enzymatic and nonenzymatic processes (Moreau et al., 2010). The nonenzymatic reduction of nitrites to NO can happen effectively and quickly in situations with high nitrate availability, low pH, or strong reducing environments. In the seed apoplast, nitrite is a crucial nonenzymatic source, particularly in acidic and reducing environments. Reduced NO synthesis that is dependent on NR and nitric oxide-associated protein 1 (NOA1) is confirmed to improve dormancy and abscisic acid (ABA) sensitivity in *Arabidopsis* mutant studies (Lozano-Juste & León, 2011). This suggests how NO can encourage early growth and germination under harsh environmental conditions. NO also acts as a key signaling molecule regulating root growth and development. It affects root growth through either promotive or inhibitory mechanisms, depending upon different parameters such as concentration and various developmental stages. Previous studies show that primary root elongation is often restricted by exogenous NO donors or by NO-overproducing mutants, while primary root growth is typically accelerated in stress situations or in mutants with low endogenous NO levels (Sun et al., 2017). NO synthesis in roots is carried out via NR, nitrite:NO reductase, and various oxygen-sensitive enzymatic pathways (Fig. 15.1). Under hypoxic or anoxic environmental conditions, NO helps roots to remain functioning in aerenchyma development and maintain cellular redox balance (Gupta et al., 2020). Interestingly, root NO also manages the relationship between nitrogen-fixing bacteria and mycorrhizal fungi, which helps roots in nutrient acquisition,

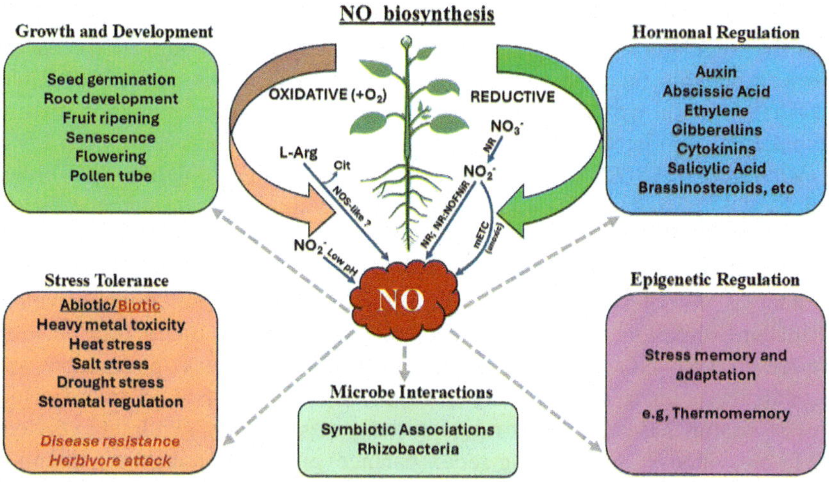

FIGURE 15.1 NO production and its diverse regulatory functions play crucial roles in controlling plant growth, development, and adaptive responses to environmental cues (*L-Arg,* L-arginine; Cit, Citrulline; NR, nitrate reductase; NOS-like, nitric oxide synthase-like; mETC, mitochondrial electron transport chain (anoxic); NOFNiR, nitric oxide-forming nitrite reductase; NO, Nitric oxide).
Some part of this figure was generated with the help of image illustrator BioRender (BioRender.com).

various stress tolerances, and root-microbe communication (Stöhr & Stremlau, 2006). NO also helps in regulating the formation of lateral roots (LR) and root hair formation (RHF). In *Arabidopsis* and lettuce, NO donors increase RHF, whereas NO scavengers or NO-deficient mutants inhibit RHF. NO, which frequently acts through auxin signaling pathways, builds up in LR primordia, encouraging LR emergence and elongation (Lombardo et al., 2006). NO is known to inhibit leaf senescence by inhibiting chlorophyll degradation and maintaining viablethylakoid membranes. It also protects rice leaves from ABA-induced senescence by decreasing oxidative stress and maintaining antioxidant levels (Hung & Kao, 2003).

NO is also an important regulator of ABA-induced stomatal regulation. It is observed that the NR double mutant *nia1/nia2* of *Arabidopsis* cannot induce NO synthesis in response to ABA and H_2O_2.. This suggests the involvement of an additional intermediate in NO–ABA crosstalk (Bright et al., 2006). The concentration of NO affects variably: at low concentrations, it promotes leaf expansion, but at higher concentrations, it inhibits growth (Brito et al., 2021). Moroever, NO in combination with reactive oxygen species (ROS), helps to maintain the water balance of leaves by regulating the production of ABA (Xing et al., 2004).

NO also plays a crucial role as a key regulator in flowering and reproduction in plants. Elevated NO levels delay flowering by inhibiting the expression of CONSTANS (CO) and GIGANTEA (GI), both of which are positive regulators of the floral transition (Krasylenko et al., 2010). At the same time, this increase in NO content raises the levels of FLOWERING LOCUS C, a known floral repressor. Flowering is regulated by NO in relation to circadian regulators, *CCA1, TOC1 (CIRCADIAN CLOCK ASSOCIATED 1, TIMING OF CAB EXPRESSION 1),* and light-responsive genes, *chlorophyll a/b binding protein (CAB), CO,* and *GI*, thereby linking environmental factors to the control of flowering (He et al., 2004). In reproductive organs, NO is also generated by pollen grains and is involved in the pollen stigma signaling system. It promotes appropriate fertilization by directing pollen tube development and direction through cGMP-mediated pathways. Interaction between pollen-derived NO and stigma-generated ROS initiates pollination and establishes cellular compatibility (Prado et al., 2004). NO is an important key regulator that delays fruit ripening and senescence by regulating ethylene (ET) biosynthesis, maintaining redox balance, and modulating fruit ripening-associated genes. Besides the 1-methylcyclopropene, which is known to block ET perception (Zhang, Ma, et al., 2020). It also acts as an inhibitory molecule for ET production through silencing of 1-carboxylic acid oxidase genes *ACO1*, *ACOH2*, and *ACO4*, including S-nitrosation of methionine adenosyl transferase, decreasing 1-aminocyclopropane-1-carboxylic acid (a precursor of ET) conversion to ET (Hussain et al., 2022). For preventing oxidative damage of fruit tissues, NO also controls defense genes and antioxidant enzymes through balancing ROS and RNS. In various fruits like banana, tomato, pepper, apple, and strawberry, NO slows down the process of ripening, also altering carotenoid and flavonoid metabolism; this alteration increases the postharvest shelf life of various fruit plants (Corpas et al., 2018). Moreover, in *Lycium barbarum* fruit, NR-derived NO, which suppresses the formation of anthocyanin biosynthesis, also interacts antagonistically with ABA (Li et al., 2019). In a nutshell, NO serves as an important signaling molecule that controls many aspects of plant growth and development, including seed germination, root, shoot development, flowering, and senescence. It also regulates various developmental and physiological processes and modifies redox signaling via S-nitrosation and nitration.

15.2 Role of nitric oxide in stress tolerance

Over the past few years, NO has emerged as an important key signaling molecule in various physiological events, especially in response to biotic and abiotic stress (Shi et al., 2012). NO plays a dual and complex role in plant responses to cadmium (Cd) toxicity. Under Cd stress, NO often enhances plant tolerance by reducing Cd accumulation, by promoting Cd sequestration in the cell wall, and by regulating Cd uptake, transport, and gene expression during detoxification (Wang et al., 2023). It also mitigates oxidative stress by directly scavenging ROS and activating key antioxidant defense enzymes.

However, NO can also intensify Cd toxicity by stimulating Fe-uptake genes with increasing Cd absorption by initiating a cellular pathway similar to that initiated upon iron deprivation (Besson-Bard et al., 2009). Similarly, under arsenic (As) and lead (Pb) stress, NO enhances plant tolerance by increasing reduced GSH and phytochelatin production, maintaining the membrane integrity, and reducing metal transfer from roots to shoots (Mostofa et al., 2015). Further studies focused on NO synthesis pathways, target molecules, endogenous signaling mechanisms, and their interaction with hormones will help us to understand how NO modulates Cd stress tolerance in plants (Meng et al., 2022). Moreover, NO also plays a crucial role in plant tolerance to heat stress by regulating antioxidant defenses, stress signaling pathways, and photosynthetic efficacy. In various species, incorporating exogenous NO donors like sodium nitroprusside (SNP), S-nitroso-N-acetylpenicillamine, and diethylenetriamine nitric oxide adducts has been shown to increase heat tolerance (Lee et al., 2008). On the other hand, these defensive effects are reversed by NO scavengers such as 2–4-carboxphenyl-4,4,5,5-tetramethylimidazoline-1-oxyl-3-oxide, 2-phenyl-4,4,5,5-tetramethylimidazoline-1-oxyl 3-oxide, and nitro-L-arginine methyl ester (Parankusam et al., 2017). In *Triticum aestivum* and *Phragmites communis*, NO alleviates heat-induced oxidative stress by stimulating antioxidants and methylglyoxal detoxification pathways (Karpets et al., 2011). Endogenous NO also promotes the heat shock protein accumulation and the activation of heat shock transcription factors in *Arabidopsis thaliana* (Hossain et al., 2010). Overall, it is emphasized that NO acts as a key redox regulator that combines hormonal, antioxidant, and metabolic signals, allowing plants to tolerate a variety of abiotic challenges while maintaining physiological stability and production.

Drought stress is another major abiotic stress factor that affects global crop productivity. NO is reported to influence DNA methylation and regulate drought-responsive genes and antioxidant-related genes, including the transcription factors (Nabi et al., 2019). In multiple plant species, NO also increases ROS scavenging by activating key specific antioxidant enzymes in salinity-induced damage. For instance, NO treatment leads to increased activities of SOD, APX, and CAT (superoxide dismutase, ascorbate peroxidase and catalase), diminishing the mitochondrial efflux of H_2O_2 and O_2^-, and preventing malondialdehyde (MDA) buildup (Zheng et al., 2009; Ahmad et al., 2016). It also reduces lipid peroxidation, maintaining membrane stability and impacting the photosynthetic efficiency. NO was also reportedly found to be crucial in brassinosteroid (BR)-dependent *alternative oxidase (AOX)* activity. This study illustrated the synergistic interaction with BRs, triggering H_2O_2-mediated NO synthesis, which is important in maintaining photosystem II efficiency and AOX activity (Arfan et al., 2019). Interestingly, exogenous NO applications demonstrated increased ABA accumulation during drought conditions, an effect that NO scavengers usually prevent, emphasizing its bidirectional impact on ABA signaling (Singhal et al., 2021). It enhances tolerance against salt stress in various plant species by regulating the expression of plasma membrane H^+-ATPase, which plays an important role in maintaining the ratio of K^+:Na^+ and reducing cytoplasmic Na^+ levels. This scenario has been observed in reed calluses and in *Arabidopsis*, where *Atnoa1* mutants with impaired NO production exhibit higher Na^+/K^+ ratios and increased salinity sensitivity (Nabi et al., 2019). NO plays a pivotal role in redox signaling, which controls plant physiological and molecular responses, helping to mitigate various abiotic stresses such as drought and flooding. Under drought conditions, NO interacts with ABA signaling to cause stomatal closure, which reduces transpiration and water loss (García-Mata & Lamattina, 2002). Furthermore, NO also promotes the accumulation of osmoprotectants such as proline and soluble sugars, maintaining membrane stability, and also plays a role in the S-nitrosation-based control of drought-responsive proteins (Tanou et al., 2012). Under flooding or hypoxic conditions, NO acts as a critical signaling molecule, promoting aerenchyma production, adventitious root growth, and maintaining cellular redox balance via modulating enzyme activities such as NR and hemoglobins (Hbs) (Planchet et al., 2005). Furthermore, exogenous application of NO donor SNP mitigates NaCl-induced inhibition of seed germination and root growth in yellow lupine seedlings (Singhal et al., 2021). Collectively, NO thus acts as a flexible mediator, integrating hormonal and redox signals to improve plant tolerance to both water- and oxygen-stressed situations.

15.3 Nitric oxide and plant–microbe interactions

The symbiotic relationship between plants and microbes is fundamental to many ecological and physiological processes that are essential for plant health and productivity. The rhizosphere is not only a medium for plant nutrition and anchorage, but it is also a dynamic ecosystem inhabited by diverse microbial communities. Within this environment, plants continuously interact with beneficial and pathogenic microbes via an intricate communication network that enables them to distinguish allies from enemies. This molecular network is mediated by specialized metabolites and root exudates, which regulate gene expression in the plant, the microbe, or both, which shapes the outcome of their interaction. These interactions can lead to variable changes like increased plant growth, suppression of soil-borne pathogens, improved nutrient uptake, formation of biofilms, and greater microbial colonization of the soil (Li et al., 2013; Mommer et al., 2016; Sasse et al., 2018). NO occupies a central position in this network among the various other physiological molecules that are involved. As plants possess multiple sources of NO, as previously mentioned, they are primarily produced through the activity of NR and NOS-like pathways (Shah et al., 2023). Being such a key signaling molecule, NO thus mediates various physiological processes, including plant growth, development, and defense responses in plants (Pande et al., 2021).

15.3.1 Nitric oxide in symbiotic associations

Symbiosis is a complex living association involving dynamic changes in the genome, metabolism, and signaling networks of different interacting organisms (Varma et al., 2020). Here, we detail inputs on plant-microbe symbiosis, which represents one of the most intricate and widespread forms of biological interaction. Based on its spatial and temporal production, NO appears to play multiple roles in the symbiotic process. It contributes to plant-microbe symbiotic interactions through multiple mechanisms, including facilitating root colonization, modulating root exudate profiles, enhancing nitrogen (N_2) fixation, and promoting biocontrol activity (Di Palma et al., 2013; Wang & Ruby, 2011). Moreover, leguminous plants form symbiotic associations with soil bacteria known as rhizobia, leading to the development of specialized organs called nodules. Within these nodules, differentiated bacteroids enclosed by plant-derived membranes in infected cells convert atmospheric N_2 into ammonium (NH_3), providing a usable nitrogen source to support the growth of the host plant (Ma & Chen, 2021). Although long recognized as a toxic molecule that inhibits N_2 fixation in mature nodules, NO also functions as an important signaling molecule that regulates the expression of genes involved in nodule formation, cell division, and various metabolic pathways (Meilhoc et al., 2011). Further findings also suggest that NO contributes metabolically to maintain cellular energy balance, which is necessary for N_2 fixation under low-oxygen conditions (Signorelli et al., 2020). Thus, NO has been shown to exert both stimulatory and inhibitory effects on symbiotic N_2 fixation.

In legume rhizobia symbiosis, NO plays key roles in bacterial infection, nodule formation, and the functioning of mature nodules. Previous findings further revealed that elevated NO levels within *Medicago truncatula* root nodules can induce premature nodule senescence (Meilhoc et al., 2013). NO has been demonstrated to play a positive role during the stages of infection, root nodule formation, and the functioning of mature nodules (del Giudice et al., 2011; Horchani et al., 2011). Overall, NO appears to participate in seemingly opposing processes during symbiosis, suggesting that its concentration and spatial distribution critically determine its function (Signorelli et al., 2020). Thus, maintaining an optimal NO level in nodules is essential for enabling its beneficial roles while preventing premature senescence (Meilhoc et al., 2013). Hbs are considered key regulators of NO levels, as evidence indicates that plant globins not only facilitate oxygen transport but also influence development and morphogenesis by locally modulating endogenous NO concentrations within specific tissues (Hebelstrup et al., 2013). Initial evidence for the presence of NO in nitrogen-fixing nodules was obtained through the detection of NO bound to leghemoglobin (Lb), the principal hemoprotein found in legume nodules (Kanayama & Yamamoto, 1991; Maskall et al., 1977; Mathieu et al., 1998; Meakin et al., 2007; Sanchez et al., 2010).

The presence of a functional NO regulatory system in both mycorrhizal and root nodule symbioses suggests that Phytoglobin1-mediated NO regulation plays a common and conserved role across diverse plant-microbe associations (Fukudome & Uchiumi, 2024). An increasing number of studies have shown that NO is generated during the establishment of mutualistic associations between plants and fungi (Calcagno et al., 2012; Espinosa et al., 2014; Martínez-Medina, Pescador, Fernandez, et al., 2019; Gupta et al., 2014). Although the precise functions of NO in these plant fungal interactions are not yet fully understood, a few findings indicate that maintaining a finely balanced NO concentration is essential for the proper regulation of mycorrhizal symbiosis (Martínez-Medina, Pescador, Fernandez, et al., 2019; Martínez-Medina, Pescador, Terrón-Camero, et al., 2019). Localized NO accumulation in roots also facilitates fungal colonization and enhances nutrient uptake, particularly phosphorus. Such NO-mediated signaling fosters nutrient-efficient crop systems and reduces fertilizer dependence. Thus, NO facilitates the establishment of symbiotic relationships between plant roots, rhizobia, and arbuscular mycorrhizal fungi, thereby enhancing the plant's ability to acquire essential nutrients, such as nitrogen and phosphorus, from the soil (Shah et al., 2023). Therefore, maintaining NO homeostasis is critical for symbiotic efficiency, including optimal nitrogen economy, which is a key aspect of sustainable agriculture.

15.3.2 Role of nitric oxide in beneficial rhizobacteria and defense

NO plays a central role in plant interactions with beneficial rhizobacteria, enhancing growth, nutrient uptake, and stress tolerance in plants. Plant growth-promoting rhizobacteria (PGPR) exert their effects through both direct and indirect mechanisms, with NO acting as a key signaling mediator in many of these processes. Studies have shown that NO enhances the establishment and functional efficiency of PGPR under saline conditions, suggesting that modulating NO levels could improve soybean growth and performance during salinity stress (Vaishnav et al., 2016). Notably, an NO-overproducing mutant of *Pseudomonas fluorescens* exhibited greater effectiveness in forming beneficial interactions compared to the wild-type strain (Wang et al., 2001). Endogenous NO production in *Azospirillum brasilense* Sp245 has also been reported to play a key role in biofilm formation and promotion of root growth in host plants (Di Palma et al., 2013). Additionally, combined treatment with the NO donor SNP and consequently *Pseudomonas simiae* inoculation significantly enhanced soybean growth and salt stress tolerance (Vaishnav et al., 2016). These findings highlight NO as a promising tool for rhizosphere modulation, with applications in sustainable biofertilizers and biostimulants.

Beyond rhizobacteria, NO is also a critical mediator of plant defense against various pathogens. Reactive nitrogen species, including NO, nitrite, nitrate, and peroxynitrite, possess antimicrobial properties and contribute to the pathogen-killing activity of activated plant immune responses (Bogdan, 2015). An early burst of NO in plant cells following pathogen attack acts as a signaling molecule in defense responses, contributing to basal resistance across a diverse range of plant–pathogen interactions (Asai & Yoshioka, 2009; Mur et al., 2006; Schlicht & Kombrink, 2013; Zeier et al., 2004). Moreover, NO plays a pivotal role in plant defense mechanisms against various pathogen invasions by its involvement in both basal resistance and hypersensitive response (HR). At the time of pathogen infection, a rapid NO burst signifies the active role of NO in early defense signaling in plants (Delledonne et al., 1998). NO also plays a role in resistance activation, which is elaborated by the suppression of HR against *Pseudomonas syringae* pv. harboring the *avrRpm1* gene when NO generation is repressed by NOS inhibitors (Shi et al., 2012). NO regulates the expression patterns of defense-related genes, including pathogenesis-related protein 1 (PR1) and phenylalanine ammonia-lyase (PAL), which play important regulatory roles in salicylic acid (SA)-mediated signaling and phenylpropanoid biosynthesis, respectively. Interestingly, NO regulates chalcone synthase gene expression, resulting in promoting flavonoid biosynthesis and lignin deposition, which are key components for structural barriers against pathogen interaction with the plant cell (Krasylenko et al., 2010). Previous studies also suggest that inhibiting NO synthesis increases pathogen growth, confirming its role in R gene-mediated resistance against *Pseudomonas syringae*. Microarray analyses in *A. thaliana* revealed that NO donors such as SNP, NOR-3 ((Z)-1-(N-Nitroso-N-methylamino)-2-hydroxy-1-propene), and S-Nitrosoglutathione (GSNO) upregulated defense-related genes like *peroxidase*, *glutathione S-transferase*, *PR-proteins*, and *AOX1a*, which enhanced oxidative stress tolerance and activated defense responses (Hung & Kao, 2003). Additionally, NO also regulates nonexpresser of pathogenesis-related genes 1 (NPR1)/TGA1-dependent signaling, where SA promotes NPR1 nuclear translocation, and S-nitrosation of NPR1 and TGA1 increases their interaction and activates downstream PR genes (Lindermayr et al., 2010).

During herbivore attacks, NO accumulates rapidly at the feeding site, where it combines with ROS to produce a localized oxidative burst that activates defense-related genes and secondary metabolite pathways. Therefore, NO is a multifunctional signaling molecule that helps plants defend against herbivorous insects and pathogens. NO also controls the production of jasmonic acid (JA) and SA, two important defense hormones, impacting both local and systemic defensive responses (Johnston, 2023). In herbivore-infested tissues, NO promotes tissue healing by inducing proteinase inhibitors, polyphenol oxidases, and phenylpropanoid pathway enzymes (Kumar & Ohri, 2023). Furthermore, NO is essential for systemic acquired resistance (SAR), a long-lasting, broad-spectrum immune response that occurs following localized infection or tissue injury. In this context, NO works in tandem with SA to modulate the redox status and activate the master regulator NPR1 via S-nitrosation-dependent conformational changes, allowing for nuclear translocation and the subsequent induction of PR genes (Moreau et al., 2010; Wilson et al., 2023). Moreover, NO also regulates the expression of SAR-associated signaling molecules such as azelaic acid, pipecolic acid, and methyl salicylate, which aid in systemic signaling to distal tissues (Shah & Zeier, 2013). Through these interactions, NO serves as a key molecular integrator, connecting local defense events during herbivory or pathogen infection to systemic immune responses, ensuring coordinated activation of both jasmonate- and salicylate-dependent defense pathways that collectively improve plant resilience. Moreover, NO also contributes to the HR response by modulating defense gene expression, activating MAP kinase cascades, and promoting SA-dependent SAR (Asai et al., 2010). Together with ROS, NO forms peroxynitrite ($ONOO^-$), which modifies proteins and lipids to amplify defense signaling (Mur et al., 2013). However, NO can have dual effects; excessive or prolonged NO production may suppress defense or facilitate pathogen colonization by S-nitrosating key immune regulators (Borrowman et al., 2023). Thus, balanced NO generation is essential for effective immunity without the detrimental metabolic consequences. Emerging evidence suggests that plants with optimized NO signaling recruit beneficial microbes, which enhance nutrient cycling, suppress pathogens, and improve tolerance to abiotic stresses. In-depth studies of NO signaling in rhizobacteria and defense may therefore become a promising strategy for designing sustainable and microbiome-enriched crops.

15.4 Nitric oxide and plant hormones

NO is emerging as an important gaseous free radical that is a central molecular regulator in different physiological processes. Different studies have described the molecular mechanisms that elucidate the role of NO in these processes and their interaction with other plant hormones. Interactions between NO and antioxidants, as well as various key plant hormones, further expand the signaling role of NO in plant growth and development under both normal and stress conditions. The NO-hormonal crosstalk fine-tunes the balance between growth and defense mechanisms, thereby improving plant resilience to environmental stressors. For instance, NO regulates leaf senescence by modulating the activity of key phytohormones, including ABA, ET, and JA (Hussain et al., 2022). Moreover, NO can also promote root nodule formation by stimulating the cytokinin (CK) signaling pathway while simultaneously modulating or interfering with auxin and ABA signaling pathways (Signorelli et al., 2020). Often, NO acts synergistically or antagonistically with these hormones, forming highly intricate signaling networks. Increasing evidence suggests that nearly all major phytohormones are influenced by NO through its effects on their biosynthesis, perception, metabolism, transport, and downstream signaling pathways (Freschi, 2013). The following section briefly highlights some of the pivotal roles of NO and its interactions with these key plant hormones.

15.4.1 Nitric oxide and auxin

Functionally, NO works in coordination with the hormone auxin to regulate numerous plant processes, such as root organ formation, root gravitropism nodule development, initiation of cell division, induction of embryogenic cells, and enhancement of NR activity (Pande et al., 2021). These processes directly contribute to improved efficiency of nutrient and water acquisition essential for sustainable crop production under different stresses. Under various stress conditions, studies conducted on *Arabidopsis* plants or with SNP treatment have shown that NO can influence auxin homeostasis by regulating its biosynthesis, degradation, conjugation, distribution, and signaling pathways, while also promoting auxin influx through upregulation of the *AUXIN-RESISTANT 1 (AUX1)* gene (Liu et al., 2018). Further evidence suggests that NO functions as a downstream signaling molecule promoting auxin-induced lateral and adventitious root formation. It can stimulate the synthesis of cGMP via guanylyl cyclase to enhance adventitious root development or facilitate root growth by activating the mitogen-activated protein kinase signaling pathway (Pagnussat et al., 2003).

Auxin-dependent root responses diminish significantly when NO is excluded (Correa-Aragunde et al., 2006; Pagnussat et al., 2004). The distribution of endogenous NO in roots has been shown to regulate auxin, leading to its asymmetric distribution (París et al., 2018). This spatial regulation of NO also contributes to the root gravitropism. The application of gravitational forces, together with the combined action of NO and auxin, has been shown to elevate cGMP levels, which govern the regulation of gravitropic bending. Interference with such processes in plants occurs with blocked auxin transport or when the NO activity is inhibited (Hu et al., 2005). Furthermore, a study on CAB protein underexpressed 1/NO overproducer 1 (*cue1/nox1*) mutants displaying altered NO levels demonstrated that elevated NO concentrations can inhibit auxin transport and signaling through a PIN1-protein-dependent mechanism. This inhibition indeed reduces the root meristem activity in such a scenario (Fernández-Marcos et al., 2011).

The crosstalk between NO and auxin becomes even more intricate under certain metal-induced stress conditions. Under Cd exposure (Yuan & Huang, 2016), NO disrupts auxin synthesis, transport, and signaling, impairing root meristem development during mercury stress (Chen et al., 2015). Exogenous NO application enhances auxin transport that improves stress tolerance under iron deficiency (Sun et al., 2017). NO also diminishes auxin levels, thereby inhibiting root elongation under Cd and arsenic exposure (Piacentini et al., 2020; Ronzan et al., 2018). Therefore, NO plays a role in mitigating harmful effects on auxin homeostasis and root growth. NO is also described to influence auxin biosynthesis, transport, and signal transduction. Indeed, it is known to modulate key enzymes YUCCA and tryptophan aminotransferase of *Arabidopsis* involved in auxin biosynthesis and affect the distribution of PIN transporters, altering the distribution of the auxin gradient (Fernández-Marcos et al., 2011; Sánchez-Vicente et al., 2021). At the signaling level, NO promotes S-nitrosation of auxin receptors and transcription factors, such as the Transport Inhibitor Response/Auxin Signaling F-Box proteins and Auxin Response Factors (Jing et al., 2023; Terrile et al., 2012), thereby regulating auxin sensitivity. Enzymes such as NR are proposed as a primary source of auxin-induced NO production involved in specific root responses (Kolbert & Erdei, 2014). Taken together, all these studies describe the potential role of NO as a regulator of auxin biosynthesis, transport, and signaling, managing plant responses under normal and stressful conditions.

15.4.2 Nitric oxide and abscisic acid

Among different plant hormones, ABA is also shown to interact with NO to coordinate stress perception and adaptive signaling. NO and ABA display an active role via crosstalking through intricate signaling pathways triggered by environmental stresses, guiding adaptive responses (Sun et al., 2021; Ullah et al., 2022). Under certain abiotic stress responses, elevated ABA levels initiate signaling cascades that regulate antioxidant enzyme activity, chlorophyll metabolism, gene expression, and stomatal movement. NO also acts as a mediator in the signaling mechanism that enhances these ABA-mediated responses in plants (Neill et al., 2008). The symbiotic interaction between NO and the hormone ABA includes its pivotal function in mobilizing plant defense mechanisms in stress responses. For instance, NO collaborates with ABA to induce drought tolerance and enhance resilience to nutrient deficiency by modulating Crassulacean acid metabolism gene expression (Mioto & Mercier, 2013) and activating defense-related pathways (Khan et al., 2019). In *Arabidopsis*, NO has been shown to upregulate *abscisic aldehyde oxidase 3* and *9-cis-epoxycarotenoid dioxygenase 3*, two key genes in ABA biosynthesis, thereby increasing drought tolerance (Khan et al., 2019). A central aspect of NO–ABA crosstalk lies in the regulation of stomatal movement, a key process in water-use efficiency and stress adaptation. NO is also strongly associated with ABA-mediated stomatal closure in *Arabidopsis* (Bright et al., 2006). During stress conditions, ABA induces NO synthesis in guard cells, which triggers stomatal closure through calcium release and activation of S-type anion channel 1 (SLAC1-type) channels (Gayatri et al., 2013).

Furthermore, NO also modulates the impact and duration of ABA signaling through regulatory mechanisms. For instance, NO generated through ABA signaling promotes S-nitrosation of SnRK2.6, further inhibiting its kinase activity and limiting the excessive ABA responses (Wang et al., 2015). Impaired GSNO reductase activity leads to GSNO accumulation, which reduces ABA sensitivity, forming a negative feedback loop that in turn fine-tunes ABA signaling (Wang et al., 2015). Beyond its role in signaling, NO also affects ABA metabolism, including ion transport. It induces ABA catabolism, maintaining a balance between growth and stress response, and directly impacts the calcium-independent outward-rectifying K^+ channels, likely occurring through specific PTMs (Sokolovski & Blatt, 2004).

15.4.3 Nitric oxide and ethylene

NO and ET generally exhibit a negative correlation from a physiological perspective in plants. The application of exogenous NO delays the senescence of both vegetative and reproductive organs by suppressing the biosynthesis of ET enzymes (Manjunatha et al., 2012). Both NO and ET are produced during plant development, including stress situations, and they regulate the level of each other. The interaction between both NO-ET influences key growth processes such as fruit ripening and de-etiolation. Under certain abiotic stress conditions, however, interactions can be either antagonistic,or synergistic (Kolbert et al., 2019; Manjunatha et al., 2012). ET production is regulated by NO (Mukherjee, 2019) and is partially known to inhibit ET biosynthesis and slow ripening (Wang et al., 2025). Conversely, ET can induce NO biosynthesis by activating NR and NOS-like enzyme activities (Xu et al., 2017). There is also emerging evidence that suggests that NO and ET, together participate synergistically in UV-B-induced ethylene synthesis in maize seedlings and instomatal closure, highlighting their coordinated role in stress responses (Wang et al., 2006; Kolbert et al., 2019).

In mature green tomatoes, inhibition of the synthesis of NO can cause a reduction in ET release and delay ripening by suppressing the enzyme activities of 1-aminocyclopropane-1-carboxylate synthase (ACS) and 1-aminocyclopropane-1-carboxylate oxidase (ACO) (Yang et al., 2016). This is also followed by the downregulation of *Solanum lycopersicum* ACC synthase 2/4 and *Solanum lycopersicum* ACC oxidase 1/3, including calcium-dependent protein kinase genes (Yang et al., 2016). Overall, NO thus acts as a negative regulator of different processes, including ET biosynthesis, delaying fruit ripening and senescence, particularly under hypoxic conditions.

15.4.4 Nitric oxide and cytokinins

NO is also known to closely interact with CKs to regulate different levels of plant growth, development, and stress responses. The influence of NO on CKs is perceived on metabolism at different levels, including biosynthesis, signaling, and downstream gene expression. However, it can stimulate the activation of CK and benefit by restoring meristematic defects observed in the *noa1* mutant (Shen et al., 2013). Consequently, elevated NO levels can also suppress CK signaling in plants (Li et al., 2013). At the molecular level, S-nitrosation of histidine phosphotransfer proteins such as AHP1 (Arabidopsis Histidine-containing Phosphotransmitter 1) inhibits the CK phosphorelay, weakening CK responses toward it (Feng et al., 2013). Interestingly, these attributes that *AHP1* mutants show insensitivity toward NO-mediated repression of phosphorylation allow them to escape the inhibitory effect of NO.

Moreover, chemical interactions between NO and CKs can further refine this regulation. For instance, CKs can interact chemically with NO to modulate its cellular concentration and may suppress or scavenge NO levels, forming nitrosylated CKs that have altered physiological activity (Feng et al., 2013; Li et al., 2013). In a developmental context, NO acts downstream of CK signaling as a second messenger, mediating the transcriptional activation of cyclins like CYCD3;1 and further promoting CK-induced cell proliferation and callus formation (Shen et al., 2013). Induction of CK during photosynthetic adaptation to drought stress correlates with elevated NO levels and NR activity, highlighting their coordinated role in abiotic stress responses (Shao et al., 2010). Beyond their developmental role, CKs are also important players in responses implicated during plant immunity. For example, enhanced CK levels strengthen resistance to infection and modulate effector-triggered immunity in plants (Naseem et al., 2015). Through these diverse interactions, NO in fact modulates these CK signaling responses via posttranslational mechanisms such as S-nitrosation of AHP1 proteins. Though CKs can influence NO availability, they usually create it through a dynamic feedback system that maintains and balances growth and defense.

15.4.5 Nitric oxide and gibberellins

Gibberellin (GA) and NO are two vital signaling molecules that also play interconnected or antagonistic roles in regulating plant development and stress adaptation. Indeed, GA is known to regulate seed dormancy and germination, with the ABA–GA balance determining the transition between these two states in crops such as wheat (Nguyen et al., 2022) and rice (Gong et al., 2022). Elevated GA levels generally reduce dormancy and promote germination (Castro-Camba et al., 2022). However, NO functions upstream of GA signaling, influencing GA biosynthesis, perception, and its transduction (Bethke et al., 2007; Lozano-Juste & León, 2011). For instance, exogenous NO or SNP treatment decreases the endogenous levels of GA (Lozano-Juste & León, 2011). Moreover, NO also suppresses GA signaling through the S-nitrosation of the RGA (Repressor of GA1-3), a key member of the DELLA protein family, thereby maintaining a balance between growth and stress responses (Chen et al., 2022). This study displays the negative regulation of GA signaling by NO acting antagonistically. Furthermore, other genetic evidence also supports a mutually antagonistic relationship between GA and NO, as seen in the GA-deficient *Arabidopsis* mutant *ga1-3,* which exhibits significantly higher NO accumulation during photomorphogenesis compared to the wild-type (Lozano-Juste & León, 2011). NO also influences GA-mediated processes involved in seed germination, including cereal aleurone longevity, activation of GA-regulated MYB transcription factor, and amylase synthesis (Beligni et al., 2002; Wu et al., 2013). Interestingly, combined GA and NO applications show neither additive nor antagonistic effects on some germination-related processes (Zhang et al., 2005), highlighting the condition-dependent nature of their interaction. Beyond germination, GA also alleviates Cd toxicity by attenuating Cd-induced NO accumulation, which subsequently downregulates the expression of the metal transporter iron-regulated transporter 1 and limits Cd uptake (Zhu et al., 2012). This mechanism illustrates the negative regulatory interplay between GA and NO that is crucial for maintaining plant homeostasis during metal stress.

15.4.6 Nitric oxide and salicylic acid/jasmonic acid

SA and NO are key signaling molecules essential for activating plant immune responses under abiotic stresses and biotrophic pathogen attacks (Rai et al., 2021). Exogenous applications of SA and NO have been reported to enhance water regulation, photosynthetic efficiency, and overall crop productivity (Shahid et al., 2019). Under stress conditions, the accumulation of SA and NO modulates the expression of defense-related genes that stimulate the production of some secondary metabolites and antimicrobial compounds (Sarma et al., 2025). Indeed, NO also activates the SA-regulated defense pathway and stabilizes *NPR1* through redox regulation, facilitates *PR* gene expression, and contributes to SAR (Durrant & Dong, 2004; Graska et al., 2023; Gupta et al., 2014; Lindermayr et al., 2010; Klessig et al., 2000; Martínez-Medina, Pescador, Fernandez, et al., 2019; Moreau et al., 2010; Pande et al., 2022; Zhou et al., 2000). Additionally, increased concentration of NO can delay MeJA-induced leaf senescence following the low concentration of SA treatment (Ji et al., 2016). Collectively, these findings emphasize the synergistic crosstalk between SA and NO, which helps to fine-tune plant defense and stress adaptation.

Moreover, JA and NO also interact closely to regulate plant defense and stress responses. NO has been shown to trigger activation of JA signaling by activating key JA biosynthesis enzymes, such as Allene Oxide Synthase and Lipoxygenase 2, that initiate a cascade of defense responses (Huang et al., 2004). Furthermore, NO-mediated activation of the ascorbate-GSH cycle induced upon exogenous JA application confers drought tolerance in wheat seedlings, demonstrating a synergistic crosstalk between NO and JA signaling (Shan et al., 2015). Moreover, an enzyme such as Cis-(+)-12-oxo-phytodienoic acid reductase is involved in JA biosynthesis, and its expression is reported to be enhanced by NO (Corpas & Barroso, 2018), indicating a synergistic relationship between these two. Recent novel studies reveal a complex regulatory network that links NO_3^- and JA signaling through the central control via Jasmonate ZIM-domain protein 1, which helps plant adaptation to fluctuating environmental conditions (Pu et al., 2025). These findings indicate that NO can either synergize with or antagonize JA-dependent responses depending on the content of the stress type in plants.

15.4.7 Nitric oxide and brassinosteroids

BRs are an important class of steroidal phytohormones impacting various plant physiological and metabolic processes. Although evidence on NO and BR interactions is limited, several inputs highlight their crosstalk in regulating plant growth and stress responses. It is known that BRs induce NO production primarily through the NR and NOS-like pathways (Tossi et al., 2013), with nitrate reductase 1 (NIA1) being the major contributor to BR-induced NO generation through the NR pathway. This interactive network of BR and NO modulates different physiological processes, including antioxidant defense, photosynthesis, osmotic adjustment, and ion channel regulation (Hu et al., 2021; Kaya et al., 2020). Moreover, NO induces transcription factors that induce activation of different genes involved in BR-mediated processes (Altamura et al., 2023; Castillo et al., 2018). For example, Zhang et al. (2011) demonstrated that nanomolar concentrations of BRs rapidly increase NO levels in leaf mesophyll cells, suggesting the role of NO as an intermediate component in BR-induced ABA biosynthesis in maize leaves. BRs also promote adventitious root formation through the upregulation of enzyme NR (Li et al., 2019). Interestingly, the interaction of BRs and NO can produce opposite effects on chlorophyll content and photosynthesis, with BRs enhancing these traits while NO leads to their reduction (Hayat et al., 2010). A recent study also reported the role of BRs via their interaction with NO toward the response of rice root systems to arsenic stress (Piacentini et al., 2023). Treatment with 24-epibrassinolide increased the endogenous NO, initiating the BR-induced antioxidant defense system in pepper (*Capsicum annuum* L.) plants during water stress (Kaya et al., 2019). Exogenous application of BR in cucumber displayed the endogenous role of NO via brassinolide induction on adventitious root development (Li et al., 2020). Overall, these reported studies emphasize BR–NO crosstalk in plants, where the role of NO is both a mediator and modulator of BR responses, highlighting its role in coordinating growth, development, and stress adaptation in plants. Future studies will lead to elucidating the in-depth molecular cross-talk between NO and BRs elicited in such responses.

15.5 Role of nitric oxide in epigenetic regulation

Plants usually also respond to various environmental stresses by modulating gene expression patterns and establishing stress memory through epigenetic mechanisms such as DNA methylation, histone modifications, and small RNA activity (Naaz et al., 2025). These specific patterns of gene expression that define each cell type are dynamic and can change over time and in different locations. In response to environmental stimuli or developmental cues, genes may be selectively silenced or reactivated to ensure the precise control of different cellular processes (Wang and Sung, 2024). In recent years, it has become evident that key mechanisms regulating transcriptional activity involve modulating DNA accessibility via changes in chromatin structure (Wang and Sung, 2024). DNA methylation and histone modifications are the primary mediators of these epigenetic modifications.

Recent studies have identified that many phenotypic alterations that previously attributed solely to NO production may, in fact, result from downstream epigenetic regulatory processes (Naaz et al., 2025). There is also increasing evidence that indicates that NO not only functions as a signaling molecule but also exerts a profound influence on the epigenome (Zeng et al., 2025). It is known to modulate a wide range of histone PTMs and induces DNA adduct formation, reprogramming chromatin architecture and its accessibility. Throughout these mechanisms, NO in turn reshapes epigenetic landscapes that control gene expression networks, ultimately influencing cell fate, developmental pathways, and differentiation (Bovee et al., 2017).

One of the key modifications is that NO has an influence on epigenetic regulation by inhibiting the histone deacetylase complex, thereby enhancing histone acetylation (Kaya & Adamakis, 2025). This is followed by promoting a more relaxed chromatin configuration conducive to transcriptional activation. In addition to acetylation, both DNA and histone methylation are pivotal in controlling chromatin dynamics and gene expression (Kaya & Adamakis, 2025). Notably, in this context, several methyltransferases and demethylases have been identified as redox-sensitive enzymes across biological systems, suggesting that NO may modulate their activity (Ageeva-Kieferle et al., 2019). Integrative RNA sequencing and chromatin immunoprecipitation sequencing analyses revealed that NO facilitates a metabolic reprogramming from a growth and development perspective toward the stress-responsive pathways (Ageeva-Kieferle et al., 2021).

Another recent study also demonstrated that NO regulates the expression and function of *ARGONAUTE 4 (AGO4)*, which is a key component of the RNA-directed DNA methylation (RdDM) pathway, controlling *Arabidopsis* stem cell function in shoot (Zeng et al., 2023). However, mutations in components of this RdDM pathway have been reported to cause meristematic abnormalities and reduced response of the stem cell system to NO signaling. Notably, the stem cell-inducing transcription factor WUSCHEL was found to interact directly with AGO4 in an NO-dependent manner (Zeng et al., 2023). This interaction provides mechanistic insight into how these factors cooperate to alter DNA methylation patterns. The application of SNPs to apricot fruits was shown to induce both hypo- and hypermethylation events in DNA, thereby modulating postharvest physiology (Jiang et al., 2025). These results provide an epigenetic framework for understanding how DNA methylation contributes via NO-mediated postharvest regulation of fruit quality. Some studies have also highlighted the role of NO in thermomemory, the ability of plants to remember and then adapt to previous heat stress exposures (dos Santos et al., 2022). NO can also modify chromatin-associated proteins through S-nitrosation, impacting their stability and function, which finally influences epigenetic pathways (Corpas et al., 2014). Furthermore, NO may regulate the expression of key epigenetic modifiers, including DNA methyltransferases and HDACs, which are critical for thermomemory development (Naaz et al., 2025; Hickok et al., 2013). A recent comprehensive review by Naaz et al. (2025) provides an overview of epigenetically regulated genes and biomarkers modulated by NO under heat stress that confers thermotolerance in plants.

Overall, NO is strongly associated with the induction of specific epigenetic modifications, including its alterations in the histone proteins and DNA methylation patterns. These NO-influenced epigenetic changes contribute to the establishment of a persistent molecular memory of such stress conditions, which reinforces the role of NO as a central gaseous regulator of stress adaptation in plants (Socco et al., 2017).

15.6 Role of nitric oxide in postharvest preservation

Postharvest preservation is a major research focus in horticultural science, aimed at developing efficient and safe technologies to maintain quality, extend shelf life, and reduce spoilage, thereby incurring economic losses. Increased experimental evidence demonstrates the functional role of NO in positive regulation, impacting the postharvest performance of fruits and vegetables via multiple physiological and molecular mechanisms. Moreover, NO, indicated as a positive effector of a signaling molecule, displays an active role in delaying senescence and postharvest fruit quality (Zheng et al., 2017). For instance, exogenous NO or its donors delay ripening and senescence through inhibition of ET biosynthesis/signaling genes (*ACS/ACO*), suppressing respiration and softening, stabilizing color, including soluble solids in many fruits (Li et al., 2022; Zhu et al., 2021). NO also increases the antioxidant-related defense genes such as (*SOD*, *CAT*, *APX*, and *GSH*) lowering ROS and lipid peroxidation, including sustenance of membrane integrity that reduces browning and loss of quality during cold or ambient storage (Cai et al., 2020; Zhu et al., 2021). Indeed, NO also affects sugar/energy metabolism,polyamine and GABA-shunt pathways, including senescence/defense proteins through S-nitrosation, all of which contribute to the balancing of postharvest metabolism and stress responses (Li et al., 2022; Zhang, Cao, et al., 2020). According to Granella et al. (2022) and Cai et al. (2020), NO fumigation under low-O_2 or ultra-low-O_2 conditions provides broad-spectrum control of insects, mites, and fungi (such as *Cydia pomonella*, *Drosophila suzukii*, *Frankliniella* spp., and *Aspergillus flavus*) (Granella et al., 2022). It also has strong antimicrobial effects when controlled NO_2 formation is influenced and frequently requires shorter exposures than conventional fumigants. Because NO oxidizes to $NO_2/NO_2^-/NO_3^-$, its practical application necessitates careful control of O_2, its dose, temperature, and termination (N_2 flush); when protocols include inert flushing, residual nitrite/nitrate levels drop to acceptable levels and product quality is maintained (Granella et al., 2022). Lastly, when commodity-specific protocols and residue safeguards are put in place, NO becomes a multifaceted, sustainable tool for reducing postharvest losses when it works in cooperation with other signals such as H_2S, H_2O_2, ABA, melatonin, and MeJA to enhance its ability to delay senescence, reduce chilling injury, and inhibit decay.

15.7 Conclusion

NO acts as an important coordinator of plant physiological and molecular networks, integrating hormonal, redox, and transcriptional regulation to optimize growth and stress adaptation. Its regulatory influence on gene expression, including epigenetic control, and its interactions with key phytohormones underscore its important role in the maintenance of cellular homeostasis and plant development. Expanding our understanding of NO-mediated signaling may facilitate the development of innovative agronomic practices and biotechnological strategies aimed at enhancing crop resilience and yield stability. Deciphering the multifaceted roles of NO provides valuable insights into plant resilience mechanisms and offers promising strategies for improving crop productivity and stress tolerance. Overall, NO acts as a major integrator between hormonal and redox networks, allowing plants to coordinate growth and respond effectively to various environmental cues. Harnessing the overall potential of NO in this context can also promote and impact resource efficiency and environmental management in crop production, thereby advancing the overarching goals of sustainable agriculture for the future.

AI disclosure

During the preparation of this work, author(s) used the ChatGPT (Open AI) tool to improve the English language of some sections in this chapter. After using this tool/service, the author(s) review(s) and edit(s) the content as needed and take(s) full responsibility for the content of the publication.

References

Ageeva-Kieferle, A., Georgii, E., Winkler, B., Ghirardo, A., Albert, A., Hüther, P., Mengel, A., Becker, C., Schnitzler, J. P., Durner, J., & Lindermayr, C. (2021). Nitric oxide coordinates growth, development, and stress response via histone modification and gene expression. *Plant Physiology, 187*(1), 336–360. https://doi.org/10.1093/plphys/kiab222, https://academic.oup.com/plphys/article/187/1/336/6277902?login=true.

Ageeva-Kieferle, A., Rudolf, E. E., & Lindermayr, C. (2019). Redox-dependent chromatin remodeling: A new function of nitric oxide as architect of chromatin structure in plants. *Frontiers in Plant Sciences, 10*, 625. https://doi.org/10.3389/fpls.2019.00625.

Ahmad, P., Abdel Latef, A. A., Hashem, A., Abd Allah, E. F., Gucel, S., & Tran, L.-S. P. (2016). Nitric oxide mitigates salt stress by regulating levels of osmolytes and antioxidant enzymes in chickpea. *Frontiers in Plant Science, 7*. https://doi.org/10.3389/fpls.2016.00347 (MAR2016).

Altamura, M. M., Piacentini, D., Della Rovere, F., Fattorini, L., Falasca, G., & Betti, C. (2023). New paradigms in brassinosteroids, strigolactones, sphingolipids, and nitric oxide interaction in the control of lateral and adventitious root formation. *Plants, 12*(2), 413. https://doi.org/10.3390/plants12020413.

Arfan, M., Zhang, D.-W., Zou, L.-J., Luo, S.-S., Tan, W.-R., Zhu, T., & Lin, H.-H. (2019). Hydrogen peroxide and nitric oxide crosstalk mediates brassinosteroids induced cold stress tolerance in *Medicago truncatula*. *International Journal of Molecular Sciences, 20*(1). https://doi.org/10.3390/ijms20010144.

Asai, S., Mase, K., & Yoshioka, H. (2010). Role of nitric oxide and reactive oxide species in disease resistance to necrotrophic pathogens. *Plant Signaling and Behavior, 5*, 872–874. https://doi.org/10.4161/psb.5.7.11899.

Astier, J., & Lindermayr, C. (2012). Nitric oxide-dependent posttranslational modification in plants: an update. *International Journal of Molecular Sciences, 13*(11), 15193–15208. https://doi.org/10.3390/ijms131115193.

Asai, S., & Yoshioka, H. (2009). Nitric oxide as a partner of reactive oxygen species participates in disease resistance to nectrotophic pathogen *Botryis cinerea* in *Nicotiana benthamiana*. *Molecular Plant-Microbe Interactions: MPMI, 22*(6), 619–629. https://doi.org/10.1094/MPMI-22-6-0619.

Beligni, M. V., Fath, A., Bethke, P. C., Lamattina, L., & Jones, R. L. (2002). Nitric oxide acts as an antioxidant and delays programmed cell death in barley aleurone layers. *Plant Physiology, 129*(4), 1642–1650. https://doi.org/10.1104/pp.002337.

Besson-Bard, A., Gravot, A., Richaud, P., Auroy, P., Duc, C., Gaymard, F., Taconnat, L., Renou, J.-P., Pugin, A., & Wendehenne, D. (2009). Nitric oxide contributes to cadmium toxicity in *Arabidopsis* by promoting cadmium accumulation in roots and by up-regulating genes related to iron uptake. *Plant Physiology, 149*(3), 1302–1315. https://doi.org/10.1104/pp.108.133348.

Bethke, P. C., Libourel, I. G. L., Aoyama, N., Chung, Y. Y., Still, D. W., & Jones, R. L. (2007). The *Arabidopsis* aleurone layer responds to nitric oxide, gibberellin, and abscisic acid and is sufficient and necessary for seed dormancy. *Plant Physiology, 143*(3), 1173–1188. https://doi.org/10.1104/pp.106.093435, http://www.plantphysiol.org/cgi/reprint/143/3/1173.pdf.

Bogdan, C. (2015). Nitric oxide synthase in innate and adaptive immunity: An update. *Trends in Immunology, 36*(3), 161–178. https://doi.org/10.1016/j.it.2015.01.003.

Borrowman, S., Kapuganti, J. G., & Loake, G. J. (2023). Expanding roles for S-nitrosylation in the regulation of plant immunity. *Free Radical Biology & Medicine, 194*, 357–368. https://doi.org/10.1016/j.freeradbiomed.2022.12.009 Epub 2022 Dec 10. PMID: 36513331.

Bovee, R. C., Hickok, J. R., Vasudevan, D., & Thomas, D. D. (2017). Mechanisms of epigenetic regulation by nitric oxideIn *Nitric oxide: Biology and pathobiology* (Third Edition). United States: Elsevier Inc, 255–270. http://www.sciencedirect.com/science/book/9780128042731.

Bright, J., Desikan, R., Hancock, J. T., Weir, I. S., & Neill, S. J. (2006). ABA-induced NO generation and stomatal closure in *Arabidopsis* are dependent on H_2O_2 synthesis. *Plant Journal, 45*(1), 113–122. https://doi.org/10.1111/j.1365-313X.2005.02615.x.

Brito, A. E.de A., Albuquerque, G. D. P.de, Palheta, J. G., Teixeira, J. S. S., Palheta de Sousa, D. J., Teixeira, K. B. S., Nogueira, G. A.dos S., Neto, C. F.de O., Viégas, I.de J. M., & Okumura, R. S. (2021). Nitric oxide as a way to mitigate copper (Cu) toxicity in the emergence and initial growth of maize seedling (*Zea mays* L.). *Australian Journal of Crop Science, 15*, 82–87. https://doi.org/10.21475/ajcs.21.15.01.2805 (15(01):2021).

Cai, H., Han, S., Yu, M., Ma, R., & Yu, Z. (2020). Exogenous nitric oxide fumigation promoted the emission of volatile organic compounds in peach fruit during shelf life after long-term cold storage. *Food Research International, 133*, 109135. https://doi.org/10.1016/j.foodres.2020.109135.

Calcagno, C., Novero, M., Genre, A., Bonfante, P., & Lanfranco, L. (2012). The exudate from an arbuscular mycorrhizal fungus induces nitric oxide accumulation in *Medicago truncatula* roots. *Mycorrhiza, 22*(4), 259–269. https://doi.org/10.1007/s00572-011-0400-4.

Castillo, M. C., Coego, A., Costa-Broseta, Á., & León, J. (2018). Nitric oxide responses in *Arabidopsis* hypocotyls are mediated by diverse phytohormone pathways. *Journal of Experimental Botany, 69*(21), 5265–5278. https://doi.org/10.1093/jxb/ery286, http://jxb.oxfordjournals.org/.

Castro-Camba, R., Sánchez, C., Vidal, N., & Vielba, J. M. (2022). Plant development and crop yield: The role of gibberellins. *Plants, 11*(19). https://doi.org/10.3390/plants11192650, www.mdpi.com/journal/plants.

Chen, L., Sun, S., Song, C. P., Zhou, J. M., Li, J., & Zuo, J. (2022). Nitric oxide negatively regulates gibberellin signaling to coordinate growth and salt tolerance in *Arabidopsis*. *Journal of Genetics and Genomics, 49*(8), 756–765. https://doi.org/10.1016/j.jgg.2022.02.023, http://www.journals.elsevier.com/journal-of-genetics-and-genomics/.

Chen, Z., Zhang, L., & Zhu, C. (2015). Exogenous nitric oxide mediates alleviation of mercury toxicity by promoting auxin transport in roots or preventing oxidative stress in leaves of rice seedlings. *Acta Physiologiae Plantarum / Polish Academy of Sciences, Committee of Plant Physiology Genetics and Breeding, 37*(9), 194. https://doi.org/10.1007/s11738-015-1931-7.

Corpas, F. J., Alché, J. D., & Barroso, J. B. (2013). Current overview of S-nitrosoglutathione (GSNO) in higher plants. *Frontiers in Plant Science, 4*, 126.

Corpas, F. J., Begara-Morales, J. C., Sánchez-Calvo, B., Chaki, M., & Barroso, J. B. (2014). *Signaling and communication in plants, reactive oxygen and nitrogen species signaling and communication in plants*. Springer International Publishing, 267–281. https://doi.org/10.1007/978-3-319-10079-1_13.

Corpas, F. J., & Barroso, J. B. (2018). Peroxisomal plant metabolism—An update on nitric oxide, Ca^{2+} and the NADPH recycling network. *Journal of Cell Science, 131*, jcs202978. https://doi.org/10.1242/jcs.202978.

Corpas, F. J., Palma, J. M., Del Río, L. A., & Barroso, J. B. (2009). Evidence supporting the existence of l-arginine-dependent nitric oxide synthase activity in plants. *New Phytologist, 184*(1), 9–14. https://doi.org/10.1111/j.1469-8137.2009.02989.x.

Corpas, F. J., Palma, J. M., & Elzenga, J. T. M. (2018). Nitric oxide on/off in fruit ripening. *Plant Biology, 20*(5), 805–807. https://doi.org/10.1111/plb.12852.

Correa-Aragunde, N., Graziano, M., Chevalier, C., & Lamattina, L. (2006). Nitric oxide modulates the expression of cell cycle regulatory genes during lateral root formation in tomato. *Journal of Experimental Botany, 57*(3), 581–588. https://doi.org/10.1093/jxb/erj045.

del Giudice, J., Cam, Y., Damiani, I., Fung-Chat, F., Meilhoc, E., Bruand, C., Brouquisse, R., Puppo, A., & Boscari, A. (2011). Nitric oxide is required for an optimal establishment of the *Medicago truncatula–Sinorhizobium meliloti* symbiosis. *New Phytologist, 191*(2), 405–417. https://doi.org/10.1111/j.1469-8137.2011.03693.x.

Delledonne, M., Xia, Y., Dixon, R. A., & Lamb, C. (1998). Nitric oxide functions as a signal in plant disease resistance. *Nature, 394*(6693), 585–588. https://doi.org/10.1038/29087.

Di Palma, A. A., Pereyra, C. M., Ramirez, L. M., Vázquez, M. L. X., et al. (2013). Denitrification-derived nitric oxide modulates biofilm formation in *Azospirillum brasilense*. *FEMS Microbiology Letters, 338*, 77–85.

dos Santos, T. B., Ribas, A. F., de Souza, S. G. H., Budzinski, I. G. F., & Domingues, D. S. (2022). Physiological responses to drought, salinity, and heat stress in plants: A review. *Stresses, 2*(1), 113–135. https://doi.org/10.3390/stresses2010009, https://www.mdpi.com/journal/stresses.

Durrant, W. E., & Dong, X. (2004). Systemic acquired resistance. *Annual Review of Phytopathology, 42*(1), 185–209. https://doi.org/10.1146/annurev.phyto.42.040803.140421.

Espinosa, F., Garrido, I., Ortega, A., Casimiro, I., Álvarez-Tinaut, M.ª C., & Balestrini, R. (2014). Redox activities and ROS, NO and phenylpropanoids production by axenically cultured intact olive seedling roots after interaction with a mycorrhizal or a pathogenic fungus. *PLoS One, 9*(6), e100132. https://doi.org/10.1371/journal.pone.0100132.

Feng, J., Wang, C., Chen, Q., Chen, H., Ren, B., Li, X., & Zuo, J. (2013). S-nitrosylation of phosphotransfer proteins represses cytokinin signaling. *Nature Communications, 4*(1). https://doi.org/10.1038/ncomms2541.

Fernández-Marcos, M., Sanz, L., Lewis, D. R., Muday, G. K., & Lorenzo, O. (2011). Nitric oxide causes root apical meristem defects and growth inhibition while reducing PIN-FORMED 1 (PIN1)-dependent acropetal auxin transport. *Proceedings of the National Academy of Sciences of the United States of America, 108*(45), 18506–18511. https://doi.org/10.1073/pnas.1108644108Spain, http://www.pnas.org/content/108/45/18506.full.pdf+html.

Freschi, L. (2013). Nitric oxide and phytohormone interactions: Current status and perspectives. *Frontiers in Plant Science, 4*. https://doi.org/10.3389/fpls.2013.00398.

Fukudome, M., & Uchiumi, T. (2024). Regulation of nitric oxide by phytoglobins in *Lotus japonicus* is involved in mycorrhizal symbiosis with *Rhizophagus irregularis*. *Plant Science (Shannon, Ireland), 340*, 111984. https://doi.org/10.1016/j.plantsci.2024.111984 Epub 2024 Jan 14. PMID: 38220094.

García-Mata, C., & Lamattina, L. (2002). Nitric oxide and abscisic acid cross talk in guard cells. *Plant Physiology, 128*(3), 790–792. https://doi.org/10.1104/pp.011020.

Gayatri, G., Agurla, S., & Raghavendra, A. S. (2013). Nitric oxide in guard cells as an important secondary messenger during stomatal closure. *Frontiers in Plant Sciences, 4*, 425. https://doi.org/10.3389/fpls.2013.00425 PMID: 24194741.

Gong, D., He, F., Liu, J., Zhang, C., Wang, Y., Tian, S., Sun, C., & Zhang, X. (2022). Understanding of hormonal regulation in rice seed germination. *Life (Chicago, Ill.: 1978), 12*(7), 1021. https://doi.org/10.3390/life12071021.

Granella, S. J., Bechlin, T. R., Christ, D., & Coelho, S. R. M. (2022). A potential role of nitric oxide in postharvest pest control: A review. *Journal of the Saudi Society of Agricultural Sciences, 21*(7), 452–459. https://doi.org/10.1016/j.jssas.2021.12.002.

Graska, J., Fidler, J., Gietler, M., Prabucka, B., Nykiel, M., & Labudda, M. (2023). Nitric oxide in plant functioning: metabolism, signaling, and responses to infestation with ecdysozoa parasites. *Biology, 12*(7), 927. https://doi.org/10.3390/biology12070927.

Gupta, K. J., & Igamberdiev, A. U. (2011). The anoxic plant mitochondrion as a nitrite: NO reductase. *Mitochondrion, 11*, 537–543. https://doi.org/10.1016/j.mito.2011.03.005.41.41.

Gupta, K. J., Kaladhar, V. C., Fitzpatrick, T. B., Fernie, A. R., Møller, I. M., & Loake, G. J. (2022). Nitric oxide regulation of plant metabolism. *Molecular Plant, 15*(2), 228–242. https://doi.org/10.1016/j.molp.2021.12.012, https://www.cell.com/molecular-plant/home.

Gupta, K. J., Mur, L. A., & Brotman, Y. (2014). Trichoderma asperelloides suppresses nitric oxide generation elicited by *Fusarium oxysporum* in *Arabidopsis* roots. *Molecular Plant-Microbe Interactions, 27*, 307–314.

Gupta, K. J., Mur, L. A. J., Wany, A., Kumari, A., Fernie, A. R., & Ratcliffe, R. G. (2020). The role of nitrite and nitric oxide under low oxygen conditions in plants. *New Phytologist, 225*(3), 1143–1151. https://doi.org/10.1111/nph.15969.

Hayat, S., Yadav, S., Ali, B., & Ahmad, A. (2010). Interactive effect of nitric oxide and brassinosteroids on photosynthesis and the antioxidant system of *Lycopersicon esculentum*. *Russian Journal of Plant Physiology, 57*(2), 212–221. https://doi.org/10.1134/s1021443710020081.

He, Y., Tang, R. H., Hao, Y., Stevens, R. D., Cook, C. W., Ahn, S. M., Jing, L., Yang, Z., Chen, L., Guo, F., Fiorani, F., Jackson, R. B., Crawford, N. M., & Pei, Z. M. (2004). Nitric oxide represses the *Arabidopsis* floral transition. *Science (New York, N.Y.), 305*(5692), 1968–1971. https://doi.org/10.1126/science.1098837.

Hebelstrup, K. H., Shah, J. K., & Igamberdiev, A. U. (2013). The role of nitric oxide and hemoglobin in plant development and morphogenesis. *Physiologia Plantarum, 148*, 457–469. https://doi.org/10.1111/ppl.12062.

Hickok, J. R., Vasudevan, D., Antholine, W. E., & Thomas, D. D. (2013). Nitric oxide modifies global histone methylation by inhibiting jumonji C domain-containing demethylases. *Journal of Biological Chemistry, 288*(22), 16004–16015. https://doi.org/10.1074/jbc.m112.432294.

Horchani, F., Prévot, M., Boscari, A., Evangelisti, E., Meilhoc, E., Bruand, C., Raymond, P., Boncompagni, E., Aschi-Smiti, S., Puppo, A., & Brouquisse, R. (2011). Both plant and bacterial nitrate reductases contribute to nitric oxide production in *Medicago truncatula* nitrogen-fixing nodules. *Plant Physiology, 155*(2), 1023–1036. https://doi.org/10.1104/pp.110.166140.

Hossain, K. K., Itoh, R. D., Yoshimura, G., Tokuda, G., Oku, H., Cohen, M. F., & Yamasaki, H. (2010). Effects of nitric oxide scavengers on thermoinhibition of seed germination in *Arabidopsis thaliana*. *Russian Journal of Plant Physiology, 57*(2), 222–232. https://doi.org/10.1134/s1021443710020093.

Hu, D., Wei, L., & Liao, W. (2021). Brassinosteroids in plants: Crosstalk with small-molecule compounds. *Biomolecules, 11*, 1800. https://doi.org/10.3390/biom11121800.

Hu, X., Neill, S. J., Tang, Z., & Cai, W. (2005). Nitric oxide mediates gravitropic bending in soybean roots. *Plant Physiology, 137*(2), 663–670. https://doi.org/10.1104/pp.104.054494.

Huang, X., Stettmaier, K., Michel, C., Hutzler, P., Mueller, M. J., & Durner, J. (2004). Nitric oxide is induced by wounding and influences jasmonic acid signaling in *Arabidopsis thaliana*. *Planta, 218*(6), 938–946. https://doi.org/10.1007/s00425-003-1178-1.

Hung, K. T., & Kao, C. H. (2003). Nitric oxide counteracts the senescence of rice leaves induced by abscisic acid. *Journal of Plant Physiology, 160*(8), 871–879. https://doi.org/10.1078/0176-1617-01118, www.urbanfischer.de/journals/jpp/p_physio.htm.

Hussain, A., Shah, F., Ali, F., & Yun, B. W. (2022). Role of nitric oxide in plant senescence. *Frontiers in Plant Science, 13*. https://doi.org/10.3389/fpls.2022.851631, https://www.frontiersin.org/journals/plant-science.

Ji, Y., Liu, J., & Xing, D. (2016). Low concentrations of salicylic acid delay methyl jasmonate-induced leaf senescence by up-regulating nitric oxide synthase activity. *Journal of Experimental Botany, 67*, 5233–5245. https://doi.org/10.1093/jxb/erw280.

Jiang, Y., Tao, J., Zuo, J., Jing, Z., Wang, Y., Bai, C., Fernie, A. R., Zheng, Y., & Wu, C. (2025). Comprehensive analysis of DNA methylome and transcriptome reveals the epigenetic regulation of nitric oxide treatment in delaying apricot fruit senescence. *Plant Journal, 123*(4). https://doi.org/10.1111/tpj.70387, http://onlinelibrary.wiley.com/journal/10.1111/(ISSN)1365-313X.

Jing, H., Yang, X., Emenecker, R. J., Feng, J., Zhang, J., Figueiredo, M. R. Ad, Chaisupa, P., Wright, R. C., Holehouse, A. S., Strader, L. C., & Zuo, J. (2023). Nitric oxide-mediated S-nitrosylation of IAA17 protein in intrinsically disordered region represses auxin signaling. *Journal of Genetics and Genomics, 50*(7), 473–485. https://doi.org/10.1016/j.jgg.2023.05.001, http://www.journals.elsevier.com/journal-of-genetics-and-genomics/.

Johnston, G. A. (2023). Uncoupling plant growth and defense through phytohormone crosstalk modification (Master's thesis). Colorado State University, Fort Collins, CO.

Kanayama, Y., & Yamamoto, Y. (1991). Formation of nitrosylleghemoglobin in nodules of nitrate-treated cowpea and pea plants. *Plant and Cell Physiology, 32*(1), 19–24.

Karpets, Y. V., Kolupaev, Y. E., & Yastreb, T. O. (2011). Effect of sodium nitroprusside on heat resistance of wheat coleoptiles: Dependence on the formation and scavenging of reactive oxygen species. *Russian Journal of Plant Physiology, 58*(6), 1027–1033.

Kaya, C., & Adamakis, I.-D. S. (2025). Redox-epigenetic crosstalk in plant stress responses: The roles of reactive oxygen and nitrogen species in modulating chromatin dynamics. *International Journal of Molecular Sciences, 26*(15). https://doi.org/10.3390/ijms26157167.

Kaya, C., Ashraf, M., Alyemeni, M. N., & Ahmad, P. (2020). Nitrate reductase rather than nitric oxide synthase activity is involved in 24-epibrassinolide-induced nitric oxide synthesis to improve tolerance to iron deficiency in strawberry (Fragaria × annassa) by up-regulating the ascorbate-glutathione cycle. *Plant Physiology and Biochemistry, 151*, 486–499. https://doi.org/10.1016/j.plaphy.2020.04.002.

Kaya, C., Ashraf, M., Wijaya, L., & Ahmad, P. (2019). The putative role of endogenous nitric oxide in brassinosteroid-induced antioxidant defence system in pepper (*Capsicum annuum* L.) plants under water stress. *Plant Physiology and Biochemistry, 143*, 119–128. https://doi.org/10.1016/j.plaphy.2019.08.024.

Khan, M. N., Alamri, S., Al-Amri, A. A., Alsubaie, Q. D., Al-Munqedi, B., Ali, H. M., Singh, V. P., & Siddiqui, M. H. (2021). Effect of nitric oxide on seed germination and seedling development of tomato under chromium toxicity. *Journal of Plant Growth Regulation, 40*(6), 2358–2370. https://doi.org/10.1007/s00344-020-10212-2, http://www.springer.com/life+sci/plant+sciences/journal/344.

Khan, M., Imran, Q. M., Shahid, M., Mun, B.-G., Lee, S.-U., Khan, M. A., Hussain, A., Lee, I.-J., & Yun, B.-W. (2019). Nitric oxide- induced AtAO3 differentially regulates plant defense and drought tolerance in *Arabidopsis thaliana. BMC Plant Biology, 19*. https://doi.org/10.1186/s12870-019-2210-3.

Klessig, D. F., Durner, J., Noad, R., Navarre, D. A., Wendehenne, D., Kumar, D., Zhou, J. M., Shah, J., Zhang, S., Kachroo, P., Trifa, Y., Pontier, D., Lam, E., & Silva, H. (2000). Nitric oxide and salicylic acid signaling in plant defense. *Proceedings of the National Academy of Sciences of the United States of America, 97*(16), 8849–8855. https://doi.org/10.1073/pnas.97.16.8849.

Kolbert, Z., Barroso, Brouquisse, Corpas, Gupta, Lindermayr, Loake, Palma, Petřivalský, & Wendehenne (2019). Hancock, A forty year journey: The generation and roles of NO in plants. *Nitric Oxide - Biology and Chemistry, 93*, 53–70. https://doi.org/10.1016/j.niox.2019.09.006.

Kolbert, Z., & Erdei, L. (2014). Involvement of nitrate reductase in auxin-induced NO synthesis. *Plant Signaling & Behavior, 3*(11), 972–973. https://doi.org/10.4161/psb.6170.

Kolbert, Z., Feigl, G., Freschi, L., & Poór, P. (2019). Gasotransmitters in action: Nitric oxide-ethylene crosstalk during plant growth and abiotic stress responses. *Antioxidants, 8*(6), 167. https://doi.org/10.3390/antiox8060167.

Kolbert, Z., Lindermayr, C., & Loake, G. J. (2021). The role of nitric oxide in plant biology: Current insights and future perspectives. *Journal of Experimental Botany, 72*(3), 777–780. https://doi.org/10.1093/jxb/erab013.

Krasylenko, Y. A., Yemets, A. I., & Blume, Y. B. (2010). Functional role of nitric oxide in plants. *Russian Journal of Plant Physiology, 57*(4), 451–461.

Kumar, D., & Ohri, P. (2023). Say "NO" to plant stresses: Unravelling the role of nitric oxide under abiotic and biotic stress. *Nitric Oxide: Biology and Chemistry / Official Journal of the Nitric Oxide Society, 130*, 36–57.

Lee, U., Wie, C., Fernandez, B. O., Feelisch, M., & Vierling, E. (2008). Modulation of nitrosative stress by S-nitrosoglutathione reductase is critical for thermotolerance and plant growth in *Arabidopsis*. *The Plant Cell, 20*(3), 786–802. https://doi.org/10.1105/tpc.107.052647.

Li, C., Yu, W., & Liao, W. (2022). Role of nitric oxide in postharvest senescence of fruits. *International Journal of Molecular Sciences, 23*(17), 10046.

Li, G., Qin, B., Li, S., Yin, Y., Zhao, J., An, W., Cao, Y., & Mu, Z. (2020). LbNR-derived nitric oxide delays lycium fruit coloration by transcriptionally modifying flavonoid biosynthetic pathway. *Frontiers in Plant Science, 11*. https://doi.org/10.3389/fpls.2020.01215.

Li, G., Zhao, J., Qin, B., Yin, Y., An, W., Mu, Z., & Cao, Y. (2019). ABA mediates development-dependent anthocyanin biosynthesis and fruit coloration in *Lycium* plants. *BMC Plant Biology, 19*(1). https://doi.org/10.1186/s12870-019-1931-7, http://www.biomedcentral.com/bmcplantbiol/.

Li, Xg, Zhang, T. L., Wang, X. X., Hua, K., Zhao, L., & Han, Z. M. (2013). The composition of root exudates from two different resistant peanut cultivars and their effects on the growth of soil-borne pathogen. *International Journal of Biological Sciences, 9*(2), 164–173. https://doi.org/10.7150/ijbs.5579, http://www.biolsci.org/v09p0164.pdf.

Lindermayr, C., Saalbach, G., & Durner, J. (2005). Proteomic identification of S-nitrosylated proteins in *Arabidopsis*. *Plant Physiology, 137*(3), 921–930.

Lindermayr, C., Sell, S., Müller, B., Leister, D., & Durner, J. (2010). Redox regulation of the NPR1-TGA1 system of *Arabidopsis thaliana* by nitric oxide. *The Plant Cell, 22*(8), 2894–2907. https://doi.org/10.1105/tpc.109.066464.

Liu, M., Zhang, H., Fang, X., Zhang, Y., & Jin, C. (2018). Auxin acts downstream of ethylene and nitric oxide to regulate magnesium deficiency-induced root hair development in *Arabidopsis thaliana*. *Plant and Cell Physiology, 59*(7), 1452–1465. https://doi.org/10.1093/pcp/pcy078, http://pcp.oxfordjournals.org/.

Lombardo, M. C., Graziano, M., Polacco, J. C., & Lamattina, L. (2006). Nitric oxide functions as a positive regulator of root hair development. *Plant Signaling & Behavior, 1*(1), 28–33.

Lozano-Juste, J., & León, J. (2011). Nitric oxide regulates DELLA content and PIF expression to promote photomorphogenesis in *Arabidopsis*. *Plant Physiology, 156*, 1410–1423.

Ma, Y., & Chen, R. (2021). Nitrogen and phosphorus signaling and transport during legume-rhizobium symbiosis. *Frontiers in Plant Sciences, 12*, 683601. https://doi.org/10.3389/fpls.2021.683601 PMID: 34239527.

Manjunatha, G., Gupta, K. J., Lokesh, V., Mur, L. A. J., & Neelwarne, B. (2012). Nitric oxide counters ethylene effects on ripening fruits. *Plant Signaling & Behavior, 7*, 476–483.

Martínez-Medina, A., Pescador, L., Fernandez, I., Rodríguez-Serrano, M., García, J., Romero-Puertas, M. C., & Pozo, M. J. (2019). Nitric oxide and nonsymbiotic hemoglobin 1 are regulatory elements in the *Solanum lycopersicum–Rhizophagus irregularis* mycorrhizal symbiosis. *New Phytologist*.

Martínez-Medina, A., Pescador, L., Terrón-Camero, L. C., Pozo, M. J., Romero-Puertas, M. C., & Brouquisse, R. (2019). Nitric oxide in plant–fungal interactions. *Journal of Experimental Botany, 70*(17), 4489–4503. https://doi.org/10.1093/jxb/erz289.

Maskall, C. S., Gibson, J. F., & Dart, P. J. (1977). Electron-paramagnetic-resonance studies of leghaemoglobins from soya-bean and cowpea root nodules. Identification of nitrosyl-leghaemoglobin in crude leghaemoglobin preparations. *Biochemical Journal, 167*, 435–445.

Mathieu, C., Moreau, S., Frendo, P., Puppo, A., & Davies, M. J. (1998). Direct detection of radicals in intact soybean nodules: Presence of nitric oxide-leghemoglobin complexes. *Free Radical Biology and Medicine, 24*(7-8), 1242–1249. https://doi.org/10.1016/S0891-5849(97)00440-1, www.elsevier.com/locate/freeradbiomed.

Meakin, G. E., Bueno, E., Jepson, B., Bedmar, E. J., Richardson, D. J., & Delgado, M. J. (2007). The contribution of bacteroidal nitrate and nitrite reduction to the formation of nitrosylleghaemoglobin complexes in soybean root nodules. *Microbiology (Reading, England), 153*(2), 411–419. https://doi.org/10.1099/mic.0.2006/000059-0.

Meilhoc, E., Blanquet, P., Cam, Y., & Bruand, C. (2013). Control of NO level in rhizobium-legume root nodules: Not only a plant globin story. *Plant Signal Behav, 8*(10). https://doi.org/10.4161/psb.25923 PMID: 23962798.

Meilhoc, E., Boscari, A., Bruand, C., Puppo, A., & Brouquisse, R. (2011). Nitric oxide in legume-rhizobium symbiosis. *Plant Science, 181*(5), 573–581. https://doi.org/10.1016/j.plantsci.2011.04.007.

Meng, Y., Jing, H., Huang, J., Shen, R., & Zhu, X. (2022). The role of nitric oxide signaling in plant responses to cadmium stress. *International Journal of Molecular Sciences, 23*(13), 6901. https://doi.org/10.3390/ijms23136901.

Mioto, P. T., & Mercier, H. (2013). Abscisic acid and nitric oxide signaling in two different portions of detached leaves of *Guzmania monostachia* with CAM up-regulated by drought. *Journal of Plant Physiology, 170*, 996–1002.

Mommer, L., Kirkegaard, J., & van Ruijven, J. (2016). Root–root interactions: Towards a rhizosphere framework. *Trends in Plant Science, 21*, 209–217. https://doi.org/10.1016/j.tplants.2016.01.009.

Moreau, M., Lindermayr, C., Durner, J., & Klessig, D. F. (2010). NO synthesis and signaling in plants - where do we stand? *Physiologia Plantarum, 138*(4), 372–383. https://doi.org/10.1111/j.1399-3054.2009.01308.x.

Mostofa, M. G., Fujita, M., & Tran, L.-S. P. (2015). Nitric oxide mediates hydrogen peroxide-and salicylic acid-induced salt tolerance in rice (*Oryza sativa* L.) seedlings. *Plant Growth Regulation, 77*(3), 265–277.

Mukherjee, S. (2019). Recent advancements in the mechanism of nitric oxide signaling associated with hydrogen sulfide and melatonin crosstalk during ethylene-induced fruit ripening in plants. *Nitric Oxide: Biology and Chemistry / Official Journal of the Nitric Oxide Society, 82*, 25–34.

Mur, L. A., Carver, T. L., & Prats, E. (2006). NO way to live; the various roles of nitric oxide in plant–pathogen interactions. *Journal of Experimental Botany, 57*, 489–505.

Mur, L. A. J., Mandon, J., Cristescu, S. M., Harren, F. J. M., & Prats, E. (2011). Methods of nitric oxide detection in plants: A commentary. *Plant Science, 181*(5), 509–519. https://doi.org/10.1016/j.plantsci.2011.04.003.

Mur, L. A. J., Prats, E., Pierre, S., Hall, M. A., & Hebelstrup, K. H. (2013). Integrating nitric oxide into salicylic acid and jasmonic acid/ethylene plant defense pathways. *Frontiers in Plant Science, 4*. https://doi.org/10.3389/fpls.2013.00215, http://journal.frontiersin.org/Journal/10.3389/fpls.2013.00215/pdf.

Naaz, S., Pande, A., & Laxmi, A. (2025). Nitric oxide-mediated thermomemory: a new perspective on plant heat stress resilience. *Frontiers in Plant Science, 16*. https://doi.org/10.3389/fpls.2025.1525336.

Nabi, R. B. S., Tayade, R., Hussain, A., Kulkarni, K. P., Imran, Q. M., Mun, B. G., & Yun, B. W. (2019). Nitric oxide regulates plant responses to drought, salinity, and heavy metal stress. *Environmental and Experimental Botany, 161*, 120–133. https://doi.org/10.1016/j.envexpbot.2019.02.003, www.elsevier.com/inca/publications/store/2/6/7.

Naseem, M., Kaltdorf, M., & Dandekar, T. (2015). The nexus between growth and defence signalling: Auxin and cytokinin modulate plant immune response pathways. *Journal of Experimental Botany, 66*(16), 4885–4896. https://doi.org/10.1093/jxb/erv297.

Neill, S., Barros, R., Bright, J., Desikan, R., Hancock, J., Harrison, J., Morris, P., Ribeiro, D., & Wilson, I. (2008). Nitric oxide, stomatal closure, and abiotic stress. *Journal of Experimental Botany, 59*(2), 165–176. https://doi.org/10.1093/jxb/erm293.

Nguyen, T. N., Tuan, P. A., & Ayele, B. T. (2022). Jasmonate regulates seed dormancy in wheat via modulating the balance between gibberellin and abscisic acid. *Journal of Experimental Botany, 73*, 2434–2453. https://doi.org/10.1093/jxb/erac041.

Pagnussat, G. C., Lanteri, M. L., & Lamattina, L. (2003). Nitric oxide and cyclic GMP are messengers in the indole acetic acid-induced adventitious rooting process. *Plant Physiology, 132*(3), 1241–1248. https://doi.org/10.1104/pp.103.022228 PMID: 12857806.

Pagnussat, G. C., Lanteri, M. L., Lombardo, M. C., & Lamattina, L. (2004). Nitric oxide mediates the indole acetic acid induction activation of a mitogen-activated protein kinase cascade involved in adventitious root development. *Plant Physiology, 135*(1), 279–286. https://doi.org/10.1104/pp.103.038554, http://www.plantphysiol.org/.

Pande, A., Mun, B. G., Lee, D. S., Khan, M., Lee, G. M., Hussain, A., & Yun, B. W. (2021). NO network for plant–microbe communication underground: A review. *Frontiers in Plant Science, 12*. https://doi.org/10.3389/fpls.2021.658679, https://www.frontiersin.org/journals/plant-science.

Pande, A., Mun, B. G., Rahim, W., Khan, M., Lee, D. S., Lee, G. M., Al Azzawi, T. N. I., Hussain, A., Kim, C. K., & Yun, B. W. (2022). Phytohormonal regulation through protein S-nitrosylation under stress. *Frontiers in Plant Science, 13*. https://doi.org/10.3389/fpls.2022.865542, https://www.frontiersin.org/journals/plant-science.

Parankusam, S., Adimulam, S. S., Bhatnagar-Mathur, P., & Sharma, K. K. (2017). Nitric oxide (NO) in plant heat stress tolerance: Current knowledge and perspectives. *Frontiers in Plant Science, 8*. https://doi.org/10.3389/fpls.2017.01582, http://journal.frontiersin.org/article/10.3389/fpls.2017.01582/full.

París, R., Vazquez, M. M., Graziano, M., Terrile, M. C., Miller, N. D., Spalding, E. P., Otegui, M. S., & Casalongué, C. A. (2018). Distribution of endogenous NO regulates early gravitropic response and PIN2 localization in arabidopsis roots. *Frontiers in Plant Science, 9*. https://doi.org/10.3389/fpls.2018.00495.

Piacentini, Della Rovere, Lanni, Cittadini, Palombi, Fattorini, Cecchetti, Altamura, & Falasca (2023). Brassinosteroids interact with nitric oxide in the response of rice root systems to arsenic stress. *Environmental and Experimental Botany, 209*. https://doi.org/10.1016/j.envexpbot.2023.105287.

Piacentini, D., Ronzan, M., Fattorini, L., Della Rovere, F., Massimi, L., Altamura, M. M., & Falasca, G. (2020). Nitric oxide alleviates cadmium- but not arsenic-induced damages in rice roots. *Plant Physiology and Biochemistry, 151*, 729–742. https://doi.org/10.1016/j.plaphy.2020.04.004.

Planchet, E., Gupta, K. J., Sonoda, M., & Kaiser, W. M. (2005). Nitric oxide emission from tobacco leaves and cell suspensions: Rate limiting factors and evidence for the involvement of mitochondrial electron transport. *The Plant Journal, 41*, 732–743. https://doi.org/10.1111/j.1365-313X.2005.02335.x.

Prado, A. M., Porterfield, D. M., & Feijó, J. A. (2004). Nitric oxide is involved in growth regulation and re-orientation of pollen tubes.

Pu, W., Wang, W., Du, R., Xie, D., & Shan, X. (2025). Reciprocal regulation of nitrate and jasmonate signaling by JASMONATE ZIM DOMAIN PROTEIN 1 coordinates plant growth and defense. *New Phytologist, 248*(6), 3040–3052. https://doi.org/10.1111/nph.70609.

Rai, K. K., Pandey, N., Rai, N., Rai, S. K., & Pandey-Rai, S. (2021). Salicylic acid and nitric oxide: Insight into the transcriptional regulation of their metabolism and regulatory functions in plants. *Frontiers in Agronomy, 3*. https://doi.org/10.3389/fagro.2021.781027, www.frontiersin.org/journals/agronomy#.

Ronzan, M., Piacentini, D., Fattorini, L., Della Rovere, F., Eiche, E., Riemann, M., Altamura, M. M., & Falasca, G. (2018). Cadmium and arsenic affect root development in *Oryza sativa* L. negatively interacting with auxin. *Environmental and Experimental Botany, 151*, 64–75. https://doi.org/10.1016/j.envexpbot.2018.04.008.

Sanchez, C., Gates, A. J., Meakin, G. E., Uchiumi, T., Girard, L., Richardson, D. J., Bedmar, E. J., & Delgado, M. J. (2010). Production of nitric oxide and nitrosylleghemoglobin complexes in soybean nodules in response to flooding. *Molecular Plant-Microbe Interactions, 23*, 702–711.

Sánchez-Vicente, I., Lechón, T., Fernández-Marcos, M., Sanz, L., & Lorenzo, O. (2021). Nitric oxide alters the pattern of auxin maxima and PIN-FORMED1 during shoot development. *Frontiers in Plant Science, 12*. https://doi.org/10.3389/fpls.2021.630792.

Sarma, H., Ramchiary, C., Sen, B., Daimary, M., & Prasad, R. (2025). Integration of nitric oxide signaling in plant stress responses: Unveiling its role in enhancing crop tolerance to abiotic stress. *Discover Plants, 2*(1). https://doi.org/10.1007/s44372-025-00102-w.

Sasse, J., Martinoia, E., & Northen, T. (2018). Feed your friends: Do plant exudates shape the root microbiome? *Trends in Plant Science, 23*, 25–41. https://doi.org/10.1016/j.tplants.2017.09.003.

Schlicht, M., & Kombrink, E. (2013). The role of nitric oxide in the interaction of *Arabidopsis thaliana* with the biotrophic fungi, Golovinomyces orontii and *Erysiphe pisi*. *Frontiers in Plant Science, 4*. https://doi.org/10.3389/fpls.2013.00351, http://journal.frontiersin.org/Journal/10.3389/fpls.2013.00351/pdf.

Sen, S. (2010). S-nitrosylation process acts as a regulatory. *American Journal of Plant Physiology, 5*(3), 122–132.

Shah, J., & Zeier, J. (2013). Long-distance communication and signal amplification in systemic acquired resistance. *Frontiers in Plant Science, 4*. https://doi.org/10.3389/fpls.2013.00030.

Shah, S., Chen, C., Sun, Y., Wang, D., Nawaz, T., El-Kahtany, K., & Fahad, Shah (2023). Mechanisms of nitric oxide involvement in plant-microbe interaction and its enhancement of stress resistance. *Plant Stress, 10*, 100191. https://doi.org/10.1016/j.stress.2023.100191.

Shah, S., Chen, C., Sun, Y., Wang, D., Nawaz, T., El-Kahtany, K., & Fahad, S. (2023). Mechanisms of nitric oxide involvement in plant-microbe interaction and its enhancement of stress resistance. *Plant Stress, 10*, 100191. https://doi.org/10.1016/j.stress.2023.100191.

Shahid, M., Imran, Q., Hussain, A., Khan, M., Lee, S., Mun, B., & Yun, B.-W. (2019). Comprehensive analyses of nitric oxide-induced plant stem cell-related genes in *Arabidopsis thaliana*. *Genes, 10*(3), 190. https://doi.org/10.3390/genes10030190.

Shan, C., Zhou, Y., & Liu, M. (2015). Nitric oxide participates in the regulation of the ascorbate-glutathione cycle by exogenous jasmonic acid in the leaves of wheat seedlings under drought stress. *Protoplasma, 252*, 1397–1405. https://doi.org/10.1007/s00709-015-0756-y.

Shao, R. X., Wang, K. B., & Shangguan, Z. P. (2010). Cytokinin-induced photosynthetic adaptability of *Zea mays* L. to drought stress associated with nitric oxide signal: Probed by ESR spectroscopy and fast OJIP fluorescence rise. *Journal of Plant Physiology, 167*, 472–479. https://doi.org/10.1016/j.jplph.2009.10.020.

Shen, Q., Wang, Y. T., Tian, H., & Guo, F. Q. (2013). Nitric oxide mediates cytokinin functions in cell proliferation and meristem maintenance in *Arabidopsis*. *Molecular Plant, 6*(4), 1214–1225. https://doi.org/10.1093/mp/sss148.

Shi, H. T., Li, R. J., Cai, W., Liu, W., Fu, Z. W., & Lu, Y. T. (2012). In vivo role of nitric oxide in plant response to abiotic and biotic stress. *Plant Signaling and Behavior, 7*(3), 437–439. https://doi.org/10.4161/psb.19219, http://www.landesbioscience.com/journals/psb/article/19219/?download_pdf=true.

Signorelli, S., Sainz, M., Tabares-da Rosa, S., & Monza, J. (2020). The role of nitric oxide in nitrogen fixation by legumes. *Frontiers in Plant Science, 11*, 521. https://doi.org/10.3389/fpls.2020.00521 PMID: 32582223.

Singhal, R. K., Jatav, H. S., Aftab, T., Pandey, S., Mishra, U. N., Chauhan, J., Chand, S., Indu, Saha, D., Dadarwal, B. K., Chandra, K., Khan, M. A., Rajput, V. D., Minkina, T., Narayana, E. S., Sharma, M. K., & Ahmed, S. (2021). Roles of nitric oxide in conferring multiple abiotic stress tolerance in plants and crosstalk with other plant growth regulators. *Journal of Plant Growth Regulation, 40*(6), 2303–2328. https://doi.org/10.1007/s00344-021-10446-8, https://www.springer.com/journal/344.

Šírová, J., Sedlářová, M., Piterková, J., Luhová, L., & Petřivalský, M. (2011). The role of nitric oxide in the germination of plant seeds and pollen. *Plant Science, 181*(5), 560–572. https://doi.org/10.1016/j.plantsci.2011.03.014.

Socco, S., Bovee, R. C., Palczewski, M. B., Hickok, J. R., & Thomas, D. D. (2017). Epigenetics: The third pillar of nitric oxide signaling. *Pharmacological Research, 121*, 52–58. https://doi.org/10.1016/j.phrs.2017.04.011, http://www.elsevier.com/inca/publications/store/6/2/2/9/3/1/index.htt.

Sokolovski, S., & Blatt, M. R. (2004). Nitric oxide block of outward-rectifying K+ channels indicates direct control by protein nitrosylation in guard cells. *Plant Physiology, 136*(4), 4275–4284. https://doi.org/10.1104/pp.104.050344.

Stöhr, C., & Stremlau, S. (2006). Formation and possible roles of nitric oxide in plant roots. *Journal of Experimental Botany, 57*(3), 463–470. https://doi.org/10.1093/jxb/erj058.

Sun, H., Feng, F., Liu, J., & Zhao, Q. (2017). The interaction between auxin and nitric oxide regulates root growth in response to iron deficiency in rice. *Frontiers in Plant Sciences*. https://doi.org/10.3389/fpls.2017.02169.

Sun, H., Lei, C., Xu, J., & Li, R. (2021). Foliar uptake and leaf-to-root translocation of nanoplastics with different coating charge in maize plants. *Journal of Hazardous Materials, 416*, 125854. https://doi.org/10.1016/j.jhazmat.2021.125854.

Sun, H., Tao, J., Zhao, Q., Xu, G., & Zhang, Y. (2017). Multiple roles of nitric oxide in root development and nitrogen uptake. *Plant Signaling & Behavior, 12*(1), e1274480.

Tanou, G., Filippou, P., Belghazi, M., Job, D., Diamantidis, G., Fotopoulos, V., & Molassiotis, A. (2012). Oxidative and nitrosative-based signaling and associated post-translational modifications orchestrate the acclimation of citrus plants to salinity stress. *The Plant Journal, 72*(4), 585–599. https://doi.org/10.1111/j.1365-313x.2012.05100.x.

Terrile, M. C., París, R., Calderõn-Villalobos, L. I. A., Iglesias, M. J., Lamattina, L., Estelle, M., & Casalongué, C. A. (2012). Nitric oxide influences auxin signaling through S-nitrosylation of the *Arabidopsis* TRANSPORT INHIBITOR RESPONSE 1 auxin receptor. *Plant Journal, 70*(3), 492–500. https://doi.org/10.1111/j.1365-313X.2011.04885.x.

Tossi, V., Lamattina, L., & Cassia, R. (2013). Pharmacological and genetical evidence supporting nitric oxide requirement for 2,4-epibrassinolide regulation of root architecture in *Arabidopsis thaliana*. *Plant Signaling & Behavior, 8*, e24712. https://doi.org/10.4161/psb.24712 pii.

Ullah, A., et al. (2022). Role of nitric oxide in drought stress. In *Nitric oxide in plant biology*. Elsevier, 197–210. https://doi.org/10.1016/B978-0-12-818797-5.00030-3.

Vaishnav, A., Kumari, S., Jain, S., Varma, A., Tuteja, N., & Choudhary, D. K. (2016). PGPR-mediated expression of salt tolerance gene in soybean through volatiles under sodium nitroprusside. *Journal of Basic Microbiology, 56*(11), 1274–1288. https://doi.org/10.1002/jobm.201600188, www.interscience.wiley.com/jpages/0233-111X.

Varma, A., Tripathi, S., & Prasad, R. (Eds.). (2020). *Plant microbe symbiosis*Springerhttps://doi.org/10.1007/978-3-030-36248-5.

Wang, P., Du, Y., Hou, Y. J., Zhao, Y., Hsu, C. C., Yuan, F., Zhu, X., Tao, W. A., Song, C. P., & Zhu, J. K. (2015). Nitric oxide negatively regulates abscisic acid signaling in guard cells by S-nitrosylation of OST1. *Proceedings of the National Academy of Sciences of the United States of America, 112*(2), 613–618. https://doi.org/10.1073/pnas.1423481112, http://www.pnas.org/content/112/2/613.full.pdf+html.

Wang, W., & Sung, S. (2024). Chromatin sensing: Integration of environmental signals to reprogram plant development through chromatin regulators. *Journal of Experimental Botany, 75*(14), 4332–4345. https://doi.org/10.1093/jxb/erae086.

Wang, X., Du, H., Ma, M., & Rennenberg, H. (2023). The dual role of nitric oxide (NO) in plant responses to cadmium exposure. *Science of the Total Environment, 892*. https://doi.org/10.1016/j.scitotenv.2023.164597.

Wang, X., Wen, H., Suprun, A., & Zhu, H. (2025). Ethylene signaling in regulating plant growth, development, and stress responses. *Plants, 14*(3), 309. https://doi.org/10.3390/plants14030309.

Wang, Y., Feng, H., Qu, Y., Cheng, J., Zhao, Z., Zhang, M., Wang, X., & An, L. (2006). The relationship between reactive oxygen species and nitric oxide in ultraviolet-B-induced ethylene production in leaves of maize seedlings. *Environmental and Experimental Botany, 57*(1-2), 51–61. https://doi.org/10.1016/j.envexpbot.2005.04.009.

Wang, Y., & Ruby, E. G. (2011). The roles of no in microbial symbioses. *Cellular Microbiology, 13*, 518–526.

Wang, Y., Yang, Q., Tosa, Y., Nakayashiki, H., et al. (2001). Nitric oxide-overproducing transformants of *Pseudomonas fluorescens* with enhanced bio-control of tomato bacterial wilt. *Journal of General Plant Pathology, 71*, 33–38.

Wilson, S. K., Pretorius, T., & Naidoo, S. (2023). Mechanisms of systemic resistance to pathogen infection in plants and their potential application in forestry. *BMC Plant Biology, 23*(1), 404.

Wu, M., Wang, F., Zhang, C., Xie, Y., Han, B., Huang, J., & Shen, W. (2013). Heme oxygenase-1 is involved in nitric oxide- and cGMP-induced α-Amy2/54 gene expression in GA-treated wheat aleurone layers. *Plant Molecular Biology, 81*(1-2), 27–40. https://doi.org/10.1007/s11103-012-9979-x, https://link.springer.com/journal/11103.

Xing, H., Tan, L., An, L., Zhao, Z., Wang, S., & Zhang, C. (2004). Evidence for the involvement of nitric oxide and reactive oxygen species in osmotic stress tolerance of wheat seedlings: Inverse correlation between leaf abscisic acid accumulation and leaf water loss. *Plant Growth Regulation, 42*(1), 61–68. https://doi.org/10.1023/B:GROW.0000014894.48683.1b.

Xu, X. T., Jin, X., Liao, W. B., Dawuda, M. M., Li, X. P., Wang, M., Niu, L. J., Ren, P. J., & Zhu, Y. C. (2017). Nitric oxide is involved in ethylene-induced adventitious root development in cucumber (*Cucumis sativus* L.) explants. *Scientia Horticulturae, 215*, 65–71. https://doi.org/10.1016/j.scienta.2016.12.006, www.elsevier.com/inca/publications/store/5/0/3/3/1/6.

Yang, Y., Zheng, Y., Liu, C., Chen, L., Ma, J., Sheng, J., & Shen, L. (2016). Inhibition of nitric oxide synthesis delayed mature-green tomato fruits ripening induced by inhibition of ethylene. *Scientia Horticulturae, 211*, 95–101. https://doi.org/10.1016/j.scienta.2016.07.026, www.elsevier.com/inca/publications/store/5/0/3/3/1/6.

Yuan, H.-M., & Huang, X. (2016). Inhibition of root meristem growth by cadmium involves nitric oxide-mediated repression of auxin accumulation and signalling in *Arabidopsis*. *Plant, Cell & Environment, 39*(1), 120–135. https://doi.org/10.1111/pce.12597.

Zeier, J., Delledonne, M., Mishina, T., Severi, E., Sonoda, M., & Lamb, C. (2004). Genetic elucidation of nitric oxide signaling in incompatible plant–pathogen interactions. *Plant Physiology, 136*, 2875–2886.

Zeng, J., Zhao, X., Liang, Z., Hidalgo, I., Gebert, M., Fan, P., Wenzl, C., Gornik, S. G., & Lohmann, J. U. (2023). Nitric oxide controls shoot meristem activity via regulation of

DNA methylation. *Nature Communication, 14*(1), 8001. https://doi.org/10.1038/s41467-023-43705-1.

Zeng, J., Zhao, X.'A., & Lohmann, J. U. (2025). The hidden architects: Nitric oxide and redox dynamics in plant stem cell homeostasis. *Bioessays: News and Reviews in Molecular, Cellular and Developmental Biology, 47*(10). https://doi.org/10.1002/bies.70048.

Zhang, A., Zhang, J., Zhang, J., Ye, N., Zhang, H., Tan, M., & Jiang, M. (2011). Nitric oxide mediates brassinosteroid-induced aba biosynthesis involved in oxidative stress tolerance in maize leaves. *Plant and Cell Physiology, 52*(1), 181–192. https://doi.org/10.1093/pcp/pcq187.

Zhang, H., Shen, W. B., Zhang, W., & Xu, L. L. (2005). A rapid response of β-amylase to nitric oxide but not gibberellin in wheat seeds during the early stage of germination. *Planta, 220*(5), 708–716. https://doi.org/10.1007/s00425-004-1390-7.

Zhang, J., Ma, Y., Dong, C., Terry, L. A., Watkins, C. B., Yu, Z., & Cheng, Z.-M.(M.) (2020). Meta-analysis of the effects of 1-methylcyclopropene (1-MCP) treatment on climacteric fruit ripening. *Horticulture Research, 7*(1). https://doi.org/10.1038/s41438-020-00405-x.

Zhang, W., Cao, J., Fan, X., & Jiang, W. (2020). Applications of nitric oxide and melatonin in improving postharvest fruit quality and the separate and crosstalk biochemical mechanisms. *Trends in Food Science & Technology, 99*, 531–541. https://doi.org/10.1016/j.tifs.2020.03.024.

Zheng, C., Jiang, D., Liu, F., Dai, T., Liu, W., Jing, Q., & Cao, W. (2009). Exogenous nitric oxide improves seed germination in wheat against mitochondrial oxidative damage induced by high salinity. *Environmental and Experimental Botany, 67*(1), 222–227. https://doi.org/10.1016/j.envexpbot.2009.05.002.

Zheng, X., Hu, B., Song, L., Pan, J., & Liu, M. (2017). Changes in quality and defense resistance of kiwifruit in response to nitric oxide treatment during storage at room temperature. *Scientia Horticulturae, 222*, 187–192. https://doi.org/10.1016/j.scienta.2017.05.010.

Zhou, J. M., Trifa, Y., Silva, H., Pontier, D., Lam, E., Shah, J., & Klessig, D. F. (2000). NPR1 differentially interacts with members of the TGA/OBF family of transcription factors that bind an element of the PR-1 gene required for induction by salicylic acid. *Molecular Plant-Microbe Interactions, 13*(2), 191–202. https://doi.org/10.1094/MPMI.2000.13.2.191, http://apsjournals.apsnet.org/loi/mpmi.

Zhu, L., Yang, R., Sun, Y., Zhang, F., Du, H., Zhang, W., Wan, C., & Chen, J. (2021). Nitric oxide maintains postharvest quality of navel orange fruit by reducing postharvest rotting during cold storage and enhancing antioxidant activity. *Physiological and Molecular Plant Pathology, 113*, 101589. https://doi.org/10.1016/j.pmpp.2020.101589.

Zhu, X. F., Jiang, T., Wang, Z. W., Lei, G. J., Shi, Y. Z., Li, G. X., & Zheng, S. J. (2012). Gibberellic acid alleviates cadmium toxicity by reducing nitric oxide accumulation and expression of IRT1 in *Arabidopsis thaliana*. *Journal of Hazardous Materials, 239-240*, 302–307. https://doi.org/10.1016/j.jhazmat.2012.08.077.

Index

Note: Page numbers followed by "*f*" and "*t*" refer to figures and tables, respectively.

O

P

Q

R

Printed and bound by CPI Group (UK) Ltd, Croydon, CR0 4YY
22/09/2026
02220473-0007